T0205771

doing mathematics

Convention, Subject, Calculation, Analogy

Second Edition

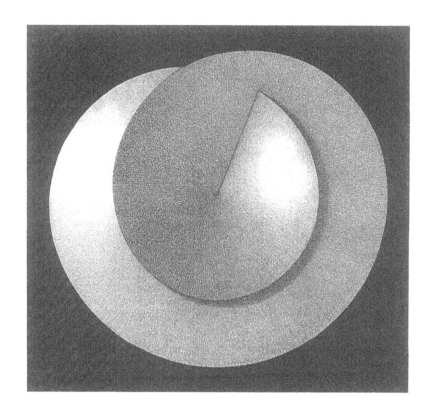

Die Riemann'sche Windungsfläche erster Ordnung

Vergl. Seite 64 – 71

Eschebuch & Schaefer Leipzig

FIGURE P.1: A Riemann surface for $z^{1/2}$ (Neumann, *Vorlesungen über Riemann's Theorie der Abel'schen Integrale*, 1884)

doing mathematics

Convention, Subject, Calculation, Analogy

Second Edition

Martin H Krieger

University of Southern California, USA

World Scientific

NEW JERSEY · LONDON · SINGAPORE · BEIJING · SHANGHAI · HONG KONG · TAIPEI · CHENNAI

Published by

World Scientific Publishing Co. Pte. Ltd.
5 Toh Tuck Link, Singapore 596224
USA office: 27 Warren Street, Suite 401-402, Hackensack, NJ 07601
UK office: 57 Shelton Street, Covent Garden, London WC2H 9HE

Portions of this book, reprinted herein with permission of the publishers, appeared in somewhat different form in: Martin H. Krieger, "Making a Paradigmatic Convention Normal: Entrenching Means and Variances as Statistics," *Science in Context* 9 (Winter 1996): 487–509 (Cambridge University Press); Martin H. Krieger, "Apocalypticism, One-Time Events, and Scientific Rationality," in, S.S. Hecker and G.-C. Rota, eds., *Essays on the Future, In Honor of Nick Metropolis*, Boston: Birkhäuser, 2000, pp. 135–151, ©2000 Springer-Verlag; Martin H. Krieger, "The Mathematics of the Ideal City in the Nineteenth Century," in, M. Leary, B. Dermody, and A. Markusen, eds., *Envisioning Cities and Regions: A Celebratory Anthology on Planning by University of Minnesota Faculty*, Minneapolis: Hubert H. Humphrey Institute of Public Affairs, University of Minnesota, 2000, pp. 309–316.

C. N. Yang, "The Spontaneous Magnetization of a Two-Dimensional Ising Model," *Physical Review* 85 (1952): 808–816, ©1952 American Physical Society. Excerpt from "On the Dirac and Schwinger Corrections to the Ground-State Energy of an Atom" by C. Fefferman and L.A. Seco in *Advances in Mathematics*, Volume 107, 1–185, copyright 1994, Elsevier Scientific (USA), reproduced with permission from the publisher. A. Weil, *Oeuvres Scientifiques*, volume 1, pp. 244–255, ©1978 Springer-Verlag; J. Leray, *Selected Papers*, volume 1, pp. 61–64, ©1998 Springer-Verlag. Figures 1.1, 3.11, courtesy of Springer Verlag; 1.3, 3.5, Koninklijke Nederlandse Akademie van Wetenschappen and the Brouwer Archive; 3.3, 3.7, 3.13, 5.4, Dover Publishing Company; 3.4, Heldermann Verlag; 3.12, A.K. Peters, Ltd.; 3.14, American Mathematical Society; 4.1, N.2, N.3, American Physical Society.

Library of Congress Cataloging-in-Publication Data
Krieger, Martin H.
    Doing mathematics : convention, subject, calculation, analogy / by Martin H Krieger (University of Southern California, USA). -- 2nd edition.
        pages cm
    Includes bibliographical references and index.
    ISBN 978-981-4571-83-8 (hardcover : alk. paper) -- ISBN 978-981-4571-84-5 (pbk. : alk. paper)
    1. Mathematics--Research. I. Title.
    QA11.2.K75 2015
    510--dc23
                                2014043918

British Library Cataloguing-in-Publication Data
A catalogue record for this book is available from the British Library.

Cover illustration is a montage of Figures 2 and 3 of L. E. J. Brouwer, "Zur Analysis Situs," *Mathematische Annalen* 68 (1910): 422–434.

Printed in Singapore

To the memory of my parents, Louis and Shirley Krieger,

my dear friend Miriam Brien and my cousin Marion Herbst,

and the memory of Abraham Polonsky and Gian-Carlo Rota.

Contents

Figures

Preface

Doing Mathematics focuses on the work of mathematics and mathematicians, and the work of those who use mathematics in the physical sciences and the social sciences. Still, I have been assured by some lay persons that the book is readable with suitable skipping.

In this second edition, I have tried to deepen and clarify the text. I have come to understand more of some of the mathematics and the examples, and so I have been able to better discern my themes. In the last decade there have been remarkable advances, and some of them are relevant to the discussion. The Prolog epitomizes these themes.

In the case of the Ising model, there is a great deal of rigorous mathematical reformulations that may well be useful for understanding its analogy with other parts of mathematics.[1]

I shall be describing some ways of doing mathematical work and the subject matter that is being worked upon and created. I shall argue that the *conventions* mathematicians adopt, the *subject* areas they delimit, what they can *prove* and *calculate* about the (physical) world, the *analogies* that work for mathematicians, and the known *tools and techniques* they borrow from a Library of Mathematics— all depend on the mathematics, what will work out and what won't. And the mathematics, as it is done, is shaped and supported, or not, by convention, subject matter, calculation, analogy, and tools. These features correspond to chapter 2 on means and variances as conventional statistics, chapter 3 on the subject of topology, chapter 4 on strategy, structure, and tactics in long apparently "messy" proofs, chapter 5 on analogy in and between two programs for research, those of Robert Langlands (1936–), and go back to Richard Dedekind (1831–1916), in number theory, and of Lars Onsager (1903–1976) in statistical mechanics, and chapter 6 on some of the tools in that Library and how they are improved when they are loaned out. The examples I shall use are drawn from applied mathematics (and mathematical physics) as well as from pure mathematics.

Mathematics is done by mathematicians located in a particular culture (chapter 6), where mathematicians may choose to work on some problems and not others, in some ways and not others. But what they can in the end prove depends on the mathematics. And by "depends on" I mean that only some possible statements are both true and provable given our current imaginations and resources, that only some definitions are fruitful, and that only some models of the physical world are in accord with how it behaves, given what we know now.

In effect, the Library of Mathematics has a wide variety of volumes, but not all we might wish for, and some of the time borrowed volumes are defaced in inventive ways and become even more useful. We shall notice that some of the time, what the mathematicians need is just what the physicist have been doing in their everyday work (albeit in a not so rigorous or general fashion). Or, what the physicists need has been developed on its own by the mathematicians and is available in the Library. Mathematicians also borrow from the Library, and so fields of mathematics deeply influence other fields. For example, curves representing algebraic expressions are then understood using deep algebraic methods (algebraic geometry).

Along the way, I am interested in saying something about what is really going on in the mathematics I describe. (And in some places, I have been rather more tentative and less successful, to be sure.) In saying what we might mean by "what is really going on," I shall argue that what is really going on is really going on only when it works in a proof or derivation and, ideally, we can show in detail just how and why it does the work. And "when it works" is itself a judgment; while, "showing in detail just how" is an interpretation of what is going on. Usually, we discover what is really going on from multiple perspectives on the same subject matter, different roads to exposition and proof, so that what at first seems miraculous and amazing (and it really is) is eventually shown to be manifest from the right points of view. We have "an identity in a manifold presentation of profiles," to use the term of art from philosophy.

Moreover, in this sort of work there is an intimate interaction of ideas and calculation, of insight and computation, of the conceptual and the algorithmic. Perhaps "proofs should be driven not by calculation but solely by ideas," as Hilbert averred in what he called Riemann's principle.[2] But, in fact, mathematical work is an interplay of both. So a combination of ingenuity, mathematical maturity, and a willingness to calculate and invent along the way is seen in Charles Fefferman's work, in the paper by C. N. Yang we shall discuss, in a series of papers by T. T. Wu and collaborators, and in Rodney Baxter's various exact solutions of lattice models in statistical mechanics. Moreover, mathematical rigor is in the end about ideas and the world; it is philosophical in that rigor often reveals aspects and counterexamples and cases we would not have otherwise been aware of. Rigor also allows mathematicians to be sure of their work, since it is error displaying, as Frank Quinn has argued.[3] And in order to implement ideas, one must calculate, theorize, and prove.

As for rigor and mathematical niceties, one often encounters comments such as,

The mathematical calculations that lead to exact results for quantities like the spontaneous magnetization have a complexity to them that many physicists writing on the subject feel obscures the essential physics.[4]

yet some of the essential physics is revealed by that complexity. But we also find,

"The two-dimensional Ising model is a Free Fermion." Remarks to this effect are commonplace in the physics literature, although for mathematicians it sounds like a cross species identification.[5]

and

The two-dimensional Ising model is nothing but the theory of elliptic curves.[6]

There are in fact many such indentifications and "nothing but"s, and it is in their variety that tells us more of what is going on.

The Topical of Table of Contents (next page) indicates where various particular examples appear in the text. It also indicates the main theme of each chapter, and a leitmotif or story or fairy tale that I use to motivate and organize my account. The reader is welcome to skip the fairy tale. My purpose in each case is to express poignantly the lesson of the chapter.

As for the leitmotifs or stories, some are perhaps part of the everyday discourse of mathematicians when they describe the work they do, often under the rubric of "the nature and purpose of mathematical proof." Sometimes, the mathematician is said to be climbing mountains, or exploring and discerning peaks in the distance that might be worthy of ascent. Or, perhaps the mathematician is an archaeologist. Having found some putative mathematical fact, the mathematician is trying to connect it with mathematics that is known already in the hopes of greater understanding.[7] Perhaps the mathematician tries to explore different aspects of some mathematical object, finding out what is crucial about it, getting at the facts that allow for its variety of aspects and appearances, in effect a phenomenological endeavor. In any case, here I have tried to describe rather more concretely some of mathematicians' work, without claiming anything generic about the nature of mathematics or mathematicians' work itself.

Again, my goal is to provide a description of some of the work of mathematics, a description that a mathematician would find recognizable, and which takes on some reasonably substantial mathematics. Yet I hope it is a description that lay readers

FIGURE P.2: TOPICAL TABLE OF CONTENTS

Chapter	Story	Theme	Examples*
2. Means and Variances	Anthropologist Studying a Culture	Conventions in Practice	Scaling
3. The Fields of Topology	Sibling Sub-Fields in Tension	A Subject or Field Defined in Practice, Algebra Everywhere	Ising Matter
4. Strategy, Structure, and Tactics in Proof	Interpreting Works of Art and Craft	Ideas and Calculation	Matter, Scaling
5. Syzygies of Research Programs	The Witch *is* the Stepmother	Analogy	Theta Functions, Ising, Scaling, Langlands, and Onsager
6. Mathematics *In Concreto*	People and Cities	Mutual Embeddedness of Mathematics and World	The City, the Embodied Prover, God as an Infinite Set

*Ising = The Two-Dimensional Ising Model; Langlands = Dedekind-Langlands Program in Number Theory; Matter = The Stability of Matter; Onsager = Onsager Program in Statistical Mechanics; Scaling = Ising in an Asymptotic Regime, or Automorphic Modular Functions, or the Central Limit Theorem.

might find familiar enough, as well, especially in terms of generic notions such as convention, subject matter, strategy and structure, and analogy. As a consequence of addressing these very different sorts of readers, the level of technical detail provided and the assumptions I make about what the reader knows already vary more than is usual. I hope that the simplifying images or the technical detail will not put off readers at each end of the spectrum.

I have been generous in repeating some details of several of the examples, at the various places where an example is employed. Each chapter may be read independently; and, sometimes, sections of chapters, of differing degrees of technical difficulty, may be read independently. It is noteworthy that some of the work I describe is better known by reputation than by having been read and assimilated, either because it is long and technical and hard to understand, or because it is deemed to be part of "what everybody knows," so few now feel compelled to take a close look at it.

Notionally, and somewhat technically, one recurrent and perhaps unifying substantive object is the central limit theorem of statistics: a sum of "nice" random variables ends up looking like a Gaussian or bell-shaped curve whose variance is the sum of the components' variances. The central limit theorem provides a model of scaling, the $\sqrt{N}$ growth of the Gaussian distribution; it is the large-N or large time (that is, asymptotic) story of the random walk on a grid or lattice; it turns out to be the foundation for a description of an ideal gas (the Maxwell-Boltzmann law of molecular velocities); and it provides a description of diffusion or heat flow, and so is governed by the heat equation (the laplacian or, in imaginary time, the Schrödinger equation), one of the partial differential equations of mathematical physics.[8]

If the Gaussian and the central limit theorem is the archetypal account of independent random variables, it would appear that matrices whose elements are random numbers or variables, and probability distributions of their basic symmetries (their eigenvalues), those distributions also determined by distant relations of the trigonometric functions, the Painlevé transcendents, are the archetype for measures of the connection between them (that is, correlation functions) and the extreme statistics (such as the maximum) of strongly dependent random variables.[9] These distributions appear as the natural asymptotic limit of correlation functions of nuclear energy levels or of zeros of zeta- and L-functions (in each case the separation between adjacent levels or zeros), in lengths of longest increasing sequences in random permutations, and in correlations of spins within a crystal lattice.

All of these themes will recurrently appear, as connected to each other, in our discussions.

Another unifying theme is set by the question, "Can you hear the shape of a drum?" That is, can you connect the sound spectrum of an object with its geometry, namely, its zeros with its symmetries?[10] (More concretely, can you connect the trace of a matrix with the determinant (a volume element) of another: the sum of the eigenvalues of one matrix with the product of the eigenvalues of another.) How is the local connected to the global, why is there regularity in both frequency and in scale size? Again, the connections will be multifold and wide-ranging.

And recurrently, we shall see that phenomena of geometry or topology or the calculus are mirrored in algebra, and so the alebraicization of much of mathematics, along the way transforming algebra itself.

More generally, there are *why* and *how* questions: Why is there *regularity* (in the combinatorial numbers, for example), seen as scaling and in fourier coefficients of nice functions? Why group *representations*, those matrices? What are the *symmetries* that underlie the relationship of counting to scaling? Why do we have *asymptotics* and these asymptotic forms?[11]

How does the algebra do the combinatorics (the motivation of Kac and Ward's famous paper)? And how does the algebra of counting lead to scaling or automorphy? More generally, how and why are objects that package combinatorial information exhibit nice scaling behavior. (And to show how does not really answer the why question.)

Descartes begins his *Rules for the Direction of the Mind* (1628),

> Whenever people notice some similarity between two things, they are in the habit of ascribing to the one what they find true of the other, even when the two are not in that respect similar. Thus they wrongly compare the sciences, which consist wholly in knowledge acquired by the mind, with the arts, which require some bodily aptitude and practice.[12]

Descartes is warning against what he takes to be incorrect analogy, and he is advocating a "universal wisdom," which allows the mind to form "true and sound judgments about whatever comes before it." I shall be arguing that whatever we might understand about universal wisdom, in actual practice our work is particular and concrete, that the temptation to analogy and to comparing the sciences and the arts is just the way we do our work.

After the introduction, *the chapters might be read out of order.* I have tried to make them self-contained for the most part. For those who prefer to start with the index, notes, and bibliography, I have tried to make them lead to the main points of the book.

As for studies of science, my analysis has been driven by the particular examples I am using, rather than by the demands of theory, sociology, philosophy, or historiography.[13] I imagine the analyses may prove useful for social and philo-sophical studies of science and mathematics. But other than my claims about convention, subject, calculation, and analogy, I have not tried to do such studies. Chapter 6, about the city, the body, and God, goes a bit further, but each of these topics demands a separate monograph, at least, if the chapter's suggestions were to be thoroughly worked out.

Rather than employing mathematics or physics to instantiate conventional philosophic (or sociological) problems, the cases we discuss suggest another set of problems: How are mathematics and physics made useful for each other? What is it about the physicist's world that makes it amenable to mathematical technologies, and how is that world so adjusted? What does it mean to have a precise mathematical description or definition of a physical phenomenon? How and why do the details in a mathematical analysis reveal physics that is otherwise not so appreciated? How is having many different solutions or proofs of the same problem useful and revealing? How do ideas and calculation support each other? How is mathematics' dynamic character fed by both applications and internal developments? How do ugly first proofs or derivations have an inner beauty? Just how does analogy actually work in these areas?

My particular choice of mathematics reflects some ongoing interests: elementary statistics (and what is not learned by students); statistical mechanics; the proof of Fermat's theorem by Ribet, Serre, Frey, and Wiles and Taylor in the 1990s, with the Langlands Program in the background; and, the recurrent appearance of scaling (or automorphy and renormalization), elliptic functions, and elliptic curves in many of these subjects. Some of this is mathematics in the service of physics. Although I cannot be sure, I doubt whether these particular choices affect my descriptions of how mathematicians do their work.[14]

As for technical knowledge of mathematics: Again, I have tried to write so that the professional scientist will find what I say to be correct mathematically and of practical interest. Yet the general reader or the philosopher should be able to follow the argument, if not always in detail then in its general flow. To that end, I have employed a number of signals:

Technically, I have sometimes flagged the more technical material by the device I used to begin this sentence. (In other cases, I have used brackets or parentheses to mark off technical material.)

***If three stars precede a paragraph, I have there provided a summary of what has gone before or a preview of a larger segment of the argument.

As for notation, of course the notation is consistent within any single example. But I have reused letters and symbols, with new meanings in different examples. The index points to the definitions of the letters and symbols.

ACKNOWLEDGMENTS

From the first edition: Abraham Polonsky and Gian-Carlo Rota encouraged my work and understood just what I was trying to do. Jay Caplan, Eric Livingston, Sam Schweber, and Robert Tragesser have been abiding readers and friends. I am grateful for my undergraduate education at Columbia College and my mathematics teachers there: J. Eels, Jr., A. Dold, and J. van Heijenoort. Colin McLarty and Craig Tracy helpfully answered my questions. The usual disclaimers apply, a fortiori.

A sabbatical leave from the University of Southern California allowed me time to write. And my students, colleagues, and deans have provided a fertile environment for my work. Over the last decades, I have received external financial support from the National Humanities Center, the Russell Sage Foundation, the Exxon Education Foundation, the Zell-Lurie professorship at the University of Michigan, the Lilly Endowment, the Hewlett Foundation, and the Haynes Foundation, support that was for other books and projects. But I would not have been able to write this one without having done that work.

My son, David, always wants to know what is *really* going on. He continues to ask the best questions, and to demand the most thoughtful answers.

More recently, I have received support from the Haynes Foundation, the Kauffman Foundation, and the Price Charities, and again it was for other books and projects. I teach in a school of urban planning and public policy. I keep finding that the notions I develop in this book enrich my ability to teach our students.

I am grateful to my physicians at the University of Southern California's Keck School of Medicine and Hospital for their care for me and my son. My son David is now more than ten years older than when I wrote the first edition. He is still my toughest questioner.

Prolog

It will be useful to preview our recurrent themes and examples: just what is *really* going on in the mathematics; how do ideas and calculation interact; how do the subjects of mathematics change as we learn more; ugly proofs have an inner beauty; analogy is a destiny we embrace; and, rigor and details are substantively informative. Substantively: we can hear the shape of a drum since sound spectrum and geometry are related; periodicity and self-similarity appear to accompany counting-up; local facts and global characteristics are systematically connected; and, you create mathematical objects so that those objects add up. In a bit greater detail:

1. When we ask *what is really going on in the mathematics*, or when we ask what is behind the various proofs of a theorem (or the various solutions of the Ising model), we are seeking what the phenomenological philosophers call *an identity in a manifold presentation of profiles*. Just what is the source of all the various seeming tricks and methods that make them work in these contexts? Put differently, how can the objects we are studying allow for such a variety of presentations. Most of the time, we have partial answers, and so fail to discern that identity as fully as we might hope.[1]

2. *Ideas and calculation* play against each other, for ideas without calculation are at best informed speculation, and calculation without ideas may or may not get anywhere or make sense. Moreover, devices and tricks employed along the way convey *meaning* and information, even if they appear jury-rigged or convenient. Again, what is really going on?

   Moreover, we might think of *mathematicians as master machinists,* using esoteric devices to make machinery that is sometimes adopted by engineers (physicists, for example) to fabricate what they want. The machinists are to some extent autonomous, tinkering and inventing, to some extent dependent on market demand.

3. *Fields of mathematics are dynamic,* changed by what we learn and can prove, what notions we invent or discover that turn out to be fruitful, and what happens in other fields of mathematics or physics (or computer

science,...). Ideas and visions may transform a field, if they work out. And we may forget useful mathematics and examples along the way. Notions from one field may find use and meaning in another area, and most notions are present in more than one area or complex.

4. *Ugly proofs have an inner beauty*, revealed after subsequent proofs show us more of what is going on (in those proofs, and more generally). And often, initial proofs are ugly.[2]

5. *Analogy is destiny*, but that destiny and just what is analogized to what depend on what we do with the analogy. Analogy may be to other mathematics, to physics, or to everyday life.

6. *Details and fine points matter*. Rigor—just what demands those details and fine points—allows us to learn more of what is going on, and also to reveal stuff we had not noticed. The exceptional case we have to work around may well be quite revealing. In effect, we are philosophical analysts, seeking meaning through differential comparison of cases.

As for substantive themes:

A. Often, we can *hear the shape of a drum*.[3]

So an object may be known by its audible sound (and so its frequency components) and/or by its everyday appearance. Spectrum and geometry, equation and curve, particle and field, the discrete and the smooth, algebraic object and topological space,...are intimately connected, albeit the connection may be of a different sort for each pair. So, an algebraic construction can stand in place for a topological or other sort of space; and we can often use a geometrical object or space to stand in for what appear to be collections of numbers or equations or an algebraic object

B. *Periodic Regularity and Self-Similarity or Scaling Symmetry* would seem to accompany various efforts at counting and enumeration.

Brownian motion—a sum of random moves—looks the same at a very wide range of scales. Partition functions (or L-functions) that package information (say, about the number of solutions to an equation, modulo a prime, p), or their close relatives, exhibit a scaling symmetry (self-similarity, "automorphy"). Think of the central limit theorem and Gaussians growing as $\sqrt{N}$. The periodicity reflects a regularity in the packaged numbers, and we want to understand the source of that regularity.

C.   Often, there is a connection between *local facts and global characteristics.*

There is a connection between locally-seen regularities (as in the number of mod p solutions to equations) and harmonic analysis (that is, a nice transform of something).

We may discover obstructions to those connections, and those obstructions are deeply informative. What we often have is a hierarchy, at each level fully adequate—as in the Standard Model of particle physics and its effective field theories.

D.   We are looking for *individuals that add up*, one way or the other.

Usually they add up as in arithmetic, or linearly, as in a vector space. What are the right parts, the right variables, the right degrees of freedom? Complex interactions should be the sum of two-body interactions, invariant to the order of adding them up—again, as in arithmetic. And that addition would seem to lead to canonical asymptotic forms, such as the Gaussian.

A and *B* say that the sum of the spectrum is related to a volume ("Weyl asymptotics"). *B* says that partition functions that package combinatorial information have asymptotic forms that are self-similar. *C* says that partition functions, as global objects, are something like Fourier transforms of properties of local objects. And *D* is about the canonical form of those partition functions.[4]

I shall be *describing* how mathematics and mathematicians work, rather than theorizing or philosophizing—although these descriptions may help in theoretical and philosophical work.[5] Again, much of this is the legacy of Riemann and Maxwell. But I have no good answers to why group representations are so useful here, or why there are the periodic regularities people notice in counting-up, or why permutations are related to self-similarity. Kant would counsel that some things are beyond scientific knowledge. Husserl would suggest that we are discovering or uncovering various aspects of a phenomenon, that identity in a manifold presentation of profiles. I suspect that fifty years from now, or maybe just a few years from now, we'll have better answers and just as frustrating questions.

1

Introduction

I want to provide a description of some of the work that mathematicians do, employing modern and sophisticated examples. I describe just how a *convention* is legitimated and established, just how a *subject* or a field of study comes to be defined, just how organization and *structure* provide meaning in the manipulations and calculations performed along the way in an extensive proof or derivation, and just how a profound *analogy* is employed in mathematical work. These just-hows are detailed and particular. So that, when we demand a particular feature in a rigorous proof, such as uniform continuity, there is lots to be learned from that.

Ideally, a mathematician who reads this description would say: "That's the way it is. Just about right." Thus, laypersons might learn something of how mathematics is actually done through this analytic description of the work. Students, too, may begin to appreciate more fully the technical moves made by their teachers.

To say that mathematicians prove theorems does not tell you much about what they actually do. It is more illuminating to say what it means in actual practice to do the work that is part of proving theorems: say, that mathematicians devise *structures* of argument that enable one to prove theorems, to check more readily the correctness of those proofs, and to see why what is being shown is almost surely the case.[1] Mathematicians might be said to construct notions and definitions that are fruitful, and theories that then enable them to prove things. The organization of those theories, or those structures of proof or demonstration or those notions, may often be shown to correspond to facts about the world. The sequencing and waypoints say something about physical or mathematical objects, or so we may convince ourselves. Whether those proofs look logical or narrative is a stylistic choice that varies among mathematicians and cultures.

> In Burnside's classical style of writing [ca. 1897], Theorem x means Theorem x together with the discussion around it. In other words, although everything is proved, only some of the conclusions are called theorems. This "classical" style is quite different from the so-called "[E.] Landau Style" of Satz-Beweis-Bemerkung [Statement-Proof-Remark]. In the classical style, you first discuss things and then suddenly say that you have proved such and such; in other words, the proof precedes the statement.[2]

One might provide a description of writing fiction or poetry similar to the one I provide here for mathematics, emphasizing convention, genre, grammar and prosody, and analogy. My claim is simple: In the doing of mathematics one uses interpretive and rhetorical devices that are peculiar to mathematics, yet of the sort seen elsewhere. And, these devices are an essential part of the mathematics, for they are mathematics as such.[3]

Hence, one of my ancillary purposes is to better understand the mathematics as mathematics, to better read and understand difficult and complex work, to get at what is really going on in a proof or a derivation.

It is also true that mathematics is done by people, whose intuitions, examples, and ideas stem from and resonate with their own culture. To illustrate this fact, in chapter 6 I speculate on the resonances between nineteenth-century cities and some of Riemann's mathematical ideas, between the fact that we have bodies and the techniques of algebraic topology, and between religious accounts of God's infinitude and transcendence and mathematicians' notions of infinity. This speculation is intended to be suggestive. I make no claim that mathematics is reducible in any useful sense to something else, or the other way around. Rather, it is a reminder that even our most abstract endeavors are embedded in history and society. That truth should appear within history is no challenge to the facticity or truth of mathematics. It is important to understand not only the history of the mathematics and how what we now know came to be elucidated, but also to appreciate the actual process of achieving insight through new proofs, reformulations, and contrastive examples, and how we take mathematical objects as objects for us (that facticity). This was of enormous interest to mathematicians such as Hermann Weyl (1885–1955), influenced as he was by the philosopher and phenomenologist and mathematician Edmund Husserl (1859–1941) in the early part of the twentieth century.[4]

I am concerned with analyzing the particular features of particular fields of mathematical activity, rather than discussing what is conventionally claimed to be philosophy of mathematics. I focus on the concreteness of the examples, no matter

2

how abstract the mathematics that is instantiated in those examples. Mathematicians usually report they are making discoveries; that proofs are a means of discovery and of checking the truth; that they are getting at the truth through various new proofs and formulations; and that even when they invent new objects there is a sense in which the objects are already-there in the material the mathematicians possessed already. I take these comments as a fiducial report, notwithstanding the fact that many philosophers of mathematics subject those reports to criticism.

I shall not at all address claims about mathematics being the language of science, or that having a language still does not provide a deeper explanation. In such instances, mathematics may be one of the main languages employed. But actual discourse is in a vernacular that mixes formal mathematics, physical and spatial intuitions, and everyday metaphorical speech. It is perhaps no more surprising that mathematics is suited to describing the physical realm than it is that ordinary everyday metaphorical speech is so suited. And the connections between mathematics and the physical sciences are as surprising as the connections between the fields of mathematics.[5]

Finally, I have focused on mathematics as published in journals and books, or codified in textbooks, and on the network of ideas in that mathematics. From our retrospective viewpoint, we might understand what is going on in the mathematics and what the mathematicians were doing. This is quite different from history of mathematics, for historians are concerned with what the scientist and their contemporaries understood, what they saw themselves as doing, and with the nature of influences and developments, and so are obliged to look more widely at additional documentary material.

We shall work out our ideas about mathematical practice through four examples of mathematical work: (i) means and variances in statistics, (ii) the subject of topology, (iii) classical analysis employed in the service of mathematical physics, and (iv) number theory's "learning to speak Riemannian." I take my warrant for these studies from Hermann Weyl, who writes about the practical need in mathematics for an historical-philosophical understanding.

> The problems of mathematics are not problems in a vacuum. There pulses in them the life of ideas which realize themselves in concreto through our human endeavors in our historical existence, but forming an indissoluble whole transcending any particular science.[6]

3

That "life of ideas" is, for my purposes, the legacies of Riemann and of Maxwell. And what we discover is that the concrete, the particular, and the exceptional in mathematics are made to carry along the abstract, the arbitrary, and the norm, and vice versa.

I have chosen my case studies because of their power to illuminate actual practice. But, notionally, they might be linked by the question: "Can one hear the shape of a drum?" If you listen to a drum, and hear its resonant tones, can you infer something about the size and shape of that drumskin? This is a beautiful and archetypal problem that epitomizes much of our substantive mathematical discussion.[7] More generically, can you connect the sound spectrum to the geometry? The answer is a qualified Yes: For example, given our ordinary experience, we know that a bigger drum has deeper tones. More technically, you can hear its area, its perimeter, and whether its drumskin has punctures or holes. More generally, however, the answer is No, for there are inequivalent shapes with the same sound spectrum.

A drum's loudness is proportional to the number of resonant tones, essentially the story of the central limit theorem of statistics, the subject of chapter 2. That the holes in the drumskin have consequences for the resonant tones is just the connection between spatial facts and set-theoretic facts and the calculus that is the foundation for topology, and which is also a source of topology's internal tensions, the theme of chapter 3. The connection between the shape and the sound of the drum turns out to model the balance between the electrical forces among the electrons and nuclei and the electrons' angular momentum within ordinary matter, a balance that ensures that such matter does not implode, the story of chapter 4. And the connection between the shape, the sound, and an accounting or partition function which encodes the resonant frequencies and their loudness, is just the analogy explored in chapter 5.[8]

"Weyl's asymptotics," as all these facts are sometimes called, is a Christmas tree on which the various themes of this book might be arrayed. And it is Weyl (in 1913) who reworks Riemann's ideas and makes them the foundation for much of modern mathematics. Hence, this is a book about the legacy of Riemann (and of Maxwell), and of Weyl, too.

Another theme is the algebraicization of mathematics. Topology becomes algebraic, both point-set and combinatorial. And more generally, a logic of analogy, functorial relationships (a picture of A in B) that mirror both objects and the relationships between them, model the systematics of analogy.

In chapter 6, I describe several themes common to both nineteenth-century urban development and nineteenth-century mathematics. My thesis here is not about influence of one domain upon the other; rather, both domains appear to draw on the same sorts of ideas. I begin with a description of what I call the Library of Mathematics, and how mathematicians and physical scientists add volumes, borrow volumes, and sometimes usefully deface volumes they have borrowed and then return them in a transmogrified form. Then I describe how our actual bodies are employed in mathematical work (without at all suggesting that proverbial Martians would have different mathematics than ours). And, finally, I survey the theological and philosophical tradition concerning the relationship of God's infinitude to notions of the mathematical infinite. Again, the lesson in this chapter is that we need not be reductionist or relativist when we place mathematics within its larger culture. And, that historical-philosophic analysis can be helpful for understanding the mathematics as mathematics, and just how technical details matter.

II

The following survey of the next four chapters highlights their main themes, indicates relationship of the studies to each other, and provides an account of the argument absent of technical apparatus.

CONVENTION

Means and variances and Gaussians have come to be natural quantitative ways of taking hold of the world.[9] Not only has statistical thinking become pervasive, but these particular statistics and this distribution play a central role in thought and in actual practice: in the distribution of the actual data; the assumed distribution of that data; or, the distribution of the values of the measured statistic or estimator. Such conventions are entrenched by statistical *practice*, by deep mathematical *theorems* from probability, and by *theorizing* in the various natural and social sciences. But, entrenchment can be rational without its being as well categorical, that is, excluding all other alternatives—even if that entrenchment claims to provide for categoricity.

I describe the culture of everyday statistical thought and practice, as performed by lay users of statistics (say, most students and often their teachers in the natural and applied social sciences) rather than by sophisticated professional statisticians.[10] Professional statisticians can show how in their own work they are fully aware of the limitations I indicate, and that they are not unduly bound by means-variance thinking about distributions and data (as they have assured me); and, that much of their work involves inference and randomization, which I barely discuss. Still, many if not most ordinary users of statistics in the social sciences and the professions, and even in the natural sciences, are usually not so aware of the limitations in their practice.

A characteristic feature of this entrenchment of conventions by practice, theorems, and theorizing, is its highly technical form, the canonizing work enabled by apparently formal and esoteric mathematical means. So, it will prove necessary to attend to formal and technical issues.

We might account for the naturalness of means and variances in a history that shows how these conventions came to be dominant over others. Such a history shows just how means and variances as least-squares statistics—namely, the mean minimizes the sum of the squares of the deviations from the midpoint, more or less independent of the actual distribution—were discovered to be good measures of natural and social processes, as in errors in astronomical observation and as in demographic distributions. There is now just such a literature on the history of statistics and probability.[11] Here, however, I shall take a different tack, and examine the contemporary accepted ahistorical, abstract, technical *justifications* for these conventions, justifications which may be taken to replace historically and socially located *accounts*. More generally, these conventions come to be abstractly justified, so that least-squares thinking becomes something less of an historical artifact and rather more of an apparently necessary fact. Socially, such justifications, along with schematized histories, are then used to make current practice seem inevitable and necessary. Put differently: What might be taken as a matter of Occam's razor—for are not means and variances merely *first* and *second* moments, averages of x and x^2, and so are just good first and second order approximations, so to speak?—requires in fact a substantial armamentarium of justification so that it can then appear as straightforwardly obvious and necessary.

One might well have written a rather more revolutionary analysis, assuming a replacement theory had more or less arrived and that means and variances and their emblematic representative, the Gaussian, had seen their day. In a seminal 1972 paper by the mathematician Peter Huber, much of what I say here about the artificiality of means and Gaussians is said rather more condensedly, and a by-then well worked out alternative is reviewed.[12] Huber refers to the "dogma of normality" and discusses its historical origins. Means and Gaussians are just mutually supporting assumptions. He argues that since about 1960 it was well known that "one never had very accurate knowledge of the true underlying distribution." Moreover, the classical least-squares-based statistics such as means were sensitive to alterations or uncertainties in the underlying distribution, since that distribution is perhaps not so Gaussian, and hence their significance is not so clear, while other statistics (called "robust") were much less sensitive.

Now if one were to take, say, Fredrick Mosteller and John Tukey's 1977 text on statistics as the current gospel, then the alternatives hinted at here and by Huber are in fact fully in place. As they say, "Real distributions often straggle a lot compared to a normal distribution."[13] Means and variances and Gaussians are seen to be an idiosyncratic and parochial case. With modern computational and graphical capacities, and the technical developments of robust statistics, we are in a new era in which we can "play" with the data and become much more intimately acquainted with its qualities, employing sophisticated mathematics to justify our modes of play.[14] I should note that obituaries for John Tukey (2000) suggest much wider practical influence for his ideas than I credit here.[15] Moreover, probability distributions much broader than the Gaussian do play an important role in the natural and the economic sciences. And computation allows for empirical estimates of variation (bootstrap, resampling).

THE FIELDS OF TOPOLOGY

The third chapter describes some of the fundamental notions in topology as it is practiced by mathematicians, and the motivations for topological investigations. Topology comes in several subfields, and their recurrent and episodic interaction is the leitmotif of this chapter.[16] Such a philosophy of topology is about the mathematics itself.[17] What are the ideas in this field, and how do they do the work? (So in a first book on topology, the Urysohn Lemma, the Tietze Extension Theorem, and the Tychonoff Theorem are so described as embodying "ideas."[18])

Topology might be seen as the modern conception of continuity, in its infinitesimal and its global implications. The great discovery was that once we have a notion of a neighborhood of a point, the proverbial epsilons and deltas are no longer so needed.[19] A second discovery was that we might approximate a space by a tessellation or a space-frame or a skeleton, and from noting what is connected to what we can find out about the shape of that space. And the notion of space is generalized, well beyond our everyday notions of space as what is nearby to what. A third was that a space might be understood either in terms of neighborhoods *or* tessellations or, to use an electromagnetic analogy, in terms of localized charges *or* global field configurations, sources *or* fields. In general, we use both formulations, one approach more suitable for a particular problem than the other. And a fourth was that we might provide an algebraic account of continuity.

7

In the practice of topology, some recurrent themes are: (1) decomposing something into its putative parts and then showing how it might be put together again; and, how those parts and their relationships encode information about the object; (2) employing diagrams to make mathematical insights and proofs obvious and apparently necessary; and, (3), justifiably treating something in a variety of apparently conflicting ways, say that Δx is zero and also nonzero (as in computing the derivative), depending on the context or moment—themes more generally present in mathematics.[20] Another recurrent theme is the connection of local properties to global ones, when you can or cannot go from the local to the global, and the obstructions to doing so. A much larger theme is the algebraicization of mathematics, so that mathematical problems from analysis or geometry or topology come to be expressed in algebraic and formal terms. And this theme has profound implications for the practice of topology and, eventually and in return, for the practice of algebra.

FIGURE 1.1: Brouwer's indecomposable plane, from "Zur Analysis Situs," *Mathematische Annalen* (1910).
FIGURE 1.2: Isles of Wada, from Yoneyama, "Theory of continuous sets of points," *Tôhoku Mathematics Journal* (1917)

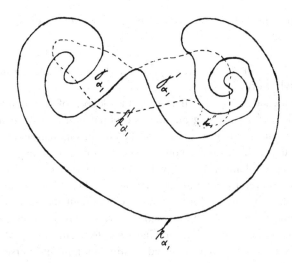

FIGURE 1.3: Proof diagram, Figure 5, from Brouwer, "Continuous one-one transformations of surfaces in themselves," *KNAW Proceedings* (1909).

We might better appreciate the tension between set theory and algebra in topology by reviewing some of the work of L.E.J. Brouwer (1881–1966). In 1909, Brouwer develops a beautiful and remarkable example, an "indecomposable continuum," accompanied by an extraordinarily handsome two-color diagram (as originally printed in the journal).[21] Here, a curve splits the plane into three parts, of which the curve is the common boundary; it is a continuum that is not the union of two sub-continua. A way of presenting this example (the "Isles of Wada"), begins:

> Suppose that there is a land surrounded by sea, and that in this land there is a fresh lake. Also, suppose that, from these lake and sea, canals are built to introduce the waters of them into the land according to the following scheme. . . . On the first day a canal is built such that it does not meet the sea-water [top map of Figure 1.2]. . . . The end point of this canal is denoted by L_1.
>
> On the second day, a canal is built from the sea, never meeting the fresh water of the lake and canal constructed the day before, and the work is continued . . .

> On the third day, the work is begun from L_1, never cutting the
> canals already built, . . . [22]

Diagrams such as in Figure 1.3 appear in his 1910 and earlier "Cantor-Schönflies" style work.[23]

Brouwer then developed a notion of the degree of a continuous map or function, the n in z^n (a winding number, the number of times a map winds the circle around itself, z^n doing it n times, or the curl or circulation, or the index or charge). And, using "fantastically complicated constructions" he proves all of the then outstanding problems in topology, including "invariance of dimension," that dimension is preserved by continuous mappings (1911).[24] His notebook in which he begins to develop the notion of what we now call the Brouwer degree is entitled by him, "Potential Theory and Vector Analysis" (December 1909–January 1910), in effect, going back to Gauss and to Riemann and Maxwell.[25]

Brouwer's 1910 proof that a closed continuous non-selfintersecting line in the plane divides the plane into two disjoint parts (the Jordan curve theorem) employs set-theoretic topological devices (curves that are deformable into each other, "homotopy"), and combinatorial devices (those tessellations, a "linear simplicial approximation").

However, Brouwer never employs the algebraic technology developed earlier by Poincaré for revealing the invariant features of an object in various dimensions and their relationship (namely, homology and chain complexes). It took some time for these algebraic technologies to be incorporated systematically into topological research, so that eventually combinatorial device would be interpreted through algebraic machinery, Betti numbers becoming Betti (or homology) groups. (The crucial figure is Emmy Noether (1882–1935) and her (1920s) influence on P. Alexandroff and on H. Hopf, and independently in L. Vietoris's work.) For until 1930 or so "set theory lavished beautiful methods on ugly results while combinatorial topology boasted beautiful results obtained by ugly means."[26]

Here, we shall be following the Riemannian path of Felix Klein (1849–1925) and Hermann Weyl. Eventually, the various paths lead to a homological algebra—the wondrous fact that the homology groups project topology into algebra. Just how and why algebra and topology share similar apparatus, why there is a useful image of topology in algebraic terms and vice versa—defines a tension of the field, as well as one unifying theme: decomposition or resolution of objects into simpler parts. Another theme is how an algebraic account of a topological space in terms of functions that define the points of that space (C* algebras), provides an algebraic model of a geometric point.

The obstructions to such decomposition, the holes in objects and the like, are expressed "dually" in what is called cohomology, the kinds of functions that can live on a space. For example, in elementary calculus when you integrate a function from a to b the path you follow does not matter; the value of the integral is the difference of the antiderivative evaluated at b minus its value at a. But, in a plane, the holes you enclose by such a path may make a difference, the road you travel from a to b, its tolls, its hills and valleys, will determine your total travel cost. And going around in a circle sometimes, as on a Möbius strip, does not get you back to where you began.

In textbook form, this history might be summarized as follows:

> The basic idea of homology comes from Green's theorem, where a double integral over a region R with holes is equal to a line integral on the boundary of R.

Namely, it is a matter of Advanced Calculus: Linear boundaries enclose planar surfaces; studies of the boundary integrals tell you about the surfaces or insides. Moreover,

> Poincaré recognized that whether a topological space X has different types of "holes" is a kind of connectivity.

But then algebra begins to have a role.

> What survives in this quotient [of cycles over actual boundaries] are the n-dimensional holes, and so these [quotient or homology groups] measure more subtle kinds of connectedness. For example $H_0(X)=0$ means that X is pathwise connected. . . . A key ingredient in the construction of homology groups is that the subgroups of cycles and boundaries can be defined via homomorphisms; there are boundary homomorphisms ∂ . . . with ker∂ [kernel—what is mapped into zero] are the cycles and im∂ [image of the map] are the boundaries . . .

And this mode of decomposition has already been developed:

11

> An analogous construction arose in invariant theory a century ago. . . .
> we seek "relations on relations" (called syzygies). . . . Hilbert's theorem
> on syzygies says that if this is done *n* times . . . The sequence of . . . free
> modules so obtained can be assembled into an exact [roughly, a nicely
> decomposing] sequence . . .[27]

As preparation for what is to come, at the end of the chapter there is an appendix
on the Ising model of ferromagnetism (a permanent magnet). I should note once
more, there is some deliberate repetition among the discussions of examples, in
order to make each discussion reasonably self-contained.

IDEAS AND CALCULATION

Some calculations and proofs appear magical at first. Technical details that are
apparently unavoidable in the first proofs seem to be pulled out of air. In part,
you do not see the piles of paper on which unproductive paths are worked out;
you only see what actually worked. You do not fully appreciate the personal
toolkit of techniques possessed by a mathematician, tools that have worked for
her in the past. Yet in retrospect those calculations turn out to build in the
deepest ideas and objects, in effect what you must do (your jury-rigged devices)
turns out to be done for a good reason, and the objects you uncover have a life of
their own. But again, you only understand that after further proofs and
calculations by others. Again, if some particular device is needed it is likely that
that device will point to objects and properties you only discover latterly.

On the other hand, ideas need to be made concrete through calculations and
the invention of technical devices. To make precise what you might mean by
unique factorization of numbers will lead to ideal numbers and algebraic
numbers. Again, we rarely hear about ideas that are not productive in proof,
theory, and calculation.

In effect, we have the phenomenologist's "identity in a manifold
presentation of profiles." An initial proof or calculation is then supplemented by
many others, and in time one has a sense of what is really going on. An idea,
realized in various contexts and calculations and examples becomes something
real for us.

> Lenard and I began with mathematical tricks and hacked our way through a forest of inequalities without any physical understanding. (Freeman J. Dyson, on their proof of the stability of matter.[28])

Chapter 4 describes several long, complex, often apparently obscure proofs that apply classical analysis to mathematical physics: proofs of the stability of matter and of the existence of permanent magnets (technically, the latter are derivations of the spontaneous magnetization and correlations in the two-dimensional Ising model of ferromagnetism).[29] In particular, the original, early proofs and derivations have been notorious for their difficulty. I am also concerned with the sequence of such proofs, from preliminary efforts that show what might be done and set the problem, to a breakthrough that solves the problem, and then subsequent work that simplifies the proof or derivation, or finds an entirely new approach, or more clearly reveals the physics and the mathematics, and so allows for a more perspicuous derivation, or explains how and why the original breakthrough worked.

We shall examine the Dyson and Lenard (1967–1968) work on the stability of matter (that matter composed of nuclei and electrons won't collapse and then explode), Federbush's (1975) reformulation of the problem, Lieb and Thirring's vastly improved solution (1976), and the elaborate program of Fefferman and collaborators (1983–1996) to achieve even better bounds, as well as to rigorously prove the asymptotic formula for the ground state energy of an isolated atom. We shall also examine Yang's (1952) derivation of the spontaneous magnetization of the Ising model, and then Wu, McCoy, Tracy, and Barouch's derivation (1976) of the asymptotic correlation functions.

I am especially interested in how such lengthy and complicated proofs are exposited or presented, both to the reader and in effect to the original authors themselves, showing just what has been done. So I shall be referring to the sections and subsections of these papers (in their being divisions of the exposition) as providing some of the vital meaning of the papers. A long and complicated proof or a derivation is staged as an orchestrated series of larger movements, each with its own logic and sub-staging.

Of course, papers that are seen in retrospect to be nicely organized, may well be difficult to follow initially absent subsequent developments and insights. Here is C.N. Yang (1922–) on his reading of Lars Onsager's 1944 paper that solves the Ising model:

> The paper was very difficult because it did not describe the strategy of the solution. It just described the detailed steps, one by one. Reading it I felt I was being led by the nose around and around until suddenly the result dropped out. It was a frustrating and discouraging process. However, in retrospect, the effort was not wasted.[30]

It helps enormously to have someone (perhaps even the author) show you what is really going on, or why the paper is organized just the way it is—say at the blackboard or in a seminar presentation—at least the author's best understanding, even if they do not really understand why it works. Writing up a paper for publication may make it all too neat, hiding the often still-inchoate insights and entrees that inform the effort.

Still, I believe that any of these proofs when studied at the right level of detail, usually reveals a strategy and a structure that makes very good sense, even if at the most detailed levels one is clearing a path through an uncharted jungle, and at the surface the various parts seem to hang together but they do not make sense except that they do achieve the proof. It may not be apparent just what is that right level of detail. Subsequent work by others often provides the appropriate point-of-view and suggests that right level of detail.

At the end of the most tortuous of paths, there is often a comparatively simple solution, although sometimes it is a mixed reward. Onsager ended up with expressions in terms of elliptic functions and hyperbolic geometry. Dyson and Lenard ended up with the simple inequality they were searching for, albeit with a very large constant of proportionality. (Lieb and Thirring's was about 10^{13} times smaller, of order 10.) Fefferman and Seco found simple numerical fractions for the coefficients of an asymptotic series. Yang's derivation of the spontaneous magnetization provided in the end a simple algebraic expression. And Wu, McCoy, Tracy, and Barouch found that the asymptotic correlation functions were expressible in terms of Painlevé transcendents, distant relations of the Bessel and elliptic and trigonometric functions. These are hard-earned saving graces. They are temptations, as well, that would appear to say that if we really understood what was going on we should be able to achieve these simple solutions rather more straightforwardly.

We might still ask, Why do elliptic functions (or Painlevé transcendents, or Fredholm determinants, or Toeplitz matrices, or Tracy-Widom asymptotic distributions) appear in these derivations. Subsequent work may suggest a very good reason, often about the symmetries of the problem that are embodied in these mathematical objects. Or, what was once seen as an idiosyncratic but effective technique, turns out to be generic for a class of problems of which this one is an exemplar (technically, here about nonlinear integrable systems and their associated linear differential equations). Still, someone had to do the initial solving or derivation, that *tour de force* or miracle, to discover that nice mathematical object X at the end, so that one might begin to ask the question, "Why does X appear?"

The applications of classical analysis may still demand lengthy and intricate calculations, endemic to the enterprise. But our assessment of those calculations, and their estimates and inequalities, can be rather more generous than Dyson's. Dyson and Lenard's work shows an enormous amount of physical understanding, whatever the size of the proportionality constant. This is evident in the organization of the proof. Moreover, the aesthetics are more complicated than terms such as clean or messy might suggest. For we might see in the inequalities some of the harmonies of nature:

> Since a priori estimates [of partial differential equations] lie at the heart of most of his [Jean Leray's] arguments, many of Leray's papers contain symphonies of inequalities; sometimes the orchestration is heavy, but the melody is always clearly audible.[31]

ANALOGY AND SYZYGY

> ... members of any group of functions [U, V, W, ...], more than two in number, whose nullity is implied in the relation of double contact [namely, $aU+bV+cW=0$, a, b, c integers] , ... must be in syzygy. Thus PQ, PQR, and QR, must form a syzygy. (J.J. Sylvester, 1850[32])

$$P \text{———} Q \text{———} R$$

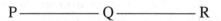

FIGURE 1.4: Three points in syzygy, as Sylvester might refer to them.

When three heavenly bodies, say P, Q, and R, are in line, they are said to be in syzygy—in effect, yoked together. In mathematics, a syzygy has come to mean a relation of relations. So, for example, the Hilbert Syzygy Theorem (1890), referred to earlier in the quote about homology theory, says that one can decompose a certain kind of object into a finite number of finite simply-composed parts, so that there are no residual relations or syzygies among them.

More generally, a syzygy is an analogy of analogies. Much modern work in mathematics forms relations of relations through forming functors of functors (functors being simultaneous transformations of spaces and mappings), structure-preserving ("natural") transformations among functors, and more generally pursues a philosophy of functoriality and syzygy.

In chapter 5 I want to describe how one quite self-conscious analogy is employed in mathematics, what is nowadays called the Langlands Program in number theory and representation theory—and which in earlier forms has borne the names of Riemann (1826–1866), Dedekind (1831–1916) and Weber (1843–1912), Hilbert (1862–1943), Artin (1898–1962), and Weil (1906–1998), among many others. I suggest how that analogy, which I shall call the Dedekind-Langlands Program, is analogous to another analogy, which I shall call the Onsager Program in mathematical physics, with an equally long list of names. In one sense this analogy of analogies is not too surprising, given that in each case one has a partition function that packages some combinatorial information about an object such as the integers or a crystal. One then studies analytical properties of that function to discover general properties of the associated object. In order to do so, one studies an analogous mathematical object that also yields that function. However, the details of the syzygy are rather more precise and informative than we might at first suspect.

The role of analogy in mathematical discovery and proof is a recurrent theme in writing about mathematics.[33] Analogy may be intuitive or formal, at the level of a single fact or between substantial theoretical structures. I came to write about analogies, and analogies of analogies, from recurrently bumping into them in quite specific contexts. Earlier, in *Doing Physics* (1992, 2012), I called them tools in a toolkit, models as analogies, which then appear ubiquitously. In my subsequent work on how mathematics is employed in mathematical physics (*Constitutions of Matter*, 1996), I had come to see in rough outline further connections between the various kinds of mathematics and the varieties of physical models of matter.[34] What was then striking to me was my subsequent reading of André Weil's (1940) description, drawn from Richard Dedekind's work of the 1880s, of a triple of parallels or analogies among different fields of mathematics: geometry, function theory, and arithmetic—respectively: Riemann surfaces and analytic functions, algebraic functions, and algebraic numbers. Weil called these parallels "un texte trilingue," or a Rosetta stone (he called two of the languages Riemannian and Italian). Weil's triple deliberately echoed Riemannian themes, linking curves to the complex spaces in which they reside, and algebraic

16

themes linking curves to algebraic function theory modeled on the abstract theory of algebraic numbers (due to Dedekind and Heinrich Weber, 1882).[35]

More to my point, Weil's triple paralleled the triple that was discernible in the mathematical physics describing a gridwork of interacting atoms, what is called lattice statistical mechanics, exemplified by the variety of solutions to Ising model.[36] Before me was a relation of relations, an analogy of analogies, a syzygy.

Roughly at the same time I was reading about elliptic curves—in part to try to understand a bit of the then recent proof of Fermat's Theorem by Andrew Wiles. By way of some bibliographic guidance provided by Anthony Knapp, in his book *Elliptic Curves* (1992), I then came upon the Langlands Program and saw how it was the contemporary version of Dedekind's analogy.[37] I cannot claim to understand much of the technical features of the Langlands Program. But I believe that the Dedekind-Langlands and the Langlands programs do provide for an analogy with the Onsager Program in mathematical statistical mechanics (and others have so indicated in the scientific literature).[38]

What the mathematicians find generically, the physicists would, for their particular cases, seem to have taken for granted. And what the physicists find, the mathematicians have come to expect. The syzygy illuminates its various elements, leading to a deeper understanding of what is going on in the mathematics and what is going on in the physics. Such relations and syzygies reflect both formal mathematical similarities and substantive ones. For a very long time, scientists (whom we now classify mostly as physicists) and mathematicians have borrowed models and techniques from each other. And, of course, many potential syzygies do not in fact work out.

We do not know the meaning of an analogy, just how something is analogous to something else, until we see how it is worked out in concrete cases. Still, there is excitement to be found in the prospect of discovery, in drawing these diagrams of relationship. Often, one is figuring them out on one's own, or hears about them orally. Rarely are these written down and published in the mainline scientific literature, for they appear so tenuous yet tempting, so unprovable for the moment even if it would seem that they must be true. A diagram I drew for myself about the Ising model, the Onsager program, and the Langlands program—before reading Weil—made it possible for me to be struck by the syzygy. (See note 34.) As Weil points out, this is only the beginning of the mathematics. It is that beginning I am describing here.

III

In 1960, the distinguished theoretical physicist Eugene Wigner published an article, "The Unreasonable Effectiveness of Mathematics in the Natural Sciences," which concludes, "The miracle of the appropriateness of the language of mathematics for the formulation of the laws of physics is a wonderful gift which we neither understand nor deserve. We should be grateful for it and hope that it will remain valid in future research and that it will extend, for better or for worse, to our pleasure, even though perhaps also to our bafflement, to wide branches of learning."[39]

There is a substantial literature discussing and explaining Wigner's claim. His article is subtle and wide-ranging, pointing out the value of new concepts from mathematics, and concerned with the uniqueness of physical theories. But it is the title that has crystallized responses to it.

Now, there is another effectiveness that strikes me as at least as "unreasonable," namely, *the unreasonable effectiveness of physics in mathematics*, the appropriateness of the language of physics for the formulation and foundation of mathematical concepts and theories, and for suggesting new directions for the development of those theories. (I will use "physics" here implicitly referring to natural and social science as well.) Physicists invent mathematical devices to do their work, or discover them in the Library of Mathematics (as Wigner suggests) to do new work: whether they be Newton's calculus; Maxwell's coupled set of partial differential equations (which eventually become matters of differential geometry and topology); Heaviside (1850–1925) and Dirac's (1902–1984) delta function (and eventually Schwartz's distributions); Heisenberg's matrix mechanics for quantum theory (leading to developments in operator theory); Schroedinger's partial differential equation, or more recently, Witten and others' topological field theories.

What is remarkable is that the mathematicians can then take these practical devices and new uses of their concepts, and not only make them rigorous but some of the time make them the foundation for a rich field of mathematics. In part, as I indicate above, the physicists are actually borrowing some mathematics that is already extant. But their use of the mathematics extends it in unexpected directions, and those directions are then articulated by mathematicians into rich theories. Also, presumably, the physicists are in part driven by physical phenomena or analogies to known phenomena, so that it is the actual world as physicists understand it that then leads to these articulations.

Feynman's path integral approach to quantum mechanics (originally suggested by a remark by Dirac), the various groups (including nonabelian or noncommutative groups) used by Wigner and others to understand the realm of particles and symmetries, the scaling symmetry methods developed by Kenneth Wilson, the discovery that the otherwise unemployed Painlevé transcendents (descendants of the sines and cosines) play a natural role in some physical problems (by Wu, McCoy, Tracy, and Barouch),...—all demanded and fruitfully received further mathematical development and rigorization. The Standard Model of the elementary particles, and more generally, quantum field theories, have not only borrowed from the Library of Mathematics. They have, as well, presented problems to mathematicians that are proving fruitful and interesting within mathematics itself.

Two further examples, we discuss in later chapters: Charles Fefferman and collaborators' (pure mathematicians, all) derivation of the ground state energy of atoms in terms fractional powers of their number of electrons (Z^n), not only makes rigorous earlier derivations by physicists and extends those derivations to higher powers of Z, it develops as well innovative methods of classical analysis. Here a distinguished mathematician takes on the terrific rough-and-ready work of physicists, makes it rigorous, extends that work, and advances the mathematics itself. Second, in derivations of the properties of the Ising model of a phase transition, and related models, Rodney Baxter and collaborators (all physicists) have developed many ingenious methods of solution that demand deeper work by mathematicians (concerning analyticity and integrability) to make sense of their meaning and why they work.

Some mathematics is developed without any influence from the physicists, perhaps most of it. But the demands of the geneticists, physiologists, engineers, computer scientists, and others in the natural sciences, also provide impetus. The usual line derived from G. H. Hardy is that some of pure mathematics will never have connection to any application in the natural sciences is often denied by actual applications. But I am quite willing to believe that lots of mathematics, even if it is eventually applied, is autonomously developed by the mathematicians.

Also, I like to think that the physicists and other natural scientists, and social scientists as well, provide the mathematicians with rough-and-ready methods and theories, begging for rigorization and generalization. With luck, that effort launches interesting mathematics that is above and much beyond the original impetus. That may well eventually serve the physicist's or other scientists' needs, or not.

None of this is surprising when we think of other human activities. Some material or method is discovered by prospectors or inventors. Eventually, that material or method is used for practical manufacture of things we find useful and convenient. In effect that material or method becomes what economists call a resource. And scarcity and invention may then suggest the need for other materials or methods that until then did not seem of value.

I remain agnostic about the miracles referred to by Wigner. Rather, historical study will suggest how varied are the influences between mathematics and physics: how often mathematics is not of use to scientists, how often scientists search for mathematics and make do with what there is, and how often physicists' rough-and-ready inventions do not lead to any rich mathematical theories, where "how often" should be read as "how often or not."

I am also struck by the effectiveness of mathematics in the mathematical realm, developments in algebra proving useful in topology, for example. Again, some mathematics does not prove useful to other subfields.

IV

Wigner begins his article with the following:

> There is a story about two friends, who were classmates in high school, talking about their jobs. One of them became a statistician and was working on population trends. He showed a reprint to his former classmate. The reprint started, as usual, with the Gaussian distribution and the statistician explained to his former classmate the meaning of the symbols for the actual population, for the average population, and so on. His classmate was a bit incredulous and was not quite sure whether the statistician was pulling his leg. "How can you know that?" was his query. "And what is this symbol here?" "Oh," said the statistician, "this is pi." "What is that?" "The ratio of the circumference of the circle to its diameter." "Well, now you are pushing your joke too far," said the classmate, "surely the population has nothing to do with the circumference of the circle."

Now the population has nothing to do with the circumference of the circle. Rather, it is a statistical measure of the population that leads to the use of pi. As we shall see, again and again, when we try to characterize objects in terms of abstractions, we find pi-like and other mathematical objects.

Now let us move forward fifty years.

Two friends from college meet after being out of touch for twenty years. Bernie Dedelands has become a mathematician doing number theory and representation theory. Larry Yangster has become a physicist doing statistical mechanics. When Bernie asks Larry to tell him about his current work, Larry mentions that he is working on a statistical counting problem that has three main methods of calculating its partition function, the physicists' way of packaging counting information about a system. The natural logarithm of the partition function is proportional to the thermodynamic free energy that the chemists so value. Bernie remarks that it is amazing that the packaging function could be so related to something a chemist would be interested in. Then, Larry tells Bernie how in order to solve the problem one might create a matrix that does the counting work, or one might look at the symmetries of the system, or one might focus on the analyticity of the solution of a functional equation and a Riemann surface that goes with an elliptic curve. In fact, although the proofs are not always there, Larry mentions that he has an intuition for when one might get away with the assumption of analyticity, and his work using functional equations to solve the problem are well-known among the physicists. It's how he got tenure.

Larry also mentions it would appear that the matrices might be seen as group representations, their trace or group character being the partition function and the group itself is often parametrized by the argument u of elliptic functions, $sn(u,k)$, that group members commute if they have the same modulus, k, even if they have different arguments, and that there are connections between modulus k and modulus $1/k$ groups members' matrices (in effect low and high temperatures). And there is an equation that gives an account of the commutativity of the matrices, and it is intimately related to those elliptic functions.

(The elliptic functions are special in that $f(k)$ and say $f(1/k)$ are more generally related. Elliptic functions are said to be "modular," in the sense that, $f((az+b)/(cz+d)) = (cz + d)^m f(z)$, if ad-bc=1, so that $f(-1/z) = z^m f(z)$, and $f(z+1) = f(z)$ or perhaps $= q \times f(z)$.)

Larry goes on to mention that he is able to compute the partition function through counting or through modularity, whether it be through the relevant counting matrices or the group representations of the symmetries of the matrices or the functional equation.

Larry then mentions that there is something called universality, so that if two systems have similar symmetries, the crucial features of the partition functions turn out to be the same, even if the exact details of the systems differ substantially.

At this point, Bernie has to interrupt. He says, You must be kidding. You tell me that a physical system of this sort exhibits modularity!

Larry then says, Oh, we physicists have been working on this for the last seventy years, and what you mathematicians call modularity is pervasive in particle physics and condensed matter physics. We call it the renormalization group, and through a variety of specific studies we have discovered how such a scaling symmetry, that is what we call it, actually occurs. When you go from one microscopic scale to a less microscopic one, such scaling begins to appear.

Bernie, would still seem to be in a trance and at the same time it would be appear that he has the experience of déjà-vu all over again (ala Woody Allen). He says, Are you telling me you have three different ways of doing your computation: arithmetically, functionally, and using symmetries; that you get analytic continuation and functional equations for free; that group representations appear naturally in terms of the counting technologies, and there are other groups reflecting what you call the modularity of elliptic functions; that you have equations parametrized by those elliptic functions and their variables; and that you have connected counting to the symmetries of elliptic functions? And, that you have something you call universality, namely that the partition function (or something related to it) would seem to be quite independent of the details of the system, as long as the symmetries are the same?

Larry says, you've got it just right.

Then Bernie says, well I got it just right because you are describing some of the main features of my own work: counting functions (we call them L-functions), defined either by counting or by symmetries (what we call automorphies), that we might find analytic continuation and functional equations by one method but not the other, that what you call universality seems to be what we call functoriality. And those elliptic or modular functions have a natural role in our theories. They are not curiosities but central to our work and our understanding. And that matrices or group representations are ubiquitous in our work. More amazingly, there is an analogy described over almost 150 years ago by Dedekind, trying to make the work of Riemann more rigorous, and more recently made more explicit by André Weil—an analogy between the arithmetic of counting and number theory, the algebra of function theory, and the geometry of Riemann and the Riemann surface. How in the world did you physicists get all of this stuff? You must have been reading contemporary mathematics. You are pretending to be ignorant, you cheat!

Larry then says, Everything I know about mathematics is likely well before 1950. I don't know anything about your business, really! I am not speaking in tongues when I talk about my work! What happened was that my colleagues kept calculating stuff in many different ways from about 1940 on, and we kept discovering these features. I am proud to say I discovered some of them, but I must admit that the early work has in it much of what we now see more explicitly.

How in the world did you mathematicians find those connections? Keep in mind that we found these connections as a consequence of brute-force calculations, done by many different people. There was no overarching theory. But you mathematicians would seem to have an overarching conception, what you sometimes call a "program" or a philosophy, and a pattern of analogy that would seem to be only sketchily supported, from what you say.

And Bernie says, You are right. Physics is such a peculiar subject. You take whatever mathematics you need, typically well after the mathematics is settled and of no research interest. You use it unsystematically, and then you find these phenomena and connections. We mathematicians have grand theories and programs, set by visionary colleagues, that guide or redeem our work. Your work seems so scattered, even if eventually you can tell a coherent story and even then often you don't much care about that.

Larry then says, How can mathematicians make such grand pronouncements when they have so little evidence, following gurus who are surely smart, who may see more widely than most other scientists? But they are human and are perhaps no more reliable than other leaders.

Larry and Bernie are pleased by the effectiveness of mathematics in mathematics, the effectiveness of physics in physics. The fields would seem to use whatever they have at each stage to move on, synthesize what they have learned, and move further along. But they are awestruck by the remarkable confluence of two lines of work, one motivated by a simple physical model, one motivated by a grand vision. They agree that they are not much struck by what Wigner called the unreasonable effectiveness of mathematics in the natural sciences, or even the unreasonable effectiveness of physics as a natural science to concretely model otherwise quite grand mathematical visions. Rather, it is that meeting ground of two different groups of scientists, albeit on very different levels of abstractions, effectively ignorant of each other's work for 150 years in mathematics, and 70 years in physics.

They both know that they will have to get down to the nitty-gritty of actual formulas and algebra and group theory to check out if the correspondence is actual, for so far all they have is talk—talk that is striking nonetheless. For it is such talk that encourages scientists to pursue analogies, to see how well they might work out.

Larry can't resist rubbing it in to Bernie, for the mathematicians always lord it over the physicists, their lack of rigor, their ersatz proofs, their ideas that never quite work out. So he then says, You mathematicians are often blind to what is needed, going off on your own paths, the mathematics not there when the physicists need it. Here is the physicist Barry McCoy:

> This model is out of the class of all previously known models and raises a host of unsolved questions which are related to some of the most intractable problems of algebraic geometry which have been with us for 150 years... machinery from algebraic geometry which does not exist.
> ...
> In one approach the problem is reduced to the evaluation of a path ordered exponential of nonabelian variables on a Riemann surface. This sounds exactly like problems encountered in nonabelian gauge theory but, unfortunately, there is nothing in the field theory literature that helps. In another approach a major step in the solution involves the explicit reconstruction of a meromorphic function from a knowledge of its zeroes and poles. This is a classic problem in algebraic geometry for which in fact no explicit answer is known either. Indeed the unsolved problems arising from the chiral Potts model are so resistant to all known mathematics.[40]

And then Larry gives the coup de grâce. It's not only Newton and Fourier who in the course of doing their work, invented mathematics that is the core of your subject. In a different paper, McCoy then says, about the physicist R. J. Baxter:

> The search for solutions of the star triangle equation has been of major interest ever since and has led to the creation of the entirely new field of mathematics called "Quantum Groups." The Baxter revolution of 1971 is directly responsible for this new field of mathematics ...
>
> Mathematics is no longer treated as a closed finished subject by physicists. More than anyone else Baxter has taught us that physics guides mathematics and not the other way around. This is of course the way things were in the 17th century when Newton and Leibnitz invented calculus to study mechanics....[41]

And, the distinguished mathematician Robert Langlands remarks,

> By representation theory we understand the representation of a group
> by linear transformations of a vector space. Initially, the group is finite,
> as in the researches of Dedekind and Frobenius, two of the founders of
> the subject, or a compact Lie group, as in the theory of invariants and
> the researches of Hurwitz and Schur, and the vector space finite-
> dimensional, so that the group is being represented as a group of
> matrices. Under the combined influences of relativity theory and
> quantum mechanics, in the context of relativistic field theories in the
> nineteen-thirties, the Lie group ceased to be compact and the space to
> be finite-dimensional, and the study of *infinite-dimensional*
> *representations* of Lie groups began. It never became, so far as I can
> judge, of any importance to physics, but it was continued by
> mathematicians, with strictly mathematical, even somewhat narrow,
> goals, and led, largely in the hands of Harish-Chandra, to a profound
> and unexpectedly elegant theory that, we now appreciate, suggests
> solutions to classical problems in domains, especially the theory of
> numbers, far removed from the concerns of Dirac and Wigner, in whose
> papers the notion first appears.
>
> Physicists continue to offer mathematicians new notions, even
> within representation theory, the Virasoro algebra and its
> representations being one of the most recent, that may be as fecund.
> Predictions are out of place, but it is well to remind ourselves that the
> representation theory of noncompact Lie groups revealed its force and
> its true lines only after an enormous effort, over two decades and by one
> of the very best mathematical minds of our time, to establish rigorously
> and in general the elements of what appeared to be a somewhat
> peripheral subject. It is not that mathematicians, like cobblers, should
> stick to their lasts; but that humble spot may nevertheless be where the
> challenges and the rewards lie.[42]

V

In sum, what I hope to provide here is a more realistic, recognizable account of
some of mathematical work, an account focusing on convention, subject matter,
strategy and structure and calculation, and analogy. And the lessons are simple

and worth repeating: The meaning of the mathematics is not at all obvious, if we are asking what it says about the world (including mathematics), if only because the world including mathematics proves refractory to being so described mathematically. The subjects of mathematics, and its fields, are not foreordained; rather, they depend on what we can conceptualize and prove, what the world will allow as true. The staging of a mathematical analysis, whether it be a computation or a more abstract proof, is just that, a staging that demands as much as we can know about how the world works, so that what we are trying to say stands out from the formalism and the nitty-gritty work. And, we rarely know what to prove until we have a range of examples, analogies, and half-baked proofs that indicate to us what is likely to be true, in detail, in the specifics, in the concreteness. My aim is to provide a rather more robust account of mathematical practice, one that mathematicians will readily recognize and lay persons might appreciate. Mathematical practice is somewhat like rather more generic human activities. But, of course, it is, in the end, in the particulars, in the actual work, mathematics as such, having its own irreducible integrity.[43]

2

Convention

How Means and Variances are Entrenched as Statistics

Convention and Culture in Mathematics; The Central Limit Theorem. *The Establishment of Means and Variances*; Justifying Means and Variances; Exceptions and Alternatives to Means-Variance Analysis. *Theorems from Probability as Justifications for Means-Variance Analysis*; Itô, Lévy, and Khinchin: A Technical Interlude; Defending Randomness from Impossible Longshots and Coincidences. *Making Variances Real by Identifying Them with Enduring Objects and Other Features of the World*; Variances as the Canonical Account of Fluctuation; Noise; Einstein and Fisher.

> Everyone believes the law of errors [the normal law]. The mathematicians imagine it is an observational fact, and the experimentalists imagine that it is a mathematical theorem. (Poincaré[1])

I

CONVENTION AND CULTURE IN MATHEMATICS

How do mathematical ways of thinking of the world become conventionalized, taken as the way the world really is? Surely, they have to model the world adequately. But we then justify those models using mathematical characterizations of the model, and then apply those characterizations to the world. Such conventions are not arbitrary. But the arguments for their necessity are likely to be superseded if and when new models are needed—for what was once just the way the world is, may no longer be the case (even if the now-former models are still approximately true).

For most users of statistics, and presumably for most purposes, means and variances (the average, and the square of the standard deviation) are treated as the statistics of choice. Although alternatives are proposed, means and variances have proved remarkably resilient to challenge. How is such entrenchment of a convention maintained? Again, the convention has to model the world adequately, but that is apparently not probative. There are both mathematical and cultural accounts of the convention. The overlap and ramification of those accounts makes a convention appear all but inevitable, just the way the world is. And, in time, in the matter of course, a new convention might arise, and it, too, will be inevitable. How is that air of inevitability maintained?

It might be useful to think of ourselves as anthropologists trying to understand the practices of a culture—here, a culture of randomness as practiced in probability, statistics, and decision theory. Anthropologists and archaeologists find great variety among the world's cultures. But in any single culture they find a much-reduced variety. They show how the particular everyday practices and rituals of a culture or society are intimately linked in a network of practices and meanings that make the particulars of a culture apparently necessary within that culture. Perhaps the world could be otherwise, but in this place and time and society the culture is structured in this particular way, and for many interlinked reasons. A ritual or a technical practice that a society treats as universal and obvious is grounded in both empirical facts and in conceptual thoughtways.[2] These interlinked reasons, facts, thought-ways, and objects allow a culture to take its own practices as natural conventions, and to marginalize anomalies and discordant cases.[3] If the culture is coherent, it accounts for the details of a variety of particular and complex cases. So are conventions maintained in the mathematical realm as well.[4]

Mathematics is rich with possibilities or models for how the world might be structured and how it might develop dynamically. Only some of these models are applied widely and usefully—or at all; for many practical purposes, less than ideal models prove adequate. Other models may well become important in time, as new phenomena are encountered, or new techniques make some possible mathematical models more usable and applicable. Technologies such as computation and graphic software, or social media and GPS-equipped smartphones, provide new opportunities and problems for analysis. And those problems, such as computing standard errors for complicated variance functions (resampling methods such as the bootstrap, for example), may make essential use of new technologies and they demand from mathematics justifications for their employment.

There are many areas in which modern notions and computers are crucial, well beyond least-squares (or what was well known as of say 1950). In so far as statistics is a decision science, many scientific decisions involve very many comparisons rather than one or two (as in micro-arrays in biology); false discovery rates; concrete facts such as a particle's mass is always zero or positive; and use of very large data sets, in part to estimate confidence and to take advantage of your knowledge (what is called "empirical Bayes").[5]

Moreover, if we are dealing with random variables which have substantial mutual dependence, the asymptotic distribution may not be a Gaussian. The Wigner semicircle law for random matrices, and the Tracy-Widom distribution (see chapter 4) play much the same role as everpresent distributions as does the Gaussian.

On the mathematical side, one of the recurrent efforts is to better characterize mathematical objects. They may be understood in a variety of ways, the varied aspects giving a rounded picture of the object. Or, the objects may be specified ever more precisely so that they are distinctively or "categorically" characterized (although there might well be several such distinctive characterizations). If we adopt a mathematical model of the world, it may then be possible to characterize the model and so, by implication, the world in this distinctive way. So Dirac (1927) might employ a first degree partial differential equation to describe relativistic electrons, and then say that the electrons are "four-component spinors"—spin up and down, positive and negative electrons. We might then justify our model by saying that the world is just this distinctive way, or at least it is very closely approximated by just this distinctive way.

THE CENTRAL LIMIT THEOREM

The Gaussian distribution or the normal distribution or the bell-shaped curve is among other things the asymptotic limit or limiting shape referred to in the Central Limit Theorem ("Central" here means important, playing a central role): sums of nice enough random variables end up (asymptotically) looking like a suitable Gaussian. Two basic assumptions are built into probabilistic conceptualizations that eventually lead to the Central Limit Theorem: for lack of anything better, we might assume that random variables are (statistically) independent of each other; and, again for lack of anything better, we might assume that whatever is not forbidden will happen, and it will happen as often as it might otherwise be expected to happen (what I shall call plenitude[6]). Pure independence and plenitude might be called noise. The Gaussian distribution, as an asymptotic approximation to the distribution of the fraction of heads in a limiting infinite sequence of presumably independent unbiased coin tosses, namely the Bernoulli process (or the binomial theorem), exemplifies those assumptions as well as the Central Limit Theorem.

Of course, it may be unwise to depend on assumptions of independence and plenitude. In extreme situations, say with real coins and substantial bets, subtle dependences and biases are likely to exhibit themselves and it is worth discovering them. And even if all goes well, there will be apparent outliers and longshots that do occur.[7] The outliers may be taken as ordinary fluctuations, and assigned to the tails or extremes of a Gaussian (or, sometimes, a Poisson) distribution. Or, it might be shown why the outliers should not be placed "under" a Gaussian; they belong to some other distribution with fatter tails ("fat" being a term of art here), so that the apparent outliers are actually more likely than we had initially thought. Or they might be trimmed off, for there is likely to be impurities in the data.

Still, the Central Limit Theorem—which says that many random processes converge rapidly and asymptotically to the Gaussian distribution—plays a large role in everyday statistics. In particular, the Central Limit Theorem justifies the crucial role that variances play in statistics, and that variances are what should be added up when we add up random variables—much as the Fundamental Theorem of the Calculus justifies tangents as the suitable stuff that should be added up for smooth (functions of) ordinary variables.[8]

The esoteric and technical properties of what we eventually take to be natural and ordinary, such as the Central Limit Theorem, are just the properties that by the way, in effect, help to show that it is necessary. Technical properties of means and variances are taken to demonstrate that means and variances are not merely conventional, but the way things are and ought to be under a wide range of circumstances.

More generally, technical features of definition and proof can be shown to refer to rather everyday phenomena; those features are not just niceties for show. Mathematical devices, employed along the way in working out a problem, are not only tricks of the trade. Eventually, they can be shown to reflect the terrain and structure of the world, everyday and mathematical. Mathematical ideas and calculation are instantiated in the actual world.

II

THE ESTABLISHMENT OF MEANS AND VARIANCES

The formulation of probability that we now take as correct developed over the last two hundred years.[9] That the reduction of data, namely statistics, should be probabilistic is a second major achievement.[10] And that decisionmaking might be thought of in those terms is a third. So has a conception of randomness been entrenched in some everyday practices, justified as the way things are, and used as a principle against which other causes or mechanisms are measured. These developments in probability and statistics were supported by and suggestive of developments in the physical sciences, namely, physical objects and processes such as the ideal gas conceived molecularly (Maxwell, Boltzmann), or the (Brownian) motion of small particles suspended in a fluid, a motion that can be modeled by a random process known by its mean and variance.

As the great statistician R.A. Fisher (1890–1962) put it, "A statistic, to be of any service, must tend to some fixed value as the number in the sample is increased."[11] And as the historians tell us, the very great discovery that launched statistics was that means taken as arithmetic averages—"the combination of observations"—were in general more reliable than were individual measurements. Combining measurements, weighting them by their inverse error squared, would provide more stable measures.[12] Most of the time averages could be made to tend to some fixed value, and variances were good measures of how reliable were those averages. Means are good statistics or estimators. They do not move around much as the number of measurements, N, increases, compared to the variation of the individual measurements. (For means move diffusively, more like $\sqrt{N}$, rather than ballistically, like N.)

To be archetypal, all these statements go along with Gaussians as the fundamental probability distribution, where "go along with" has meant either an assumption or a consequence. Now, means can be justified by least-squares, minimizing the sum of the squares of the deviations from a point. They can also be justified as possessing good statistical properties (for example, "best linear unbiased estimators"). If the distribution of the data is Gaussian or normal, they maximize the likelihood; and, like many other estimators, the sampling distribution (needed for inference) is normal.

There are notable exceptional cases. If the underlying distribution is Cauchy $(1/\pi \times 1/(1+x^2))$—the so-called gunshot distribution, the distribution on a line of bullets from a gun that is rotating in a plane—then the variance is infinite. Averages do us little good since the proportional error does not decline with greater N. Here,

31

the median—the middle value, half the measurements above, half below—does the work of estimating the central point, and the average of the absolute value of the deviations measures the spread. The Census Bureau reports income medians, to better take into account the long upper tail of the income distribution. Sometimes, typical behavior and mean behavior are not at all the same, especially for a quantity that has a very wide range of values.[13] Thinking of the income distribution, the average is not at all typical since the average may be dominated by rarer anomalous values.

JUSTIFYING MEANS AND VARIANCES

Corresponding to the various empirical observations when the mean is typical, there developed mathematical justifications, sometimes ahead of the empirical observations, showing under reasonable conditions why means and variances were such good measures, why for example there is central bunching rather than several disparate peaks, or why rare anomalous values do not overshadow the mean. The justifications include: (1) Methods of least squares, and the method of maximum likelihood and the Gauss-Markov theorem, each of which under reasonable criteria picked out least squares and the mean and the variance as crucial. (2) Laws of large numbers, that the mean settles down as N increases. (3) Asymptotic limit theorems, about the distribution or shape of sums of random variables, which explained why means, for example, so securely and rapidly (in terms of N) settled down rather than bounced around, and why the sum distribution had to be Gaussian, fully characterized by its mean and its variance.

The historians tell us there were other possibilities developed earlier, such as using least absolute deviations and medians—namely, minimizing the sum of the absolute value of the deviations ($\Sigma |x_{midpoint}-x_i|$) rather than the sum of their squares, which can be shown to lead to the median as the midpoint (as for the gunshot distribution) rather than the mean. But those other possibilities mostly faded under the successful onslaught of least squares.[14]

Latterly, it was also noted that regression and correlation, taken as a measure of the relationship of two variables, show themselves most literally if the dependent and independent variables are expressed in terms of their standard deviations (the square-root of the variances) from their respective means. It was also noted that variances (and not standard deviations) might be analyzed into components (the "analysis of variance"). And, that means associated with each of those components were nicely related to the mean of the total once we take variances into account.[15] More generally, in adding up random variables one must "standardize," divide by the standard deviation, if one is to get the sums (of errors, say) to converge appropriately. And, in fact, as we discuss later, such norming constants are in some cases not standard deviations and $\sqrt{N}$.

So might go a condensed and selective history.

Whatever the history, nowadays most students learn about means and variances in a rather more canonical fashion in which least-squares, central limit theorems, methods of maximum likelihood, regression, and analysis of variance are catechistically invoked along the way. Devices that support that canonization include modes of calculation and visualization (such as the Pythagorean theorem), criteria for estimators, and theorems. In more advanced courses, they learn about bootstrap methods, with the proviso that they are employed because of difficulties in inverting complicated variance-covariance matrices, and the Gaussian assumption is unhelpful. Or, they learn about robust and resistant methods, such as trimmed means, when the data is likely a mixture of distributions. Or, they learn about James-Stein or shrinkage estimators, when one is simultaneously estimating more than two parameters, as in estimating the season batting average of a bunch of hitters, given their averages at All-Star time (in effect a Bayesian estimator)—yet these estimators are biased and have other tabooed features, but they are more accurate for these purposes. *Pace* such peculiar cases, the convention is retained.

As for calculation, what once proved crucial practically—before modern computing—were the closed analytic forms for computing least-squares statistics: formulas or calculator keystroke sequences, the ones memorized for the exams. There were as well lovely mechanical computing algorithms suited to large chug-a-chug electromechanical calculators.[16]

For example, for small numbers, if you put x at the far right and y toward the far left of a ten digit entry, and then square and accumulate you will get the quantities needed for doing regression: Σx^2, Σy^2, and $2\Sigma xy$. (There is a 21 digit accumulator.) And those forms and algorithms are often pointed to when statisticians want to tell themselves how least squares became entrenched, despite a perhaps more sophisticated but esoteric understanding of alternatives then available within the science.

Moreover, the development of analytic forms for small sample statistics—t, chi-square, F—and tables of their numerical values, allowed the least-squares orthodoxy and the fundamental character of the Gaussian to be maintained even in these small-N regimes. Fisher's extremely influential *Statistical Methods for Research Workers* (1925) is the canonization of all of this. Sample distributions needed for inference, for statistics that are not Gaussian and are larger-tailed, are a consequence of well-understood small-N effects (or of our interest in variances as for chi-square)—not anything fundamentally new.

In general, many a statistic would not be so readily calculated if one assumed other than Gaussians or perhaps a Poisson for the actual data (*pace* the Central Limit Theorem, which says that eventually Gaussian behavior "should" dominate) or for the predicted sample distribution. Even medians are not so easy, since you have to order the data efficiently (for large N, a nontrivial if algorithmically solvable task). Conversely, with comparatively inexpensive and powerful computational resources, such algorithms became feasible, and we can perform Monte Carlo simulations for the distribution of any particular statistic and its spread. We are freed from some of the previous practical constraints.[17] And our methods are justified by mathematical theory. But students still first learn the old convention.

Visually, if we assume the conventional Euclidean measure of distance or metric (the Pythagorean theorem: $(\text{Distance})^2 = (\Delta x)^2 + (\Delta y)^2 + (\Delta z)^2$), then geometrical diagrams and vectors may be used to show how statistical ideas and formulae have natural and seemingly inevitable interpretations, once we choose a suitable hyperspace. Statistical independence is expressed as orthogonality or perpendicularity of appropriate vectors; least squares as a principle is built into the metric; analysis of variance is effectively the Pythagorean theorem in reverse, the square of the hypotenuse decomposed into the sum of the squares of the sides of the triangle; linear regression is seen as a projection of the dependent variables onto the space of the independent variables. And, again, if we standardize, then regression as a tending toward or regressing to the mean is manifest. What was for Fisher a geometrical way of thinking, and a way of dealing with multiple integrals, becomes in one intermediate statistics text a patent justification, on page 376, of the first 375 pages of text and algebra—namely, an illustration of the various statistical objects as sides of right triangles in the appropriate hyperspaces.[18]

Other modes of thinking statistically might correspond to a different geometry, perhaps just as patent. The minimization problem of least absolute deviations, minimize $\Sigma\ |x_{\text{midpoint}}-x_i|$, can be shown to be equivalent to linear programming as a method of optimizing. Hence, we are led to the simplex algorithm and its geometry of searching for the most extreme point of an irregular multidimensional polygon.[19] And given that the simplex algorithm in actual practice scales roughly linearly with the number of variables, this is not so onerous a task. And there is a mathematical literature that shows why the simplex algorithm is so effective most of the time.

Not only are means and variances to be calculationally and visually justified, they are also to be justified in terms of phenomenological criteria, which are taken to be obvious or intuitive, or we so learn to see them as such, embodying quite reasonable hopes we might have for a statistic. Here we think of Fisher on sufficiency, efficiency, unbiasedness, and consistency (and, again, the often invoked BLUE = best linear unbiased estimator).[20] Here is Fisher in a didactic mode:

> Examples of sufficient statistics are the arithmetic mean of samples from the normal distribution, or from Poisson series; it is the fact of providing sufficient statistics for these two important types of distributions that gives to the arithmetic mean its theoretical importance. The method of maximum likelihood leads to these sufficient statistics when they exist.[21]

Fisher links practical, theoretical, and phenomenological factors. Sufficient statistics "contain in themselves the whole of the relevant information available in the data," even for small samples.[22] A distinguishing feature of some such statistics is that, having chosen the right statistic, the statistic of the combination of M samples or sets of measurements is just the statistic of the M statistics of the samples.[23] Namely, the mean of the combined M same-sized samples is just the mean of the M means of the samples; the maximum of the M samples is just the maximum of the M maxima of the samples. Such additivity, as we might call it, implies that in combining samples there is no interaction among the samples. In effect, each sample is an individual. These criteria lead to statistics which not only work well, but which also have some nice simple qualities characteristic of well-defined individuals.

We might also develop suitable phenomenological criteria for measures of spread, and then show that the variance is the ideal statistic by these criteria.[24] In each case, a set of apparently reasonable criteria is then shown to lead to a unique statistic (usually, the one that is conventionally employed). The set of apparently reasonable criteria leads to a proof that justifies the statistic as superior to others. In part, this is the aim of the mathematician, for then the statistic is well characterized by a uniqueness or categoricity proof (uniqueness up to isomorphism). But if the actual phenomenology of the data changes, say the typical and the average diverge, such a proof can no longer be so cogent for our situation, even if the proof is still formally valid.

EXCEPTIONS AND ALTERNATIVES TO MEANS-VARIANCE ANALYSIS

By the devices we have discussed, at least for most students and for many of their teachers, a particular hypostasized historical development is now entrenched technically, and as a readily invoked catechism. What was once a contingent although not at all arbitrary mode of analysis is now seen as necessary and exclusive of alternatives, justified by modes of calculation and visualization and by reasonable criteria. Alternatives that are not so calculable, visualizable, or intuitive are precluded. Since, as in all studies of the entrenchment of technology, there are alternatives that still might take pride of place, we need to see how those alternatives are accommodated, so to speak put in their places, and so just how the current conventions might be said to be conventional.

There is a variety of ways of measuring midpoint and spread, and we might classify many of them as L_p, namely in terms of the metric used in computing the sum of deviations that is to be minimized: $(\Sigma_i |x_{midpoint} - x_i|^p)^{1/p}$. L_1 is least absolute deviations, and, again, it corresponds to the median and the technology of linear programming, with an appropriate geometry and algorithm.[25] L_∞ corresponds to the mid-range, $(x_{max}-x_{min})/2$. And L_2 is least squares, corresponding to means and variances.

The L_p classification system places a variety of phenomenological notions of midpoint and spread within a single system. And it allows the not so canonical ones, the non-L_2, to have their own particular and idiosyncratic places and applications.

Least absolute deviations (L_1) is perhaps the greatest contemporary challenge to least squares and normality. For least absolute deviations is said to be less sensitive to distortions from Gaussian behavior in the underlying distribution of the data, and so it is more "robust," and less influenced by extreme or outlying data points, and so it is more "resistant" (although the median is sensitive to bunching). In effect, new criteria for point estimators, likely to favor medians, are being justified. These new criteria, robustness and resistance, challenge how we think of and interpret Fisher's criteria. We might retain our commitment to Fisher's criteria, but understand their technical meaning differently and so they now pick out different statistics. Or, perhaps the criteria themselves will no longer seem so natural, rather more about a justification of an idiosyncratic point of view about error and randomness.

It is perhaps difficult to imagine giving up criteria with names like efficiency or consistency. Yet, we do know of situations in which we might do so. Say we have an all-or-nothing bet and need merely to be close rather than right on; or, the costs of acquiring the data and computing the statistic for an inefficient, inconsistent, or biased statistic is much less than for an ideal one. Or as in the James-Stein case, we are estimating more than three parameters simultaneously. In each case, one does not so much give up on these criteria but views them as inappropriate for some particular situations. What would be more striking is if the criteria were seen as no longer universal and natural, but as particular themselves.

Much more practically, it is acknowledged that the tails of many actual distributions of data really are too large or fat. There are "too many" events in the extreme values. Again, means may not work well as estimators, being too much influenced by outliers; and, correspondingly, the outliers affect the variance even more. L_1 medians allow us to give the tails less emphasis than do L_2 averages; or, such averages might be "trimmed," taking into account only the middle 2/3 of the data. We might even examine the data in rather more detail, what is sometimes called data analysis. We would be especially concerned with apparently wild data points, no longer apriori taken as random fluctuations, and with potential heterogeneous mixtures of data sets. Thereby we might be doing just what we were conventionally told we should not be doing, making judgments about the believability of individual fluctuations, rather than leaving it to L_p formulaic methods.

There are other strategies, rather more conservative in their implications for the convention. First, we might show that our distribution is not intrinsically Gaussian, arguing that fat tails are fat for good reason, and there should be lots of observations in the tails. Perhaps, for good reason the fundamental distribution has infinite variance—which means that in repetitive measurements the variance grows as, say, N^2 or $N^{3/2}$ rather than N, and so the variance per summand is unbounded. We would need to employ the correct statistics for these distributions. We still might have a good estimator of the midpoint, but it might be an L_1 median. Second, we might show that the Gaussian distribution is just a poor approximation for our situation, as for the tails of the finite binomial, which it underestimates. "Large deviations" estimates are considerably more accurate, but here the variance is replaced by a measure of entropy as the crucial stuff that is to be added up, reminding us that statistics is concerned with information.

In each of these cases, the conventional criteria might retain their universality. But because of readily acknowledged peculiarities in the situation, means and variances are, appropriately, no longer what is to be justified by those criteria.

So alternatives are put in their place by being marginalized: classified by L_p, marked by additional criteria (resistant, robust), or treated as applicable to the idiosyncratic case and the peculiar data set. The conventional statistics retain their hegemony, and they remain readily calculable and obviously universal—rather than being particular themselves.

In the background is the possibility that the marginal classification or stigma becomes the norm: The idiosyncratic case is generic, and the peculiar data set is a commonplace: "Real data straggle a lot when compared to the normal distribution," say Mosteller and Tukey. Rather different "combinations of observations" than arithmetic means actually work well in practice, since data sets are rather more diverse and heterogeneous and straggly than Gaussians allow.[26] The combinations might even be a mixture of mean and median, and hence the L_p classification scheme no longer applies. What we once took as conventional (L_2) is now seen to be an artifact of a particular history. Whatever the justifications, and no matter how hard it is to conceive of finding means and variances as outdated orthodoxy, there are alternative conceptualizations and criteria waiting in the wings. Statisticians are aware of this possibility. Some elementary courses are based in alternative concepts (such as "data analysis"). There develops a mathematical justification for robustness or resistance, and for the methods of data analysis they encourage. The former mathematical justifications are revised because they are to be applied to the rather different kinds of data sets and statistics now taken as canonical.

New mathematical phenomena become significant, as in gene expression data, once we have comparatively large but not huge numbers of observations, each observation characterized by a radically large number of measured characteristics (rather than a large number of observations, each observation characterized by a small number of well-chosen characteristics). In such high-dimensional data analysis, it turns out that fluctuations are well controlled ("concentration of measure"). Asymptotic methods (letting the dimension go to infinity) provide rather nice results that would seem to apply even to smaller finite cases (as happens as well in the Central Limit Theorem).[27]

Calculationally, computers redefine what is calculable. Visually, Fisher's geometry of a multidimensional hyperspace representing all measurements in a sample, the set of samples a cloud of such points, is perhaps replaced by a cloud of points in a hyperspace whose dimension is the number of dependent and independent variables—and metrics may not be Euclidean, and the Pythagorean theorem no longer applies. Each point represents a measurement in the sample. Computer graphics allow us to readily visualize that cloud—what we always could do manually for x-y data, but now from many aspects in several dimensions. We may more readily try out various editing and weighting and reduction schemes. Fisher's criteria, now subject to different sorts of data sets, might well lead to different statistics, especially if new criterial modifiers like robust and resistant are added to notions such as efficiency. And new sorts of theoretical notions, such as concentration of measure, become important.

The canonical and fundamental subject matter of statistics changes from Gaussian random variables to not only the L_p story but also to justifications of more data-interactive techniques. Once-marginal subject matter is now central and primary, the Gaussian story a special case. One can provide better measures of reliability.

III

THEOREMS FROM PROBABILITY AS JUSTIFICATIONS FOR MEANS-VARIANCE
ANALYSIS

Laws of large numbers and central limit theorems may be viewed as justifications
for means and variances. They tell us about the existence, the shape, and the
stability of the sums of random variables—variables more or less identically
distributed, more or less independent of each other—such as sums of errors or
averages of measurements.[28] Laws of large numbers say that means become more
and more settled down as N increases, their diminishing fluctuation and the
persistence of that diminishment being justification for taking means as more
reliable and so real than individual measurements.[29]

Central limit theorems, in their most popular forms, say that in the end what
asymptotically survives in those sums of N random variables is a Gaussian
distribution, more or less independent of the distributions of the summands, whose
variance is the sum of the addends' variances and whose central peak rises as $\sqrt{N}$.
(This is much more than we might expect from the fact that for independent
random variables the variance of their sum is equal to the sum of their variances,
the cross-terms being zero by independence.)

"Asymptotically" is the crucial term of art here. It means that there is not only
convergence, but there is convergence to a given distribution—"asymptotically to."
Not only is there an averaging and convergence to a nice mean, the measurements
add up to a random variable whose distribution is a specific shape, a Gaussian. A
Gaussian distribution is fully characterized by its mean and its variance. The
Central Limit Theorem, in its statement of "asymptotically to a Gaussian," becomes
an assertion of the full adequacy of means and variances as statistics in this realm.

Those laws of large numbers and central limit theorems, in so far as they
become more overarching, taking into account not so identically distributed
random variables, and not so independent random variables, are even more
powerful such justifications.[30] In so far as the proofs of these laws and limit
theorems become simple and perspicuous, they show, literally, why means and
variances are all you need, over wider and wider ranges of concern.

What is perhaps not at all made obvious by these provings is why Gaussians
and means and variances are such good approximations, why the asymptotics work
so well in practice. We might learn something from higher order approximations,
as in the Berry-Esséen theorem, which say how important are non-Gaussian parts
of the summands, say third moments, to the sum distribution. And small sample
statistics, such as Student's t, which do become Gaussian for large enough N, say

just how N plays a role. But in the end what you would really want is an intuition—a combination of a picture, an insight, and an argument or proof—of why you can so often get away with just means and variances for such small N.

There are perhaps three approaches or strategies at this point:

1. Show how variances are natural for measuring rates of change in just the way that derivatives as tangents are natural, so drawing from our well worked intuitions about linearity from the ordinary differential and integral calculus: namely, the Itô calculus and stochastic integration (Kiyosi Itô, 1915–2008).

2. Show how universal is the Gaussian asymptotic limit, how peculiar the exceptions are since they all are of infinite variance, namely Paul Lévy's (1889–1971) stable distributions.

3. And, show ever more delicately what it would mean, in actual practice, to deviate from the Gaussian on the way to that utopian asymptotia, namely, Aleksandr IAkovlevich Khinchin (1894–1959) and the law of the iterated logarithm.

All of these strategies set the goodness of means-variance analysis into a wider context of justification. Ideally, one might want a mathematical structure, a theory, in which all of these strategies are buried behind technical devices that show just what is going on—something I do not know how to provide.

I take what might be called, eponymously, Itô, Lévy, and Khinchin as ways of justifying an L_2 world and taking care of the seeming exceptions to it. The Itô calculus points to variances as the tangents or first derivatives of the random world, and so variances inherit the mantle the calculus has endowed to tangents. Lévy's moves are a way to justify the orthodoxy of Gaussians and the Central Limit Theorem, by systematically labeling the exceptions.[31] And Khinchin defines further the meaning of "asymptotically Gaussian," how close any particular sum's behavior comes to the Gaussian shape, so that the sum's particular fluctuations might prove the rule rather than be exceptional. In these fashions, means and variances have received from the field of probability both their theoretical justification, through the apotheosis of the Gaussian, and their exculpation for not accounting for some deviant or extraordinary situations.

I should note that polluted and mixed data sets and outliers are not dealt with in these regimes. Robust and resistant statistical methods are meant to deal with these quite messy phenomena.

ITÔ, LÉVY, AND KHINCHIN: A TECHNICAL INTERLUDE

Here I want to elaborate a bit on some of the more technical strategies for justifying means and variances. (Some readers may want to continue with the next section at this point.)

1. *Why variances?*: As I have indicated earlier, we can show that much as in the calculus, where tangents or first derivatives are in the limit sufficient to sum up a smooth function's changes (in the limit no higher order terms are needed), here variances or second moments (averages of squared differences) are just enough to sum up the changes in a random variable. So a random walk or diffusion is characterized by its variance. First moments average to zero (by definition of the average) and no higher order moments than second may be needed to characterize the distribution of the sum of similar random increments (where I have assumed that the means of the summands are all zero).[32] In each case, the contributions of higher order terms cancel or become negligible—in the limit. This is the import of the stochastic calculus of Itô.

A version of what is called the invariance principle is another way of ensconcing variances: "Whatever" the approximative model one uses to model a random process, one ends up with the same result, a result most readily found by a model fully characterized by its variance (such as a Wiener process or continuous random walk). Variances turn out to be the whole story, given a reasonably smooth jump-free world. Here, again analogously, think of proofs of the Riemann integral being independent of or invariant to how you divide up the interval, or of derivatives being invariant to how you take the limit.

Practically, in application, proofs of the necessity of a convention may be given in a particular context. So their inevitability is perhaps more obvious and necessary, given the demands of the specific situation and the limits of our analytic capacities. For example, in finance theory the economist Paul Samuelson proves that means/variance analysis (where the variance in the price series is a measure of risk and so, under equilibrium, a measure of expected rate-of-return) is adequate for a portfolio theory of investment for what he calls compact distributions, namely, a continuous no-"big"-jump in price trading regime—essentially, the smoothness we referred to above.[33] Modern finance theory must deal with fat tails and asymmetric distributions.

2. *Why and when should we add variances, and why and when Gaussians?*:

The importance of the normal distribution is due largely to the central limit theorem. Let $X_1, \ldots, X_n$ be mutually independent variables with a common distribution F having zero expectation and unit variance. Put

41

$S_n=X_1+\ldots+X_n$. The central limit theorem asserts that the distribution of $S_n n^{-1/2}$ tends to the normal. For distributions without variance similar limit theorems may be formulated, but the norming constants must be chosen differently. The interesting point is that all stable distributions and no others occur as such limits. (Feller[34])

We know that a finite number of Gaussians add up to a Gaussian, Poissons to a Poisson. To repeat, the Central Limit Theorem tells us specifically how a large number of random variables which are not necessarily Gaussian or Poisson will add up: if the variance of each is finite and they have identical means, their sum will be asymptotically a Gaussian (or a Poisson). One need only know their variances to determine the only crucial parameter of the limiting distribution. More precisely, one scales or standardizes the sum of the random variables by $1/\sqrt{}$(Sum of the variances), and then the limit distribution is a variance-one Gaussian distribution.

But, as we have indicated, there are some random variables that do not add up in the limit to Gaussians, namely some of those of infinite variance (what Feller refers to as "distributions without variance")—although they do add up to something, more specifically something similar in distribution to themselves much as Gaussians add up to Gaussians. They do have what Paul Lévy called stable limiting distributions. And the midpoint of these distributions is not to be best characterized by a mean, and the spread is surely not distinctively characterized by their infinite variances. Lévy also showed that if the variance of each random variable is finite, then the Gaussian with its mean and variance is in fact the only story. So, much as in our discussion of L_p, exceptional cases are put within a classification scheme, here parametrized by an exponent α which is between 0 and 2 (where $N^{1/\alpha}$ characterizes the growth rate of the sum), while the Gaussian ($\alpha=2$) has a distinctive and extreme realm unto itself.[35]

Stable distributions generalize the Central Limit Theorem, concerning the normed asymptotic limit of a sum of similar random variables. If the variance of the random variable is infinite, still some of the time such limiting behavior is possible. But the parameters are not variances and the scale factors are not $\sqrt{N}$, and there are other than Gaussians as such limiting distributions. Again, arithmetic means may no longer be the point estimators of choice for the center; and variances must be replaced by other additive quantities of lower than second moment. When adding up, the scaling factor in terms of the number of addends is $N^{1/\alpha}$, which also points to the additive measure of spread for the sum distribution, much as $\sqrt{N} = N^{1/2}$, $\alpha=2$, points to the variance.

For example, there is an additive measure of Stark broadening of atomic lines in the Holtsmark distribution of fluctuating electrical fields in a plasma due to the chaotic motion of charged particles, which is scaled by $N^{2/3}$. It is a width for a Cauchy distribution (whose inverse is the radioactive lifetime), which is scaled by N, as in a Breit-Wigner distribution of particle energies in radioactive decays in nuclear physics, where total width or decay rate is just the sum of partial widths or partial decay rates. For the distribution of first passage times for a random walk, the scale factor is N^2. This is what we might expect for diffusion, where we know it will take four times as long to go twice the distance. And broad stable distributions, with power law falloff (rather than exponential falloff), play a significant role in understanding laser cooling of atoms and risk in financial markets. [36]

There is a natural characterization of each distribution, and of how it adds up in its own way. Again, having put the Central Limit Theorem in its restricted place, Lévy then re-ensconces the Central Limit Theorem and variances in a special place: If the variance of each summand is finite then the Gaussian and its variance and the conventional Central Limit Theorem is the whole story (that is, $\alpha \geq 2$). This is not a demonstration of just why means and variances are so good, why Gaussians are such good asymptotic approximations. Rather, the claim is: There is nothing else that will do at all.

3. *Fluctuations in the processes of addition*: In adding up finite-variance (σ^2) random variables of mean zero, say we just added up variances, as we are supposed to, and so obtained asymptotically a Gaussian with the appropriate width. Now let us ask just how well constrained is any particular sum of random variables to the Gaussian prediction, on its way to N being infinite? For any particular sum, we will surely expect seemingly improbable fluctuations along the way. "Infinitely" often, namely as often as we might specify, the sum will exceed (or be less than) the value given by any fixed number of standard deviations out in the tails, the sum being standardized by $\sigma\sqrt{N}$. So, in particular, the standardized sum will exceed s standard deviations "infinitely" often on its way to N infinite. That is, the actual sum will be greater than the mean plus $s\sigma\sqrt{N}$ infinitely often.

Now if we let s rise with N that turns out not to be the case. For while the standard deviation for the actual unstandardized sum will be $\sigma\sqrt{N}$, it turns out that there will still be an infinity of times (an unending number of times, that is) along the way to N infinite when the sum will almost hit the value associated with $\sigma\sqrt{N} \times \sqrt{(2 \log \log N)}$. Eventually, for N large enough, $\sqrt{(2 \log \log N)}$ will obviously be larger than any fixed s. And, there will be only a finite number of times when the sum will exceed $\sigma\sqrt{N} \times \sqrt{(2 \log \log N)}$.[37] Khinchin's law of the iterated logarithm (that is, log log) gives a sharper limit on fluctuations than does the Central Limit Theorem. Here, we have a sharp bound for fluctuations that goes as:

$$\sqrt{N} \sqrt{(2 \log \log N)},$$

rather than the weaker estimate, $\sqrt{N}$, the standard deviation in the Gaussian.[38]

The law of the iterated logarithm says how good the Gaussian asymptotic approximation really is, how good variances alone are in telling you about a random variable's fluctuations. The law of the iterated logarithm might be taken to stand at the edge of the Central Limit Theorem, reaffirming Gaussians as being the whole story, carefully confining their implications or meaning in borderline cases.

DEFENDING RANDOMNESS FROM IMPOSSIBLE LONGSHOTS AND COINCIDENCES

> The postulate of randomness thus resolves itself into the question, "Of what population is this a random sample?" which must frequently be asked by every practical statistician. (R.A. Fisher, 1922[39])

Probability and statistics defend randomness as the source of variation and idiosyncrasy, as against device and design and malice and intention. Probability and statistics are recurrently used to show how what are initially taken as peculiar phenomena are in fact merely the consequence of random processes. What if what would initially appear to be an "outrageous event" actually does occur?[40] How is an extreme longshot handled, other than to say that eventually it might have happened, so why not now? Surely some of the time means and variances and Gaussians are inadequate. But perhaps less often than it would appear.

If a purported longshot would seem to be too frequent in its appearance, much more frequent than we might once have expected, we may try to show that it is in fact a coincidence with sufficiently high probability, once that probability is properly calculated; or, perhaps, the appropriate probability distribution has much larger tails than the Gaussian's.[41] The crucial claim here is that randomness is the whole story, or at least most of it. (It helps as well to take into account potential admixtures of corrupting distributions and to excise potential contaminations in the data set.)

Alternatively, we might carve out regions where the orthodoxy of the Central Limit Theorem does not apply and provide a canonical phenomenology of what are shown to be peculiar situations, so leading to a more correct probability distribution for them. Now, some extreme or highly improbable but truly random events are not supposed to be explained away. They are outside the Gaussian Central Limit realm. Extremely, you may have to go for broke because otherwise you are doomed. There are times when such a bold go-for-broke strategy is in fact rational in probability terms—as when you need a million dollars now, anything less is worthless to you, and you put all your money on one bet.[42] Or, the events you are concerned about are extreme in their probability distribution: say, zero or one. And as you go along you can gain some apparent information on which of the extremes applies, even though you will never gain enough information: for example, as in deciding if an ongoing decimal expansion represents a rational or an irrational number.[43]

Moreover, in each of these extreme cases, while the mathematical and formal requirements of probability and statistics readily apply, in actual ordinary life the analysis offered does not reliably model the world. If we relax the assumptions—and one would have to do that to apply the analysis to actual situations—their results do not then follow along.

In response, we might pool risks, so as to make fluctuations less problematic. Such an insurance is possible only if the seller of insurance believes, in the end, that the set of longshots, as a set, is in effect Gaussian enough (at least in the tails, although it is just here that large-deviations estimates are relevant), pooling enough varied sorts of independent risk, and having sufficient reserves and reinsurance to stay in business. And the seller might still demand an extra risk-premium dependent on their estimate of the likelihood of deliberate and devious and idiosyncratic influences, not to speak of covariances and dependence.

In general, what appear to be anomalies are shown to be ordinary fluctuations, most likely in a Gaussian—once systematic effects have been carved off. What are taken as technical moves or better models are by the way also reassertions of the authority of the core practices of the disciplines of probability and statistics concerning randomness and Gaussians. And these are emblematically about means and variances.

IV

MAKING VARIANCES REAL BY IDENTIFYING THEM WITH ENDURING OBJECTS AND OTHER FEATURES OF THE WORLD.

After statistical calculation, visualization, and phenomenology, and limit theorems from probability, what practically ensconces means and variances is their being successfully identified with what we ordinarily take as real, whether it be the average height of members of a population or a diffusion rate of a gas. Namely, means and variances can often be identified with quantities which would appear to adequately characterize a system or an object, and which do not change much. Variation that is manifestly present in the world may be accounted for by the spread of a Gaussian. And phenomena, which are taken as error and noise, may be seen as random, and so, too, might be characterized by means and variances.

Conversely, we might say that because phenomena are actually instances of means and variances, they are nicely additive and may be averaged. And the phenomena do not move around too much, only as $\sqrt{N}$. And so they are real.[44] Much has been written about means. Here I shall focus on how variances become identified with social and physical objects and systems, through the analysis of variance, and in the account of fluctuations in these objects and systems.

The historians of statistics tell us that in the early-nineteenth century observation error in astronomy becomes identified with what we would call the variance of a random process. Next, perhaps, came variation in some societal property, say the height of members of a society, which is taken to reflect a variance within another random process, although formally it is much the same as measurement error. And next comes the analysis of error, as in Sir George Airy's (1801–1892) study, in 1861, of astronomical observations, in which the error as measured by a variance is analyzed into distinct additive sources of error, each with its own contribution to the variance.[45] This procedure is then extended by the analysis of variation, as encountered in agricultural experiments, varying or randomizing the fertilizer treatments of a partitioned field. The variance as a measure of variation is seen to be the sum of several variances, each due to different actual material factors such as fertilizer or tilling, or, put perhaps more mechanistically, due to different sources of variation, as well as an error term.

The crucial move here is just that process of analysis, of breaking something down into its putative distinct and separable parts. The variation is now called a total variation, consisting of two parts: the variation from the mean plus the variation of the mean from zero or "no effect." (In fact, it is an algebraic identity of the following sort: $\Sigma x_i^2 = \Sigma(x_i - \langle x_i \rangle)^2 + \langle x_i \rangle \Sigma x_i$, given the definition of the mean, $\langle x_i \rangle$, where angle-brackets indicate an average. The algebraic identity is then endowed with substantive content and statistical or probabilistic significance.[46]) And then one endows those indirectly seen parts as different and distinct in nature, calling them the unexplained or random variation and the explained variation: Crop yields would appear to correlate with the amount of fertilizer (the explained variation from "no effect") yet there is still some variation that is not accounted for.

The next crucial move is to determine the probability distributions of the ratios of the variations, assuming the x_is are Gaussian. Is the measured improvement in crop-yield actually due to fertilizer or is it more likely to be noise or a fluctuation? The tests for significance that employ these distributions (for these explicitly "small-sample statistics"), tell if these putative parts are to be taken as real. Again, are the indirectly seen parts random fluctuations or are they persistent, much as we learned to think of means as persistent? So analysis of variance becomes a way of explicitly treating variance as something real, composed in part of actual, real, material, apparently causal parts.

A suitable geometry will make analysis of variance seem natural. Again, analysis of variance is a restatement of the Pythagorean theorem set in the correct hyperspace, the appropriate choice of hyperspace, subspaces, and projections being the powerful analytic move. And variances' adding up as they do in terms of explained and unexplained variation, is merely the meaning of each of the sides of the Pythagorean triangle. Moreover, there developed understandable algorithms for computing the relevant quantities and suitable tabular devices for keeping track of them, algorithms not unlike those used to compute regression on an electromechanical calculator. As for the tabular devices, Fisher says, "This useful form of arrangement is of much wider application than the algebraical expressions by which the calculations can be expressed, and is known as the Analysis of Variance."[47]

But algebra, geometry, and algorithm and computation are not enough. That analysis of variance would seem to be an algebraic identity, or a geometrical observation, does not make it statistics. The claim is deeper than an identity or an observation or a diagram. A set of assumptions about the statistical independence of variables and the Gaussian distribution of variation and error is needed if inference is to be possible. And, again, the correct significance tests need to be developed. Then a justified practice needs to be codified. (Fisher is the source of much of this.) Significance tests, formalism, diagram, and algorithm are effective in endowing variances with a palpable practical place in the world.

Now, we might learn what would appear to be the same information we obtain from analysis of variance by employing an apparently different method, regression with dummy variables. They are in fact formally equivalent techniques. But here the variances do not play the same central conceptual role, as parts, that they do in analysis of variance, with its explicit tabular format and its tabular and visual interpretations.[48] Variance-based notions such as heritability and the distribution of variance among several sources suggest a very different way of conceiving the world than does the formally equivalent formulation in terms of the differences of means, just what is highlighted in regression with dummy variables.[49]

***As we see again and again,[*] an empirical regularity or an algebraic fact is then practically entrenched and made conventional by means of: (1) theorems that justify the convention as inevitable, natural, and necessary, given some reasonable

---

[*] I use *** to mark passages that provide an overview of where we have been.

assumptions; (2) canonical images, diagrams, geometries, and tabular formats, so that it is manifest or obvious what is the case, and it is straightforward to do the right sort of calculation (if you have been properly trained, that is); (3) practical and resonant interpretations of formal symbolic objects that as such become useful and meaningful (so that a component of variance is taken as some actual thing, or a criterion for estimators is seen as inevitable); and, (4) canonization of justification, diagram, and meaning into a particular way of doing things, so distinguishing it from any number of other ways of laying out the facts—and so that regularity or that fact is seen as natural and readily interpreted in its detail.

VARIANCES AS THE CANONICAL ACCOUNT OF FLUCTUATION

Whole classes of phenomena may be seen as fluctuations, as such. So we often take error and variety as fluctuations. Those fluctuations might well not be random or they might not have a Gaussian distribution, and measures other than variances might better characterize them.[50] Still, much of the time apparently random fluctuation and noise may be characterized by a variance, and so a variance is identified with something quite real in the world. For chemists and physicists and finance theorists, variances often are the canonical measure of random fluctuation. In these contexts, fluctuation is a fundamental process: in essence, the contents of the second and third laws of thermodynamics, or the Heisenberg uncertainty principle, or the source and nature of the liquidity in securities markets.

Physical laws and principles sometimes say that fluctuation is ineliminable and that variance is necessarily nonzero. That variance is to be identified with something physical and palpable: the specific heat of matter; the possibility of lasing; the polarization of the vacuum (the spontaneous creation and annihilation of electron-positron pairs that then affects the value of an atomic spectral line, leading to the Lamb shift, for example). Fluctuation as noise—which may well average out to zero—does have an enduring meaning nonetheless: Its variance adds up, and that summed variance is something, such as ohmic electrical resistance and the consequent energy dissipation by heating, that we do experience as real and necessary and palpable.

More technically, variances and covariances are crucial notions in statistical physics, which includes quantum field theory, statistical mechanics, and the study of heat and of diffusion.[51] Temperature is a measure of the variance of a random system, how dynamic it is microscopically: $\langle mv^2/2 \rangle$=kinetic energy$\approx k_B T$, m being the mass and v the velocity of the atoms, k_B is Boltzmann's constant, and T is the temperature. (At room temperature, $k_B T$ equals 1/40 of an electron-volt, roughly the kinetic energy of an air molecule.) And, diffusivity is a measure of how spread out it is ($\langle x^2 \rangle$, where x is a measure of position).

Moreover, as Einstein showed, we can relate fluctuations measured by a variance, diffusion itself, to dissipations, measured by a viscosity or resistance—and the temperature (T) is the crucial link: Fluctuation = Variance $\approx$ Diffusivity $\propto$ $T/(Viscosity)$. Due to fluctuations in the atomic motions, and so random scattering of electrons off the crystal lattice of a wire, there will be a noise current, i, across the terminals of a resistor (R), allowing it to dissipate energy (ohmic heating) as fast as it absorbs energy from its thermal environment: $<i^2> \propto T/R$. Or, temperature fluctuations in an object are related to how hard it is to heat up that object, its specific heat at constant volume, c_v: $\langle (\Delta T)^2 \rangle / T^2 \propto 1/c_v$. Boltzmann's constant, k_B, appears in each of these relationships, for these are thermal fluctuations.

Put differently, variance is a measure of the rigidity of an object or an environment, large variance meaning low viscosity or low resistance, so connecting fluctuation with dissipation, and again temperature is the crucial link. For these are thermal fluctuations, larger as the temperature rises.

Fluctuation-dissipation relations in effect say that "all there is" are fluctuation processes—namely, all there is are variances. For it is through their variances that those fluctuation processes are known, and dissipation is just a consequence of fluctuation.[52] Lars Onsager's regression hypothesis (1931) says that ordinary thermal fluctuations are indistinguishable from externally caused heat shocks.[53] In settling down to equilibrium from an external shock, dissipation then occurs through those self-same fluctuations.

In each case—electrical resistance and ohmic heating, viscosity, the specific heat, or returning to equilibrium from disequilibrium—fluctuation, and a variance that measures it, accounts for the phenomenon (at least in a first approximation).

As we might infer from our discussion of stable distributions, some fluctuations are not measured by a variance. Radioactive decay, here taken as an archetypal transition between physical states, is also the consequence of a fluctuation process: quantum fluctuations between the initial state of the particle and all possible final states, some of which are the decay of the particle. And measures of radioactive lifetime, or equivalently the width of mass plots of very rapidly decaying particles, are measures of the rate of fluctuation between states and of dissipation as a decay process.

NOISE

In the natural sciences, means and variances are entrenched as conventions by means of statistical practice, mathematical theorems, and scientific theorizing. In the world of social exchange, variation exemplified—noise—comes to be seen as real and essential if that world is to work as it does. And for many processes, variances characterize that variation.

> Noise makes financial markets possible, but also makes them imperfect. . . . If there is no noise trading, there will be very little trading in individual assets. People will hold individual assets, directly or indirectly, but they will rarely trade them. . . . In other words, I do not believe it makes sense to create a model with information trading but no noise trading where traders have different beliefs and one trader's beliefs are as good as any other trader's beliefs. Differences in beliefs must derive ultimately from differences in information. A trader with a special piece of information will know that other traders have their own special pieces of information, and will therefore not automatically rush out to trade. . . .
> Noise trading provides the essential missing ingredient. Noise trading is trading on noise as if it were information. People who trade on noise are willing to trade even though from an objective point of view they would be better off not trading. Perhaps they think the noise they are trading on is information. Or perhaps they just like to trade. . . . With a lot of noise traders in the markets, it now pays for those with information to trade. (Fischer Black[54])

The economist Fischer Black's description of securities markets says that it is noise—trading on possibly false information as if it were true, or trading on no information—that makes for a liquid securities market. Presumably, it pays for those with real information to trade with those who are noise traders. Moreover, fluctuations in prices are to some substantial extent noise or random fluctuations (rather than a consequence of changes in information). Then we might treat the price of a security as a noisy or random variable, and apply the technology of option pricing developed by Black and others (the Black-Scholes formula, essentially about the diffusion of a random variable, the price), the source of much of modern financial engineering. One of the themes in this work is how security prices do not quite follow the conventional random walk.

More generally, variance—as error, as variation, as fluctuation—is a measure of the possibility of fit among the parts of society, the matchings that allow for a coordinated process.[55] Unless there is noise, there is unlikely to be matches or trades.

Consider precision manufacturing on an assembly line. If the variance in the dimensions of one element matches that of another element into which it must just fit, and their means differ by an amount allowing for, say, the flow of lubricant—given a relatively small tolerance or precision compared to the variances—then sets or stocks of individual parts will provide for interchangeability and fit. There will be in the inventory or stock of one part an individual that fits just right, namely, within that tolerance, into a given receiving part—and that will be true the next time, too (*pace* runs in a particular size, in which the need for smaller parts comes in longer runs than you stocked for, for example). Now if we want individual parts, rather than inventories, to surely fit, then the variances must be less than the tolerance. This may well require very substantial investment in surveillance, craftsmanship, or technology. The nice feature of matching sets of parts is that the tolerance can be small and the variance large given reasonable costs of measurement and storage. There is no necessity here to loosen tolerances to get a fit; enlarging the stock inventory will do. Moreover, if tolerances are small, then wear and friction is less, and fewer spares are needed, to some extent compensating for the larger stock size. (But if variances are very large compared to tolerances, it is hard to believe a part is uniform enough, within tolerance everywhere, to really fit.[56])

In sum, variance measures the degree of potential interdigitation of sets of objects. In combination with measures of cost, it determines the minimum size of those inventories given the demands of fit.

If noise and variation are crucial to the workings of modern society, it is perhaps not surprising that they are then explicitly justified as essential: real and measurable and commensurable, and conventionalized by variances, although, again, other statistics might well replace this measure. Noise and variation are, moreover, intrinsic to or the direct consequence of the actual mechanisms of social and natural processes. Variances are not only real, they are needed if there is to be an ongoing world. Were variances zero, the world would come to a halt. (But even "at" T=0 there are quantum fluctuations.) Conversely, since variances are in fact not zero, even if means are zero in absolute terms, something might even arise from nothing, the fluctuations in the vacuum producing pairs of particles.

51

EINSTEIN AND FISHER

Albert Einstein (1879–1955) and R.A. Fisher each showed how noise could be canonically tamed, its consequences elegantly epitomized.[57]

"Better than anyone before or after him, he [Einstein] knew how to invent invariance principles and make use of statistical fluctuations."[58] From the nineteenth-century kinetic theory of gasses, it was known how molecular motions in a fluid medium might be treated as fluctuations, characterized by a variance such as the diffusion constant. Einstein (1905, and contemporaneously R. Smoluchowski) showed how those fluctuations could account for apparently irreversible phenomena such as viscosity in fluids. The crucial fact is that system is in thermal equilibrium; it has a fixed temperature. Noise is real and significant here just because Einstein, following Maxwell and Boltzmann, connects it intimately to— identifies it with—actual physical molecular processes.

Perhaps Fisher's greatest contribution is his formulation of the statistical task, both philosophically and practically:

> . . . the object of statistical methods is the reduction of data. . . . This object is accomplished by constructing a hypothetical infinite population, of which the actual data are regarded as constituting a random sample. The law of distribution of this hypothetical population is specified by relatively few parameters, which are sufficient to describe it exhaustively in respect of all qualities under discussion. . . . It is the object of the statistical processes employed in the reduction of data to exclude this irrelevant information, and to isolate the whole of the relevant information contained in the data [namely, sufficiency].[59]

Fisher, in 1920–1925, develops descriptions or criteria for statistics or estimators, which we have already discussed, and in so doing he provided a canonical description of noise and variation, one which often turns out to be fulfilled by means and variances. What is crucial is just the notion of the reduction of data while retaining all the information in a sample (whether that reduction is envisioned geometrically as a projection in a hyperspace or as an efficient coding of the data by maximum likelihood parameters). But, noise is not only to be described; it is as well to be partitioned and analyzed. Driven in part by practical problems, such as "the manurial response of different potato varieties," in part by the recurrent appearance of certain statistics (z or F), Fisher took the algebraic identity that founds the analysis of variance, and he then imports meaning and statistical significance to each of its terms or parts.[60]

Einstein and Fisher might be said to epitomize two centuries of work that insisted that noise and variation are somethings real and canonically knowable. Noise and variation are the consequence of particular mechanisms and processes, such as molecular kinetics and genetic mixtures and varied environments. And, in terms of variances, those mechanisms and processes have actual effects and natural descriptions as diffusion constants, interaction rates, and heritabilities.[61]

V

Entrenchment of conventions can be rational, well founded in theory, in mathematics, and in practice. In fact, the convention is likely to be overdetermined. Namely, the entrenchment is ramified, multiply supported from many sides, and the various justifications provide mutual support for each other. In the background, there may be viable and even apparently preferable alternatives. Perhaps they will become the conventions of the future, or perhaps they once were the reigning convention. What is most curious will be to reexamine our ramified, overdetermined justifications when they are no longer quite so effective, when they are more or less abandoned, the convention they supported no longer taken as the way things are. They are presumably still true and valid, as justifications, but they no longer hold sway over us. We wonder how they could have been seen as definitive. But that of course is what we must see, albeit implicitly, if conventions and cultures change.

3

Subject

The Fields of Topology

Analysis and Algebra; The Subjects of Topology; Preview. *Before the Birth: Stories of Nineteenth-Century Precursors. General Topology*: Examples and Counterexamples, Diagrams and Constructions; Parts in Analysis; Algebraic Topology as Playing with Blocks. *Open Sets as the Continuity of Neighborhoods*; Dimension and Homology as Bridging Concepts. *The Tensions in Practice*: The Local and the Global; The Tensions in Analysis: The Hard and the Soft; Hard Analysis: The Ising Model of Ferromagnetism and The Stability of Matter. *Keeping Track of Unsmoothness and Discontinuity*; Keeping Track of the Shape Objects. *Diagrammatics and Machinery*; Pictures and Diagrams as Ways of Keeping Track of Flows and Systemic Relationships; The Right Parts; Naturality. *The Tensions and the Lessons. Appendix*: The Two-Dimensional Ising Model of a Ferromagnet; Some of the Solutions to the Two-Dimensional Ising Model.

I

ANALYSIS AND ALGEBRA

The subject matter of a mathematical field is circumscribed by the variety of objects with which it is concerned and the concepts used to describe them, and by the range and depth of what can be proved or understood about them—or we might hope to prove or understand. The linkages, the commonalities, and the differences among those objects, concepts, and methods of proof provide some of the glue that holds a field together.

54

Topology provides a lovely example of the development of a field and its subject matter. If means and variances understood probabilistically defined statistics (at least for many of its lay users), a set of tensions, accounting schemes, and archetypal examples define topology. In particular, topology is defined by a tension between its two major subfields: general or point-set topology and combinatorial or algebraic topology. They represent the tension between analysis (and the calculus and set theory) and algebra, and a tension between the local and the global, the particular features and the general symmetries. Yet, we can algebraicize general topology, while algebraic topology can fall under the purview of sets and analysis. Technically, the local may be globalized (as in p-adics and adèles), and the global may be explored locally (Hasse's principle).

In 1931 Hermann Weyl uses the tension between topology and abstract algebra—say, the real line considered as a continuous manifold and as the domain of the arithmetic operations—to describe "the intellectual atmosphere in which a substantial part of modern mathematical research is carried out."[1] He suggests that "modern axiomatics is simpleminded and (unlike politics) dislikes such ambiguous mixtures of peace and war, [so] it made a clean break between the two [fields]." We shall see how topology and abstract algebra then proceeded to invade and occupy each other's territory. Twenty years later, he says, "The connection between algebraic and set-theoretic topology is fraught with serious difficulties which are not yet overcome in a quite satisfactory manner. So much, however, seems clear, that one had better start, not with a division [of space] into cells, but with a covering by patches which are allowed to overlap. From such a pattern the fundamental topologically invariant concepts are to be developed."[2] (Such a covering then leads to the natural algebraic objects; say, a "sheaf" of stalks, each stalk an algebraic object, a ring, associated with a patch.)

As in much of mathematics, here particular sorts of pictures and diagrams are deployed to do some of the technical mathematical work almost automatically, as well as to make arguments clearer and more manifest. What seems especially characteristic of these pictures and diagrams is that they are meant to keep track of lacunae, such as discontinuities, disconnectednesses, and leftover parts, in the mathematical objects they are representing.

Now, I take it that adding-up the world out of simpler components is one way of describing the endeavor of arithmetic and the calculus, and decomposing the world into simpler parts (units or tangents) is another. Lacunae in those additions, which may reflect hiatuses: 1, 2, 3, . . . ∞, or disconnectedness or discontinuities or holes or infinities or singularities, became the theme of the calculus of Gauss and Green and of the set theory of the latter part of the 19th century.[3]

Topology may be said to arise out of traditional mathematical problems of arithmetic addition and partition and problems of geometrical structure—effectively, an arithmetic of space. One of the achievements of topology was to be able to consider nearness, continuity, and disconnectedness in single account. As for continuity, the critical observation is that continuous functions (actually, their inverses) preserve closeness or neighborhoodness properties. And a function cannot be continuous if it maps a connected set into a disconnected set. More generally, the notion of continuity is the link between two sorts of tasks: characterizing the nearbyness of various sets of points and characterizing the shape of a space or an object.

That achievement of topology—considering nearness, continuity, and disconnectedness within one account—has been recurrently riven by the fabulous tension between analysis and algebra, the one representing continuity, the other structure and discreteness. Moreover, there are family resemblances between analysis and vast swaths of mathematics, and between the algebra of topology and the rest of algebra and discrete mathematics and geometry. So a tension between continuity and disconnectedness is a very deep one indeed, its resolutions of high import.

In fact, the tension is perhaps of mythic proportions: The medieval story of Valentine and Orson is an account of twin brothers separated at their birth, one to be brought up by royalty, the other by bears.[4] Their mother (and father), their birth, and their histories are unknown to them—until, eventually, after a fight almost to the death, there is mutual recognition and, later, reconciliation. And, unlike for Oedipus's recognition scene where he blinds himself, here we have romance and comedy rather than tragedy.

As in all such stories, various of the characters barely miss recognizing each other several times before the redemptive moment, and afterward as well. Mysteriously, the brothers are deeply drawn to and influence each other whenever they do meet, their kinship unrecognized. Of course, the brothers are very different from each other, although both are brave and true and virtuous. Valentine is global in his vision, supplied with the machinery of arms and organization. Orson is local, with the keenest of sight and the capacities for opportunistic invention and craftsmanship.

Valentine is algebra and algebraic topology, while Orson is analysis and point-set topology. Again, what they share, their mother and father, is the real numbers, to be known by its arithmetic (+−×÷) and by its being a continuous line or manifold. Of course, the brothers come to depend upon each other, Orson's continuity becoming Valentine's algebra, and vice versa. Here I tell their story, the lesson being about the stranger and the brother.

Our recurrent themes will be linearization, continuity, algebraicizing, functions and the spaces on which they live, mappings, and that a space may have many topologies.

THE SUBJECTS OF TOPOLOGY

Topology is a comparatively recent branch of mathematics, having originated in the mid-nineteenth century, and become identified as such a little more than one-hundred years ago. While by rubric it is a single subject, in fact it would appear to be two subjects, point-set topology and algebraic topology, taught and practiced in almost complete independence from each other. This was not always the case. At various times some mathematicians worked seamlessly between them: Poincaré (ca. 1890), Brouwer (ca. 1910), and Vietoris (1926), for example. Perversely, we might ask, Is there a subject of topology? And the answer is not only philosophic and historical, it is as well mathematical.

We might identify a tension between local and global knowledge of the world. Do we learn about the world by looking around us, nearby, or by standing outside the world, in an olympian position, looking down and peering in? And, can local knowledge lead to global knowledge, and vice versa? Technically, point-set topology, with its neighborhoods and open sets, is the local story, while algebraic topology, with its skeletal approximation of a space, a space occupied by sources and fields, is the global story. Both subjects have "topology" in their name; that is perhaps not just an historical artifact but a substantively interesting one as well. It might be said that topology is the study of the impact of continuity, whether locally or globally, in terms of how continuous functions preserve structure.

In the nineteenth century and the first part of the twentieth century, mathematicians learned how to express both local and global features of a space in terms of the behavior of the functions that can "live" in that space—most poignantly in Riemann's surface for a function, on which the function is single valued. (See Figure P-1.) At one time, a function might have been a formula or a curve or an ordered pair of numbers (say x and y $(=f(x))$). The great discovery was that one might speak of a *mapping*, a function in functional term, namely in terms of what the function does and the properties of what it does: a Mercator projection mapping is a projection of the sphere onto a cylinder. And in so speaking one can describe both local and global features of the world.

Continuity is an extraordinarily rich concept and demand.

It turned out that the intuition of continuity could be expressed in terms of how a mapping operated upon ("open"—here think of a ball with a fuzzy surface) sets. If a continuous mapping, f, takes a set W into an open set, then W itself is open. (More correctly, if $f: X{\rightarrow}Y$, is continuous, then, if W is an open set in Y, then $f^{-1}(W)$ is open in X.) And a global feature such as the connectivity of a space—Does it have holes or not?, Is it two well-separated parts?—could be expressed in terms of invariant features of such mappings: for example, whether an integral of a smooth function around a closed path in the complex plane was non-zero, which it may well be if there are holes in the plane.

Moreover, one might study the algebra of continuous functions from the topological space, X, to the real line, $\mathbb{R}$: $f:X{\rightarrow}\ \mathbb{R}$, and thereby reconstruct the topological space (Banach-Stone theorem), if X has some nice properties. A point is represented by the set of functions that are zero on that point. (Emmy Noether saw in the combinatorial topology of her time, groups and their kernels (what maps to zero) and images.) More generally, in modern mathematics there is a recurrent mirroring of a space or a geometry with an algebra: fields and particles, or zeroes or poles, geometry with spectrum, spectrum with a ring, a curve as a picture (or a variety) or as its zeroes (its ideals).

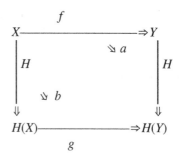

FIGURE 3.1: Diagrammatic representation of commutativity of continuity and structure: $H(f(X)) = g(H(X))$.

The second great discovery was that some properties of "good" maps could be expressed in terms of *diagrams* and algebraic rules. A map that was continuous might preserve structural features of a space, X, structural features that might be encoded in an algebraic object, say a group (say, modeled by the integers under addition), here called $H(X)$. And we might first map at the level of the space and then compute the structural feature (Figure 3.1, path *a*), or first compute the structural feature and then map at the level of the structural feature (path *b*). Following either path, we end up in the same place. Continuity and algebra are said to commute, and that fact is encoded in a diagram. Moreover, those diagrams would seem to be able to help in proving things:

> The diagrams incorporate a large amount of information. Their use provides extensive savings in space and in mental effort. . . . in the case of many theorems, the setting up of the correct diagram is the major part of the proofs.[5]

More familiar to many students are Venn diagrams, which nicely encode some of the rules of inference in logic. Diagrams encode facts about the world. The discovery here is of those facts *and* that they might be expressed diagrammatically.

A third discovery was how what mathematicians call *nilpotence* allowed one to naturally decompose objects into simpler components. So, in the calculus, a function, $f(x)$, is decomposed into its tangents: $\Delta f / \Delta x$, given that in the limit we can ignore second differences, those proportional to $(\Delta x)^2$. This basic notion of the calculus is encoded in an algebraic statement; epsilon-delta arguments become algebraic facts. Namely, for a given function or operator or variable, F's nilpotence means that $F^2 = 0$ even though F does not equal zero. Geometrically, the boundary of an object's boundary is zero: The boundary of a ball is a sphere, and the sphere has no boundary. In topology, boundary operators, in effect a difference operator Δ, allow us to decompose objects into natural parts.

$$A \to B \to C \qquad g[f(A)] = 0$$
$$ f \quad\; g$$

FIGURE 3.2: An exact sequence. The image of A, $f(A)$, is a part of B which g "annihilates," that is: $g[f(A)] = 0$. Note that $f(A)$ need not be zero.

Significantly, nilpotence can also be encoded in diagrams much as is commutativity. One says that a sequence of spaces and transformations is *exact* if a series of two consecutive transformations of an object in a space gives zero: say, $g[f(A)] = 0$. (Figure 3.2) And then one can learn to reason with such diagrams, so that for other purposes one can make use of what is known about exactness (or commutativity) in the particular cases of interest. (I should note that saying that commutativity or exactness is encoded in diagrams does not mean that they are the case because one can draw a diagram. Rather a diagram can encode those properties if they are the case. Certain diagrams of exactness do imply the exactness of other diagrams. This is what it meant to reason with diagrams.)

And a fourth discovery was a remarkable coincidence between analysis and algebra, that the description of *path-dependence*, what is called *cohomology*, is the same for each. Say we took two different routes for a trip from New York to Chicago. The mileage, difficulty of travel, and the total cost of gasoline would almost surely be different (technically, the integrals of the gasoline-cost over the paths are different). The total tolls we paid would also likely be different, and would depend on the path we took. Imagine that we wanted to arrange it so that "the total tolls + total gasoline-cost" were the same for both paths. The analysis that allows us to figure out what tolls to charge at each tollbooth, in order to take into account path-dependence, is the cohomology of the roadway system. If the cohomology is zero, the cost of travel is path independent and depends only on the endpoints.

In the realm of algebra, the path dependence is expressed by a statement to the effect that $s(g)$ times $s(h)$ does not equal $s(gh)$, where s is selects an element of a group. If we say that $s(g)s(h)=u(g,h) \times s(gh)$, then group cohomology puts consistency constraints on the u's, in effect the tolls we discussed above. And if u turns out to be the identity, then we have $s(g) \times s(h)=s(gh)$, and there is no path dependence (for the product depends only on the "end points," g and h).

Mappings and continuity, diagrams and their commutativity, nilpotence and exactness, and path-dependence and cohomology are pervasive notions in topology. They suggest how analysis and the calculus and geometry might be understood algebraically. How far should or could one go in algebraicizing the world? Quite far, indeed, it would appear. And, conversely, perhaps one could rely on continuity to understand algebra, as well.

The history of algebra and of topology is pervaded by attempts to take complex objects and decompose them into simpler ones, the complex object being some sort of sum of its simpler components. In effect, we end up linearizing the world. Namely, Complex Object = "Σ" $a_i \times A_i$ the a_i being coefficients such as integers, the A_i being the simpler components, which themselves might be decomposed. We have what appears to be a linearization (a vector space or module). That is the sense in which cohomology is said to linearize complex objects. (Decomposition may also be in terms of a sequence of objects, with maps or inclusion between them, a Russian doll model.)

PREVIEW

Surely, one might survey the tables of contents of a collection of topology textbooks to assay the subject matters of topology. The purpose here, as well as in chapter 2, is to describe the enduring motivations of a field, the ideas that are behind the chapter and section titles of the texts. The aim is to give a better account of what is really going on in a field, where "what is really going on" is a notion we have to explore along the way.

I will describe some nineteenth-century problems that are taken as the precursors to topology and its tensions, then briefly describe the two subfields in terms of archetypal themes and examples, and then offer something of a condensed history of the tensions between the two. The tension is nicely illustrated in modern mathematical analysis applied to physical problems. I close the chapter with a consideration of diagrammatics and machinery as part of a functionalist, map-oriented way of thinking that as well keeps track of lacunae such as discontinuity and disconnectedness.

That diagrammatic, map-oriented, algebraic way of thinking has been of late supplemented by computer visualization, in which the topological object itself plays a central role, rather than an algebraic abstraction of it.[6] The tension between point-set topology and algebraic topology is perhaps shifting from the world of Cantor, Poincaré, and Brouwer to the milieu of Riemann and Maxwell and those plaster models of functions found in glass cases outside of mathematics departments.

Those pretty pictures and models, and even the calculations needed to generate them, are now in tension with formalisms that make topology and analysis newly algebraic, but now with an emphasis on even more general properties of mappings. So Jean Leray (1906–1998), the expert on hydrodynamics before World War II, afterward develops "sheaf theory" to appreciate the force of cohomology (which itself represents the generic, qualitative features of fluid flow). (See Appendix C.) And, Alexandre Grothendieck (1928–2014) takes that development, that mode of algebraic abstraction, as a generic move to be applied elsewhere.

***The tensions between Valentine and Orson, between algebraic and set-theoretic topology, are actual. They constitute the world as mathematicians discover it and experience it. The tensions appear in the work as mathematics, not philosophy, not myth. The same will be true for some of our other themes, throughout the book: strategy, structure, and tactics; analogy and syzygy; following a proof with your body and with chalk; choosing apt names and diagrams; employing a philosophy of mathematics that is a generic idea of what is interesting and why, what is sometimes called a "program," such as Felix Klein's *Erlanger Programm* of the late nineteenth century (named after Klein's university and town, at the time); finding important ideas embedded in particular functions or other mathematical objects or within lengthy computations; working out theological issues through technical mathematical means; and, finding again and again that one's mathematical specifications of objects allow in others that were not anticipated, and hence the productivity of the world compared to our imaginations—all these are expressed in the technical mathematics as it is actually practiced. Beyond that, there is hot air, useful for getting a balloon off the ground but subject to evanescence.

II

BEFORE THE BIRTH: STORIES OF NINETEENTH-CENTURY PRECURSORS

Every field of inquiry tells a story of its precursors. Such a story is not meant to be good scholarly history. Rather, it motivates the scientist and provides a justification for how the field has developed. As for topology, we might provide one such fanciful account:

Much of general or point-set topology may be seen as theme and variations on the real line and the plane, and the development of our intuitions about them. Counterexamples such as Brouwer's indecomposable plane (Figure 1.1) show the limits of our intuitions.[7] Here a curve splits the plane into three parts, of which it is the common boundary; we have a continuum that cannot be the union of two sub-continua. Or, consider the fact that the limit of a series of smooth curves may not be smooth (unless there is "uniform continuity").

Correspondingly, much of combinatorial and then algebraic topology may be seen as deriving from analysis of functions of a complex variable, potential theory of fluid flows, and a structural engineering in which the algebra of the matrices that specify what is connected to what within an object (in effect, the combinatorics) supersedes those matrices themselves.[8]

Riemann (1826–1866) may be taken as the emblematic figure. His 1854 dissertation on the representation of a function by a trigonometric or fourier series sets the stage for point-set topology and its concern with peculiar subsets of the real numbers. (And so Cantor's work on infinities.) In other work, Riemann connects flows and sources with the shape of the space they inhabit, setting the stage for a combinatorial topology that unifies these phenomena. An object's shape is characterized by the slices or cuts needed to flatten it and remove any holes. Flows, charges, and shapes are also connected through what Riemann called the Dirichlet Principle, a principle of energy minimization, in effect a mechanical or a thermodynamic analogy.

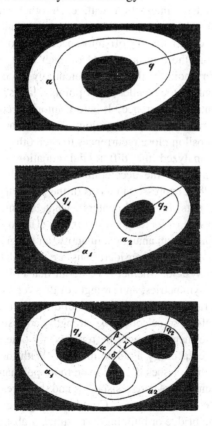

FIGURE 3.3: Riemann's slices or cuts in domains, from Riemann, "Theorie der Abel'schen Funktionen" (1857).

Nineteenth-century science and society is also a precursor to topology. Celestial mechanics, structural engineering, and electromagnetic theory are intimately connected to societal issues: the age of the universe and its long term future; the development of cities with their tall iron-skeleton buildings, railroads with their sufficiently sturdy bridges, and the growing distance between city and its supplying hinterlands; and, the electrification of life in the telegraph, the dynamo and the electric motor, and the light bulb. As for topology,

—Poincaré, in his analysis of the stability of the solar system (1885), had to account for the planets' interaction with each other and their moons. The problem could be posed in a number of ways: Does a planet continue to encircle the Sun, or does it eventually fly off into space or into the Sun itself? Is the planet's distance from the Sun at the end of each orbit constrained to a finite range or not? Is every point of that range potentially a locus or at least epsilon-close to one, or is only some smaller set of points a locus? What is the nature of the set of those points of recurrence? It is to Cantor's account of infinite sets of points and what we would call their topology that Poincaré turns to characterize those loci and their epsilon-close recurrences to each other.[9]

Poincaré also analyzed the differential equations of planetary motion, namely Newton's laws for the acceleration and velocity of the planets. The stability question is then expressed in terms of the geometry of that velocity field (its nodes, saddle points, foci, and centers), what becomes the "index" of that velocity field, a measure of its torsion or vorticity. The topological shape of that velocity field is defined by an analytical quantity, an integral around the origin.

In each case, topological considerations are crucial for understanding the stability of the solar system, whether it be about that Poincaré set of recurrence points, or the shape, symmetries, and strength of that velocity field.

—In designing railway bridges and tall buildings, structural engineers had to balance cost with safety, so selecting the strength and distribution of supporting members. Any single member might be replaced by others; and a collection of members might be replaced by a single member. With the development of iron and steel structures these issues became even more poignant. Very similar issues arose for electrical circuits, where resistances and the wires linking them to other such resistances are the members.[10] Could the engineers develop a form of analysis in which the bridge or building were suitably abstracted or skeletonized, so that one might understand the structure in a canonical and automatic way. And then having done so, "throw away" any particular part of the skeletal structure and analyze the consequences. Such a "simplicial approximation" (say, in terms of triangles) is one foundation for topology, a combinatorial topology of the members linked into substructures and superstructures. (Of course, these problems had arisen much earlier, the paradigmatic story being that of medieval and early modern gothic cathedrals (the thickness of their walls, their flying buttresses), each structure to be higher and better able to let the light in, or to fall down if too weak.)

—The theory of equations also set up two problems, which may be taken to be the conceptual origins of point-set and algebraic topology, respectively. The first is the nature of the solutions of polynomial equations, given that their coefficients are integer or rational or real or complex or polynomials themselves. Just how much of the real line or the complex plane is covered by solutions of various sorts of equations? And second, given a system of simultaneous equations, and given the nature of the coefficients, can we develop a systematic procedure of elimination in order to solve those equations. How and when do we find ourselves in trouble, when the algorithms do not work?

—The legacy of Ampère, Faraday, Gauss, Kirchoff, and Maxwell was a unified and mathematical account of electricity and magnetism in terms of interrelated fields that pervade all of space, those lines of force, and sources such as charges and currents considered as fluids. Electrical phenomena were a consequence of the interaction and the intersection of those charges and currents with those fields. Currents were sources of fields; so a line integral of a field around a current is determined by that current itself. These formulas which connect fields, sources, and the shape and connectivity of objects and spaces were for Maxwell topological considerations. (Maxwell was aware of J.B. Listing's and Riemann's work.) And they are the precursors of algebraic topology.

In 1915 Arthur D. Little developed an industrial or chemical engineering that analyzed complex manufacturing systems into "unit operations" such as drying or distillation. The manufacturing system would be represented by a linked network of boxes, one for each unit operation, much as an organizational chart represents a bureaucracy. What proves crucial for electrical or chemical engineering are notions of flows and their conservation, of network and node, and notions of hierarchical decomposition.

Contemporaneously, mathematicians start developing formalisms for canonically specifying how to decompose objects into their natural discontinuity-free and hole-free parts. The diagrams of modern algebraic topology express the fact that the singularities (cuts, holes, discontinuities—what I shall call *fissures*), and the general spatial configuration of the flows or parts, are related much as sources and fields are related in electromagnetism, or sources and flows are related in fluid dynamics or in processing plants.

—Finally and speculatively, we might note that much of nineteenth century philosophy is a conversation with Kant (1724–1804), an attempt to say how we constitute the apparent world. That concern with constitution, and so I think also with construction, leads to the idea that mathematical concepts are accessible to us, in and as we construct or constitute them. And so Brouwer constructs the continuum in great detail (as does Weyl under Husserl's influence). And Brouwer also constructs (simplicial) approximations and mappings basic to topology's understanding of dimension. Constructing the world as we constitute it is for Brouwer the crucial task for foundations of mathematics and for topology.

These vignettes of nineteenth century mathematics and its applications in natural science and engineering may be seen to motivate and justify the subject matter of topology by displaying prominent precursors for that subject matter. They are not so much history as origin story or folktale. Next, we shall examine themes and concrete examples central to topology's understanding of itself as a discipline and a mode of inquiry: point-set topology's archetypal examples; analysis and composition as an arithmetic, in effect playing with blocks; and fissures, diagrammatics, decomposition, and "hard" analysis.

III

We might define topology thematically in terms of a set of concepts instantiated in particular archetypal examples. So we might say that topology is a way of formalizing notions of nearness, convergence, continuity, smoothness, and the shape of the world, and showing that the world can be decomposed or cut up into nice parts—in each case, in particular canonical ways.

GENERAL TOPOLOGY: EXAMPLES AND COUNTEREXAMPLES, DIAGRAMS AND CONSTRUCTIONS

Set theory developed to better understand collections of mathematical objects, their relationship to the foundations of mathematics, fourier analysis, and the kinds of infinities mathematics might encompass. Point-set topology or "general topology" studies how the various sets fare when they are mapped by continuous functions and, it turned out, how the sets' structures so discovered are about their local properties or systems of neighborhoods, what is called their topology, and

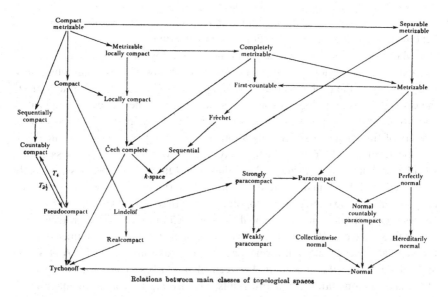

Relations between main classes of topological spaces

FIGURE 3.4: "Relations between main classes of topological spaces," Engelking, *General Topology* (p. 508).

there may be several such, each defining a topological space; yet that local notion has structural and global information. In Kuratowski's topology text, set theory and point-set topology fully overlap in a general account of functions, and not only continuous ones, functions being sets themselves.

The usual notions of topology almost always depend on intuitions from the real line and the plane, and on the wide applicability of the idea of a point and a space (so that in function-spaces a point is a function). An account of point-set topology includes its origins in problems of adding up and of decomposition (for example, the relationship of a function and its fourier series representation, Cantor's original set-theoretic problem), continuity, mappings, peculiar curves in the plane, conceptions of the infinite, and nonuniqueness (or the lack of categoricity) in our specification of sets. Various surprising cases and theorems, examples, and counterexamples ramify those notions, so leading to new concepts, such as a set's compactness or connectedness. Eventually, the various notions are codified. The concepts allow for a rich description of the world, and for interesting theorems connecting the various conceptions. A *topology* is a structure that endows a set with a simple but powerful notion of nearbyness, by describing a distinguished collection of its subsets (the "open" sets) and how they behave under finite intersections and infinite unions. So the descriptions become more articulate, the theorems more interesting.

Many of the demonstrations in point-set topology employ peculiar curves and examples, and tricks and mechanisms, as in Brouwer's work (although once one appreciates the problems there is perhaps less peculiarity than initially meets the eye). Point-set topology is sometimes caricatured as being about peculiarities and idiosyncrasies, "pointless sets" or "pathetic sets of points," obsessed with matters of fine distinctions, classification, and hierarchy: in short, "botanical," to use the term of art. Yet a science may gain power in its faithfulness to the world's richness and variety. And that faithfulness can be systematically organized and be founded on deep principles (much as is botany). So the chapters in one text on general topology are titled: Introduction, Topological spaces, Operations on topological spaces, Compact spaces, Metric and metrizable spaces, Paracompact spaces, Connected spaces, Dimension of topological spaces, and Uniform spaces and proximity spaces.[11] And in its appendix it charts the "Relations between main classes of topological spaces," "Invariants of operations, and Invariants and inverse invariants of mappings." These various facts about the topological world can be systematically organized, and powerful rubrics or species encompass a diverse range of facts. Not only do we discover the variety of the world, we also discover ways of putting it into an order.

Another textbook will begin by specifying a list of fiducial examples:

> Now it is all too easy in studying topology to spend too much time dealing with "weird counterexamples." . . . But they are not really what topology is about. . . . there is a fairly short list which will suffice for most purposes. . . .
> —the product of the real line with itself, in the product, uniform, and box topologies.
> —the real line in the topology having the intervals [a,b) as a basis.
> —the minimal uncountable well-ordered set.
> —the closed unit square, in the dictionary order topology.[12]

A space may be "endowed" with various topologies, various distinct notions of neighborhood or nearness.

As for tricks in the demonstrations, the Urysohn Lemma that allows for a continuous function between two disjoint sets, depends on putting down a topography whose heights vary smoothly and which incorporate the intermediate space. In the Tychonoff Product Theorem, the choice of appropriate sets is

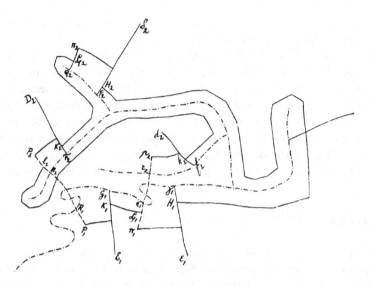

FIGURE 3.5: Proof diagram from Brouwer, "Continuous one-one transformations of surfaces in themselves, 3rd communication," *KNAW Proceedings* 13 (1911)

Theorem 24.2: *Suppose one is given the commutative diagram*

$$
\begin{array}{ccccc}
& \varphi & & \psi & \\
0 \to & A \to B \to & C & \to 0 \\
& \downarrow\alpha \quad \downarrow\beta & & \downarrow\gamma & \\
0 \to & A' \to B' \to & C' & \to 0 \\
& \varphi' & & \psi' &
\end{array}
$$

in which the horizontal sequences are exact sequences of chain complexes, and α, β, γ *are chain maps. Then the following diagram commutes as well:*

$$
\begin{array}{cccc}
\varphi_* & \psi_* & \partial_* & \\
\to H_p(A) \to H_p(B) \to & H_p(C) \to H_{p-1}(C) \to \\
\downarrow\alpha_* \quad \downarrow\beta_* & \downarrow\gamma_* & \downarrow\alpha_* \\
\to H_p(A') \to H_p(B') \to & H_p(C') \to H_{p-1}(C') \to \\
\varphi'_* & \psi'_* & \partial'_*
\end{array}
$$

FIGURE 3.6: Diagrams typical of algebraic topology, adapted from Munkres, *Elements of Algebraic Topology* (1984)

justified by a counterexample that shows the need for a maximal enlargement that would appear to avoid the counterexample.

The diagrams and constructions are in general wondrous. Without knowledge of the trial examples and the conventional constructions, the specific particular diagrams of a proof would appear to come out of nowhere and be ad hoc tricks, jury-rigged to do the nitty-gritty work. Brouwer ingeniously constructed the required apparatus, as can be seen in the figures in his papers. An exposition of set theoretic topology of the plane might be epitomized by its diagrams. (The diagrams in algebraic topology, mechanical and formal, represent a very different style of ingenuity and a different sort of argument.)

PARTS IN ANALYSIS

Topology's concern with the right sort of parts into which an object might be decomposed, or conversely from which it might be built up, is prefigured in the effort to define the right parts for analysis in the calculus or in fourier (or frequency) analysis.

The great achievement of the calculus of Newton, Leibniz, and Riemann was to show how to decompose processes and objects into summable flows or infinitesimal changes, the sum invariant to the exact decomposition—using either geometric (Newton) or algebraic (Leibniz) insights: Add up the tangents or slopes or changes, in finer and finer partitionings of the interval, and you will get just one answer. (As we saw in chapter 2, the great achievement of nineteenth-century statistics, embodied in the Central Limit Theorem, was to show that what needs to be added up in this probabilistic realm are variances, in effect, second-order differences, since the first-order ones cancel out.)

The decomposition problem proved to be not only of mathematical but also of philosophical interest. For Bishop George Berkeley (1734), Leibniz's infinitesimals were a problematic mode of decomposition, their employment— sometimes as zero, at others as nonzero—was both ersatz and inconsistent. Infinitesimals were perhaps rather more mysterious than God, much as Newton's universal Gravity was suspected to be an occult force since it would appear to act at a distance. For Karl Marx (1860s), the "mystification" of infinitesimals is a dialectical process of "unity and identity of opposites": the decomposition of a variable into x and $x + \Delta x$, where Δx is nonzero; then doing some algebra on the difference, $f(x + \Delta x) - f(x)$; then dividing by Δx; and, eventually Δx is set equal to zero.[13] The possibility of a mathematical decomposition and reconstruction provides for Marx a model for an ontology or a societal process, and for how that process, too, is mystified, its true operation deliberately hidden from us. Modern nonstandard analysis, an algebraic view of the calculus, would appear to disenchant these philosophic accounts. It prescribes the arithmetic of these infinitesimals, so that it no longer appears to be arbitrary or mysterious when they are to be treated as finite nonzero, and when they are to be treated as ignorable and zero (namely, those nilpotent nonzero quantities, whose square is zero).

Fourier analysis as linearization is pervasive in the various attempts to work out mathematical decompositions that reflect symmetries or the orderliness of a space or a system (when those symmetries are expressed in terms of the algebra of certain operators, namely, as groups or rings).[14] So there are the frequency components of signals or music, reflecting our notions of pure single-frequency tones. (Technically, there is a group over which that decomposition takes place—an element of the fourier group being $g_x = \exp ikx$.) Again, modern set theory and analysis, the legacy of Cantor and of Lebesgue, arose out of the problem of determining the distinctiveness of a fourier decomposition of a function. Two functions that differ on a "small" set of points will have the same fourier coefficients. The problem was to characterize the size or the measure of those small "sets of uniqueness."[15]

Heisenberg's matrix mechanics makes sense once one works with a fourier-like decomposition of a process, essentially those pure frequency tones or discrete atomic energy levels, so that frequency shifts or transitions are readily discerned. Or, using the right variables ("action-angle variables"), planetary orbits can be defined as sums of various period-of-revolution or frequency components. In each case, symmetries of the system are expressed through the harmonies of nature.

Finding the right parts of the sort I have been describing is an achievement; then the world may be added-up, literally. But, if there is path-dependence or jumps, then discrete compensating add-ins are required and a very different kind of topology is needed. And, sometimes the parts are components, meant to be put together as are pieces of a puzzle, or the bricks in a construction, or the factors of a number.

ALGEBRAIC TOPOLOGY AS PLAYING WITH BLOCKS

An object may be built out of discrete parts fitted together, a game of LEGO or TINKER TOY, whether the parts be algebraic or the structural members of a

71

building.[16] So, sets are the union, intersection, or the Cartesian product, etc., of other sets of a perhaps more elementary sort. Conversely, an object may be decomposed into simpler discrete parts. There are "composition sequences" for groups in terms of groups of a smaller size. Integers have a unique factorization in terms of primes (yet there are many ways of partitioning them additively: $24 = 2 \times 2 \times 2 \times 3 = 20 + 4 = 11 + 13 = 2^2 + 2^2 + 4^2 = \ldots$). And the real numbers may be described in terms of sets of natural numbers (as limiting decimal expansions).

The various modes of discrete decomposition provide models for how combinatorial topology and its algebraicized version, algebraic topology, might analyze an object into simpler discrete ones. (In effect, objects that can be so understood are the objects of algebraic topology.) Conversely, in combinatorial and then algebraic topology we want to compute features of complex objects in terms of features of the parts used to construct them. Given the properties of each of the objects that we combine or paste together, what are the properties of the construction? (Technically, for example, there is the Mayer-Vietoris account of the relation of the homology of the sum of two topological spaces to that of their union and intersection as sets. And, there are Künneth formulas which give the homology of the Cartesian product of spaces in terms of a convolution of the homologies of the individual spaces.)

Blocks allow for an enormous variety of constructions and deconstructions. Many of those constructions are beautiful abstractions of objects we ordinarily encounter, while others are surprising inventions. So, too, are the constructions of algebraic topology, their success being a matter of what we can actually construct or deconstruct, and what we might learn from such constructions. The payoff from such an analysis is a canonical accounting of the shape and properties of objects and spaces, and an explanation of why apparently different objects may be said to have the same shape or not—given a criterion for sameness (such as, the objects' being continuously deformable into each other).

***Some of the themes I have so far employed to characterize topology are: adding-up the world out of parts, infinitesimal and discrete; linearization; the relationship of the local to the global; mappings, diagrams, nilpotence, and path dependence; and construction and constitution and deconstruction. These themes are given concreteness and flesh through particular examples and pictures and diagrams, often drawn from our everyday and not-so-everyday intuitions about the real line and the plane (and, as we shall see, everyday objects such as doughnuts and blocks). For the problems of topology arose in the context not only of mathematics but of mathematics' application to the natural sciences.

Newton and Leibniz, and Gauss, Riemann and Maxwell, are perhaps the archetypal figures. Still, any such abstract listing as this one is of necessity schematic, its particular embodiment within a subject or a field to be given concretely in the field's history, problems, and codifications. The rest of this chapter is an exploration of some of those concrete facts.

***Again, topology may be described in terms of a tension between analysis and algebra, continuity and structures (section IV); a tension between the local and the global (section V); or an accounting scheme for keeping track of continuity and connectedness, algebraic objects, and flows (sections VI and VII). More precisely, those schemes keep track of lacunae and fissures: the absence of continuity or connectedness; the failure of nice decompositions of algebraic objects; the nonconservation of stuff, and the sources, sinks, and dissipations in flows; and, more generally, the play of local peculiarities and global structure. Much of topology may be characterized by the interchangeability or commutativity of algebra and continuity, in that the order of group operations and structures and of continuous mapping might be interchanged: namely, the compatibility of neighborhoods, structures, and maps. These themes and schemes are often represented by pictures and diagrams, epitomizing detailed calculations and providing natural and canonical ways of abstractly visualizing topological objects. Finally, topology has a readily recognized set of examples and counterexamples, a legacy of a tradition of proving and instructing.

IV

Topology is the geometry of continuity. (Alexandroff and Hopf, *Topologie*, 1935[17])

[For Felix Klein] a geometry is determined by a group of transformations . . .
. . . The widest group of automorphisms one can possibly envisage for a continuum consists of all continuous transformations; the corresponding geometry is called topology. (Hermann Weyl[18])

Topology is the branch of mathematics that studies continuity: . . . those properties of objects which are invariant under topological maps . . . [It also studies] equations. It begins with the definition of topological spaces; these are abstract spaces in which the following notions are meaningful: open and closed sets of points, and their mappings. It continues with the introduction of new algebraic-geometric objects: complexes, groups and rings. Set theoretic topology is the part of topology which employs only the following operations: union, intersection, and closure of sets of points. . . . Algebraic topology (or combinatorial topology) is the part of topology that employs algebraic-geometric notions . . .

My initial plan was to develop a theory of equations and mappings which applied directly to topological spaces. . .

I introduce, beside the notion of a covering, which belongs to set theoretic topology, a more flexible notion, that of a "couverture," which belongs to algebraic topology.

. . . the methods by which we have studied the topology of a space can be adapted to the study of the topology of a mapping. (J. Leray, 1945–1946[19], See Appendix C.)

A selective and condensed history of topology will allow us to appreciate how the subject of topology is constituted, and to sketch the tension between point-set and algebraic and structural notions, the recurrent question within the field being, Is topology one field or two, or more?

OPEN SETS AS THE CONTINUITY OF NEIGHBORHOODS

Recalling some of what we have already discussed, by the end of the 19th century, considerations of the uniqueness of the fourier series decomposition of a function and considerations of complex analysis on variously connected domains, develop into: (1) an account of continuity of functions, and of compactness, connectedness, and convergence and limit points, all of which characterize various nice sets of points—what I shall call the *cons*; and (2) an account of decomposing a space into nice parts, or slicing up a space associated with a function (Figure 3.3) so that both space and function can be smooth and without fissure or discontinuity or multiple-valuedness. The notion of a neighborhood, as embodied in an open set, is supple enough to express many of these subtleties of analysis and geometrical structure. Continuity of functions—expressed by means of properties of mappings of open sets—does much of the work we expect from topology, point-set and algebraic. More than one topology can define a space and so the relevant notion of continuity and continguity.

Hausdorff (1914) was able to encode, independent of metric and number, the notion of neighborhood—what it means to be nearby, local, that there is no action at a distance—by means of finite intersections and infinite unions of sets, and so deliberately redefining the topology of a space as its systems of open neighborhoods.[20] And what we might mean by continuity turns out to be that points that are in each other's neighborhoods should be mapped into points that are as well in each other's neighborhoods. More precisely and correctly, if we demand that if a set is mapped into an open set, it too is open, we have in fact provably achieved our goal. So, continuity is encoded into what becomes an algebra of mappings and sets.[21] And we have a topological space.

In their *Topologie* (1935), Alexandroff and Hopf initially define topology as the geometry of continuity, in that it is the geometry that remains the same or is invariant under continuous mappings. Alexandroff and Hopf then followed "the tendency to algebraicize topology by means of a group theoretic foundation."[22] So topology might as warrantedly have been called *Stetigkeitsalgebra*, the algebra of continuity. By page 205 of *Topologie*, Betti groups rather than Betti numbers enter the discourse.

At the same time, the skeletal structure of a space (a tessellation or triangulation, in effect, a space frame), the wherewithal of a combinatorial topology and the source of an algebraic topology, is being replaced by point-set topological notions of a covering by open sets, due to Alexander, Vietoris, and Čech, so linking combinatorial and point-set notions.[23] The trivium of analysis (and continuity), geometry, and algebra are beginning to meet harmoniously. To learn about the world, we ask: what are the open sets or coverings; or, what are the natural units of combined objects, or complexes; or, what are the structures and mappings? Later, a topological space, say a space frame, need not be conventionally smooth or nicely connected, but merely a set of linkages.

Alexandroff and Hopf want to present "Topologie als ein Ganzes,"[24] (topology as a whole) acknowledging the tension between point-set and combinatorial topology. They do touch all the fields, although the connections among them is not at all so clear, the whole not quite apparent. But this is only to be the first of their three projected volumes (and, it turns out, the only one that appeared). And, Tietze and Vietoris, in their 1929 *Enzyklopädie der Mathematischen Wissenschaften* survey of connections among the various branches of topology, suggest:

> Our discussion of the three areas of topology (general, *n*-dimensional continuous manifolds, and combinatorial) will not begin with the most abstract, but first with point sets in an *n*-dimensional space of real

numbers [*n*-tuples of reals]. In strict systematic sequence, this section should come after the one on general topology.[25]

Tietze and Vietoris sequentially connect the various branches of topology, point set to general to *n*-dimensional to combinatorial and back, but the chain's links do not there appear to be compellingly strong.[26] Further developments are required.

Algebra as a Unifying Force. By 1935 combinatorial topology has become self-consciously algebraic. The algebraist Emmy Noether had already (ca. 1926) taught several men ("Noether's boys") that the combinatorial methods that involved incidence matrices indicating what was connected to what, what was incident to what, which defined objects or spaces, and their boundaries as well, in terms of a skeletal or polygonal approximation, also generated algebraic objects, groups, which themselves might be studied. Alexandroff says in 1935, memorializing Noether:

> In the summers of 1926 and 1927 she went to the courses on topology which Hopf and I gave at Göttingen. She rapidly became oriented in a field that was completely new for her, and she continually made observations, which were often deep and subtle. When in the course of our lectures she first became acquainted with a systematic construction of combinatorial topology, she immediately observed that it would be worthwhile to study directly the groups of algebraic complexes and cycles of a given polyhedron and the subgroup of the cycle group consisting of cycles homologous to zero; instead of the usual definition of Betti numbers and torsion coefficients, she suggested immediately defining the Betti *group* as the complementary (quotient) group of the group of all cycles by the subgroup of cycles homologous to zero. . . . These days [eight years later] it would never occur to anyone to construct combinatorial topology in any way other than through the theory of abelian groups; . . .[27]

And, apparently independently, Leopold Vietoris made use of such groups in 1926.[28]

The tension between the subfields or the approaches to topology is still alive today. So a text on general topology concedes that:

The proof that the space $\mathbb{R}^n$ has topological dimension n requires a deeper insight into the structure of this space; by the nature of things, some combinatorial arguments must appear in it. To preserve the uniformity of arguments used in this book [on general topology], which are all of point-set character, . . . [we shall employ the well known] Brouwer fixed-point theorem . . . [which, for its proof] requires some combinatorial or algebraic arguments . . . [29]

The recurrent sticking point is a notion of dimension. O. Veblen's 1922 *Analysis Situs* deliberately has no point-set topology, except for an unavoidable part concerned with dimension and Brouwer.

And, again, a recurrent passing remark of the last fifty years is how the subfields are at last being unified:

The unification of the two areas of interest has been under way for a generation and is still not complete. Even today [1961], one hears of point-set topologists as distinguished from algebraic topologists. (Hocking and Young[30])

Presumably, any such unification should be mathematically fruitful, and not just a marriage of convenience. What could you see more clearly, or be able to prove much better, under such a theory?[31]

DIMENSION AND HOMOLOGY AS BRIDGING CONCEPTS

Alexandroff and Hopf's admittedly provisional unification is the topology of polyhedrons (rather than, say, of point sets or of complexes more general than polyhedrons, or rather than a unification of set theory and algebra).[32] Polyhedrons allowed for the crucial unifying notions: the "invariance of dimension" and homology groups, which account for problems that bridge the fields of point-set and algebraic topology.

In about 1900, a problem that presented itself was the invariance of dimension under a continuous one-to-one transformation. Cantor had shown that dimension need not be invariant under a one-to-one mapping. The number of points on the line or in the plane are the same. And Peano's space-filling curve is conversely disturbing. The central figure is Brouwer, the great conceptual and proof innovator in topology from 1909–1914. He employs his notion of the degree of a mapping, in effect, the n of z^n, to show that there cannot be a one-to-one continuous mapping between an m-dimensional and an n-dimensional real space, $\mathbb{R}^m$ and $\mathbb{R}^n$, if m and n are different—thereby proving the topological invariance of dimension.[33] Combinatorial methods, and approximations of a space by a tessellation or a polyhedron (a "simplicial approximation"), provide a method for studying how, for example, a line is covered by closed intervals, the crucial feature of Brouwer's and also of Lebesgue's analysis of dimension. So combinatorial methods do set-theoretic work.

Homology offered another unification, where by homology we might mean the number of holes in a space, the word homology coming from the fact that different approximations of the space still counted the same number of holes. Homology is a bridging concept because it is both about the shape of the space (the province of combinatorial and algebraic topology) and about what is invariant under continuous transformations. Earlier, for Poincaré, the shape and connectedness of objects may be turned into a combinatorial and then algebraic problem, once you realize that shape and connectedness is a matter of the distinct paths you can follow continuously from here to there in a space, and the combinations of those paths ("homotopy"). Eventually (1920–1945), it was understood just how the discrete methods of combinatorics and algebra might be said to commute or be compatible with considerations of continuity and the cons more generally.

The problems of dimension and homology demanded the deeper understanding of continuity provided by point-set topology and of the shape of a space provided by algebraic topology.

Developments in topology then transform analysis, geometry, and algebra. If algebra explicitly enters topology to characterize homology in about 1926 or 1927 (and in a sense much earlier in Poincaré), it is then realized that homology's dual, cohomology (called such by Whitney in about 1935), which studies the functions that can live on a space, is formally like group extensions (Mac Lane, 1942). Namely, given a group D, defined as X/H, and a group H, what could be X? Topologically, given a doughnut, D, and its holes (H), find out the shape (X) of the baker's original dough. In effect, find the X that solves the equation, X _minus or modulo_ H _equals_ D.

And then those "homological methods" transform algebra itself (ca. 1945):

> The new methods and concepts which I have just mentioned [of algebraic topology] are topological in nature, and were introduced to solve topological problems. What was totally unexpected in 1940 is that the methods of algebraic topology could be bodily transplanted to a host of mathematical situations in analysis and algebra. . . .

> Homological algebra . . . started in fact as a kind of glorified linear algebra [matrices], by introducing concepts such as the Ext and Tor functors, which in a way measure the manner in which modules over general rings misbehave [that is, just where division may not be possible] when compared to the nice vector spaces of classical linear algebra; and the similarity with homology groups, which tell us how much a complex deviates from being acyclic [or hole free], is now a commonplace. (Dieudonné[34])

A proviso: In characterizing a field of mathematics, whatever tensions I describe, whatever subject matter I discern, they are in the end subject to revision due to actual developments within the mathematical field and in its applications to other mathematical fields and to the natural and social sciences: what we understand; what we believe is true; what we can prove; and, what we can do with the mathematics. The descriptive characterizations of topology (its tensions, its objects) must be appreciated within actual practice, again within the field and in its applications. The characterizations only notionally, albeit usefully, have an independent abstract meaning. But the abstractions and characterizations could not legislate for the field or for the mathematics as it turns out. That depends on the mathematics.

V

THE TENSIONS IN PRACTICE: THE LOCAL AND THE GLOBAL

Within topology, there is a tension between local and global analysis: understanding an object in terms of its parts or what happens close to each of its points, and understanding an object in terms of its more general characteristics as a whole. Do we understand a fluid flow in terms of its velocity at each point, or in terms of the general features and shape of its flow-lines and of its sources, sinks, and boundaries—or both? The local is often signified by epsilon-delta arguments, and by an "uncertainty principle" that connects the value of a function with its derivative. The global is often signified by the spanning or the tessellations of a space or a flow, so as to characterize its shape, and by decompositions of that tessellation into simpler components.

It is one of the recurrent themes of topology that the local and the global are to be connected. They are "duals" of each other, complementary ways of getting at "the topology" of some object. That is the legacy of Fourier, Riemann,

Weierstrass, and Cantor—a belief that global and local features are intimately connected with each other, that Valentine and Orson are brothers, yet each has its own dominion.

So applied mathematics is concerned both with matters of form and structure and symmetry and with matters of sharp limits and bounds and singularities, what is sometimes called soft and hard analysis, respectively.[35] If topology is in any reasonable sense a reworking of the basic notions of the calculus, then it would not be surprising that topology's culture deeply influences the applications of mathematics. There is an interplay of global and combinatorial matters with sharp estimates and bounds. Whether we are studying quantum field theory or biological molecules, topology is everpresent.

Again, whatever the tension, whatever the objects that are employed to define a field of mathematics, those tensions and objects depend on the mathematics and on what you can do with it. If new tensions and objects are more fruitful, or if new notions of a tension or an object are called for, the field's definition will change. No field or science could be adequately characterized by terms such as fissure, diagrammatic, decomposition, or blocks. Only in those terms' particular realization, in the actual mathematics itself—the facts, the proofs, the examples—do such terms begin to make sense, essentially as pointers to what is especially important in such practices. But in so far as topology informs applied mathematics—in combinatorial, algebraic, set-theoretic, and function-theoretic ways—ways in which "topology" may be hardly mentioned at all, these terms or notions are now so deeply entrenched, we might see their validation.

THE TENSION IN ANALYSIS: THE HARD AND THE SOFT

> Let δ, ε be two very small numbers, the first chosen such that . . . (Cauchy, *Calcul Infinitésimal*, Leçon 7[36])

> [P]hysics consists of all the consequences of the basic laws that have to be unearthed by hard analysis. (W. Thirring[37])

As general topology developed we might see it as responding initially to the needs of analysis: the world of limits and of Cauchy's and Weierstrass's epsilon-delta (ε-δ) arguments.[38] (Epsilon-delta goes, "For all ε, there exists a δ (both positive), such that if $|x| < \delta$, then $|y(x)| < \varepsilon$, where $\delta = \delta(\varepsilon)$." Note that we define the open set in the image space and then *ask* for an open set in the original

80

space.) Epsilon-delta prescribes a practical and rigorous way to implement the notion of a neighborhood and build in notions of continuity and differentiability, open and closed sets, accumulation points, and compactness.[39] It also provides an account of sharp results or bounds: $A \leq c \times B$, where c is as small as possible—"best possible" since you have in your back pocket a case that is otherwise precluded that just violates the inequality. In such "hard" analysis, best-possible may demand higher order approximations and delicate inequalities.[40] (Note, however, that rigor is not the same as hard analysis, although those who do hard analysis may often speak about the need for rigor.)

An archetypal basic result might be called an uncertainty relation, connecting the gradient or derivative of a function to its height or size, namely $|\operatorname{grad} f| \geq c / |f|$), much as the Heisenberg uncertainty principle connects momenta or velocities, which are the gradients, to uncertainty or variance in position, a norm, $|f|$. So fourier analysis also connects the momentum and spatial representations of a function.[41]

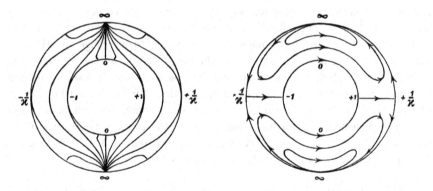

FIGURE 3.7: Possible streamings or flows in an annulus, as pictured by Klein, *On Riemann's Theory of Algebraic Functions and Their Integrals* (1893).

Correspondingly, as topology became combinatorial and algebraic, it responded to the needs of geometry. So there developed a soft or global analysis, in which the model objects are knots, paths, and generic configurations of field and flow in hydrodynamics and electromagnetism (as in Gauss, Maxwell, Riemann, and Felix Klein). It is a global topology.

Such a topology is committed to composition (pasting things together, adding things up) and to decomposition, and to characterizing the shape of objects and spaces in those terms. The topology provides an accounting scheme in terms of stocks and flows, sources and fields, black boxes and inputs-and-outputs, holes and boundaries, and edges and faces and vertices. Imbalances and leftovers are expressed as mismatches and fissures and singularities and toll charges, to be displayed graphically, pictorially, and diagrammatically, as well as algebraically.

For a complicated object in a higher dimensional space, the fissures (or nonzero homologies) appear at different dimensions. So the diagrammatics of (homological) algebra are set up to keep track of such decomposition and fissure and mismatch, as a decomposition sequence of objects and mappings.

In so far as possible, global and discrete facts are to be expressed locally and infinitesimally. Adding-up and actual flows are the featured moves in the vector calculus. The gradient, curl, and divergence operators point to the right stuff that needs to be added up locally; and they are employed to express the conservation laws of momentum, angular momentum or vorticity, and energy and charge. And the integrals of these operators (Gauss, Green, Stokes) lead to global information. In electromagnetism, potentials (and "gauge invariance") may be seen to be a way of expressing the conservation of charge, effectively embodied in Kirchoff's second law; and the potential itself is a means toward making things as path-invariant as can be, in effect Kirchoff's first law.[42] (Put differently, cohomology measures conservation of source through flow, while homology measures nodal features and so characterizes the network. And these are complementary and in effect equivalent formulations.) Here is such a description of the electromagnetic and gravitational fields:

Define this field and "wire it up" to the source in such a way that the conservation of the source shall be an automatic consequence of the "zero boundary of a boundary" [$d^2=0$, nilpotence, an exact sequence]. (Misner, Thorne, and Wheeler[43])

By its devices, the topology builds in the physics or the geometry, the global facts or conservation laws instantiated locally.

Some of the time what is at stake in the distinction between hard and soft are very different ways of approaching a problem. Do we calculate, in a virtuosic fashion, or can we draw a picture from which we can read off the result?

By studying the hydrodynamical limit for Hammersley's process we show by fairly "soft" arguments that lim . . . This is a known result, but previous proofs relied on hard analysis of combinatorial asymptotics.[44]

This [Buffon needle] problem can be solved by computations with conditional probability [as in Feller's treatise]. It is, however, more instructive to solve it by another method [an additive set-functional on certain sets of lines in the plane], one that minimizes the amount of computation and maximizes the role of probabilistic reasoning.[45]

In soft analysis, the right setup or example may lead to the relevant insight and proof, although it may be quite difficult to come up with that right simple picture, and a good deal of technical machinery and virtuosic calculation may be needed along the way.

But only some of the time can you avoid hard analysis if you want deep results that are not in their nature generic. Sharp limits usually demand complicated and perhaps lengthy and arduous computations, and even numerical calculation, while employing a toolkit of inequalities, asymptotic approximations, and generalized functions ("distributions") and special functions.[46] And you must apply that apparatus sophisticatedly. You might hope to show that your sharp limits are generic, but that is often just that, an unfulfilled hope.

As we shall see in chapter 4, the tension we have identified between the hard and the soft is often expressed as a tension between calculation and ideas, computation and concepts:

> I have tried to avoid Kummer's elaborate computational machinery, so that here too Riemann's principle may be realized, according to which proofs should be impelled by thought alone and not by computation. (Hilbert, *Zahlbericht*, 1897[47])

Sometimes and tendentiously, one distinguishes between blind computation and insightful thought. But actual computation is rarely blind, and really monumental calculations only become doable when they are endowed with insight as well as inventiveness and perseverance. What appears as a miracle usually reflects enormous amounts of calculation that did not work out and is not displayed or mentioned, the actual experience of the scientist with similar problems and so a personal bag-of-tricks, and the good fortune to have gotten someplace with the work.

Technology complements ingenuity. Mechanical generic methods may substitute for ad-hoc inventions. Historically, much of what Leibniz and Newton achieved in the calculus was already known by special and idiosyncratic analytic and geometric constructions. Their achievement was to systematize and unify the variety of methods and tricks and examples into a single theory.

Again, Berkeley's (1734) critique of the Newtonian and the Leibnizian calculuses argues that in fact they are still rather ad hoc: fluents and infinitesimals sometimes considered nonzero while at others set equal to zero. And if one does not do it just right, that is, treat infinitesimals as nonzero or zero as the case may be, one won't get the right answer. (Again, Robinson's nonstandard analysis does provide a canonical and consistent way of doing this, albeit 250 years later.)

HARD ANALYSIS: THE ISING MODEL OF FERROMAGNETISM AND THE STABILITY OF MATTER

The distinction between hard and soft, or between intense computation and generic modeling, is a useful one for describing different sorts of mathematical endeavors, if it is not taken too rigidly or seriously. An excursus into mathematical physics will display how the distinctions appear in practice, although the subject of topology appears only briefly. (Chapter 4 discusses the following problems in much greater detail, focusing on matters of strategy, structure, and tactics in computations and proofs. The appendix to the current chapter provides a survey of the Ising model.)

FIGURE 3.8: The Two-Dimensional Ising lattice, where J is the coupling between horizontally (row σ) and vertically (rows σ and σ') adjacent spins, has a hamiltonian: $H = \Sigma_i -J\{\sigma_i\sigma_{i+1} + \sigma_i\sigma_i'\}$. Spins have values $+1$ and -1. If $J > 0$, then alignment of spin directions is preferred, and hence there is the possibility of ferromagnetism. The statistical mechanical partition function is $\Sigma \exp{-\beta H}$ (β is the inverse temperature, H is the energy) and the sum is over all horizontal and vertical linkages and all possible spin configurations. And it is also a problem in algebra, namely, group representations of the deepest symmetries.

The Ising Model. We can discern the shifts between hard and soft analysis in studies of the thermodynamic properties of the two-dimensional Ising model (named after E. Ising, 1925, and earlier proposed by Ising's advisor, Lenz, 1920, albeit in one dimension). Here, it is a model of a two-dimensional ferromagnet, in epitome represented as planar grid of interacting dipoles or spins, σ_i, with stylized local dipole-dipole interactions preferring alignment of the spins' directions.

The Ising lattice in two-dimensions was the first mathematical model that actually modeled a phase transition: a discontinuous or, in this case, a non-analytic transition of the free energy at the critical point in the quite realistic limit of an infinite lattice. It is actually a good model of some physical systems (such as a "lattice gas"), and provides the archetype for a whole range of lattice models, that is discrete models. of field theories. And it turns out to be remarkably rich mathematically.[48] The task is to compute the statistical mechanics partition function, a way of packaging in one function the number of macroscopically equivalent microscopic systems for each value of the temperature and density, say. The logarithm of the partition function is proportional to its thermodynamic free energy.

The solution of the two-dimensional Ising model, namely its partition function, turns out to be a problem in algebra or in graph theory and combinatorial topology. It is also a problem in analysis and point set topology, in effect about the scaling symmetries of the partition function and about the locus of its zeros.

Lars Onsager's original solution (1944) for the partition function is an amalgam of algebra, analysis, and hyperbolic geometry. His student Bruria Kaufman (1949) reformulated the solution in terms of spinors, the mathematical objects originally employed by Dirac to represent spin-½ electrons, so making Onsager's algebraic methods more transparent to his contemporaries. Lee and Yang (1952) pointed out the significance of the locus of the zeros of a partition function; that is where a phase transition might take place. In the graph-theoretic, combinatorial, topological vein, Kac and Ward (1952) ingeniously constructed a matrix whose determinant counts up the interactions among the spins, in terms of the number of closed polygons on a lattice, and so they compute the lattice's statistical mechanical partition function and demonstrate how Kaufman's (and Onsager's) algebra does its work. The next moment in this back-and-forth progression was a rather less ad hoc appearing piece of mathematics, as well as being quite deep physically: the Ising lattice understood as a free fermion quantum field theory (Schultz, Mattis, and Lieb, 1964; Samuel, 1972), now emphasizing algebraic rather than combinatorial techniques, with objects that resembled the Cooper pairs of superconductivity. And, roughly at the same time, using a relative of the determinant, the Pfaffian, the combinatorics was systematized into an algorithm that is as automatic and natural as is the analysis or the field theory (Hurst and Green, 1960, 1964; Kastelyn, 1963; Montroll, Potts and Ward, 1963), in effect the Wick product of quantum field theory. During the 1960s, Domb and Sykes and collaborators developed cluster expansions for the partition function, in effect a perturbation theory, and then, using Padé approximants, explored the zeroes and other features of the partition function. Lieb's use of the Bethe Ansatz (1967, following Yang and Yang, 1966; also, McGuire (1964) and Sutherland, 1967), to solve more complex models (the ice model), showed how it all could be reduced to a sum of two-body interactions. Baxter (1970s), then showed how analysis (the partition function as a function of a complex variable) could be used to derive the partition function once we paid attention to its scaling and its algebraic symmetries incorporated into a functional equation—what is now called the Yang-Baxter equation and what Onsager called the star-triangle relation and the engineers called the Y-Δ transformation. (I should note that Onsager's original work echoes everywhere in this progression.)

Subsequently, Wu (with McCoy, Tracy, Brouch, and their collaborators) go from automatic devices to dogged and detailed symbolic manipulation, to solve much more difficult problems concerning correlations and random impurities. They use Kastelyn's, and Montroll, Potts, and Ward's Pfaffians to compute the combinatorics, and they employ a kind of fourier transform and some very special special-functions (Wiener-Hopf techniques and Painlevé transcendents, respectively) to achieve their final result. In the end, those problems still demand great ingenuity. In the matter of course, eventually, what we might call softer algebraic and geometric techniques show how to make these moves as well (Sato, Miwa, Jimbo, and collaborators). And so in a sense we are both advanced in our ability to calculate and back where we started from, for great ingenuity and device is needed again. Yet, there is as well a payoff in the variety of quite specific analogies with quantum field theory that are suggested along the way by the various techniques.[49]

The Stability of Matter. From the fact that ordinary bulk matter (say a gram of material) in our everyday world does not implode (that is, the electrons and nuclei collapsing into each other, eventually leading to an explosion), we know that ordinary matter is stable. To rigorously prove that this is the case turned out to be a formidable problem in hard analysis in mathematical physics—although after much subsequent work, the proof was considerably simplified. At first, it was not clear just how Coulomb forces and quantum mechanics would be sufficient for a proof, although presumably everyone believed they ought to be sufficient.

Once mathematical physicists could provide a proof of the stability of matter, they then could warrantedly say that the stability of matter depended on an uncertainty principle (and quantum mechanics); the fact that roughly equal positive and negative charge densities cancel each other's effects at long distances ("screening"); and, that electrons are fermions which obeyed the Pauli exclusion principle (one particle per state). Without these features, the proof would fail, and in fact one could show that matter would be unstable. The proof provided a philosophical analysis of "the stability of matter," just what could be meant by that notion, and an account of what is essential for us to have our ordinary everyday world.

Freeman Dyson and Andrew Lenard's (1967–1968) mathematically rigorous proof of the stability of matter is generally acknowledged to be an exemplary tour de force of hard analysis in mathematical physics. Technically, one wants a lower bound, E_*, for the energy per particle (E/N) of bulk matter, $E \geq E_*N$, where $E, E_* \leq 0$.

> [I]f N charged non-relativistic point particles belonging to a finite number of distinct species interact with each other according to Coulomb's law, and if the negatively charged particles of each species satisfy the exclusion principle, then the total energy of the system cannot be less than ($-AN$), where A is a [positive] constant independent of N. The practical meaning of this theorem is that chemical or electrical reactions, in a lump of matter containing N electrons, can never yield more than a bounded quantity of energy per electron. Nuclear and gravitational interactions between the particles are not considered. The theorem is obviously false if gravitational interactions are included. (Dyson[50])

In another tour de force of hard analysis, Joel Lebowitz and Elliott Lieb (1969) used the Dyson-Lenard bound to prove the existence of a bulk matter or thermodynamic limit: $\lim_{N\to\infty} E/N \to$ Constant, an asymptotic limit, that is, about the *form* of the equation: $E \approx$ Constant $\times N$, rather than a lower bound.[51] Lieb and Walter Thirring's new proof of the stability of matter (1975), also mathematically rigorous, used rather more physics to set things up. So it provided a much sharper bound by a factor of perhaps 10^{13} over Dyson-Lenard, and with somewhat less (tour de) force. It focuses on generic physical features of bulk matter. And in just this complementary sense, it may be said to be "softer."

> Lenard and I began with mathematical tricks and used them to hack our way through a forest of inequalities without any physical understanding. [This is surely unfair to Lenard and Dyson.] Lieb and Thirring began with a physical idea, that matter is stable because the classical Thomas-Fermi model of an atom is stable, and then went on to find the appropriate mathematical language to turn their idea into a rigorous proof. (Dyson[52])

For the stability of matter, one has to show not only that isolated atoms held together by Coulomb forces ($\sim 1/r^2$) are stable, but also that an Avogadro's Number of atoms in bulk are stable as well. As Dyson points out, if the attractive forces were stronger than repulsive forces by one part in a million, there would not be stability of matter, since attractions of next-nearest neighbors would dominate over the repulsions, and there would be collapse. The generic problem, both for the stability of matter and for the thermodynamic limit, is to understand how to take into account all these just-about-canceling forces among the nuclei and the electrons.

Lenard started working on the stability-of-matter problem in the context of plasma physics. He was inspired by an observation by Michael Fisher and David Ruelle (1966), who pointed out there was as yet no rigorous proof of the stability of matter. Notably, earlier work on the thermodynamic limit had not dealt with the $1/r$ problems of the Coulomb potential (it had used a "tempered" potential which (albeit realistically) smeared out the nuclear charge, and it fell off more rapidly with distance than $1/r^2$). Eventually, Lenard was joined by Dyson, and they achieved a proof. One can discern in all derivations the importance of figuring out how to build into the formalism a proto-crystalline structure, in effect that each atom was within its own nice little cube.[53]

The later Lieb-Thirring 1975 proof was of a very different sort than the Dyson-Lenard 1967 proof. They realized that one could employ a model of atoms in which atoms did not bind to each other (there is no chemistry in this model): the Thomas-Fermi model of an atom, in effect a $Z \rightarrow \infty$ approximation.[54] That turns the N-body problem back into a one-atom problem. (Since Thomas-Fermi model atoms do not bind to each other, we can consider one atom at a time. Their binding energy, the energy needed to pull the electrons and nucleus apart, suitably modified, can be shown to provide a lower bound to the binding energy of actual atoms in bulk.)

Whatever the derivation, one of the main tasks is to learn to sequence the problem in physically meaningful and mathematically doable stages. Again, both Dyson and Lenard's work and Lieb and Thirring's work are pieces of hard analysis. But in return for more detailed physics in the latter effort one does obtain a much better constant with a lot less computation. But it is still not yet best possible. (We actually know some of the measured binding energies of isolated atoms, or those computed by means that are quite accurate, the Hartree-Fock approximation. These are quite close to the empirical binding energy per atom in bulk.[55] See chapter 4.)

Charles Fefferman and collaborators (1983–) have taken on the problem of proving the existence of a gas phase of hydrogen atoms or molecules, namely a regime of chemical potential (say, density) and temperature in which N electrons and N protons form a hydrogen gas, with very high probability, rather than a crystal or an ionized plasma. This is a many-body problem par excellence. In order to make it into a few body problem, even more physics is needed to guide the formulation, including a physically reasonable, mathematically useful definition of a gas (namely, the mean distance between an electron and a proton is roughly of atomic scale, yet there is lots of room between atoms so that atoms themselves do not much interact with each other). The idea is to use a good approximation to the solution, those noninteracting atoms, and then bound or control the correction term, this being the task of the hard analysis, which in this case is part of a monumental endeavor involving "gruesome details."[56] (Ideally, eventually, we might want these details to be about the actual physics in the model, not mathematical technology. And that involves not only improved proving, but a deeper understanding of what we are doing in a calculation.)

Along the way, in order to find a regime of chemical potential and temperature that is a gaseous phase, Fefferman needs a much sharper value for the stability-of-matter lower bound of the binding energy per particle, E_*. He and his collaborators have subsequently devoted enormous effort to obtaining a rigorous derivation of a sharper E_* (so far, only partially successful), and then of the binding energy of a single isolated atom.[57] In the latter effort, they must find a way to take into account the correlations of an atom's electrons with each other, and to estimate the energies or eigenvalues of the atom's hamiltonian.[58] The problem is to bound the first few derivatives of the wave function, hence bounding the kinetic energy.[59] "The result obtained is quite sharp."[60]

It would seem, at least for the moment, that if we want sharp bounds, and we must have them if we are to account rigorously for some everyday phenomena such as a gas, there will be no soft road to the solution. There is no simple generic picture. Moreover, yet more physics and more hard analysis are needed still to squeeze down the constant E_* so that it will allow us to account for those everyday phenomena.

What we might take as physically and intuitively obvious about our world—that matter is stable, that there are phase transitions, that pushing on a system will decrease its volume ("thermodynamic stability"), that there are gases—is not at all so obvious once we have to be quite clear about what we mean and to rigorously prove it. In return, we get a much better sense of what is essential about atoms that makes possible the everyday world (quantum mechanics and an uncertainty principle, Coulomb forces, and electrons' being fermions so they obey the Pauli exclusion principle).

These phenomena, magnetization and stable matter, are the consequence of a very large number of almost-canceling interactions.[61] What is vital in their analysis is the tension between the local and the global: hard and more local methods, featuring epsilon and delta, complement soft and global and often algebraic/combinatorial methods.[62] The interesting side-problem is to figure out how to chop up space into nice cubes (perhaps containing and contained in other cubes) or other such volumes, and then to put space together again.

VI

Topology was driven by the phenomena to become a field concerned with keeping track of path dependence, fissures or cracks or holes, and dissipations. Complex analysis provided ways of handling these phenomena in two dimensions, while topology provided more general ways: An integral from a to b in the complex plane sometimes depended on the path from a to b. Important spaces were neither smooth nor simply connected. There were discontinuities, poles, cuts, and holes in spaces. And there were fluid-like flows that did not obviously conserve that fluid; we sensed that we were missing something.

Topology provided an account of path dependence and ways of compensating for it, so integrals might be single valued. Topology provided accounts of discontinuity and disconnectedness that showed that such fissures might be an artifact of a peculiar perspective on a higher-dimensional much nicer space, or that they were a matter of adding together several otherwise nice spaces. And, topology provided systematic ways of keeping track of dissipation as algebraic accounts of the "lack of exactness."[63] Similar phenomena appeared in algebra.

If local properties of objects and spaces—path independence, smoothness, and conservative flows—did not extend globally, topology provided an account of obstructions to that extension, the discontinuities and the inexactitudes.[64]

KEEPING TRACK OF UNSMOOTHNESS AND DISCONTINUITY

A typical course in point-set or general topology studies what I called the *cons*: continuity of maps; properties of sets such as compactness, connectedness, convergence, and completeness, all of which may be preserved by good-enough maps in the right sort of spaces (continuous maps and nice topological spaces, respectively); countability and separability and coverings; and, the peculiar sets of points which may or may not possess these properties—what are called counterexamples. (So, a book entitled *Counterexamples in Topology* can convey a good deal of the central content of the subject.)

Notionally, I shall take *cons* to refer to realms where there is robust integrity and wholeness, and you can just add things up willy-nilly. Continuum limits exist and so derivatives may be defined and the physicist's partial differential equations are meaningful and may have nice solutions. (Analysis eventually learned to allow for lacuna, in generalized functions, microlocal analysis and singular integrals.)

(More technically, and relevant to some of the physical examples in chapter 4: When the points are functions in a topological function space, then there are the physicists' infinite volume limits or "statistical continuum limits."[65] The functional differential equations of statistical mechanics and of quantum field theory, in which the field at each position, f_x, is itself an independent variable, make sense, and there exist nice solutions representing bulk matter. The observed quantities are well-defined averages over the field or measures of correlation among the field quantities. Moreover, notions of uniform convergence (that is, the rate of convergence is in the end independent of the point within the interval) guarantee that when we add up functions the properties of the addends, for example smoothness, are reproduced in the sum. There are no surprises. With the absence of uniformity, the limit of analytic functions need not be analytic, so allowing in mathematical statistical mechanics for those non-analytic critical points where there are phase transitions.)

Many of the cons properties are inherited from the calculus, part of "the rebuilding of the foundations of calculus achieved during the 19th century."[66] The intermediate value theorem leads to connectedness, the uniform continuity and maximum value theorems lead to compactness.[67] (Countability and separation properties do not come from the calculus.) We might say that general topology shows how to make a function continuous (what is called the induced topology), and more generally it shows how nice properties such as compactness survive or not in restricted and combined spaces and mappings (subspaces, sums, Cartesian products, inverse limits, function spaces, and quotients).

But, the precision craftsmanship of the world of epsilon and delta, the algorithmic way of thinking about nearbyness, otherwise represented by open sets, neighborhoods, and nice topologies, cannot account for fissures, for holes and discontinuities and jumps and path-dependence. Here, the manner and order in which we group and add things, and take limits, would seem to matter. That local story of continuity and smoothness, the derivative, and the global story of paths and the calculus of variations, has to deal with situations that do not add up so nicely. Topology must provide ways of accounting for mismatches, breaks, and twists, and for path-dependence.

Cantor's and Lebesgue's great insight was that some sorts of fissure might not matter much at all, such as on small sets (such as a countable infinity of points), those "sets of uniqueness" that do not affect the fourier coefficients. Lebesgue provided a way of adding things up even if the derivatives were infinite at these points. In Lebesgue's adding-up, one partitions the image space into horizontal slices rather than the vertical rectangles of the familiar Riemann integral.

> Lebesgue explained the nature of his integral by an image that is accessible to all. "Let us say that I must pay a certain sum; I search in my pocket and take out coins and bills of various values. I hand them over in the order in which they present themselves until I pay the total of my debt. That is Riemann's integral. But I could work in a different fashion. Having sorted all my money, I put the coins of the same values together, and the same for the bills, and I make my payment as a series of payments each with money of the same sort. That is my integral."[68]

Unavoidable fissures, discontinuities, breaks, holes, and cracks might be accounted for explicitly, something like MONOPOLY's "Collect $200 when you pass GO," or a toll booth. Riemann's moduli, here a $2\pi n$ increment, are such $200 rewards or $7 tolls. Topology prescribes compensations or corrections to path-dependent sums so that they are then path-independent or invariant. Similarly, Maxwell's electromagnetism, with its scalar and vector potentials (and their gauge invariance), prescribed how to incorporate time and space path-dependence.[69]

Alternatively, imagine playing on a spiral board so that you always know how many times you have gone around. Your assets depend in part on the number of times you have gone around. Your location on a flat playing board at any particular moment gives no direct indication of that number, only the spiral level does. A paradigm is Riemann's surfaces, and his prescription for making a multiple-valued function such as the square root ($\sqrt{4}=\pm2$) single valued: enlarge the domain of the function into separate branches, positive and negative roots, say, suitably pasted together. (Figure P.1) Hermann Weyl then provides an account of Riemann surfaces that leads to a combinatorial/algebraic topology, initially (1913) in terms of functions of a complex variable.

In each case, the idea is to make the world smoother and more continuous by systematically dealing with fissures. If there is a function that is multiple-valued or has discrete jumps, one makes it single-valued and smooth on a larger domain, or extends the dimension of a domain. Dimension may become a variable itself. If there are singularities, they are to be given charges or degrees or indexes that epitomize their effect on their surroundings; more complicated measures, in homologies, take into account the holes of various dimensions. If there is periodicity, one arranges the space so that periodicity is built in. If there is path dependence, one devises compensating factors such as gauges or potentials that account for that path-dependence.[70] And, if there is linear dependence, so that a space is actually smaller than it would appear, one develops modes of revealing that dependence.

Again, the notion of an open set grounds the topological enterprise, encoding what we mean by a neighborhood and by continuity. The infinite union and finite intersection of such sets is also open; and, a continuous map that yields an open set is a mapping of an open set. In general, open sets are not nicely or finitely covered by their subsets (they need not be "compact"), and so at their boundary, which in general need not be in the set, there may well be peculiar features not at all nicely related to tamer features within the set. So, in effect, there is no continuous (or "homeomorphic") way of getting rid of fissures. We have to acknowledge them as such.

More generally, topology connects local features to global ones, various cracks and holes in the ice to the possibility of doing figure-eights on the pond. Numbers and groups are associated to shapes and forms. The spatial nature of the world is encapsulated into neighborhoods and fissures. So are related the *cons* and the pastings. And, the ways we systematically account for neighborhoods and fissures, those algebraic structures, turn out to be more generally useful.

KEEPING TRACK OF THE SHAPE OF OBJECTS

We might infer the shape of an object by keeping track of its local discontinuities and fissures. But, we might also keep track of its shape in a more global geometrical way, describing the structure of relationships of the members that compose the object or are combined to form it. Again, think of the skeletal structure of a body or a building. Notably, many objects cannot be flattened-out without taking them apart, breaking some of the linkages among the structural members. This fact, generalized, turns out to be a key insight in understanding objects topologically.

Say we have a skeletal structure that approximates an object, the tessellation or triangulation we mentioned earlier. We might decompose that structure into parts, each of a specific dimension, and then decompose those parts into subparts, again each of a specific dimension. So Poincaré developed a combinatorial topology in the 1890s. That process of decomposition may be expressed algebraically in terms of (homology) groups, pointing to the invariances in structures, features independent of just how we do the triangulation: for example, that a doughnut has a hole. That algebra turns out to have its own pictures, actually diagrams, which help to keep track of the decomposition. So we have gone from an object, to its schematic picture, to its skeleton, to an algebra and its diagrams, all in the service of keeping track of the shape of that object.

Or, we might stay with the object, but then try to cut it apart, flaying it out. Eventually it may be flattened out, and then presumably one has to decompose that flattened-out version into its natural parts. The object becomes known and pictured through the cuts and slices needed to decompose it, and conversely the pastings needed to put it back together. An algebra of cutting, slicing, and pasting leads to an algebraic representation of the original object. And there will be a diagrammatic representation of that algebraic account of the object.

In each case, algebraic machinery becomes the device of choice for keeping track of the shape of an object. New sorts of pictures, diagrams of the algebraic decomposition of the object, come to represent everyday objects or their tesselations. The payoff of this sort of abstraction is a canonical and powerful way of keeping track of fissures.

A third way of keeping track of the shape of an object is to ask what kinds of functions can comfortably reside on it. Rather than the object itself, it is the inhabiting functions that are front and center. More physically, the shape of an object is accounted for by the fields and fluid flows it accommodates, with whatever discontinuities they possess. So, for example, we can fit the function $\theta(x)$=arccos x/r onto a circle of radius r in the x-y plane, where θ is the radial angle. The function will be single valued if we allow for a -2π jump at $x=r$, $y=0$, If that circle had a fissure at $\theta=0$, the function mirrors the topology of the object.

Technically, this is just what we do when we patch things together in analytic continuation in complex variable theory: One pastes together regions, within each of which, and even across some of the boundaries between, adding-up of changes in the function is straightforward. For there, the functions are smooth and without fissure. And then one prescribes what to do at the unpatchable boundaries, bridging the fissures.

For Riemann and for Poincaré the task was to find out what was equivalent to what: which objects or spaces were equivalent to each other, perhaps modulo some compensating add-in that was a measure of lacuna or fissure. What was for Riemann the more general problem of "Abelian integrals" and algebraic functions, saying what kinds of functions can reside on a surface, became a technique for distinguishing inequivalent surfaces.

The algebraic technology of decomposition and reconstruction, whether the algebraic objects are derived from skeletal structures, literal cutting and pasting, or the analysis of fluid flows, is much the same for algebra's account of decomposition and reconstruction of some of its own objects into simpler ones and relationships among those simpler objects. And so we might use the same algebraic topology methods for algebra itself. And just as there might be many skeletal structures that span the same object, there might be many ways of decomposing an algebraic system into simpler objects and relations among them, although often we may show they are in some sense "the same."

What the algebra of topology provides in each case is a means of extracting the common information in those various different decompositions.

VII

DIAGRAMMATICS AND MACHINERY

Diagrams are the practical means of organizing the work at hand.[71] A tabular arrangement or a diagram can make an algorithm more natural and automatic, as in double-entry bookkeeping, analysis of variance in statistics, or commutative diagrams in algebraic topology. Diagrams and machinery can be abstracted from one context, and for good reason they may be applied to new contexts. The formal machinery for keeping track of fluid flows, the vector calculus of Gibbs, is also a means for keeping track of the decomposition of algebraic objects into simpler parts. For, just as derivatives measure velocities or flows, they also in effect transform objects into their boundaries. So algebraic topology keeps track of flows and of decomposition-into-parts.

Good diagrammatics should be familiar, modular, and appropriately natural, and of course they have to work. That is, they have to be true to the world and the problems we take as important, or the problems we learn to take as significant because the machinery handles them.

Diagrams would appear to do the work at hand automatically, whether it be showing a possibility or a counterexample, proving a theorem, displaying a flow, taking apart an object, or prescribing an algorithm. Brouwer's "fantastically complicated constructions"[72] and their diagrams (Figures 1.1, 1.3, 3.5) allow him to demonstrate visually what he means or to push through a proof. A schematic representation of an object might show how it is put together out of simpler parts or moves. So a diagram in a Boy Scout handbook teaches you how to tie a knot. A series of pictures of Klein bottles and Möbius strips in the making, shows how to construct them. A set of rules and gluings with particular orientations, as in paper models put together by gluing tab A to tab B, so constructs an innertube out of a flat piece of rubber by gluing its edges together. The diagram may be of a sequence of objects, and mappings among them, animatedly proving the Snake Lemma (see chapter 6), laying bare a process of calculation and proof, and suggesting analogously true facts.

We may learn to read off from a diagram the structure of a proof or a calculation, as in high school plane geometry, or the physicist's Feynman graphs, or the logic instructor's Venn diagrams for Boolean algebra. Of course, the student or the scientist has to learn to read and to draw those diagrams in the canonical fashion. And some of the time, what is read off is just not the case, even in the best of practices. So the science must learn to read better the implications of its diagrams or learn to draw them in a new way.

At various levels of abstraction, pictures and diagrams, and algorithmic modes of calculation that may be read off from those diagrams, provide powerful epitomes of systemic relationships.

Historically, Kirchoff's two laws (1845, 1847) have been an important model of how sources and flows are imaged in a circuit diagram and calculated from that diagram: the conservation of charge and current; and, the sum of voltage drops in a circuit is zero (the conservative nature of the electric field when there is no changing magnetic field). An electrical circuit diagram, with its symbols for batteries and resistances, indicates where currents have to balance (everywhere, but in particular at the nodes). It displays or projects out the sources and the dissipation of energy at each point, so that energy is demonstrably conserved. The resistive loads are discrete elements which exhibit Joule heating, i^2R, where i is the current and R the resistance. The currents in the wires and the batteries and generators are in effect streamlines in a fluid-flow diagram.

Kirchoff's insights, now abstracted, inform the proof of the Four Color Theorem through Heesch's notion of "discharging."[73] Attaching suitable "charges" to the nodes of a graph that represents the coloring of a map, one employs what is called a discharging algorithm (rearranging the charges so that the system is "discharged") to try to get to an equilibrium, knowing that the net charge here is positive, by definition. Obstructions to such discharging lead to candidates for problematic, possibly non-four-colorable map configurations. These "unavoidable" configurations are then shown to be "reducible," to be four-colorable.

In the design of an oil refinery or other such flow-through processing plants, the chemical engineer describes the plant in terms of unit operations and materials flows, operations and flows that conserve materials and energy. One works with abstract pictures of the industrial plant, namely, flow charts or box-and-arrow diagrams. What is an actual sequence in time and space of chemical species and thermodynamic states, and transformations among them, is now seen in another object, a diagram linking states and materials and flows. We have a picture of the processing plant in a flow diagram, a picture of A in B.[74]

Felix Klein, in his little book on Riemann's theory of algebraic functions (1882), pictures sources and flows in particular spaces (Figure 3.7), so linking the shape of the underlying space to a solution of a partial differential equation

$$A \rightarrow B \rightarrow C \rightarrow D \rightarrow E$$
$$\downarrow \quad \downarrow \quad \downarrow \quad \downarrow \quad \downarrow$$
$$A' \rightarrow B' \rightarrow C' \rightarrow D' \rightarrow E'$$

FIGURE 3.9: Parallel exact sequences, A and A', mirroring each other.

for the flow of a field or a fluid. Or, sprinkling iron filings on a sheet of paper pierced by a current-carrying wire, we see another such flow.[75]

Algebraic topology's diagrams abstract from Kirchoff and Riemann on conservation and flow, and from the algebraists' techniques for solving sets of simultaneous equations and for representing groups in terms of their basic generators and the relations among them. The diagrams detail relationships among the various objects, in various dimensions or in various spaces. Often, these are parallel linear sequences, linked ladder-like to each other. (Figure 3.9) Ideally, along each linear sequence there is a conservation law in operation, what is called exactness (as in an "exact differential"): in effect, that the boundary of a boundary is zero (much as a surface might be thought of as a derivative of a volume, or much as a flow might be a derivative of a field). And the exactness of one sequence implies the exactness of the parallel sequence. And the parallel sequence projects out features of the objects in the first sequence.

In general, detailed and complicated calculation of particular cases and much proving has taken place, before these various diagrammatics are invented and codified. So, on the way to understanding the homology of what we now call Eilenberg-Mac Lane spaces and the homology of spheres, over a ten-year period Samuel Eilenberg and Saunders Mac Lane covered piles of sheets of paper with "mountains of calculations" to find a topological space with an appropriate associated algebraic object (its fundamental group). By studying the topology of that space (its cohomology) one could find out about the algebraic object. Subsequently, they then found a way of expressing their discovered understanding in a comparatively calculation-free way.[76] Richard Feynman calculated a wide variety of physical processes to test out his diagrammatic methods, which represented a physical process in terms of flows and interactions, each diagram leading to a formula or contribution to the process.[77] Although such diagrams may look cute and simple, and visually obvious, all of these diagrams are technical and densely coded, and one has to be trained to meaningfully draw and appreciate them and to employ them in a correct calculation. Diagrams as such are never enough.

The canon of diagrams in algebraic topology prescribes an analogy between properties of one space's decomposition sequences and those of another onto which it is mapped. In school, one learns to mechanically and correctly invoke the diagram that ought do the desired work for paradigm cases, much as one learns to make circuit diagrams or Feynman graphs. In this case, the student learns to infer properties of maps and objects from properties of the maps and objects linked to them. Exactness might be made to propagate down the lines, so that in the end one has a decomposition sequence or resolution for one object derived from the decomposition sequence of another sort of object.[78]

A recurrent theme is that of flows and of conservation of flowing material (exactness, gauge invariance, the first law of thermodynamics). In part, those flows are metaphoric flows. But what redeems such a metaphor is the commonality of the apparatus, the similarity of the diagrammatics, and the suggestiveness of one diagrammatic technology for another.

It would appear that a set of diagrams might well in time be abstracted from, reconceived, and taken in terms of a new diagrammatic. So Serge Lang says, in discussing the work of Grothendieck and Dieudonné in algebraic geometry:

A theorem is not true any more because one can draw a picture, it is true because it is functorial.[79]

Functoriality is just the idea of diagrammatic parallelism, a mirroring, a picture of A in B. Remarkably, functoriality is not only a mirroring, but also a mirroring of that mirroring. At each stage, a picture or diagram is displaced by a more generic claim that that picture and structure is like another that is more general. And the claim is that that likeness is where our focus ought be. Such an analogy of analogies is called a syzygy. So Grothendieck and Dieudonné reconceive the objects of algebraic geometry, going from curves to the most generic features of those curves, which then leads to a revised notion of geometric intuition. So we might go from a picture of an object, albeit abstracted, to a diagram of its decomposition, to algebraically equivalent objects in those homology groups, to making a claim of functoriality or syzygy, and then to see that functoriality as having its own diagrammatically expressible features, which then presumably are analogous to another such realm of objects, and then to see that functoriality as having its own . . . , eventually stopping.

Of course, actual examples and facts constrain and inspire these towers of analogy. What is most striking is mathematicians' commitment to genericity and functoriality. The task becomes one of figuring out the right form of these syzygetic claims. This commitment is a practical one. Adopting more generic notions, may allow one to prove previously intractable theorems of great import and to understand better what we could already prove.

Rigor has not been abandoned in the use of pictures and diagrams. There are canonical ways of using them, derived from systematic theory. And, one learns through actual practice when a diagram encapsulates all of what is true, and does not miss out on something or sneak in nonsense.

As we shall see in chapter 5, this issue is a recurrent one. Riemann's pictorial and analytic way of thinking is then made rigorous (as they saw it) and algebraic by Dedekind and Weber (1882).

THE RIGHT PARTS

Decomposition is not only into parts, but also into the right sort of parts. So Adam Smith describes the division of labor in the pin factory at the beginning of *The Wealth of Nations* (1776). Smith justifies that division in terms of the propensity to make exchanges (to "truck and barter") and "the extent of the market," and these in relation to the manifest advantages of efficient or low-cost mass production. The factory has to be productive and a going concern.

For mathematicians, the right sort of parts are sometimes distinguished by the fact that they can be put together in a number of ways, yet one obtains the same object at the end (something not true for most factories). For example, we might take a derivative first and then a limit, or vice versa, and expect to get the same result. But this is the case only if there is uniform convergence in that limit. To say that "the diagram commutes," as a claim about the equivalence of different sequences of steps, is a substantive claim about mathematical objects and maps—even if it is presented as a formal diagrammatic fact. Other "natural" equivalences are, again, formally displayed, but actually they are about particular substantive objects and maps for which that formal display is true. One might be deceived by a diagrammatic similarity, but to be a trained professional is to be wary of such deception.

To find the right sort of parts in algebraic topology one wants to irreducibly and canonically decompose an object. Finding the right parts and their sequential relationships is called computing the homology or cohomology or homotopy groups of an object. Again, one ends up with a sequence, each step or part or sub-object connected to the next by a derivative or boundary operator, much as a volume is connected to a surface. The suitably understood composition or sum of these parts is, at least formally, the original object or an equivalent one. Imagine gluing walls together to make a room, the rooms together to make a storey, and the storeys together to make a home.

Decomposition is only sometimes easy. We learn in college algebra how to do such a decomposition, say in solving a set of simultaneous equations. But what is for the most part straightforward for fields such as the real numbers and their vector spaces, usually the only case that is taught, is not at all trivial for rings (such as the integers) and their vector-space-like structure, modules, when division is not provided for. For real number fields, decompositions and dimensions divide out or subtract out nicely. They do not do so when division is not available, as for the integers. Such divisibility or "exactness" is a given for vector spaces, an achievement for modules. (Technically, one says, "Homological algebra starts from the regrettable fact that not all modules are projective."[80] Those regrettable facts, the breaks or failures of exactness, are measured by the nonzero homology groups.[81])

For example, consider our system of roadways and toll booths: Group together equivalent paths, say those paths costing roughly the same in tolls+gasoline. Presumably, there will be several such groups. Do members of each group of paths share some quality that would seem to account for their equal cost? If we are fortunate, such residual path-dependences (differences between each of the groups of paths) are then brought to the fore. The "homology groups" tell one how to take geometrical path-dependence into account, for example counting the number of times you have encircled a lake and so you cannot avoid a larger gasoline bill.

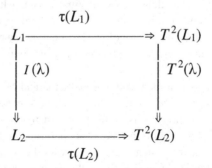

FIGURE 3.10: Commuting diagram for a natural transformation of the functors I and T^2, adapted from Eilenberg and Mac Lane, "General Theory of Natural Equivalences" (1945), p. 232.

NATURALITY

The topological decomposition factory is modular, to be described at various levels of detail (the "nth homology group"). At each stage one is disentangling syzygies, the relations of relations. Eventually, one hopes to have exhausted the sequence, nothing being left to decompose. Hilbert's theorem on syzygies in algebra (1890) says that, given certain conditions, at the end of the (provably) finite composition sequence there is a "free" module, freely (without restriction) and finitely generated by its basis elements, with no constraints or relations among them; and it, too, is finite.

The syzygy theorem leads eventually to an algebra of analogies (the homological algebra we have earlier mentioned) and a technology of decomposition. And analogies among systems interchange with or commute with internal operations of those systems: analogize and then operate, or operate and then analogize—you end up with the same object or space. And analogies of those analogies are "natural" if they, too, commute. (Of course, again, you could be misled by such analogizing to incorrect statements.)

Why "natural"? The notion of a function that transforms both objects and the mappings between objects, invented by Eilenberg and Mac Lane (1945) to account for homologies, and called by them a "functor," leads to a canonical sense of what mathematicians take as natural equivalences. Objects that were equivalent were to be equivalent intrinsically, no matter how they were represented, say independent of the basis of a matrix, for example. (The standard example is the dual of the dual of a vector space, designated as T^2 in Figure 3.10, as being naturally equivalent to the original space, while the first dual is not, since the dual depends on the chosen basis.) Naturality as a diagrammatic fact is a generalization of point-set topology's archetypal way of understanding the properties of sets and functions: the continuity of a mapping commuting with the openness of a set.

In general, it came to be taken that mathematical constructions ought survive the test of naturality. (Technically, for example, a tensor product is made to be bilinear, and an ultraproduct is made to retain total ordering, for the obvious Cartesian product will not do the required "natural" work.) Not only do we have objects, arrows between them (maps), arrows between object-arrow systems (functors between "categories"), and arrows between those (that is, transformations of functors), Mac Lane (1950) also noted that in this realm everything came in pairs. For notions of group theory often came in pairs, and if notions were expressed in diagrams the arrows would merely need to be reversed to express the notion's twin. Reversing the arrows in one object-arrow system or category led to its Doppelgänger or sibling "adjoint" category.[82]

The phrase "the diagram commutes," says that algebra and its structures commutes with or is compatible with topology, or more generally, that two categories are so compatible. There is something so formally constrained about the natural construction of an object, that no matter which path you take to make it, you end up with the same thing. Whether we take the boundary (or derivative, ∂) first and map second, or map first and boundary second, we end up with the same thing. The capacity of this system of relations of relations to then abstract to a higher level of relations of relations, the seamlessness of its processes, is striking in its ultimate productivity for mathematical work.[83] Presumably, that productivity is about the nature of the world and our inventiveness. One might want to know just why the formalism works so well, just what is behind these compatibilities. The formalisms are not necessary. The diagrams might not work out, the objects not fulfilling the diagrammatic facts. We might ask, What kinds of objects and diagrams do this work, what kinds do not? This, too, is a mathematical question.

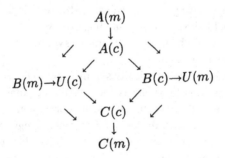

FIGURE 3.11: Kunen-Miller chart, from Judah "Set Theory of Reals" (1992).[84]

Recurrently, we encode the visible, the symmetric, the diagrammatic, and the natural (however understood) in what we take as a diagram's geometrical symmetries. For example, in the definability theory of sets (descriptive set theory, recursive functions), one has the Kunen-Miller chart connecting the various sets' properties. We are able to epitomize the mathematics in the peculiar geometric symmetries of a nice diagram.[85]

Or, consider Cichoń's diagram (Figure 3.12), a generalization of the Kunen-Miller chart, which describes the relative sizes of various measurements of the cardinality of "small" sets, the arrow here indicating "no larger than."[86]

The task is to show that the 13 arrows connecting the 10 nodes are just what is the case. The mathematical intuitions are embodied in a curious graph, not at all obvious initially, but which then becomes canonical, setting up the mathematical agenda.

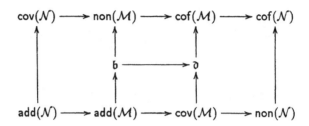

FIGURE 3.12: Cichoń's diagram, from Bartoszynski and Judah, *Set Theory: On the Structure of the Real Line* (1995). Cichoń's diagram connects the cardinal coefficients of "meager" (*M*) and "null" (*N*) subsets of the real line.

Now, it may turn out that nice symmetries in diagrams actually mislead us. The world may not go these particular ways. If the world turned out differently, presumably we would draw a different diagram. We might then warrantedly wonder how natural would be the new diagram's symmetries. Only examples and proof can provide assurance.

Diagrams need not be geometrically symmetrical to be useful and indicative. Often, in set-theoretic topology one is looking for peculiar or pathological or exceptional sets, so as to discover the limits and consequences of generic notions such as the cons, as well as to discover more general regularities. Epitomes of these discoveries are displayed in diagrams of inclusive hierarchies, or map-like overlapping realms of properties. (Figure 3.13) Sierpinski, in his (1934) study of the consequences of the continuum hypothesis, displays a possible systematic order among the various detailed consequences, some of it to be discerned by the geometrical symmetry, other parts to be discerned by various notational devices. (Figure 3.14) There is perhaps as well an implicit invitation to prove that systematic order or to discover where it fails.

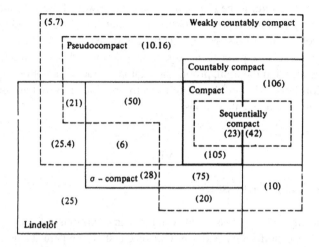

FIGURE 3.13: Hierarchies of topological properties, from Steen and Seebach, Jr., *Counterexamples in Topology.*

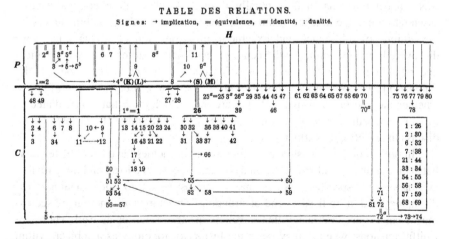

FIGURE 3.14: "Table of Relations" of Theorems, from Sierpinski, *Hypothèse du Continu* (1939, 1956).

The great inventions here are pictures or diagrams and machinery that make manifest what turns out to be true. The geometric power of the imagination is to be educated by graphic representations of mathematical objects and relations: plaster models of functions, geometric figures drawn from Euclid, commuting diagrams and exact sequences in algebraic topology. Insight into the actual mathematics, through theorem, proof, and example, leads to a notion of a natural set of relationships. The problem is to learn to draw pictures that are likely to be true or to lead to doable calculations.

VIII

THE TENSIONS AND THE LESSONS

Out of Cauchy, Weierstrass, and Cantor came notions of open sets and convergence and uniformity and ε-δ. Out of Oersted, Ampère, Gauss, Faraday, Maxwell, Riemann, and Poincaré, came notions of sources and fields with their fissures and discontinuities and path-dependences (cohomology), and the decomposition of those objects into nice parts. Out of Poincaré and Brouwer came the idea that we could understand the shape of the world in terms of what was connected to what, and eventually, through Noether and others, that what is essential here is the algebra of those incidence matrices. That algebra centers on the notions of nilpotence and exactness, an expression of conservation laws discovered in the nineteenth century, and an epitome of mathematics' understanding of solving systems of equations. These are some of the foundations of topology as discovered in the last century or so. Infinities, whether they be infinitesimal or large, are the other legacy that is articulated in topology, whether they be continua or countable infinities. (In chapter 6, we discuss these infinities further.)

In our account of the subject of topology, our themes have been the tensions between continuity and structure, analysis and algebra, the local and the global— all marked by specific pictures and diagrams meant to show analogy and disanalogy and those fissures and obstructions. Structure and function are seen to mirror each other, a strategy that develops in other sciences as well. These are fruitful tensions, whether they be in the history of topology as a subject, within analysis as a branch of mathematics, in accounts of discontinuity, discon-nectedness, and conservative flows, or in diagrammatic machinery. A pervasive theme is the linearization of the world: derivatives, cohomology, and modules or vector spaces.

In the fable, eventually Orson learns to appreciate his brother's more global and formal (algebraic) commitments, and perhaps how his own commitments to the local can be formalized as well. And Valentine learns through actual experience (computation and detailed mathematical visualization) that the particular will guide him to how the world works generically. Orson and Valentine will always be brothers. And perhaps they will not be estranged, but be in a renewing and productive tension.

As for the subject of topology, our account is surely partial and provisional. But what is pervasive and enduring is the world's variety of topological spaces and mappings; the various topologies (notions of closeby-ness) that may be endowed to a space; the taming of their discontinuities and fissures; and objects that beg to be decomposed into simpler ones. And the sign of these endeavors are pictures and diagrams that often appear to do the work of mathematics by the way. But in the end it all depends on concepts and definitions and what you can discover and prove—notions that are productive; and finding the right level of abstraction to do the work.

APPENDIX: THE TWO-DIMENSIONAL ISING MODEL OF A FERROMAGNET[*]

FIGURE 3.15: STUDIES OF THE TWO-DIMENSIONAL ISING MODEL, 1925–1985[87]

1925	Ising's dissertation, one-dimensional (1-D) lattice model.
1936	Peierls argues that a phase transition is possible for two dimensions, based on the topology of a plane vs. that of a line.
1941	van der Waerden shows that counting closed polygons on a lattice will get the 2-D partition function.
1941–43	Kramers and Wannier's solution of toy 2-D model (duality, transfer matrix); Montroll's solution of toy 2-D model ("resonance" of almost-ordered and almost-disordered eigenfunctions). Lassettre and Howe on the correlation function; Ashkin and Lamb on series expansions.
1944	Onsager's exact solution of 2-D, using Clifford algebra, quaternions, and elliptic functions. (Stephen and Mittag, 1972, develop the diagonal transfer matrix formalism hinted at in Onsager.)
1949	Onsager announces a formula for the spontaneous magnetization, but never directly publishes his method of solution.
1949	Kaufman's spinor formulation, Kaufman and Onsager on correlations.
1952	Yang derives the spontaneous magnetization. Kac and Ward develop a counting matrix (for closed polygons), whose determinant is the partition function, reproducing Kaufman's results for the partition function for a finite-size lattice.
1953	Newell and Montroll review article.
1963	Montroll, Potts, and Ward use the Pfaffian method for counting polygons on a decorated lattice applied to the two-dimensional Ising model (Hurst and Green, 1960; Potts and Ward, 1955; Kastelyn, 1963) to derive the spontaneous magnetization using the Szegő limit theorem.
1960s–	Domb, Sykes, Fisher and collaborators do high-order series expansions using cluster diagrams, and Padé approximants, to study the analytic structure of the partition function.
1964	Schultz, Mattis and Lieb show how the two-dimensional Ising model is a fermion field theory, reminiscent of Cooper pairs in the BCS theory of superconductivity. Samuel (1970) expresses the partition

---

[*]The two-dimensional Ising model plays a role in chapters 3, 4, and 5. In order to make each chapter more self-contained, within each chapter there is some overlap with what I say here.

	function in terms of an action integral of totally anticommuting (Grassmann) variables
1965–80	Fisher, Kadanoff, Wilson, and others exploit renormalization group analysis. Symmetries and functional equations of Ising lattice (Baxter). Connection to quantum field theory (Wilson, 1975; Wu, Tracy, McCoy, 1977).
1966–81	Wu, McCoy, Tracy and collaborators, using the Pfaffian method and Toeplitz determinants, solve for asymptotic correlation functions. Wiener-Hopf techniques (1966), Painlevé transcendents (1976).
1966	Yang and Yang on one-dimensional scattering, leading to Bethe Ansatz (all is two-body) and the Yang-Baxter relation. Also, McGuire, 1964; Sutherland, 1967.
1967	Lieb solves lattice models (of ice, or 6-vertex model) using Bethe Ansatz, effectively particle scattering. (Also, Lieb and Lininger, 1963.) In 1970s Baxter solves 8-vertex model.
1967	Gardner, Greene, Kruskal, Miura on solitons and Kortweg-deVries equation and nonliearity.
1973	McCoy and Wu, *The Two-Dimensional Ising Model*.
1978ff	Sato, Miwa, Jimbo use a continuum scaling of a fermion field theory ("holonomic quantum fields") to directly get Painlevé transcendents in correlation functions, as a consequence of symmetries of fermions.
1982	Baxter, *Exactly Solved Models in Statistical Mechanics*.
1985	The continuum Ising lattice at the critical point as a conformal field theory (Friedan, Qiu, and Shenker; Cardy; Fonseca and Zamolodchikov, 2003).
1985-present	The conformal field theory at T_c may then be perturbed by a magnetic field and by a shift away from T_c, to develop in an eight-particle massive field, having E_8 symmetry. Baxter solves more complex models, including the Potts model, using a variety of new techniques. There is progress on correlation functions, the spontaneous magnetization, and the susceptibility, scaling from the critical point, using form-factor expansions, Toeplitz methods, and cluster expansions—discovering interesting features of their analytic structure (poles, branch cuts). Mathematicians (Langlands, Smirnov) begin to make rigorous renormalization and scaling.

The Ising lattice of interacting spins and its various solutions may be used to model ferromagnetism (permanent magnetization, as in an iron bar)—as well as a gas of particles whose positions are restricted to points on a lattice (a "lattice gas"), a quantum field theory, and a neural net. It is one model, at least in its two-dimensional version, that may well have "399" exact solutions—as one paper playfully suggested (providing a particularly lovely solution that made maximal use of the symmetries of the model).[88] Historically and originally, Henrik Kramers and Gregory Wannier (1941) and Elliott Montroll (1941), were able to provide a model solution, a derivation of the partition function for a toy lattice in two dimensions (infinite in one of its dimensions, and having a small number of rows in the other).[89] (Ising's dissertation, in 1925 under Lenz's guidance, was for one dimension. It does not exhibit a nonzero-temperature phase transition.) In 1944, Lars Onsager provided an exact solution for the infinite lattice, in part based on the earlier insights. Onsager's solution retains still a reputation for opaqueness, perhaps now much less deserved. While one can follow each step, the general scheme was initially not so obvious.[90] (The schematic history I provide here leaves out some of the most ingenious contributions. My purpose here is to help the reader appreciate how I use the Ising model as an example.)

There developed at least three sorts of exact solutions to the Ising model in two-dimensions. There is already in Kramers-Wannier and in Onsager, an algebraic solution to do the combinatorics (and the trace of a matrix that would seem to be a group representation), which in the end employs automorphic or modular functions, in particular, the elliptic functions. Kac and Ward developed the first combinatorial solution, in part to show how the Kaufman's algebra (and Onsager's) did the work. And in Baxter (and earlier in Yang, 1952) one solves a functional equation derived from symmetries of the model, with suitable assumptions about analytic continuation. The doubly-periodic elliptic functions, the natural successors to the trigonometric or circular functions, appear in different places in each sort of solution, marking the algebra and modularity, the combinatorics, or the analyticity: namely, the elliptic functions' automorphy and their addition formulas (much as $\sin(x+y)$ is expressible using $\sin x$ and $\sin y$); the fact that the fourier coefficients of the elliptic theta functions are of combinatorial significance; and, the elliptic functions' being known by their periodic poles and zeros. (See Figure 5.1.)

The elliptic functions are doubly periodic, here reflecting two symmetries of the lattice. First, a reduced temperature, k, which is the modulus of the elliptic functions, is the right variable to characterize the orderliness of the system (and not the temperature). Second, the relative size of the horizontal and vertical couplings among the spins (the reduced ratio, u, the argument of the elliptic function) plays a secondary role, although u also embodies a translation symmetry of the lattice. (Eventually, it is realized that k is analogous to an energy, u to a momentum.) There is symmetry for that relative size: Matrices that are employed to calculate the partition function by transferring the interaction down the line of the lattice so summing the Boltzmann weights, "transfer matrices," of different u but the same k commute with each other. That matrices of the same k but different u commute allows one to relate the partition function for a hexagonal lattice (which is made up of three-pointed "stars") to the partition function for a triangular lattice, both at the same k—and so the symmetry is called a star-triangle relation. (If k goes to $1/k$, u goes to a dual u, ku or $k(I' - u)$; in the star-triangle relation, k does not change, but u goes to $I'-u$. iI' is the imaginary half-period of the elliptic function.) It can be shown that the k and u symmetries are equivalent to Lorentz invariance in an analogous quantum field theory.[91]

The elliptic functions are said to be modular. Namely, under a birational transformation of the independent variable ($z \rightarrow (az+b)/(cz+d)$) they exhibit only a change in scale: $f(z)$ is proportional to $f((az+b)/(cz+d))$: the elliptic function will be multiplied by a number or modulus. The crucial quantities in the Ising model exhibit such modular invariance, presumably due to the symmetries mentioned above. So the partition function, Q, exhibits modularity in what was called duality: $Q(k)$ is proportional to $Q(1/k)$, again k being a reduced temperature, so we have high temperature and low temperature partition functions proportional to each other. (In contrast, again in the star-triangle relation k remains unchanged but u changes, and so in that case we are relating partition functions for the same k but different u.) And it turns out that the (eighth power of the) spontaneous magnetization is a "level 2 modular form."[92]

We might say that there are analytic, algebraic, and combinatorial solutions to the Ising model, and these are mirrored in the properties of elliptic functions. (This fact will play a large role in chapter 5.) Analogously, we might try to understand the behavior of a gas of molecules in terms of the differential equation for heat or diffusion (for which the elliptic theta functions provide solutions, as Fourier found); or, as adding up the effects of atoms whose average energy is given, and so we are adding up random variables under the regime of the Central Limit Theorem with its automorphy (the sum of gaussians is still a gaussian, with a scale factor); or, as a random walk where we are counting up the ways we might walk a certain distance in N steps, as a function of N, or the distances we might walk in N steps.

SOME OF THE SOLUTIONS TO THE TWO-DIMENSIONAL ISING MODEL

Again, given a two-dimensional regular grid (say, rectangular), place spins at each vertex, endow each spin with a discrete spin-spin interaction with each of its nearest neighbors (say, horizontal and vertical), one that favors alignment, and ask if there is, classically and statistically, a temperature below which there is a net magnetization or orientation of the spins of the lattice. The answer is, Yes.

We might classify the solutions as:

I: Analytic or topological or geometric (continuity is the theme): functional equations (item #5 below); elliptic curves and the Yang-Baxter relation.

II: Algebraic or function theoretic (automorphy is the theme): Onsager and Clifford algebra (#1); fermion field theory (#2); scattering and Yang-Baxter (#3), braid group; renormalization and automorphy (#4).

III: Combinatorial or arithmetic (regularity of the combinatorial numbers is the theme): Pfaffians (#6); zeta functions (chapter 5).

Over the years, innovative new solutions for some property then show how that solution's final result is the same as another's, even if they have different apparent forms, say one has two determinants and the other has four. Or, they show how one method of solution is equivalent to another: how the algebra does the combinatorial work; or, how the analysis does the algebraic work.

1. *Algebra and Automorphy: Onsager and predecessors.* Kramers and Wannier showed that there was a non-zero transition temperature for the Ising lattice. Moreover, they and Montroll showed that the partition function could be expressed as the trace of a matrix (the "transfer matrix"). To cover a two-dimensional lattice, Kramers and Wannier went down a helix or screw that covers a cylinder, in effect a strip, wrapped barber-pole-like around the cylinder, the transfer matrix adding-in statistical weights (exp$-\beta E$) to the partition function, atom-by-atom.[93] Kramers and Wannier noted and exploited the fact that the transfer matrix exhibited in its matrix elements a curious symmetry between high and low temperature representatives (or disorderly and orderly arrangements), locating the transition temperature at the fixed point of such an inversion symmetry. (Their normalized temperature variable, κ, is the same for the high and low temperatures whose transfer matrices were so related. Here, we shall just use Onsager's k, where again k and $1/k$ are the high and low temperatures (or vice versa).) Because of this symmetry, the partition function, Q, for a low temperature lattice, was proportional to the partition function for a high temperature lattice, the proportionality constant being merely a function of the temperature: $Q(k)=k^{-N/2}Q(1/k)$, where N is the number of atoms. (The

exponent is proportional to the "weight" of the modular function.) It was not noted as such by Kramers and Wannier in their paper, but Q's symmetry is an automorphy or modularity (as for example for the theta function: $\theta(u,\tau)$ is proportional to $\tau^{-1/2}\theta(u/\tau,-1/\tau)$). Later, Onsager took explicit concrete advantage of this modularity in his analysis. (Again, automorphy means that under the set of transformations, one ends up with an object of the same sort as one started out with: Under inversion in k, Q is automorphic since $Q(1/k)$ is a Q, here $Q(k)$, albeit multiplied by a number, the modulus.) The phase transition took place at $k=1$.

This inversion symmetry or modularity was called duality, in part recalling the duality of points and lines in projective geometry. Duality connected the partition function at a high temperature with the partition function at a low temperature, the partition function of a lattice that exhibits disorder with the partition function of a lattice that exhibits an orderliness dual to the disorder. The rectangular lattice was as well transformed geometrically, in effect rotated by $\pi/2$.

Although at this time notions of group representations were standard in quantum mechanics, there was little discussion that the transfer matrix might be a group representation, what group it might represent, or that the trace might be a group character.[94]

Using Kramers and Wannier's techniques, as well as ones developed with Mayer for the theory of imperfect gases, Montroll derived series solutions for the partition function for narrow strips of lattice, a row-to-row transfer matrix. Again employing Kramers and Wannier's techniques, Ashkin and Lamb derived a series expansion for the propagation of order in the lattice, the result that Yang eventually uses to check his exact solution.[95] The concreteness of these analyses is a nice complement to the more abstract techniques employed for the exact solution, showing more of what is going on in the solution.

Onsager showed how one might exactly compute the partition function of the two-dimensional Ising model, in effect diagonalizing a "kinetic-energy + potential-energy" operator, as one often does in a first course in quantum mechanics, although we are here in the classical realm. In order to do this, he analyzed a quaternion algebra, through which his A and B operators (defining a pseudo-hamiltonian) could be jointly diagonalized. (Bruria Kaufman (1949) showed how to use the more familiar spinor algebra to do the analysis in a perspicuous and motivated fashion.) Onsager confirmed and extended Kramers and Wannier's earlier work, and in effect showed the correctness of Montroll's insight that the eigenfunction at the critical point would be a "resonance" of almost-ordered and almost-disordered states.[96] Essentially, one builds up the critical-point solution by introducing more and more disorder into the fully ordered T=0 state (so it is almost-disordered), or introducing more and more order into the fully disordered infinite T state (so it is almost-ordered).

Onsager's algebra can be interpreted to show that the Ising model solutions are associated with an infinite number of symmetries and conserved quantities or charges (the possible orderlinesses of the spin arrangements in a row of the lattice).[97]

Onsager mentioned in passing that he understood how one might more directly employ Kramers and Wannier's duality for an exact solution. He also mentioned that there was as well what he called a star-triangle relation, a Y-Δ transformation as in electrical circuit theory, that converts a honeycomb lattice to a triangular one. And the star-triangle transformation allowed one to determine the partition function for one such triangular lattice at high temperature from that of another honeycomb lattice at high temperature, as well. (For different lattices, the actual temperatures corresponding to the same k are different, but the orderliness of the lattices is the same.) Although not explicitly mentioned by Onsager, the star-triangle relation is k-preserving, rather than duality's k-inverting, the corresponding geometric transformation of replacing triangles by three-pointed stars different from duality's turning lines into points and vice versa. He reports that he used the star-triangle relation and commutation properties of the transfer matrix to obtain a solution, for lattices such as the hexagonal and triangular two-dimensional lattice. (Diagonal transfer matrices commuted with his pseudo-hamiltonian, $B+k^{-1}A$. Note that for low temperatures and so small k, the row-operator, A, in effect orderliness, dominates.) This work was never published. Stephen and Mittag (1972) work out all of this, and derive nonlinear difference equations for the eigenvectors' Pfaffian coefficients.

2. *Algebra and fermions.* It turns out that Onsager's Clifford algebra and Kaufman's spinors,[98] employed as devices in solving the Ising model, can be shown to represent the Ising lattice as populated by fermions. They are (quasi-) particle excitations of the lattice, much as a vibration is an excitation of a string; more specifically, each particle is a patterned row of spins. At most only one of each type is excited. They are fermions.

Fermions are particles, such as the spin-1/2 electrons and protons, which obey the Pauli exclusion principle. Exclusion can be guaranteed if the fermions are described by antisymmetric functions: for two fermions, if $f(x,y) = -f(y,x)$, antisymmetry, and if $x=y$ (that is, the same state), then f=0. Or, if α is an operator that formally creates fermions in a particular state, then creating two in the same state, $α^2$, should be impossible: namely, $α \times α = 0$, even if α does not equal zero (namely, nilpotence).

114

Schultz, Mattis, and Lieb showed how to solve the two-dimensional Ising model using the language and machinery of quantum field theory: it was a field theory of fermions.[99] This was and perhaps still is the demonstration of choice for those who wanted to read a solution to the two-dimensional Ising model having the most familiar technical apparatus and the most familiar physical model. After decoupling the spins by going to another space, one could describe the Ising lattice as composed of pairs of fermions of opposite momenta (namely, Cooper pairs, as in superconductivity). Each of these fermions is a quasiparticle, an orderly collective excitation of the lattice.[100] Samuel (1970) showed how to write the partition function as an integral of anticommuting variables (the anticommutator is zero, unlike for fermions), drawing from intuitions from quantum field theory.

At the critical point, which is at a finite temperature, there is an infinite specific heat since an infinity of (almost) zero-energy degrees of freedom or fermions are excited. Again, the combinatorics of interest are of nonoverlapping, closed polygons as in a self-avoiding random walk, and the fermion rules ($\alpha \times \alpha = 0$) in effect prevent counting any edge twice.

These fermions are spatially nonlocal particles since each is actually a patterned row of spins. So their interaction with an external magnetic field is not at all straightforward, their status as observable particles being questionable.[101] The true observables must be the bosons that are the spin variables. Still, the fermions might be said to interact with the thermal field, the specific heat being accounted for in their terms.[102]

3. *Algebra, Scattering, and Yang-Baxter.* The commutation rules for the transfer matrix, and their expression in terms of relations among the coupling constants of the lattice, are called the Yang-Baxter relations.[103] They appear in work on quantum mechanical scattering in one dimension: *n*-particle scattering can be expressed in various equivalent ways as a sum of two-particle scatterings. (More generally, one-dimensional quantum mechanical systems--one space, and time--are formally identical to the two-dimensional lattice model, with various sorts of external fields in one-dimension corresponding to different two-dimensional lattice interactions.) They also appear explicitly and ubiquitously in Baxter's work. The star-triangle relation of Onsager, and of Wannier, is one form of the Yang-Baxter relation (actually preceding it by two decades). These commutation relations of the transfer matrices are also reminiscent of the rules for the building up a knot or a braid in terms of basic moves.[104]

Lieb (to whom Baxter expresses his indebtedness) employed a scattering interpretation of the lattice's symmetrical arrangements of atoms (a sum of products of plane waves, the "Bethe Ansatz") in order to solve a lattice model of ice. Again, Lieb and collaborators had earlier described the Ising lattice has being a fermion field theory, so a particle interpretation was perhaps quite congenial.

4. *Algebra, Automorphy and Renormalization.* Duality allows one to compute the partition function for high temperatures in terms of the partition function for low temperatures. At very low temperatures, spins are mostly aligned, their thermal energy unable to overcome their mutual magnetic forces. At low temperatures, presumably only a few spins are unaligned, and so the computation of the partition function is still straightforward (as Montroll realized). For very high and merely high temperatures, duality provides for a complementary account. So, the remaining and very hard problem is the intermediate temperatures, near the critical point. Ideally one wants a way of analytically continuing the known partition function from very high or very low temperatures to the computationally less tractable critical point. Series expansions, exciting 1 spin, then 2 spins, then 3 spins, . . . , allow one to do this approximately (using Padé approximants). The series expansions are an approximation to the combinatorial problem, as well as to the scaling symmetries near the phase transition point.

The lessons that were drawn from modularity and from these expansions were that: we might analytically approach the critical point from the simpler and more tractable very low or very high temperatures; as we approach the critical point, more and more complex and much larger spin configurations or fluctuations play a role, while the smaller ones still are present; the partition function scales up or down, perhaps as k^v, for very large or very small k, where v will slowly depend on k; and, automorphy suggests that as one scales up (a Gauss-Landen transformation of k), the form of the partition function should remain the same.

A characteristic physical feature of the Ising lattice near the critical point is that its correlation length (the distance over which correlations fall by $1/e$) becomes infinite. (It is finite and small at high or low temperatures.) In order to control the infinity of degrees of freedom, Kadanoff suggested spins be grouped into blocks composed of blocks of spins. And one then treats the new lattice (with its reduced degrees of freedom) much as the original lattice, modifying its hamiltonian as little as possible from the original's. Wilson showed how this might be done in a more exact way, which involved a transformation of the form of the hamiltonian to take into account multispin interactions at large scales. What saves the day is that many of the extra terms become less significant as the scale changes, and only a large but finite number of terms are needed for any reasonable degree of accuracy (say 1%). There is an automorphy that is perhaps less precise algebraically, but rather more exact computationally. Wilson's themes are: "the search for analyticity," that is, finding analytic or smooth functions that characterize the system at and near phase transition points, even if we expect nonsmoothness in the usual thermodynamic potentials at the critical point;[105] phenomena that depend on many different scales or an infinity of degrees of freedom; and, as we mentioned earlier, the "statistical continuum limit" which appears as one averages out fluctuations at each scale of length or momentum.[106]

5. *Analyticity and Functional Equations: Baxter.* Baxter has effectively exploited the algebra of the transfer matrix, using the matrix's algebraic or commutation rules or symmetries to find perspicuous solutions to the Ising model, and exact solutions to other more complex lattice models.[107] Again, transfer matrices, characterized by a k and a u (whatever the horizontal and vertical coupling constants of the lattice, $\{K_i\}$) commuted if they had a common k, as Onsager knew.[108] k is effectively an energy or temperature. We might say that the collections of atoms represented by different transfer matrices with the same k were in an equilibrium, even if their actual temperatures differed, in that they exhibited the same degree of orderliness. Onsager appreciated that k and u correspond to a modulus and an "argument" of an elliptic function; Baxter makes full use of this fact.

These commutation rules, some reasonable symmetries reminiscent of those of quantum mechanical scattering and the S-matrix, and assumptions about smoothness or analyticity (which then often lead to solvability), then lead to a functional equation for the eigenvalues of the transfer matrix, and so to the partition function as a trace of that matrix.[109] Or, they can lead to a functional equation for the correlation function of neighboring spins.[110]

If we attend to k's definition in terms of the interactions, actually the Boltzmann factors, for each k there is an elliptic curve parametrized by u. The Yang-Baxter relation is in effect addition on that "spectral" curve. Hence, Maillard says that the Ising model is "nothing but" the theory of elliptic curves. (See note 6 of the Preface)

6. *Combinatorics and Pfaffians.* The task of computing the partition function was equivalent to computing the combinatorics of closed polygons on a lattice. Kac and Ward (1952) showed how Onsager's 1944 solution, as reformulated and extended by Bruria Kaufman (1949) in which she employed the algebra of spinors, actually if implicitly counted up the right polygons, so that there was now a combinatorial or graph-theoretic account. The anticommuting algebra of Kaufman's spinors, as well as Onsager's Lie algebra of the figurative kinetic and potential energies that appear in his model, actually

enumerate the polygons. Kac and Ward jury-rigged a matrix whose determinant explicitly more or less did the counting work. Subsequently, a series of combinatorial solutions were developed using Pfaffian forms, in effect determinants of antisymmetric matrices, echoing the Wick product in quantum field theory. Montroll, Potts, and Ward used this procedure to compute the permanent or "spontaneous" magnetization, which was first computed by Onsager and first published by Yang. Wu, McCoy, and Tracy and their collaborators have shown, in several tours de force of mathematical analysis, how this approach may be employed to calculate asymptotic correlation functions.[111] It turns out that the asymptotic correlation functions, for large lattices close to the critical point (the scaling region), are expressed in terms of Painlevé transcendents, the natural successors to the trigonometric, elliptic, and Bessel functions.[112]

The highly symmetric (Toeplitz) matrix that often appears in the calculations of the spontaneous magnetization and of correlation functions naturally leads to the Painlevé transcendents.[113] Namely, one is taking a determinant, $|1+A|$, where 1 is the unit matrix and A is Toeplitz (and so $1+A$ is Toeplitz). This is a Fredholm determinant, just what appears in the solutions of integral equations. Fredholm determinants are naturally associated with solutions of total systems of differential equations (such as "second-order, nonlinear, ordinary differential equations with nonmovable branch points"), of which Painlevé's equations are some of the most elementary. Moreover, the Toeplitz matrix is in effect a convolution operator, and might be seen as a way of operating on the vector of spins for a row.

Or, as it is said, integrable nonlinear systems have associated with them linear problems. Sato, Miwa, and Jimbo showed how to derive the asymptotic correlation function by treating the Ising lattice as a continuum scaling field theory directly, represented by a massive Dirac equation. The symmetries inherent in the fermion quasiparticles lead naturally to the equation for the Painlevé transcendent, as an integrability condition, and this also turns out to give the correlation function as well.

The Painlevé transcendents, Fredholm determinants, and scaling, linked as they are in the Ising model correlation functions, seem as well to lead to apparently universal asymptotic distributions (Wigner semicircle, Gaudin-Mehta-Dyson, Tracy-Widom) that appear recurrently: as energy level densities in nuclear physics, as correlations of zeros of the zeta function, as extreme properties of strongly correlated random variables. The model system is a random matrix and the properties of its eigenvalues. It may turn out that all the various methods of solution of the Ising model are ways of pointing to one of these distributions, much as the various ways of thinking about independent random variables lead to the gaussian.

More generally, even if a particular mathematical object does not appear in a solution, one may still end up with the same sorts of phenomena. The correlation function for two spins in general position (that is, they are in different rows and columns, and not on the main diagonal) would seem not to be expressible in Toeplitz form, but the final expression is much as if it were so expressible. And then, of course, someone shows how it can be expressed in something that looks like the Toeplitz form (what Yamada, 1986, calls a "generalized Wronskian" form).

4

Calculation

Strategy, Structure, and Tactics in Applying Classical Analysis

Deciphering Strategy, Tactics, and Structure in Mathematics Papers. Rigorous and Exact Mathematical Physics; C.N. Yang on the Spontaneous Magnetization of the Ising Lattice; Figuring Out Some of Yang's Paper; Strategy and Structure in Proofs of the Stability of Matter; Some of the Mathematics and the Physics; Mathematical Techniques. *Staging, Sequencing, Demonstrating*; Why All These Details?; New Ideas and New Sequences: The Elliott Lieb-Walter Thirring Proof. *Charles Fefferman's Classical Analysis*; 1. The Stability of Matter; 2. Matter as a Gas; 3. Stability of Matter with a Good Constant; The Lessons, So Far; More on Fefferman-Seco; The Fefferman-Seco Argument, A Technical Interlude. *"Why Are These Mathematical Objects Useful in Describing This System?"*; Tactics Ascendant: Unavoidable Algebraic Manipulation and Lengthy Computational Proofs; The Wu-McCoy-Tracy-Barouch Paper; Illuminating Wu-McCoy-Tracy-Barouch Through – Strategies, Parallels, and Resonances; Why Wiener-Hopf and Toeplitz?; Figuring Out More of What is Really Going On; Clues Along the Way: Identifying the Formal and the Abstract With Substantive Meaning.

It is a fact that beautiful general concepts do not drop out of the sky. The truth is that, to begin with, there are definite concrete problems, with all their undivided complexity, and these must be conquered by individuals relying on brute force. Only then come the axiomatizers and conclude that instead of straining to break in the door and bloodying one's hands one should have first constructed a magic key of such and such shape and then the door would have opened quietly, as if by itself. But they can construct the key only because the successful breakthrough enables them to study the lock front and back, from the outside and from the inside. (Hermann Weyl[1])

I

Deciphering Strategy, Tactics, and Structure in Mathematics Papers

Mathematical work may be said to have conceptual, manipulative, and formal aspects, in effect, strategy, tactics, and structure. Strategy is the larger meaning of what you are doing, the relationship of means to ends. Even with a good strategy, you still need to be able to manipulate the formulae and organize the argument. Ideas are not enough. Tactics are the line-by-line computations you perform, and the tricks you use that do the work of the moment even if you are not sure why they work so well. Tactics are connected to matters of strategy and structure by the various simplifications, groupings of terms, and other such ordering, that makes a formula or a construction meaningful and useful for the next step in the argument. And structure is the overall plan of the argument, often to be displayed in a diagram or list or chart, a plan that says just how a proof is a proof of what it claims to be proving, how the tactics fulfill the strategy. If you want to know what is going on in a proof or a demonstration or a derivation, neither the line-by-line details nor the structure will tell you enough. You need the strategy as well.

When you read a long and complex paper, you are both checking the various steps and figuring out how the paper and its argument is organized, on your way to guessing and assessing the strategy being employed. The paper's introduction may tell you the strategy and the structure, and even some of the distinctive tactical moves, but they only begin to make sense when you start to read the details. Exceptionally, the work is so close to work you have already done that you can readily infer what is going on or ought to be going on. And then you are figuring out how to prove this part of the paper before reading it. So Andrew Wiles' draft of his extensive paper on Fermat's Last Theorem engendered improvements by the experts as they checked it over. By these various modes of reading, we might gain insight, and see what is true as necessarily the case.

Discerning the structure of the argument, and even that of the paper, may not be so easy. Even if the author tells you the plot or the structure in the introduction, it may be hard to figure it out. And technical apparatus and elaborations in the argument may temporarily hide that structure. In my experience, one of the virtues of review articles is that they give another account of the structure and the argument, one that might be more apt for some readers. And then, again, when you go back to the original paper you can see how its description in the introduction is in fact correct and helpful, how the authors meant exactly what they said.

For difficult papers, the reader, and even the author, begins to understand what is really going on only after substantial subsequent work over the decades has simplified the argument and pointed to peculiarities in the original which are as well pregnant clues. The author was neither sleepwalker nor magician. She had motivated her argument with a strategy. Still, it seemed almost miraculous that it worked at all. But, in retrospect, she was groping through the obscurity of a good enough proof for an olympian view in which the proof and its argument actually made sense.[2]

Ideas and theories about what is really going on emerge after multiple readings, and after the subsequent work that makes clear that apparent connections are generic rather than opportunistic.

Moreover, there is a substantial gap between knowing what must be really going on in the physics or the mathematics, and making that into proof or a derivation. Yet having at least some sort of suspicion of what is really going on is almost surely needed if we are to provide a stronger proof or derivation. However, in the end, *what is really going on is really going on only when it works in such a proof or derivation and, ideally, we can show in detail just how it does the work.* Otherwise, there may well be surprises in the actual proving; or, in fact, you may be wrong about what is really going on. So our experience teaches us.

Reading a once (and still) difficult paper, we are much like architectural historians who are trying to understand cathedrals and worship, from Byzantium on. One discovers that the surviving structures are sturdy for good reasons, modern structural engineering and materials science principles confirming the logic of their design and build. One can also observe how the design changes over the centuries, and becomes more transparent, members being thinned and more effectively employed. New materials are discovered to be usable. Of course, there are many examples that have redundant elements—especially early on. And there are few surviving examples of marginally stable structures.

If we are good historians, we also appreciate how people lived in and used these cathedrals. Much of what is obscure to us and hard to acknowledge, was clear and obvious to contemporaries, not even needing to be noted. What they did by craft, by ritual, and by trial-and-error, we now try to see systematically.

I shall illustrate these claims about the organization and meaning of mathematical work by examining specific cases in some detail, the only way I know of to lend support to a description of what people (here mathematicians) actually do. Most of this chapter is an account of the strategy, tactics, and structure of several lengthy scientific papers: How do the aspects of strategy, tactics, and structure work together to encompass mathematical work? In particular, we shall examine in much greater detail than we could in chapter 3, some long and complex proofs or derivations in mathematical and statistical physics concerned with the existence of our everyday world of bulk matter (the stability of matter problem) and with how a piece of iron might be permanently magnetizable (the two-dimensional Ising model).

These initial papers are not only lengthy, their calculations are complex and not very transparent. Why not focus on the more recent, simplest, most physically and mathematically transparent calculations and papers for the phenomena that concern me here? Why focus on the earlier papers?

I am interested in showing how strategy, tactics, and structure work in the earliest original work, when the mathematical and physical insights are much less clear, when the calculation is much more involved and intricate, when one does not yet know ahead of time that a result could be proved, and one does not already know at least one way to achieve the result. Moreover, it turns out that these papers are rather more motivated and readable than was initially appreciated, that motivation now more readily seen in light of subsequent work. There are insights within these early papers that only much later come to be appreciated as essential rather than accidental.

***Before going further, it may be useful to put this chapter's argument in light of where we have gone already and where we will be going. We have examined particular mathematical fields and practices to better understand mathematical work. Statistics' employment of the Gaussian distribution shows one way conventions are established, maintained, and justified. The development of the field of topology, in particular its on-and-off division into two fields, shows how subjects become established as a consequence of mathematical work. In chapter 5, number theory's use of devices from geometry and analysis, and the inter-connections between analysis, algebra, and arithmetic, show one way analogy plays a productive role in doing mathematical work. And in chapter 6, I describe a relationship between difficult mathematics and everyday images and common-places, a theme that is instantiated more technically in the current chapter.

Recurrent themes are asymptotic approximations, linearization, diagrammatics, and the canonical decomposition of objects. And under the rubrics of "what is really going on" in a proof, and the notion of a research program, we attend to the generic themes and goals, the Holy Grails, in mathematical work.

Along the way, I indicate a number of analogies. I view them as suggestive and heuristic, rather than exact, unless they are worked out in detail and all the peculiarities are accounted for (which for the most part has not been done). Now, if an analogy is pressed too hard, too early, it will surely disappear. And once the analogy is worked out fully, it loses its power. I would hope these analogies are illuminating, and with revision they prove to be correct technically. If not, they will have to be abandoned.

II

RIGOROUS AND EXACT MATHEMATICAL PHYSICS

> There are numerous reasons why it is not so trivial to render the calculation in the present paper rigorous... The author is unable to answer these questions in general. (T. T. Wu, 1966)

Imagine a long, cumbersome proof or derivation, apparently essentially long, filled with "extensively explicit calculations" in which "details are formidable," and there are, for example, "forests of hyperbolic functions" and jungles of formulas along the way, not to speak of "baroque" hypotheses.[3] Some of this work, especially in mathematics applied to the physical world, is approximative but with unproven error bounds; some of it is correct but sprinkled with quite plausible but unproven mathematical assumptions. Here, I shall be concerned with work that is either exact or rigorous or both. As for "exact," the claim is that the result is not an approximation but in fact is the correct solution. (It might be an approximation, but with correct error bounds, although this is not what people usually mean by "exact.") R.J. Baxter entitles his book, *Exactly-Solved Models in Statistical Mechanics* (1982), even if some of the derivations have deliberate mathematical lacunae. As for "rigorous," here we are careful and clear about what is proven and what is yet to be proven, paying attention to mathematical fine points. Whether the solution is exact or an approximation, the proof or derivation itself has no handwaving and the like. David Ruelle entitles his book, *Statistical Mechanics, Rigorous Results* (1969). Wu's pioneering 1966 work, from which I quote above, had a number of lacunae, of which he is quite aware. The paper by Wu, McCoy, Tracy, and Barouch (1976) that we describe later provides an exact asymptotic solution, but further work needed to be done (by Wu, McCoy, and Tracy and their collaborators) to make it rigorous. What is characteristic of modern presentations, as in Barry Simon's *Statistical Mechanics of Lattice Gases* (1993) or in Elliott Lieb's work, is the attempt to pay careful attention to matters of rigor.[4]

The attention to mathematical rigor proves to be of physical interest. Being careful and precise about what we are doing is not merely for show. So, for example, if a lower bound on a quantity may be shown to be sharp, there are cogent examples demonstrating that without certain assumptions the bounds are crossed. The fine points of detail, needed to make a proof or derivation rigorous, point to crucial features of the physical situation—although we might not have been explicit about those features until we paid attention to mathematical details and rigor. It is in this sense that I shall call mathematical physics philosophical. It reveals what we mean by what we are saying and doing.

Again, usually after a period of decades, ways are found to articulate these complicated structures into rather simpler ones, or new proofs are found that are manifestly briefer and more transparent. Mathematical technologies are developed that allow for a more straightforward proof, so you are in effect standing on the shoulders of predecessors. Or, theories are developed for which the original theory is a straightforward consequence.

For example, one can describe how a piece of mathematical physics, Lars Onsager's 1944 derivation of the solution to the two-dimensional Ising model, was conceptually and formally simplified over the years. (See also, pp. 84-86, and the appendix to Chapter 3.) The algebraic formalism that Onsager employed is now seen as generic rather than ad hoc.[5] The original paper, formidable as it might be taken to be, can make not only formal step-by-step mathematical sense, but also physical sense in its formal structure and in its details. (Or, so I believe is the case, having devoted a good part of an earlier book to that paper and its resonances.) Yet, still, there remain obscurities, tempting hints, and curious facts in the paper.

Perhaps the first simplification of Onsager's paper was provided by Bruria Kaufman. Kaufman (1949) derived the partition function as a function of finite lattice size. The paper illuminates Onsager's Lie algebra by using in its stead the then more familiar formalism of spinors. Kaufman and Onsager (1949) then derived the short-range order, or correlation between two nearby spins in the lattice, employing those spinor techniques.

Spinors were known to describe the (spin-½) electron, as a particle represented by an antisymmetric wavefunction (a "fermion"). In 1927, Dirac developed them (as the four-component objects) in the course of solving what we now call the Dirac equation for a relativistic electron. My impression is that for physicists of 1949 spinors were rather better known mathematical objects than the algebras (or the elliptic functions) that Onsager deployed in the 1944 paper. Still, these papers, too, are formidable in their calculations. Kaufman and Onsager's used what appear to be novel mathematical objects (at least for their contemporary physicists): highly symmetric matrices ("Toeplitz matrices" in which the elements in each right-downpointing diagonal row are the same), in effect a convolution operator; and determinants of those matrices, eventually in the limit of infinite size. Yet, the penultimate formula for the correlations in Kaufman and Onsager's paper, in terms of those determinants, would appear to be means of figuratively adding up probability amplitudes (as in quantum mechanics), as in the constructive and destructive interference of light, but here leading to order or disorder. Notionally, each of the amplitudes is a contribution to the magnetization from each of the various different paths connecting the two spins of interest.[6]

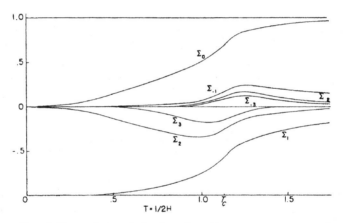

FIG. 4. Dependence of the functions Σ_a upon temperature. At low temperatures all Σ_a are negligible, except for Σ_1 which is $\sim +1$. At high temperatures only Σ_0 is significant.

FIGURE 4.1: Values of elements of the Toeplitz matrices as a function of temperature, whose determinants give the correlation function in the Ising lattice. From Kaufman and Onsager, "Crystal Statistics. II" (1949). Σ_1 is ~ −1 at low temperatures, as in the diagram.

Kaufman and Onsager's work on the correlations among atoms in the two-dimensional Ising model is notable not only for its mathematical techniques, but also for their actually plotting the numerical values of some of the elements of those highly symmetric matrices whose determinants gives the correlations among molecules, showing which terms are important at each temperature. So, at very low or very high temperatures just one of the diagonal rows of the matrix (whose elements are Σ_1 and Σ_0, respectively) is nonzero, reflecting the behavior of the phase angle of the amplitudes.[7] They plot as well the numerical values of the determinants at the critical temperature (decreasing to zero or no correlation, for large distances between molecules, their Figure 6).

In these papers subsequent to Onsager's 1944 paper, simplification and insight is provided by a more familiar formalism, by a poignant analogy (say, of a two-dimensional classical statistical mechanics system to a one-dimensional quantum system), and by actual quantities—all of which suggest how the complicated machinery produces a result.

There are by now bodies of work in mathematical physics even more arduous if not more difficult to follow than Onsager's 1944 paper. It is just this sort of difficulty that is the subject here. Namely, I want to examine the work of Charles Fefferman and his collaborators (1983–1997) on proving the stability of matter and the ground state energy of an isolated atom, and the work of T.T. Wu, Barry McCoy, and Craig Tracy and their collaborators (1966–1981) on solutions of the two-dimensional Ising model in the scaling regime, when the size of the lattice goes to infinity and the temperature is close to T_c. These are lengthy technical papers or series of papers, with enormous amounts of formal setup and calculation, mostly drawn from classical analysis. Again, there are now simplified derivations of much of this work, characterized I believe by their better appreciation of the physics of each situation. So, for example, the work based on combinatorial analysis (that is, the Pfaffian method of doing the counting for the partition function), becomes simplified through work that focuses more directly on the fermionic nature of the quasiparticles in this problem (the particles that may be said to make up the excitations of the Ising lattice) and which builds in the scaling symmetry from the very start (conformal field theory and its perturbation). But, in retrospect, this is perhaps not so surprising given the power of Kaufman's spinor formulation and of Schultz, Mattis and Lieb's (1964) fermion field theory formulation of the Ising model (and, again, in the background is the Dirac equation for the electron, considered as a spinor), as well as the ubiquitous appearance of modular functions in this problem. Except, of course, we have no reason to be sure ahead of time that these insights will actually enable us to do this work, this problem, this derivation. As Weyl suggests, only afterwards do we know what works and why.

As we shall see again and again, scientists find—discover, invent, create—a strategy, a physical and a mathematical meaning that goes along with a structure and with technical means and devices. And those devices are filled with meaning. In effect, the formulae and the staging of a proof are meaningful, about the world, and not only about the proof itself.

The derivations I shall be considering are "bulk matter limits," namely mathematical models of everyday matter built out of an Avogadro's Number of atoms, where Avogadro's Number (6×10^{23} molecules) is effectively infinite. In part, it is attention to taking these limits carefully and rigorously that then leads to the length and complexity of these proofs or derivations.

Technically,[*] the infinite volume limits may lead to lower bounds, $E(N) \geq E_*N$, for example, a bound on the energy needed to pull bulk matter apart into its constituents (that is, a lower bound on the binding energy of matter, E, as a function of the number of electrons, N). (Note that binding energies are negative, and therefore E_* is less than zero.) That there is a lower bound is a sign of the stability of matter. Otherwise, ordinary room temperature matter would be expected to implode much as it does in a supernova, but now unlimitedly so, the protons, neutrons, and electrons forming a glob rather than say an orderly crystalline solid. (Better put, within our model they would appear to collapse until the approximations we employ in the model are no longer applicable.)

Or, there are point limits, which say that in the limit of large volumes, V, and large numbers of particles, N, and constant density (N/V), something like the thermodynamic free energy, F, or entropy, S, makes sense and is defined. Namely, $\lim_{N,V \to \infty,\ N/V \text{ constant}} F(N,V,...)$, is well defined (that is, there is a limiting value), say in terms of the density and temperature. (The temperature is defined in terms of the mean energy per atom. Or, thermodynamically, the temperature is $\partial U/\partial S)_V$, where U is the internal energy.) These infinite volume limits are point limits, that is, for a particular density and temperature, although we want the limit to be well defined and smoothly changing over a range of the independent variables, reflecting the existence of a phase of matter.

If the thermodynamic potentials such as the free energy make sense, namely, that such limits exist, one may separate the quantitative and qualitative properties of matter, those extensive and intensive quantities such as free energy and temperature, F and T, respectively. Often, in a proof or derivation one divides space into balls or cubes of matter, and then notes that the interstitial contributions to the free energy, from the regions between the balls or from interactions of the cubes with each other, go to zero, at least as a fraction of the total, in the infinite volume limit. And so there are properties proportional to the volume or amount of matter.

---

[*]*Technically* is a flag. The next paragraph or more might be skimmed or skipped by nontechnical readers. In each case, I have tried to preview the conclusions of the technical analysis, as well as provide some of the nitty-gritty details.

Or, there are asymptotic estimates, which are statements about the form of a limit or its growth, here as N approaches infinity. For example, for the binding energy per particle for a piece of matter, $\lim_{N\to\infty} E(N,Z)/N = c(Z)$, a number, where Z is the charge of the nucleus. In this case we have an asymptotic limit linear in N, $E(N,Z)\approx cN$. Or, we can give an asymptotic approximation in $Z^{-1/3}$ for the ground state energy of an isolated atom: namely, $E(1,Z)=E(Z)= -c_7 Z^{7/3} + c_6 Z^2 - c_5 Z^{5/3} + O(Z^{5/3-1/a})$ Rydbergs. (1Ry = ¼ in these units.) The Oh notation means that $|(E_{correct}-E(Z))/Z^{5/3-1/a}| < K$, as $Z\to\infty$, K being a constant greater than zero. (Rigorously, the approximate value of $c_7 = 1.54$, $c_6 = 1$, and $c_5 = 0.54$.) The Gaussian is the asymptotic limit in the central limit theorem. Or, more generically in perturbation theory, a series expansion in the coupling constant is in fact an asymptotic expansion good only up to a certain term, when the series then begins to blow up.[8]

C.N. YANG ON THE SPONTANEOUS MAGNETIZATION OF THE ISING LATTICE[9]

Again, I shall focus on proofs and derivations before they become transparent, which is I believe still the case for some of these pieces of work. In these cases, it is obvious that ideas and concepts are not enough. There is no neat trick that gives it all away. Dogged, scrupulous, and inventive effort is required to find a proof or to follow one.

C.N. Yang (1952, see Appendix A for the paper) describes such work, concerning his derivation of the permanent magnetization of a crystal ("the spontaneous magnetization of the two-dimensional Ising model"), an $N\to\infty$, infinite volume limit.

> Full of local, tactical tricks, the calculation proceeded by twists and turns. There were many obstructions. But, always, after a few days a new trick was somehow found that pointed to a new path. The trouble was that I soon felt I was in a maze and was not sure whether in fact, after so many turns, I was anywhere nearer the goal than when I began. . . . [A]fter about six months of work off and on, all the pieces suddenly fit together, producing miraculous cancellations.[10]

This account mirrors Yang's account of his earlier reading of Onsager's 1944 paper on the solution of the two-dimensional Ising model:

FIGURE 4.2: SUBHEADS AND SUBTOPICS OF C.N. YANG, "THE SPONTANEOUS
MAGNETIZATION [M] OF A TWO-DIMENSIONAL ISING MODEL" (1952)

I: "Spontaneous Magnetization" (M), (Eqns. 1–16)
       $M = \langle max|\Lambda_2|max\rangle$, $|max\rangle \approx (\,|-\rangle + |+\rangle\,)/\sqrt{2}$
       M as an off-diagonal matrix element of a rotation (30) in spinor space (15):
       $\langle -|\Lambda_2|+\rangle$.[11] Note that $\{U, \Lambda_2\}=0$ (anticommutation), for the operator U: $U|+\rangle = |+\rangle$,
       $U|-\rangle = -|-\rangle$, for even and odd eigenvectors, respectively.

II: "Reduction to Eigenvalue Problem"
       A: (Eqns. 17–25) M as a limit of a trace of a rotation matrix,
       M=lim trace Λ' = lim trace $\Lambda_2 \Lambda_1'$, where lim Λ' = Λ = $\Lambda_2\Lambda_1$, where Λ_1 is made up
       of creation or annihilation operators for the even and odd states. In effect, Λ_1'
       restores the rotational symmetry that is broken by the even-odd matrix element.
       Eventually, taking the limit will effectively project out half the system (34).[12]
       [Employing Yang's notation, $\Lambda_2=V^{1/2}sV^{-1/2}$, $\Lambda_1'=S(T_+^{-1}MT_-)$.]
       B-E: Eigenvalues for an improper rotation.
       B: (26–33) M as a product of the eigenvalues of Λ' (which is to be block-
       diagonalized by ζ).
       C: (34–44) λ_α (Eigenvalues for the improper rotation Λ') in terms of those (l) of
       another matrix, G, which is independent of the limit.
       D: (45–59) $\xi_{\alpha\beta}$ (Eigenvectors of the diagonalizing matrix ζ, then taking the limit
       to get ξ in terms of the surrogate "eigenvector" y).
       E: Summary

III: "Limit for Infinite Crystal"
       A: (59a–69) Going to an infinite crystal, express how G (actually now D_+D_-)
       operates on vectors as a contour integral around the unit circle. It is an integral
       representation of a sum of terms, each of which is expressed in terms of the roots
       of unity. Set up the eigenvalue problem for l (IIIA): a singular integral equation in
       terms of Onsager's exp $i\delta'$.
       B: (70–75) y (surrogate eigenvector) is to be found by the same eigenvalue
       problem as IIIA but now for zero eigenvalue (IIIB). Elements of ξ in terms of
       these eigenvectors.
       C: (76–80) Second integral equation (IIIB) solved by inspection. M^4 (the fourth
       power "to eliminate the undetermined phase factor" introduced by the definition
       of the odd and even states) as product of limiting eigenvalues of rotation (IIIA).

IV: "Elliptic Transformation" [or Elliptic Substitutions] (81–88)
       $\Theta(z, =\exp i\omega) = \exp i\delta'(\omega) = cn\,u + i\,sn\,u = [(k\,-\exp i\omega)/(k\,-\exp -i\omega)]^{1/2}$. (See our
       Figure N.2 for δ' and ω.) Eigenvalues for IIIA; IIIA now expressed as an integral
       equation (Eq. 84) using the substitution.

V: "Solution of the Integral Equation (84)" (89–95)

Analyticity and pole-and-period considerations for elliptic functions give solution and eigenvalues. The product is now expressed as a q-series,[13] which is an elliptic function.[14]

VI: "Final Results." M as fourth root. (96)

NOTE: λ_α and $\xi_{\alpha\beta}$ are defined in terms of a limit of a going to infinity on the imaginary axis. Numbers in parentheses refer to equations in the original paper. Papers subsequent to Yang's inform my comments in this figure.

> The paper was very difficult because it did not describe the strategy of the solution. It just described detailed steps, one by one. Reading it I felt I was being led by the nose around and around until suddenly the result dropped out. It was a frustrating and discouraging experience. However, in retrospect, the effort was not wasted.[15]

The effort was not at all wasted given his subsequent work on the spontaneous magnetization. For now he could fully appreciate the advance provided by Bruria Kaufman's solution.

In writing up the paper, Yang developed a structure that organized the calculation, as a calculation. In retrospect, with many years of subsequent work by others, what appears to be miraculous and accidental and opportunistic turns out to become more natural, motivated, and understandable. Of course, this is not a claim that the original scientist could know all of what we now know, whatever we now read into the paper. And it is not a claim that the paper and the work were originally without motive and understanding, adequate to doing the work. However, it is a claim that we now believe we see more clearly why things went the way they did.

FIGURING OUT SOME OF YANG'S PAPER

When one examines Yang's nine-page paper (again, see Appendix A), the physics would appear to be confined to the first section. The permanent magnetization of matter, the "bulk spontaneous magnetization," is the matrix element of a rotation in spinor space (the small notional applied magnetic field) between otherwise degenerate (equal energy) even and odd states. Formally, this is similar to an anomalous Zeeman effect, degenerate atomic levels split by a magnetic field.[16]

Within a not unfamiliar structure provided by the headings of the sections ("Reduction to Eigenvalue Problem," . . .), and the ingenious device of temporarily restoring a rotational symmetry by means of an artificial limiting process, a symmetry otherwise broken by the even and odd states, the rest of the paper would appear to be matter of tricky and tedious manipulation and invention: "we notice that," "rearranging rows and columns" etc., solving an eigenvalue problem, finding the right diagonalization.[17] Yang's work draws in part from Kaufman's and Onsager's earlier work, using a spinor analysis rather than Onsager's (1944) quaternions.[18] Again, Kaufman's "commuting plane rotations" and spinor analysis were comparatively familiar to many physicists of Yang's generation, drawing as well on their intuitions about spatial rotations more generally.

So, for example, if $C = \left(\begin{smallmatrix} 0 & 1 \\ 1 & 0 \end{smallmatrix}\right)$, the 2×2 matrix, $1 + C = \left(\begin{smallmatrix} 1 & 1 \\ 1 & 1 \end{smallmatrix}\right)$ (which results from forming $|+\rangle\langle-|$), is not a rotation. Yang notes that, since $\tan i\infty = i$,

$$1 + C = \lim_{a \to i\infty} 1/\cos a \times (\cos a - iC \sin a).$$

And, since $C^2 = 1$, $(\cos a - iC \sin a) = \exp -iaC$, which is a rotation. This is Yang's artificial limiting process: $1 + C = \lim_{a \to i\infty} 1/\cos a \times \exp -iaC$.

But, for me, at first and even after a while, it was hard to see the mathematics or the physics of the paper as something conceptual rather than formal, although it is my faith that that can be seen, a faith that has in part been redeemed. Most of the time you can say what is going on, here at this line, and its connection with Yang's formal strategy. But much less can be said about a larger picture, mirroring Yang's own experience. At the end of Section II, Yang provides a summary of his calculation so far, sketching the path connecting the various mathematical objects that now need to be calculated:

> To summarize the results of this section [II]: the spontaneous magnetization ["I," which I denote by M] is given by (31), in which the λs [of section IIC] are related through Eq. (36) to the eigenvalues l of Eq. (44), and in which ξ_{11} and ξ_{12} are the first and the $(n+1)$th element of the column matrix ξ_1 [of IID] calculated through (48) from the column matrix y which in turn is determined by (54) and (56). ξ_1 is to be normalized according to (59) [IIE].

Schematically:

M=(31): (32)$\rightarrow\lambda\rightarrow$(36)$\rightarrow l$(44) and (33)$\rightarrow$ $\xi_{11},\xi_{12}\leftarrow\xi_1(48)\leftarrow$ y(54,56).
Section: IIB   IIB   IIC   IIC   IIB   IID   IID

Sections III and IV of the paper set out to fulfill this agenda.

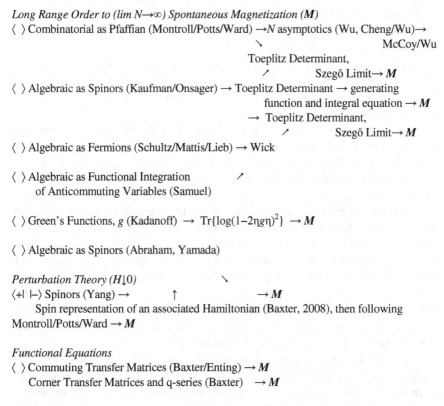

Long Range Order to (lim N→∞) Spontaneous Magnetization (M)
⟨ ⟩ Combinatorial as Pfaffian (Montroll/Potts/Ward) →N asymptotics (Wu, Cheng/Wu)→
                                            ╲   McCoy/Wu
                              Toeplitz Determinant,
                                    ╱   Szegő Limit→ *M*
⟨ ⟩ Algebraic as Spinors (Kaufman/Onsager) → Toeplitz Determinant → generating
                                  function and integral equation → *M*
                                 → Toeplitz Determinant,
                                      ╱   Szegő Limit→ *M*
⟨ ⟩ Algebraic as Fermions (Schultz/Mattis/Lieb) → Wick

⟨ ⟩ Algebraic as Functional Integration   ╱
   of Anticommuting Variables (Samuel)

⟨ ⟩ Green's Functions, *g* (Kadanoff) → Tr$\{\log(1-2\eta g\eta)^2\}$ → *M*

⟨ ⟩ Algebraic as Spinors (Abraham, Yamada)

Perturbation Theory (H↓0)   ╲
⟨+| |−⟩ Spinors (Yang) →   ↑   → *M*
    Spin representation of an associated Hamiltonian (Baxter, 2008), then following
Montroll/Potts/Ward → *M*

Functional Equations
⟨ ⟩ Commuting Transfer Matrices (Baxter/Enting) → *M*
   Corner Transfer Matrices and q-series (Baxter)   → *M*

FIGURE 4.3: Some of the various derivations of the two-dimensional Ising model spin-spin correlation function, ⟨ ⟩, and spontaneous magnetization, *M* (= lim ⟨ ⟩$^{1/2}$). Using techniques originally developed to derive the partition function, some derivations find the long-range order or correlation and then go to infinite size to get the spontaneous magnetization. Yang employed a perturbation by a small applied field, while Baxter in 1982 employed the commuting transfer matrix approach with his corner transfer matrices.

It would seem that the contemporary readers of Yang's paper could not go much further than this in understanding the calculation, even knowing the connection with Kaufman's and Onsager's work, or knowing Onsager's publicized formula for the spontaneous magnetization, the latter called a "cryptogram" by Montroll, Potts, and Ward.[19] Even Yang could not.[20]

On yet another rereading of Yang's paper, while rewriting this chapter, his sections and subheads became for me more meaningful, and even many of his moves along the way now seemed motivated in part by physics. In particular, I saw the paper as being filled with objects familiar from other subsequent derivations or other mathematical physical techniques. The annotations in Figure 4.2 reflect some of these sightings and insights.

I still do not believe I see the paper as physical. Baxter redoes Yang's paper in 2008, and again there are many familiar objects. Baxter's way of working is ingenious, but by now it is instantiated in many other of his papers—so it is still amazing but the legerdemain is familiar.

We might say that Yang's is "just" a hard eigenvalue problem. But along the way, one just has to temporarily restore a rotational symmetry (a symmetry that, in the limit, is to be broken), and the eventual appearance of a particular product is just the Szegő limit theorem. But it is just these details that might point to the physics of this problem.

I now cannot help but see the paper in the light of subsequent achievements by others (Figure 4.3), as well as in light of Kaufman and Onsager's earlier work, and what we now know of Onsager's unpublished derivation (1949), itself seen in light of subsequent work by others. Yang's paper resonates with other work, earlier and later: by Kaufman and by Onsager, of course, and Potts and Ward (1952), and Montroll, Potts, and Ward (1963), and Wu and McCoy and collaborators (1966–1981) and Baxter (1970s, 1982, 2008). And Abraham (1972) redoes Yang's derivation in quantum field theoretic language, much as Schultz, Mattis, and Lieb (1964) redid Onsager's and Kaufman's derivations.

Some of these papers approach the problem very differently than does Yang. Yang's is a perturbation theory analysis in which one allows for a notional eigenvalue-splitting magnetic field, which then approaches zero from above, effectively the derivative of the free energy.[21] Onsager took the long-distance or macroscopic limit of the short-range order of Kaufman and Onsager.

Technical devices in Yang make their appearance in one or more of these papers, often with a more intuitive physical or mathematical meaning: highly symmetric (Toeplitz) matrices (Potts and Ward; Montroll, Potts, and Ward); infinite distances to represent spontaneous magnetization expressed as limit theorems (for Toeplitz determinants, or sums becoming integrals); integral equations as integral operators that sometimes perform the Bogoliubov-Valatin transformation on Jordan-Wigner anticommuting operators, so creating Cooper pairs (Abraham, Eq. 5.12); elliptic substitutions (Onsager; Baxter); Wiener-Hopf techniques for solving an integral equation (Onsager, Wu and McCoy and collaborators, Kadanoff); analysis of elliptic functions in terms of their poles and zeros (Baxter); and q-series (Ferdinand and Fisher, Baxter, Nickel). But, does all of this yield to a simple account? Not yet.

The "astute reader" of Kaufman and Onsager (1949) might have appreciated some of these objects in Yang's 1952 paper, already, although Yang is almost surely as astute as is imaginable. Better put, he knew what he was doing, but surely not all of what he was up to, in its implications and meaning and resonances.[22]

I think that a difference in strategy from other approaches is not enough of an explanation of the difficulty in Yang's paper. In the course of working out his derivation, Yang discovered (or uncovered) a variety of formal technical features that were to prove mathematically or physically significant, recurrent in subsequent work by others (some of which were, apparently, already in Onsager's unpublished work). Again, it makes an enormous difference to know about work that appeared in the next forty years—at least for me as a reader. I then appreciate better a little more of what is going on in the paper, and the significance of that little more, even if Yang does not tell me and testifies he is not so sure, as of 1952. (And knowing the answer, as Onsager had already publicized it several years earlier (1949), or even having been shown the methods of Kaufman and Onsager by Kaufman herself, was surely still not enough.)

Technically, and reading retrospectively (again, see Appendix A for the paper itself, and see Figure 4.2 for the structure of Yang's paper): In section IIC, Yang's matrix D (Equation 63bis) is a highly symmetric (Toeplitz) matrix, doing a convolution.[23] The eigenvalue problem expressed in terms of D and p, builds in Onsager's (1944) exp $i\delta'$, and the fermion quasiparticle momentum, ω, as eventually understood in Schultz, Mattis, and Lieb's account of the Ising lattice as a field theory. (These variables have a different but equivalent interpretation in Onsager's paper, as angles and sides of a hyperbolic triangle; their trigonometric relationship becomes the energy momentum relationship for the quasiparticles. See Figure N.2.) Section III is the infinite volume limit. This leads to a product of eigenvalues that must do the work of the Szegő limit theorem for Toeplitz matrices, as Abraham suggests.[24] (Montroll, Potts, and Ward speak of their "using a [Szegő] theorem on the evaluation of Toeplitz matrices. Evidently, this is one of the methods used by Onsager himself."[25]) And the eigenvalues are of an operator that does the Bogoliubov-Valatin transformation (for some cases), producing the good particles for this system (that is, they diagonalize the transfer matrices).

Section IV's elliptic substitutions are those of Onsager (1944), and, much later, of R.J. Baxter (1982, actually 1970s), the current master of these techniques.

Section V's analysis of the poles and zeros of an integral, using features of elliptic functions, is just what Baxter does in his arguments, which are standard in this mathematical realm. And at the end of the section there is a mathematical object that may be used to express an elliptic function: a q-series, say $\Pi(1-q^{2m})^2/(1+q^{2m})^2$, where $q=q(k)$ and k is a normalized temperature (q is called the "nome"), ala Ferdinand and Fisher (1969) and Baxter.[26] The q-series expresses the series expansion for the solution in terms clusters of spins (Nickel).

Ideally, one would like to be able to explain Yang's formal strategy and make it physical or mathematical. Still, such resonances and sightings of commonalities with other work, as I have just described, need not equate with increased understanding of the overall strategy. Sometimes they do, especially if you have been working through a paper in some detail. Repeated resonances and sightings can insist on being illuminating, especially if notation and formulae are much like previous work, the typography doing typological work, recalling the similarities.

Now, some papers are deliberate attempts to clarify earlier work. Kadanoff (1966) reformulated the correlation problem in terms of Green's functions, and at various points makes explicit contact with Yang's analysis—for example, pointing out where Yang's argument implicitly depends on the Wiener-Hopf method.[27] Abraham's (1972) resonant and illuminating and "extremely elementary" account of the correlation functions and magnetization employs the fermion techniques of Schultz, Mattis, and Lieb, and then Yang's argument, and then Onsager's elliptic substitutions, in each case echoing their notation and symbols, much as Yang echoed Kaufman's (1949) and Onsager's (1944).[28] As I mentioned above, Abraham is deliberately redoing Yang's argument in quantum field theory language (although this problem is classical statistical mechanics); the repetition of symbols and equations makes clear that it is a redoing and an interpretation. The paper is in effect a translation, readable on its own for conventionally trained physicists. Abraham is aware of the net of ideas and notation he is working within, and how pulling on one corner of the net of proofs brings along the rest: "solution of the integral equation [Yang's IIIA] provides an alternative, albeit rather ghastly [very roundabout], approach to the Szegő-Kac [limit] theorem," its eigenvalues providing the terms in a suitable product. Abraham's paper is in fact about the quantum mechanical problem of a one-dimensional chain of spins, formally or mathematically identical to the classical two-dimensional Ising model (the chain's other dimension is time).[29]

So Yang's work on the spontaneous magnetization now fits more richly into a web of papers in this field and into the more common practices of theoretical physicists.[30]

It should now be possible to provide a more detailed physical account of Yang's argument as it was presented, beyond the rubrics (the anomalous Zeeman effect, an artificial limiting process, and fermion field theory) and the recognizable and familiar forms. I suspect that that physical understanding will also require a better understanding of the meaning of the Wiener-Hopf method, to which we shall return toward the end of this chapter. (But, I cannot do more now.)

***The kind of analysis I have offered here will characterize our argument more generally. In each case, the task is to appreciate the connections between strategy and structure, and to understand how tactical moves are often initially motivated by neither strategy nor structure but by localized facts on the ground, say the way the formulae are written. Only much later can those facts and tactics be seen more globally, in strategic and structural terms.

To follow a proof line-by-line does not mean the proof or an argument is convincing. One wants to see more globally how it is a proof of what it is claiming to be proving, and wants to appreciate the various tactical moves in strategic terms. Moreover, subsequent work does not obsolete its predecessors. There is a residue of insight within many of the earlier papers that is not necessarily captured by the improved and more perspicuous work.

Mathematicians do not follow the proof as a logical deduction, but rather as a series of demonstrations of how what is the case is to be shown to be the case. [31] And claims about a proof—generically, "we believe that the factor of 4π does not belong in the formula"[32]—are promises about what is to be eventually demonstrated, it is hoped for good albeit not-yet-proven reasons.

STRATEGY AND STRUCTURE IN PROOFS OF THE STABILITY OF MATTER

FIGURE 4.4: TITLES OF PAPERS THAT PROVIDE THE FEFFERMAN-SECO-CÓRDOBA PROOF OF THE ASYMPTOTIC GROUND STATE ENERGY OF A LARGE ATOM

Eigenvalues and eigenfunctions of ordinary differential operators. (161 journal pages)
The density in a one-dimensional potential. (181 pp.)
The eigenvalue sum for a one-dimensional potential. (73 pp.)

The density in a three-dimensional radial potential. (74 pp.)
The eigenvalue sum for a three-dimensional radial potential. (91 pp.)
[And, bringing it all together:]
On the Dirac and Schwinger corrections to the ground-state energy of an atom.

(185 pp.)
Aperiodicity of the hamiltonian flow in the Thomas-Fermi potential. (142 pp.)

With Córdoba:
A trigonometric sum relevant to the nonrelativistic theory of atoms. [an announcement
of results] (3 pp.)
Weyl sums and atomic energy oscillations. (62 pp.)
A number-theoretic estimate for the Thomas-Fermi density. (16 pp.)

FIGURE 4.5: SOME OF CHARLES FEFFERMAN AND COLLABORATORS' RESEARCH
PROGRAM IN THE STABILITY OF MATTER

The uncertainty principle. (1983) [stability of matter (again), atoms exist, there is
chemical binding]
The atomic and molecular nature of matter. (1985) [atoms exist as a gas]
The N-body problem in quantum mechanics. (1986) [stability of matter with a small
constant]
Relativistic stability of matter. (with de la Llave, 1986)
On the energy of a large atom. (1990–1997) [ground state energy, asymptotic formula]

FIGURE 4.6: THE STABILITY OF MATTER RESEARCH PROGRAM, IN BRIEF

Onsager: Electrostatic Interaction of Molecules. (1939)
Fisher and Ruelle: The stability of many-particle systems. (1966). There is as well
earlier work by van Hove, Lee and Yang, and Lévy-Leblond
Dyson and Lenard: The stability of matter. (1967–1968)
Lenard: Lectures on the Coulomb stability of matter. (1971)
Federbush: A new approach to the stability of matter problem. (1975)
Lieb and Thirring: Bound for kinetic energy which proves the stability of matter.
(1975) + subsequent work by Lebowitz and Lieb and collaborators on "the constitution of
matter." Lieb and collaborators, and others, on reducing the size of the Lieb-Thirring
constant.
Fefferman and collaborators (1983–1997)
Bach, and Graf and Solovej: Simplified derivation of the asymptotic form with better
bounds (1992–1994)
+ work on the relativistic stability of matter problem.

138

So far, I have discussed mostly a paper's structure rather than the meaning of that structure, its strategy. In part, Yang's actual strategy was very much influenced by tactical moves, that larger physically-meaningful strategy only weakly discerned at best. Nor, again, have I provided much of the detailed nitty-gritty of any single calculation. What I provide is something of a round-by-round rather than a blow-by-blow account. I shall quote subheads of long papers, often remaining silent about technical details concerning bounds, inequalities, and apriori estimates of differential equations—those technical details of course being at least half the essence of an achievement. I do not believe the reader should now conclude that a line-by-line analysis of a paper could make strategic sense, even though I would like to believe that is the case. I have surely not provided such evidence here (in part, and only in part, because such a detailed and technical reading is precluded economically by the very small number of readers it would allow.) And the alternative of taking on a highly simplified toy-problem never convinces the skeptic.

How do we discern strategy and structure within hundreds of pages of formula-strewn papers, rather than merely nine pages?

For example, the list of titles of seven long papers conveys the structure of the Fefferman-Seco proof of the asymptotic formula for the ground state energy of a large atom; in addition, there are three papers by Córdoba-Fefferman-Seco on the oscillating $Z^{3/2}$ next-term in the series, which we will discuss only briefly. The initial paper improves the semiclassical approximation for solutions to differential equations. First, in one dimension, and then applying those results to three dimensions, taking advantage of radial symmetry, these improved approximations are employed to obtain the charge density and the sum of the negative eigenvalues (bound-state energies) in an atom. The crucial paper, whose argument we shall review later on, is "On the Dirac and Schwinger corrections." It employs all the technical results in the other more generic papers. The paper on "Aperiodicity of the hamiltonian flow" makes sure those techniques are applicable in order to achieve the $Z^{5/3}$ term with errors that are asymptotically smaller. The Córdoba-Fefferman-Seco series will in effect shift the problem to a $Z^{3/2-\varepsilon}$ term. By staging the proof as they have, Fefferman-Seco are able to defer technical detail and generic considerations. At each level, the argument depends on more detailed results provided elsewhere, and results are made ready for employment in a larger argument. (Although (Córdoba-) Fefferman-Seco is for an isolated atom, not bulk matter, ideally the proof would provide information, the Thomas-Fermi energy, needed on the way to providing a rigorous account of the gaseous nature of matter by providing a better account of its stability, that is "stability of bulk matter with a small constant." See note 57 of chapter 3.)

These papers are part of a larger endeavor by Fefferman and collaborators, and as we shall see by many others, to provide a rigorous account of the atomic and molecular and gaseous nature of matter—which we shall review as an endeavor. Again, we might see, at each point in Fefferman and collaborators' papers, that some problems are deferred to previous or prospective work, while results are prepared for the larger account. In so far as the sequence of papers is not planned as a sequence, but merely cumulated as they are published, there is likely to be a less tight fit than in a single major endeavor.

And this endeavor is part of an even larger endeavor over several decades to give a rigorous account of the stability of matter and the existence of thermo-dynamic systems. Such a program is likely to be even more disjointed than any single group of papers by one set of authors. At the same time, the meaning and import of any particular paper is likely to be enhanced by appreciating its place in a much more extensive program of work.

Authors, either in their introduction or their conclusion, usually summarize the contents of a paper in terms of a proof's structure, and perhaps even provide motivation that allows one to appreciate the strategy employed. However, again, it is my experience that that summary only makes sense after one has read the article, paper and pencil in hand, or perhaps if one already works in that particular field and on that particular problem.

The various proofs of the stability of matter employ architectonic structures, perhaps custom made for each proof, based in either mathematical or physical design. The original proof by Dyson-Lenard (1967–1968) has such a structure (despite Dyson's claim that they were hacking through a forest of inequalities). At least, they could so describe the proof when they wrote it up for publication. Seemingly, they were driven by calculational issues (as in Yang's paper), although in fact the physics is very important in guiding them in what to prove and to calculate, and in the structure of the papers themselves. Such a structure is an obvious requirement for the exposition and organization of a lengthy and complex piece of work, if the work is to be understood as a proof that actually proves what it claims to prove, if it is to get past the journal editors and referees. But the structure also reveals a strategy, perhaps discovered latterly or along the way: the meaning of the various steps, the hierarchy of such meanings. And in that sense, the structure leads to an explanation, however dimly perceived, of what it is about the physics or the mathematics that makes the proof go through.

The strategy and structure are discovered in the course of working out the proof, the final presentation normally hiding many lengthy calculational details and mistaken paths. We are presented with a reconstruction that can be followed and leads to understanding, somehow and somewhat.

The advantage of the subsequent Lieb-Thirring (1975) proof over the Dyson-Lenard one is mainly its perspicuousness, the enormous reduction in the proportionality constant for the lower bound of the binding energy, by about 10^{13}, an extra plus. Moreover, the work of Lieb and his collaborators is illuminated as physics by how Lieb organizes its presentation in his review and survey articles (collected in his wonderful selecta, *Stability of Matter*), didactically bringing to bear the needed physics or mathematics at just the right time.

Fefferman and Seco and Córdoba (1990–1997) use comparatively esoteric technical classical analysis (at least for many mathematical-physicist readers), to get a rigorously-derived value of the ground state energy of an isolated atom, that is, $E(1,Z) = E(Z)$. (We know, empirically from chemistry or nonrigorously from calculations, that $E(1,Z)$ is only slightly less in absolute value than the absolute value of the binding energy per atom in bulk matter, $|E(N,Z)/N|$ (which is greater than the lower bound, $|E_*|$). Real atoms bind to each other; as we shall note, Thomas-Fermi-model atoms do not. This is not the stability of matter itself.) Their account of the derivation of the basic *form* for the expression for the ground state energy of an atom in terms of a series in powers of the atomic number Z (recall, $E(Z) = -c_7 Z^{7/3} + c_6 Z^2 - c_5 Z^{5/3} + O(Z^{5/3-1/a})$) has been considerably improved and made more perspicuous, and its bounds made more sharp, through somewhat different modes of working, by Bach (1992, 1993) and Graf and Solovej (1994).[33] But these improved methods do not provide the rigorous derivation of the coefficient of the (Dirac-Schwinger) $Z^{5/3}$ term, the major part of the Fefferman-Seco work.[34] Nor do the improved methods lead to an approach to the next term in the series (presumably $Z^{4/3}$, but in fact an oscillating term roughly of order $Z^{3/2}$), the term that takes into account angular momentum and atomic-shell effects. Schwinger and Englert had earlier (1985) extended Thomas-Fermi in this direction. Fefferman-Seco have a handle on this term as well, and Códoba-Fefferman-Seco (1996) have a rigorous handle on it, and an estimate of the error, $O(Z^{3/2-\varepsilon})$. In the end:

$$E(Z) = -c_7 Z^{7/3} + c_6 Z^2 - c_5 Z^{5/3} + s(Z) Z^{3/2} + O(Z^{3/2-\varepsilon}),$$

where $s(Z)$ is the oscillating term.

Still, there are about 800 pages of proof in just the Fefferman-Seco series, not to be assimilated readily even given their dividing the work into distinct parts, and the introduction and the summary provided in the "On the Dirac and Schwinger corrections" paper (for which, see Appendix B). Still, without those aids one could not segregate off or black-box much of the technical differential equations material or describe a structure a physicist might love.

SOME OF THE MATHEMATICS AND THE PHYSICS

In each of the proofs of the stability of matter there appears the crucial physics for the stability of matter: (1) Coulomb forces that can be *screened*, the attractive force of the nuclei for the electrons is canceled or screened by the repulsive force due to the cloud of those electrons around the nuclei, and that there are roughly equal positive and negative charges—so that a nucleus' charge is felt very weakly if at all a few lattice sites away from it; (2) the Heisenberg (or an improved) *uncertainty principle*, so connecting the potential energy and the kinetic energy for an atom (or "bound state"); and, (3) electrons are fermions or the Pauli *exclusion principle*, so forcing each additional electron to be less bound than its predecessors. But, in fact, that crucial physics appears somewhat differently in each proof as well. And somewhere along the line, the crystalline structure of matter, its periodic structure, must appear, too—represented by those balls or cubes I have mentioned already.

There are several notable features of these proofs: (a) Structure and strategy is crucial. Even hacking through a forest of inequalities is presented in an orderly if not wholly transparent fashion. (b) Facts matter. These are not merely conceptual endeavors, but the particular features of the particles and interactions make or break the calculations. (No fermions, no stability of matter.) (c) In time, proofs may be simplified. Intricate lengthy calculations with gruesome details may become lengthy calculations with extensive details, but these arguments may not readily simplify into conceptual ones. When they do become so simplified, as in the Federbush and the Lieb-Thirring redoing of Dyson-Lenard (or the Schultz-Mattis-Lieb redoing of Onsager), we are impressed. And, usually, new ideas are introduced into the proof. And then it turns out that the new proof may be even further simplified, while the putative new ideas may sometimes then be found to be embedded already, more or less explicitly, in the original proof. These recurrences and simplifications take no credit away from the original authors, or from subsequent improvements that are then improved upon. And, eventually, it proves possible to develop enough of the preliminary apparatus in such a way that the proof of the stability of matter comes naturally as a small section of a textbook, as in W. Thirring's *Quantum Mechanics of Large Systems*, or as part of a review article by L. Spruch on the Thomas-Fermi model, or in Lieb and Seiringer's *The Stability of Matter in Quantum Mechanics*.[35]

The crucial physical and mathematical move is localization, breaking a global object into local parts and then patching things together afterwards, a recurrent theme in modern mathematics. And there needs to be careful attention to the right comparisons, the right approximate systems, and so the right inequalities. That demands a great deal of invention, whether it be employing the Thomas-Fermi model of an atom in Lieb-Thirring, or the right sort of weakly interacting system in Fefferman-Seco.

I do not know if the particular features I have mentioned about the sequences of proofs are necessary. And perhaps, for some problems, there is no way out of intricate arguments, ever. Or, perhaps there will come along something like the Hilbert Syzygy Theorem (1890) that will change the whole field, as the Hilbert Theorem and its method apparently "demolished" classical invariant theory. However, the Hilbert Theorem did not make it easier to actually calculate invariants, and only modern algebraic geometry and Groebner bases and powerful computers do that. And so a fantasized Big Theorem or Generic Theory might not give us the proportionality constants we need. One would like to believe that eventually things will become conceptual rather than calculational. But that belief may not be fulfilled.[36]

We might also discern a physical sequential structure of proving in these endeavors. (See Figure 4.6.) First, prove the stability of a hydrogen atom; then, the stability of a larger multielectron atom; then, the stability of bulk matter by employing the Thomas-Fermi model of an atom or electrostatic inequalities; and then, the existence of well defined bulk properties, such as the free energy, what is called the thermodynamic limit, and that the system exhibits thermodynamic stability (that pushing on something does not make it grow in size), as analyzed by Lebowitz and Lieb and Thirring. Then, show there is chemical binding to form molecules, and then the possibility of a gas of hydrogen molecules (Fefferman). Along the way, rigorously derive the terms in the power series in Z (actually $Z^{-1/3}$) for the ground state energy of an isolated atom, the exact values for the coefficients being another major endeavor. Again, some of this structure is discovered in the proving, some in the writing up, and some is only discerned after a sequence of papers by many different authors has appeared, each paper perhaps depending on some of the previous papers.

Some methods may prove to be quite powerful. So one starts out from an uncertainty principle and correct volume counting in phase space for differential equations—namely, the number of eigenvalues or bound states, thereby connecting geometry to spectrum. But there are always seemingly idiosyncratic tricks, and miraculous and remarkable moves—beyond strategy or structure—that are invoked at particular moments. One such, found in the Lieb-Thirring proof, is the formal interchange of nuclear and electronic coordinates in proving an electrostatic inequality concerning the electron-electron interactions, in effect making each electron into a "nucleus" of its own "atom." Fefferman calls the trick "remarkable."[37] As Spruch describes it, "the method is imaginative enough to bring pleasure to even a jaded theorist's heart . . . [it] remind[s] one of some brilliant castling move by Bobby Fisher in his prime."[38] Making use of the Teller's No-Binding theorem of Thomas-Fermi atoms, these "atoms" no longer interact with their neighbors. It would appear that the formal-interchange trick is effectively, or leads to an inequality proved by Onsager (1939), which comes out of a classical theorem in electrostatics: that the total potential energy including the self-energies is positive, since the electrostatic field self-energy is always positive (that is, it is a positive-definite form).[39] The net effect is to convert an $N \times N$ bound of electron-electron interactions into an N bound. Graf and Solovej achieve that same reduction by means of an operator inequality.

Fefferman achieves the reduction of the $N \times N$ bound to an N bound by means of a formula that is supposed to be "well known to fourier analysts," an observation that in three dimensions $1/|x-y| = 1/\pi \iint dz\ dR/R^5\ \chi(x)\ \chi(y)$, where the integral is over all possible balls, located at point z in three-dimensional space and having radius R, and the χ are indicator functions testing if x and y, points in three dimensional space, are within each ball.[40] If there are N electrons and M nuclei of charge Z, the integral gets a term (multiplied by the wavefunction squared):

$$\{N(N-1) + Z^2M(M-1) - 2ZMN\} = \{\ [N - ZM - \tfrac{1}{2}]^2 - \tfrac{1}{4}\} - (Z^2 + Z)M$$

$$\geq - (Z^2 + Z)M,$$

the linear proportionality that we need for a lower bound.[41] In effect the balls of atoms again allow for Onsager's insight. Fefferman shows how this also leads to screening. Yet the $N \times N \rightarrow N$ trick is, again, nineteenth century electrostatics, now dressed up in rigorous and more general clothing suited to quantum mechanics and actual atoms. Lieb and Seiringer provide a derivation that would appear to be model independent. That dressing-up needs to be done rigorously to be sure it would do the work.

And it is recurrently noted that the ad hoc tricks and formalism of one generation become the theories and powerful methods of subsequent generations.

MATHEMATICAL TECHNIQUES

Besides the sequential hierarchy of proofs, powerful generic methods, and the marvelous idiosyncratic devices, there are the technical crafts of the mathematician, the professional's legerdemain. A number of technical themes recur in Fefferman's work:

(1) Again, solve global problems by local approximations, and then patch them together to achieve global solutions. So we find a division of actual space or phase space (the space of *momentum* × *position*[42]) into cubes or balls or Thomas-Fermi model atoms, ones that do not much interact with each other; or, the various schemes for locally solving differential equations and patching together solutions ("microlocal analysis" and "pseudodifferential operators"). There is a wonderful tradition of cubes and balls of various sorts, filling up all of space either ball-by-ball, or by scaling up into balls/cubes containing smaller ones, or by schemes of fitting-in smaller balls into the interstices within and between larger ones (Lebowitz-Lieb's "Swiss cheese").[43] These various schemes are not only geometrically different, but they may have different mathematical convergence properties in the limit of bulk matter.

(2) Bend cubes in phase space so that they fit much better into a particular phase space (much as we might bend and fit small frankfurters into a pan), and solve the differential equation trivially and more accurately on those bent cubes. That is, we solve differential equations by fourier analysis, so converting derivatives into multiplications. Then we modify or phase shift the fourier coefficients themselves. Put differently, first we employ a straight-line geometrical optics approximation, and then we try to gently bend the rays.[44]

(3) Employ some sort of averaging or mean-field approximation for the charge density (for example, Hartree-Fock or Thomas-Fermi).

It turns out that this is all we need for a reasonably good approximation—largely because atoms do form, and they arrange themselves regularly and perhaps periodically in space, namely one particle in each cube.[45]

Classical Analysis. Charles Fefferman's work is classical analysis of a high order, the differential and integral calculus of Newton and Leibniz now 300 years later. It has been said of this work: "These are relatively old subjects and the problems in these fields are notoriously complicated and difficult."[46] That calculus is now seen as being constituted by fourier or harmonic analysis, partial differential equations, and functions of complex variables.

There was a period, in the 1940s and 1950s, when classical analysis was considered dead and the hope for the future of analysis was considered to be in the abstract branches, specializing in generalization. As is now [1978] apparent, the rumour of the death of classical analysis was greatly exaggerated and during the 1960s and 1970s the field . . . [saw] the unification of methods from harmonic analysis, complex variables, and differential equations, . . . the correct generalization to several variables and finally the realization that in many problems complications cannot be avoided and that intricate combinatorial arguments rather than polished theories often are in the centre.[47]

Recurrent themes in classical analysis are inequalities and taking limits, choosing the right delta to go with a given epsilon, and being good at approximations, themes still found in introductory calculus texts.[48]

Although my examples are drawn from mathematical physics, there are of course other recent examples of apparently unavoidably long and complicated efforts in pure mathematics, perhaps most notoriously the Gorenstein-orchestrated classification of the finite groups (1970–1990); the Appel-Haken computer-aided proof of the Four Color Theorem (1970s); arguably, Wiles' proof of the Taniyama-Shimura conjecture for a restricted domain and so Fermat's Last Theorem (1990s); and, the Perelman-Hamilton proof of the Poincaré Conjecture (2003). (The issue in the latter two cases is about how one delimits a "single" proof vs. foundational material.) Sometimes, the basic ideas are straightforward. But, then the length of the proof is a matter of exhausting the variety of cases, given our limited capacity to decrease that variety by theoretical means. So Appel says:

It is totally maddening that none of us seem to understand reducibility well enough to prove good general theorems about useful enough classes of reducible configurations and thus computers must be used to show each individual configuration reducible. It is totally frustrating that it is becoming intuitively clear that almost any reasonable use of the discharging procedures will work and that the collection of reasonable unavoidable sets is huge.[49]

Often, as in Andrew Wiles' work, the final proofs in an endeavor stand upon very substantial efforts by others (here, Weil, Serre, Ribet, Frey, plus Mazur, Langlands, . . .) that made it possible for the lengthy proof to have its extraordinary significance. Wiles says that he began his work on the Taniyama-Shimura conjecture when he learned of Ribet's work concerning Serre's and Frey's ideas, arguably all founded in the work of Taniyama, Shimura, and Weil, which showed how the Taniyama-Shimura conjecture's proof would lead to a proof of Fermat's Last Theorem. The final parts of the proof are comparatively brief. Most of the papers' contents are developments of rather more general techniques (as is the case in Fefferman-Seco), their application to the question at hand taking up a smaller fraction of the work. As we shall see, pioneering work is often ugly, although in retrospect the deep ideas may be discerned even there.

III

> In the middle of my life's journey I found myself in a dark forest, the right
> way being lost. To tell about those woods is difficult, so tangled . . .
> (Dante, *Divine Comedy* (ca. 1300), Canto I)

STAGING, SEQUENCING, DEMONSTRATING

In a long and complicated proof, it is imperative to stage and sequence the work so that the proof becomes a demonstration. Ideally, a demonstration shows that what is true must be true. More practically, it displays the structure of the argument, showing what is at stake at each point. The structure may even indicate infelicities and problems in the proof itself, places where it might be improved. Such staging and sequencing is itself a creative act, as much as the proving of individual theorems and lemmas within that organization. Moreover, to write a review article, staging and sequencing the various efforts so that their meaning is apparent, is also a creative act. Much of the proof may remain irremediably technical and detailed. Still, the general structures and sequences employed are rather less technical, available to most readers.

Again, our case in point will be a series of proofs in mathematical physics of the stability of bulk matter: namely, rigorously proving there is a lower bound for the energy proportional to the number of particles, N. Or more quaintly, that classical bulk matter does not implode. That is, in putting together two bits of bulk matter, enormous energy (much more than chemical, much less than nuclear bomb scales) is released, then leading to an explosion.

To recall, the physics is fairly straightforward: nonrelativistic quantum mechanics (the Schrödinger equation) and Coulomb forces among the electrons and protons ($\sim q_1 q_2/r^2$, for two charges, q_1 and q_2, at a distance r from each other). Again, the delicate problems here are: (1) screening, the cancellation at long ranges

147

of attractive by repulsive forces among the charged particles (on the other hand, gravitation is not screened, although gravitation is too weak to matter here); (2) the fact that electrons are fermions, and so are represented by antisymmetric wave functions or anticommuting operators (the nucleus's being a fermion or not should not matter for stability, since there are stable atoms with both odd and even numbers of protons and neutrons); (3) the uncertainty principle that says that if the electrons are bound very close to the nucleus their kinetic energy will rise, and hence there may be a balance point of the potential and the kinetic energies (but we need a form of an uncertainty principle that excludes cases such as the electron wavefunction being bunched up in two places distant from each other); and, (4), that the Coulomb force is singular, it goes to infinity at $r = 0$. The hard part is that this is an N-body problem, where N is roughly Avogadro's Number.

Although the problem might well have been solved in 1930, since all the needed mathematics and physics were available then, the first published effort was by Onsager in 1939:

> In the electrostatic theory of crystals and liquids the existence of an additive lower bound for the electrostatic energy is generally taken for granted as a very trivial matter, and it does not appear that a rigorous proof, even for a simplified model, has ever been attempted. For an assembly of ions or even electric dipoles, a computation of the Coulomb energy by direct summation meets with considerable difficulty because the convergence is conditional at best, . . . The idea of a lower bound for the electrostatic energy *per se* cannot be separated from the assumption of a closest distance of approach between the centers of charges. We shall therefore restrict our considerations to "hard billiard ball" models of molecules and ions, and attempt no more refined representation of the short-range repulsive forces.[50]

Onsager then applies classical electrostatics to derive a lower bound in an elegant and physically intuitive way. The paper is eight pages long, and the thought-experiment, a very chemistry-sounding derivation, takes two of those pages.[51] Assuming that the particles are in fixed positions, the main theorem states, in a nice paraphrase, that "no matter how one arranges the particles, . . . the potential energy is bounded from below by the potential energy of a system of particles in which each particle interacts *with one and only one other particle*, a particle of charge equal and opposite to its own and located at the position, in the original system, of its nearest neighbor."[52] "The fictitious system always has a potential energy lower than the real Coulomb system, and—what is most essential—the number of terms out of which this fictitious potential energy is made up is N and not of the order of N^2 as for the true energy."[53] The N^2 interactions among N particles become bounded by N stylized interactions, or N self-energies, given that the total electrostatic energy is positive. Hence, there is a lower bound proportional to N.[54]

Others extended Onsager's work, now employing quantum mechanics but also employing a tempered potential that was of short range (unlike the Coulomb potential, but in effect simulating screening), and smeared out charges to avoid the $1/r$ problem. For our purposes, the crucial paper was by Fisher and Ruelle, entitled "The Stability of Many Particle Systems" (1966); it is especially concerned with the existence of the thermodynamic limit, when the energy is asymptotically proportional to N, rather than just a lower bound. (By the way they show that the long range character of the Coulomb force won't be a problem for stability.) Their posing and specification of the problem of "the stability of matter" set others to work on the problem.

Before we go into details of the "ugly" and less-ugly proofs of the stability of matter (that it won't collapse and then explode), for comic relief I want to discuss the current slimmed-down proofs due to Lieb and collaborators.[55] Essentially, one proves a series of general inequalities, sometimes concerned with the $1/r$ Coulomb potential, sometimes about nice functions. This is all ingenious but also fairly standard mathematical analysis, with many-particle systems in mind.

There are several approaches, each providing a slightly different lower bound on the energy/nucleus for hydrogen. All require a kinetic energy bound, the kinetic energy bounded by the sum of the bound state energies of the potential, an integral of the charge density to the 5/3 power. These Lieb-Thirring inequalities are in effect Weyl asymptotics, and their coefficient L affects the stability of matter by $L^{2/3}$.

One path uses a "Basic Electrostatic Inequality," as Lieb calls it. It includes Onsager's notion that the electron-nuclei interaction is bounded by the each electron's interaction with its nearest neighbor nucleus, and an estimate of the quantum mechanical exchange or correlation energy, since electrons as fermions are otherwise indistinguishable from each other (an inequality due to Lieb-Oxford, an integral of the charge density to the 4/3 power). Using the Cauchy-Schwartz inequality, the 4/3 is converted to 5/3. Then one must deal with the nearest neighbor distance in terms of a bound, and finally one optimizes to get the stability of matter with a constant of 7.29 Rydbergs/nucleus for hydrogen. Again, the binding energy of hydrogen is 1 Rydberg. As Lieb and Seiringer describe it, there are:

...three essential ingredients. One is the LT inequality, which relates the kinetic energy of fermions to their density... The second is the estimate on the indirect (or exchange) part of the Coulomb energy, which gives a lower bound on the electron interaction in terms of the classical electrostatic energy of the electron charge distribution. Finally, the electrostatic inequality of Section 5.2 relates the total classical electrostatic energy to the interaction energy of the electrons to the nearest nucleus only [ala Onsager]. In the end, one has to deal with the $\rho^{5/3}$ semiclassical energy, the negative $\rho^{4/3}$ exchange estimate, and the negative nearest-neighbor electron-nucleus interaction.

The Thomas-Fermi method, Lieb and Thirring's earlier work (which we shall get to), but now with some improvement for the exchange energy (the Lieb-Oxford inequality), gives 5.60 Ry/nucleus for hydrogen (an improvement of about a factor of four or five over the original work). A different derivation, again without Thomas-Fermi, that does not take exchange into account, in effect assumes that the electrostatic interactions cancel each other, except for the electron to nearest-neighbor nucleus interaction, in addition smearing out the electron's charge a bit, gives 30.5 Ry/nucleus for hydrogen. A semi-classical derivation of the kinetic energy inequality suggests that these methods should get about 4.90 Ry/nucleus for hydrogen.

To recall, in the early 1960s, Lenard was an applied mathematician working in plasma physics, and hence concerned with Coulomb forces between large numbers of positive and negative particles at very high temperatures. Inspired by Ruelle and Fisher's work, and employing Onsager's ideas, Lenard started to think of how he might build in the long-range Coulomb cancellations through suitable combinatorial mechanisms. (Yang told Lenard that since the problem was not "trivial," it must be "difficult.") Eventually, and by chance, he interested Dyson in the problem, and together they provided the "lengthy and complicated"[56] first proof of the stability of matter, as we now understand it, published in 1967–1968.

FIGURE 4.7: SUBHEADS OF DYSON AND LENARD'S 1967–1968 PAPERS ON THE STABILITY OF MATTER

I: 1. Introduction
     2. Statement of Results [Thms. 1, 2, 3, 4, *5* (the stability of matter)]
     3. Proofs of Theorems 1 and 2 [N^3 and N^2 lower bounds, Lem. 1]
     4. A Theorem of Electrostatics [Thm. 6 and 7, $N \times N \to N$ interactions]

5. Proof of Theorem 3 [$N^{5/3}$ lower bound, using Lems. 2, 3, 4]

6. Proof of Lemmas 2, 3, and 4 [about functions and their derivatives; the geometry of points; arithmetic of numbers]

7. Proof of Theorem 4 [N bound, all particles are fermions, Lem. 5, using Thm. 8]

8. Proof of Theorem 8 [space occupied by fermions]

9. Smooth Background Charge [Theorem 4, improved to Theorem 9, is almost but not quite Theorem 5 (point nuclear charges), the stability of matter]

II: 1. Introduction [Stability of Matter requires lots more work than Theorem 4 and 9; and now a very large proportionality constant]

2. The Plan of Attack [one particle at a time in a chopped-up space; toy calculation of only one negative particle in space, Thms. 10, 11 (cf. Lem. 9), 12 (cf. Thm. 5)]

3. Preliminary Simplifications [Now to Theorem 5, the main result, making use of almost all that has gone before.]

4. Configurational Domains with Uniform Nearest-Neighbor Separation [chopping up space into good cubes, each containing one negative particle and an arbitrary number of positives, Lem. 6]

5. Reduction of the N-body Problem to a One-Particle Problem [Theorem 9 for each cell, and then reassemble the fragments, using Lems. 7, 8, 9, Thm. 13]

6. Solution of the One-Particle Problem

7. Proof of Lemmas 7, 8, and 9

8. Informal remarks on Lemma 9

FIGURE 4.8: THE "NUMBERED" FLOW OF THEOREMS AND
LEMMAS OF THE DYSON-LENARD PROOF

Theorems

I: Argument of Part I

1	Fisher-Ruelle, $-N^3 \le E_{min}$
2	refined $-N^2 \le E_{min}$
3	$-N^{5/3} \le E_{min} \leftarrow$ 7, L4, L3, L2
4	all particles fermions, $-N \le E_{min}$ $\leftarrow$8a, L4
*5	e^- fermions $\leftarrow$ L6, 13, 8a
	=Stability of Matter
6	electrostatics—lower bound of Coulomb energy for an arbitrary system of point charges
7	potential energy of the system bounded by each particle attracted by its nearest neighbor $(N \times N \rightarrow N)$
8	space occupied by particles, given antisymmetry$\leftarrow$L5
8a	space occupied by just two particles, given antisymmetry

Argument of Part II: Single Particle
      *9   smooth background charge ←4

Toy Calculation:
      10   single negative particle, lower energy bound in a field
      11   Sobolev-type inequalities
      12   stability for a single negative particle ←L2

      13   reduction to one-particle case ←L6, 6, L7 (which is like L2), L8, L9
[This leads to Theorem 5]

Lemmas
Argument of Part I:
      L1   particle in a Yukawa potential need not bind →2
      L2   inequality connecting wave function, its gradient, and its potential
      L2a
      L3   geometrical fact about a finite set of points within a sphere
      L4   numerical or arithmetic fact
      L5   uncertainty relation of points within a sphere and functions of them

Argument of Part II:
      L6   partition of space into disjoint suitable cubes
      L7   L2 for a cube ← L2a
      L8   comparing energies and potentials for a cube vs. a sphere ←L2
      *L9   bound interaction energy using kinetic energy of negatives and potential
                  energy of positives

FIGURE 4.9: THE FLOW OF THE DYSON-LENARD PROOF, AS LEMMAS HANGING
FROM A TREE OF THEOREMS
Theorem (L=lemma)
1 [N^3 lower bound]
2 ←L1 [a refinement, N^2]
3 ←L2 + L3 (←7, 6) + L4 [improved to $N^{5/3}$]
[Toy problem: reduction to a one-particle-
in-all-of-space problem]↓
{1-particle} 12,11,10 ←9 ←4 ←8a, 8, L5 [N bound, all particles
fermions]
**5 ←13 ←(L6, L7, L8, L9) [e⁻ fermions]
↑[good cubes; configurational
domains with uniform nearest-
neighbor separation]

NOTE: Theorem 5 is The Stability of Matter.

. . . The logical structure of the proofs of the two inequalities [roughly, relating averages of density times potential to the average of the gradient of the density times the average of the gradient of the potential times the average density itself] Theorem 11 and Lemma 9, is the same, and the logic of their use in the proofs of Theorem 12 and Theorem 5 [stability for a single particle, stability for a particle in a cube] is also the same. Only the details are more complicated for the case of the cube, and the numerical coefficients are correspondingly less precise.[57]

The proof is all about inequalities and estimates, what might be called a nice application of classical analysis. The proof has an architectonic structure reflecting the argument, an argument that has not been cleaned up much. Still, it has a certain order, driven as it is by mathematical technique and device, the physics, and the need to be didactic. Theorems may need technical lemmas, but some of those lemmas will be quite physical in their meaning (such as Lemma 9), and toy calculations are employed to make a complicated argument more understandable.

The crucial theorems are enunciated first, their connection to the needed lemmas comes later. We might unfold the sequence of presentation, to display the flow and staging and structure of implication of the argument. I call this a bunch of lemmas hanging from a tree of theorems (Figures 4.8 and 4.9).

The structure and strategy are interesting in and of themselves, in part because of the tree-like mode of organization, and in part because of how much technical material concerning the details of the lemmas is deferred in the presentation to the latter parts of the work. There are parallels between a calculation for all of space, and for space chopped up into cubes. And at several points, calculations for simplified systems, toy models, are provided to elucidate the more complicated realistic model. More to the point is that there is a structure here, on several levels, not just a series of mathematical deductions, traipsing through a jungle. That structure is seen in the subheads of the papers, in the set of implications and in the flow list, the middle structure mediating between the first and the third. And that structure is in part about the physics, in both simpler and more realistic models. I have not gone into the details of each part, but those details are also carefully structured.

Dyson and Lenard's work is not only a sophisticated piece of rigorous classical analysis (at least for a physicist); it is highly orchestrated, with a great deal of attention to making the proof readable. And there is physics that guides the presentation of the mathematics and the calculations. (Dyson says that he and

Lenard think mathematically more than physically, and that the proof is not really physics, and that they didn't understand the physics of the problem, that the proof is opaque, presented in two very long papers nobody [now] reads. It is surely the case that there is a good deal of mathematical manipulation in the proof, but I believe that the proof is still much informed by physics.[58])

The Dyson-Lenard proof is constructive, giving a numerical proportionality constant for the lower bound, and it is "rigorous," that is, requiring no perturbation theory (and I believe it is mathematically sound as well). The proof is provided in two papers (I and II, Figure 4.7), each divided into sections, each section proving some fact, sometimes providing a proof of a fact that was needed earlier to push through the argument. In Figure 4.7, I have indicated the contents of each section in brackets. The numbering of theorems and lemmas is continuous between the two papers. Sections 3 to 5 of Paper I are for improved N^α bounds, sections 7 to 9 achieve stability ($\alpha=1$) for a smooth background charge. Here N is the number of electrons. As the proving goes on, and α decreases, the proportionality constant for stability becomes larger in absolute value. For $Z=1$, hydrogen, as α decreases from 3 to 5/3, the proportionality constant rapidly decreases: -1, $-2\sqrt{2}$, -165, -1600 Rydbergs. (1 Ry equals 13.6 electron-volts, roughly the binding energy of the hydrogen atom. In units where $\hbar^2/2=1$, the mass of the electron, m_e, equals 1, and the speed of light, c, also equals 1, 1 Ry=¼.)

Paper II begins with a toy calculation of one negative particle in all of space, then in sections 4 to 6 chops up space into cubes, and then shows how the cubes themselves are stable and how space might be put together again. The proportionality constant now is -4.1×10^{14} Ry, when we might hope it to be of the order -1.[59]

Dyson and Lenard describe the argument:

> In Theorem 9 of Paper I [section 9, stability subject to a smooth charge density] we have a result that almost solves the stability problem. . . . [But] we want to prove that the fermions remain stable in an arbitrary distribution of positive classical point charges,
>
> . . . [W]e succeeded in sharpening Theorem 9 only for one negative particle at a time. We are consequently driven to an elaborate and unphysical scheme of chopping up our space into cells, each containing one negative particle (cf. Paper II, Sec. 4). We then prove a sharpened form of Theorem 9 for each cell separately, with its one negative and an arbitrary number of positive particles. Finally, we reassemble the

fragments and show that stability in the individual cells implies stability for the whole space (cf. [II] Sec. 5).[60]

. . . In our whole proof of stability, Lemma 9 is the innermost core. . . set[ting] a bound to the interaction energy between negative and positive charges, the bound depending only on the kinetic energy of the negatives and on the potential energy of the positives.

. . . a [single] particle of mass m in the periodic Coulomb potential has a ground state binding energy less than 16 Ry[dbergs], . . .

FIGURE 4.10: THE STRUCTURE OF LENARD'S REFORMULATED PROOF (1971)

I: NTC Inequality (Theorem 2, Section 3)
NTC + Cubes (Theorem 3, Section 3)
Gradient inequality (4,3)
$\rightarrow$ Section 4, more NTC inequalities, and bounds dependent on
    $\sqrt{}$(Kinetic Energy) and $\sqrt{}$(Potential Energy).

II: Electrostatic inequality (5,5)
Cubic Partitioning (6,6)

III: Lower bound for energy in terms of K_1, a Coulomb-like (C) potential energy (–,7)
$K_1 K_2 T$ kinetic energy inequality (–,8)
Gradient inequality (7,9)
Antisymmetry inequality (8,9), where Theorem 7$\rightarrow$8

IV: K_2, T inequality(–,10)
K_2 (–,10)
(K_1, K_2 inequalities) $\rightarrow$ (1,11), where Theorem 1 is the Stability of Matter.

Subsequently, in 1971, Lenard polished up the proof so that it was rather more compact and transparent, at least mathematically.[61] Now it consisted of four parts: so-called NTC inequalities, connecting integrals of densities (N), gradients (T, for kinetic energy, where the momentum is the gradient of position), and a stylized Coulomb potential, $f^2/|x|$, (C), f^2 being a charge density of sorts; electrostatic inequalities; bringing these together with antisymmetry to estimate certain integrals (T, K_1, K_2); and then bringing those estimates together to get the lower bound.

It might be said that this was more a mathematician's proof, the deeper physics not so apparent. And Dyson and Lenard knew that their estimate of the proportionality constant, E_*, – 4.1×10^{14} Ry, when the expected value is of the order of 1, is "not just a joke" but also a challenge[62]: "It is likely that any significant improvement will come from a stability proof which, at least in part, depends on new ideas."[63] (Note that 4.1×10^{14} Ry is about six-million proton masses.)

Let me emphasize once more, what these charts show is that whatever one might say about these efforts compared to later ones, it is clear they are deliberately organized, that there is a physical and mathematical story to be told here. And there is masterly technique supported by a larger vision.

In 1975 Federbush showed how the Dyson-Lenard result might be derived using easier or at least more familiar methods drawn from constructive quantum field theory, mirroring Kaufman's and then Schultz, Mattis, and Lieb's redoing of Onsager's work. The fermion wave function for the electrons is replaced by anticommuting annihilation and creation operators, and so the Pauli exclusion principle appears differently than it does in earlier derivations. And the electron-electron interaction is now expressed as a correlation or two-point function, a standard object in quantum field theory. "[O]ne calculates the energy of fermions moving in a fixed distribution of bosons [nuclei here taken as bosons], and compensates this by the repulsion energy of the bosons."[64] Federbush provides two derivations, one assuming the periodic potential, the second not making that assumption but carving up space into cubes as in Dyson-Lenard. His first paper begins by announcing three Facts, the proof of those Facts being in effect a sidebar on the way to the result. The second paper has four such Facts.

Federbush does not expect his method to much affect the value of the proportionality constant as found by Dyson and Lenard, and to calculate that number he would have to pay rather more attention to cube sizes, etc., than he does.[65] Federbush believes that his real achievement was to fit the stability of matter problem "into a framework of greater flexibility."[66] (Again, in 1964, Schultz, Mattis, and Lieb showed how the Onsager solution of the Ising model might be transferred from the realms of group theory reminiscent of crystallography and of classical analysis as Onsager understood it (say, Whitaker and Watson's *Modern Analysis*, 1902, 1927), into standard quantum field theory, a problem of many fermions, the "only" way for most physicists who nowadays claim to have read a derivation.)

Technically, the argument goes as follows: Assuming a periodic potential: Using the anticommuting character of fermion operators, Federbush calculates "N_τ" inequalities, τ being the exponent on the energy/laplacian (Fact 1); then a packing inequality for the density (Fact 2). He then divides up the hamiltonian into carefully chosen parts—the boson kinetic energy, the fermion kinetic energy, the fermion

156

interaction, a compensating difference of fermion and boson long-range interactions, a screened fermion-boson interaction, a screened boson-boson interaction—in such a way that the long range forces are controlled (effectively, Lenard's original idea), bounds each of its parts, and then adds in the electrostatic repulsion of the nuclei, here taken as bosons (Fact 3). Again, the proof is presented in terms of Facts about inequalities, and then lemmas and theorems that prove both the Facts and the stability of matter.

For the infinite-volume second case, Fact 1 is already given. One then localizes the Fermi-sea energy of the electrons (Fact 4), and then localizes the positivity of the Coulomb energy (Fact 5), again breaks up the hamiltonian into convenient parts and bounds them, one-by-one, and finally adds in the interaction between neighboring atoms (Fact 6).

> . . . space is cut into a union of unit cubes, with interactions smoothed over the edges. The kinetic energy is localized not with local N_τ operators as in field theory, but by writing the free kinetic energy as a sum of local kinetic energies in cubes with Neumann boundary conditions. . . The most difficult problem is the interaction of neighboring cubes when the cutoffs are different in the different cubes.[67]

WHY ALL THESE DETAILS?

The level of detail in this chapter is perhaps deadening, and the knowledge required to follow some of the technical arguments is that of an expert, who not only possesses at least a masters degree but also has a close acquaintance with the actual literature. Still, read for rhythm, tone, and associations, the diagrams and outlines, and the quotations, are suggestive even to the layperson of how important in these endeavors are matters of strategy, structure, and organization.

Yet my claims do depend on my being faithful to the original papers. We are describing what people actually do, how they present their work and how others encounter it. I do not believe simplified examples of the work are cogent.

How could I be wrong here? What have I been demonstrating? These proofs do have a strategically meaningful structure, indicating tactical moves. Mistaking some single point in a paper is not likely to disconfirm my argument, but it will decrease my authority with experts. Misunderstanding the flow of a proof or derivation is much more likely to be fatal to an argument that structure and strategy matter. And if my analysis is not illuminating to someone who is trying to follow these papers, in detail, then I have not been as successful as I would hope to be. The detail is crucial to the claim.

My flagging of technical passages in the text is deliberate. Again, I have endeavored to mark by *** those passages which draw the lessons from the technical arguments that will follow. In particular, the dedicated skipper is invited not to skip the initial parts of section V, or other *** passages.

NEW IDEAS AND NEW SEQUENCES: THE ELLIOTT LIEB-WALTER THIRRING PROOF

In 1975, Lieb and Thirring provided the breakthrough—in both physical understanding and in arriving at an order 1 estimate $(O(1))$, actually about -8 in natural units (equal to -32 Rydbergs, a measure of atomic energies), for the proportionality constant.[68] They found that a lower bound is provided by an average or mean-field approach for the electron density, the Thomas-Fermi model of an atom: each electron moved in a field generated by the average total charge density of all of its electrons. The crucial fact was that Thomas-Fermi atoms do not bind to each other: no molecules, no chemistry! And hence one obtains the desired proportionality-to-N of the lower bound: once one proves the stability of an atom, namely, a lower bound for its energy; and, one also shows that Thomas-Fermi estimates for N electrons and their nuclei, with altered constants of proportionality, are in fact the already estimated good lower bounds for the actual N-electron Hamiltonian of matter composed of many atoms (and so the proportionality-to-N can now be employed).

Figure 4.11 provides the logic of the Lieb-Thirring proof.[69] The abiding principle is that the Thomas-Fermi atom is a good-enough approximation to actual atoms. That Thomas-Fermi atoms do not bind to each other is perhaps less important in that approximation than the fact that they account for most of the binding energy of a single atom. The averaging in step 3d turns the electron-nucleus attraction into a Thomas-Fermi electron-electron interaction, and the nucleus-nucleus interaction into the actual electron-electron interaction. The final proportionality constant depends inversely on K', the kinetic energy bound or the sum of the eigenvalues bound. There is a conjecture of Lieb and Thirring that it should achieve its classical value; currently it is temptingly close, less than two-time "too large" in absolute value.

FIGURE 4.11: THE STRUCTURE OF LIEB AND THIRRING'S ARGUMENT IN "BOUND FOR THE KINETIC ENERGY OF FERMIONS WHICH PROVES THE STABILITY OF MATTER" (1975)

Basic insight: The energy of the Thomas-Fermi model atoms in bulk, with suitably modified constants, can be shown to be a lower bound for the actual N-electron hamiltonian, H_N. And, since Thomas-Fermi atoms do not bind to each other, their (Thomas-Fermi) N-electron hamiltonian is proportional to N. Hence, we have a lower bound proportional to N, the stability of matter.

Teller's argument for No-Binding is, roughly: adding some charge to a Thomas-Fermi nucleus will raise the potential, and so building up a molecule is harder than building up two well-separated atoms.[70]

Now, along the way, we need to bound both the kinetic and the potential energies of H_N:

(A) *Kinetic Energy Bound:* The Thomas-Fermi kinetic energy, $K \times \int \rho^{5/3}$, where ρ is the Thomas-Fermi averaged charge density of the electrons in bulk matter and K is a constant, turns out to be a lower bound, for some K', for the actual kinetic energy of N fermions (the electrons) in bulk matter. "Surprisingly," a statement about the energy levels of a carefully chosen single-particle hamiltonian H_1 (for one electron in bulk matter) can tell us about the kinetic energy of N fermions for the correct H_N.[71] It is carefully chosen in that the Thomas-Fermi form appears, and surprisingly it works as well for the sum of eigenvalues.

1. The ground-state energy of any N-fermion hamiltonian is less than or equal to its kinetic energy, T, plus potential energy, V, for any antisymmetric wavefunction, ψ:

$$E_0 \leq T_\psi + V_\psi.$$

Eventually, V will be chosen to be the sum of independent- or single-particle potentials (those of H_1) of the form $\int \rho^{2/3}$.

2. If the ground state energy is to be as low as possible, the fermions should fill up the energy levels two-by-one. The ground state energy is bounded by the sum of the negative eigenvalues of H_1,

$$2 \times \text{sneg}_1 \leq E_0.$$

3. $\text{sneg}_1 \geq -L \times \int (-V)^{5/2}$ by a general argument connecting the eigenvalues to the shape of the drum. If $(-V)^{5/2}$ is made proportional to $\rho^{5/3}$ (as promised above), then we have:

$$\text{sneg}_1 \geq -L' \times \int \rho^{5/3}.$$

Moreover, V_ψ is $\int \rho V$, and so, by our choice of V, V_ψ, too, is proportional to $\int \rho^{5/3}$.

4. Substituting (1) and (3) into (2), we get:

$$K' \times \int \rho^{5/3} \leq T.$$

(Note that T can be computed here for the correct wavefunction for H_N.) This bound is just the Thomas-Fermi kinetic energy with modified constant K'. In effect, we have an uncertainty principle connecting momenta to a functional of positions. Any such principle would have a $\rho^{5/3}$ integral (or at least something with those dimensions), by dimensional analysis, given that T is an integral of the gradient squared of the wavefunction.

(B) *Bounding the N × N term by an N term:* The electron-electron repulsion's contribution to the potential energy, an N^2 term, needs to be bounded by a term proportional to N. The Thomas-Fermi approximation for the repulsion can be converted to a lower bound for H_N's repulsion. So the total repulsion term is now proportional to N. (The "trick" here is that at a crucial point one interchanges the electron and nucleus coordinates.[72])

1. The N-particle energy, E_N is a sum of the kinetic energy (T), the attractive energy of electrons and nuclei (A) and the repulsive energy of the electrons with each other (Re) and the nuclei with each other (Rn):

$$E_N = T - A + Re + Rn.$$

2. By (A) above:

$$K' \times \int \rho^{5/3} \leq T.$$

3. We make use of No-Binding for Thomas-Fermi atoms, and the interchange of electron with nucleus coordinates trick, to get Re proportional to N.

a. No-Binding for Thomas-Fermi atoms means:

$$E_N^{\mathrm{TF}} \geq N \times E_1^{\mathrm{TF}}.$$

b. Substituting the Thomas-Fermi version of equation 1 into equation 3a, but now for "heavy" electrons of mass m/γ, and for an equal number of nuclei $Z=1$, we find:

$$\gamma \, T^{\mathrm{TF}} - A^{\mathrm{TF}} + Re^{\mathrm{TF}} + Rn^{(\mathrm{TF})} \geq N \times E_1^{\mathrm{TF}} / \gamma Z^{7/3}.$$

c. A^{TF} and $Rn^{(\mathrm{TF})}$ are functions of the nuclear coordinates, which are arbitrary. So we might make them the heavy-electron coordinates (there is the same number of nuclei and heavy-electrons). Note that such an "$Rn/e^{(\mathrm{TF})}$" ($Rn^{(\mathrm{TF})}$ with electron coordinates) is a curious term, essentially the electron-electron interaction but assuming the charge distribution is pointlike (which it is not). We need to bring in the actual distribution, and so we average over the square of the wavefunction.

d. Applying 3c, and averaging equation 3b over the system, we get:

$$\gamma \, T^{\mathrm{TF}} - 2 \, Re^{\mathrm{TF}} + Re^{\mathrm{TF}} + Re \geq N \times E_1^{\mathrm{TF}} / \gamma Z^{7/3}$$

(Note how A^{TF} and $Rn/e^{(\mathrm{TF})}$ are transformed!) Now we have an inequality for Re in TF terms.

4. Substituting (2) and (3d) into (1):

$$E_N \geq [(K' - \gamma) \, T^{\mathrm{TF}} - A^{\mathrm{TF}} + Re^{\mathrm{TF}} + Rn^{(\mathrm{TF})}] + N \times E_1^{\mathrm{TF}} / \gamma Z^{7/3}.$$

5. The bracketed expression in 4 is just the Thomas Fermi energy functional for a system with electrons of mass $m/(K' - \gamma)$, and by No-Binding we can replace the bracketed expression by $N \times E_1^{\mathrm{TF}} / (K' - \gamma)$, giving:

$$E_N \geq N \times E_1^{\mathrm{TF}} \times \{1/(K' - \gamma) + 1/\gamma Z^{7/3}\}.$$

We can minimize the braced expression, and we have our result (which turns out to be inversely proportional to K').

There are four crucial physical factors that have to be made mathematically usable: an uncertainty principle; electrons are fermions; and so the kinetic energy depends on a binding energy; and, the electrostatic inequality

An Uncertainty Principle: Parenthetically, the rigorous proof of the stability of a single multielectron atom is not trivial, and involves a fair amount of classical analysis: namely, inequalities between quantities of the sorts $(\int \rho)^3$ and $\int \rho^3$, where ρ is the electric charge density. The kinetic energy is the square of the gradient of the wavefunction and the density is the square of the wavefunction itself.[73] In particular, inequalities are employed to bound the kinetic energy by functionals of the electron density, that is, integrals of the density to various powers (namely, Sobolev and Hölder and Young inequalities).

So we relate the kinetic energy to the potential energy, a functional of the gradient of the wavefunction to a functional of the wavefunction itself: an "uncertainty principle," connecting dx/dt with dx. Again, we have to show rigorously that for a single isolated atom the kinetic energy is large enough to prevent collapse due to attractive Coulomb forces.[74]

I should note that by a virial theorem that averages the potential $(-A + Re)$ and kinetic energies (T): $2T - A + Re = 0$. One can show that: $T: A: Re :: 3: 7: 1$.[75]

Fermions: That electrons are fermions, which obey the Pauli exclusion principle, must play a crucial role in the proof of the stability of matter in bulk. Were they bosons, matter would not be stable.[76] In each of the various proofs, there are either anticommuting operators, or antisymmetric wavefunctions, or one notes that there can be only one electron of a particular spin in each quantum state. Notably, according to Lenard,

> . . . [Dyson-Lenard] did not use the Pauli Exclusion Principle to its fullest, namely that no two (spinless) particles occupy the "same 1-particle state." Instead, we only used the much weaker statement that the 1-particle ground state cannot have more than one electron in it. Actually, I recall Dyson musing about how little of the full force of the Fermion-ness of the electrons is needed for stability.[77]

In Lieb-Thirring, in order to estimate the kinetic energy bound, the electron's fermionic nature comes in as "one-particle per state." The electrons occupy and so fill the energy levels, two per level (up-spin and down-spin), much as they do for an atom as we go up the Periodic Table. Similarly, if there were N fermions in a box, their (kinetic) energies must escalate with N so that there are energy states available for larger numbers of fermions. A small box, say in effect defined by the strength of the electrical forces, with lots of fermions, has quite energetic electrons—and that prevents collapse. But just how that comes to pass (or not) in actual bulk matter depends on details incorporated into a rigorous proof.

The Shape of a Drum: More precisely, the energy available to prevent collapse is, ironically, roughly proportional the sum of the binding energies for the electrons (technically, estimated by the sum of the negative eigenvalues of a single-particle energy operator). For those binding energies are then instrumental in setting the large-enough kinetic energy bound.

Parenthetically, this eigenvalue-sum problem is very much like the problem discussed in the introduction, the problem of trying to estimate the size and shape of a drum from the tones it produces. (Again, the sum of the resonant tones or the eigenvalues is proportional to the drum's area plus a term representing its perimeter plus higher order terms.) In general, it is possible to hear a good deal about the shape of most drums. And, in the current case, what you "hear" is the stability of matter. Actually, you can "see" the stability of matter: namely, the spectrum of the atoms in bulk—reflecting the negative eigenvalues or energy levels of the atoms, which are just what you must sum to prove the stability of matter. You may see the atomic lines if the bulk is a dilute gas, or the blackbody spectrum if the bulk is condensed matter.

Technically, as for the kinetic energy bound computed by Lieb-Thirring:

First (WKB or semiclassical approximation for the "Weyl asymptotics," the shape of the drum problem): Will the kinetic energy be large enough to prevent the system from collapsing? Lieb and Thirring prove a crucial inequality connecting the sum of the negative eigenvalues of a single-particle hamiltonian ("sneg$_1$") to an integral of a power of the potential, V: sneg$_1$ is approximated by, actually is greater than or equal to $-L \times \int (-V)^{5/2}$ (recall that V is negative) for the three-dimensional case. ($L^{2/3}$ is inversely proportional to K'.) They use a semiclassical or WKB approximation, in effect one state per unit of phase space. They in-effect localize each electron state, and so compute its energy at a particular radius, so to speak.[78] (As we shall see, Fefferman shows how to do the volume counting more sharply, using refined WKB estimates. In general, the semiclassical eigenvalue estimates are not very sharp, for which Fefferman provides several examples.[79]) This is step A3 in Figure 4.11.

Second (fermions): Again, Lieb-Thirring consider a stylized multielectron hamiltonian, where the electrons do not directly interact with each other but only interact through an averaged charge density (of which they are the source), the Thomas-Fermi model. The electrons fill up the various negative or bound-state energy levels, two-by-one since there are two spin states, and so the binding energy of the stylized ground state is greater than $2 \times$ sneg$_1$. This is step A2.

Third (Rayleigh-Ritz variational principle): Noting that the ground state energy for the stylized hamiltonian will be lower than the stylized-hamiltonian energy of the true wavefunction (that is, the wavefunction defined by the correct hamiltonian, H_N), they connect the true kinetic energy to the stylized ground state energy and a $V^{5/2}$ integral as in the first step. This is step A1.

Finally, these three observations are combined to get the kinetic energy lower bound. What is remarkable is that the arbitrary independent-particle potential, V, can be adjusted (that is, made proportional to $\rho^{2/3}$) so that sneg$_1$ and the potential energy are of the same functional form: $\int \rho^{5/3}$, just the form of the Thomas-Fermi kinetic energy. Presumably, this is testimony to just how good is the Thomas-Fermi model.

$N \times N \rightarrow N$: As Thirring (1994) describes it, the no-binding feature of the Thomas-Fermi model was quite attractive, since it gave one proportionality to N.[80] But their initial work was for fixed N, and $Z \rightarrow \infty$. However, what is needed is $N \rightarrow \infty$ and fixed Z. They realized that one could bound the Coulomb repulsion of the electrons (a potential energy contribution) by an effective field, an electrostatic inequality (as in Onsager's 1939 paper).

> Since, by our inequality for the Coulomb repulsion, we had already reduced the problem to noninteracting electrons in an external field, I had the feeling that once one has a formula for the sum of the binding energies in a general external potential, one should get a lower bound for the energy. Elliott [Lieb] played more on the side of TF [Thomas-Fermi] theory and noted that after bounding the exchange correction with $\int \rho^{5/3}$ [ρ=charge density] the only thing we were still missing was to show that $\int \rho^{5/3}$ actually gives a bound to the kinetic energy. . . . With a general formula on the number of bound states below an energy E we could deduce a bound on the sum of the binding energies, which in turn showed that $\int \rho^{5/3}$ gives a bound for the kinetic energy of fermions.[81]

No-Binding means we may treat each atom as discrete, enclosed within its own ball of space. The $Z=1$ and interchange-of-coordinates trick produces "atoms" centered now on the electrons. (The Thomas-Fermi charge density is arbitrary in the No-Binding theorem, not constrained to be self-consistent.) These "atoms" are just like the hard-sphere molecules in Onsager's argument or the balls that do not touch (equals No-Binding) in Dyson's version, and so perhaps it is not so surprising that one gets the proportionality to N here.

It is remarkable how powerful is the Thomas-Fermi model, combining as it does a good estimate of the binding energy, No-Binding to other molecules, and the integral of the charge density to the right power when it is needed.

It is noteworthy how much actual physics is employed here. At least ahead of time, in no sense will the mathematics itself suggest the models. Nor will our intuitions about the Pauli principle say just how it ends up being used in a particular proof, one that may employ a good deal more physics (here, Thomas-Fermi) as well.

The proofs of the stability of matter show that there is a gap between knowing what is "really" going on in the physics or the mathematics, and making that into proof or a derivation. As I indicated at the beginning of this chapter, in the end, what is really going on is really going on only when it works in such a proof or derivation and, ideally, we can show in detail, and technically, just how it does the work. And the proof itself might well reveal crucial features, for if they were not present the result would be very different.

Again, in the end, the Lieb-Thirring proof does lend itself to a much condensed and more didactic presentation once the Thomas-Fermi model is worked out. What was formidable mathematical physics in Dyson-Lenard is now (ala Lieb-Thirring) built into other presentations and into Fefferman's more general exposition of methods of solution of partial differential equations. The physical ideas and the mathematical technology can now do their work even more transparently. My exposition has been quite detailed, teasing out a sometimes magical argument. This contrasts nicely with the emphasis I placed in the discussion of the Dyson-Lenard proof on architectonic and structure.

(Although we shall now leave the work of Lieb and Thirring and their collaborators, I should note that they, too, address some of the issues that concern Fefferman and his collaborators, more than ample evidence of which is provided in Lieb's selecta, *The Stability of Matter: From Atoms to Stars* (1991, 1997), and in Lieb and Seiringer's book, the source of our more recent estimates of the proportionality constant.)

***At the halfway point in this chapter, it may be useful to indicate the lessons we might draw from these descriptions. Namely, a proof or a derivation is an organized endeavor, the organization itself indicating what is really going on in the proof (at least as it is understood at a certain time). And, the sequence of papers that provide more and better proofs and derivations are also to be organized as a staged endeavor, again to indicate what is really going on. Such structure indicates matters both of strategy and of tactics. And it allows the reader of the work (including the person who is writing it) to see what is going on, not only in the proof or proofs, but also in the world that is constituted by mathematics or by physics or both. Rigor is not merely for show, for it displays crucial features of the world. Note that "rigor" as employed by mathematical physics may not be the same as the mathematicians' criteria.

IV

CHARLES FEFFERMAN'S CLASSICAL ANALYSIS

> I had the idea that the Lieb-Thirring conjecture* [that a factor of 4π should not be there] might actually be harder than the [direct] calculation of E_*. (Charles Fefferman[82])

Charles Fefferman is primarily a mathematician, not a physicist. (Again, he is known for his contributions to a variety of fields in classical analysis.) In a series of papers with collaborators he has brought to bear his considerable technical talent to these problems of mathematical physics, taking off from Lieb and Thirring. Leaving out all work on magnetic fields, and on relativistic effects, an important theme, Fefferman's endeavors concerning the stability of matter might be grouped as follows (Figures 4.4, 4.5, 4.6):

1. Stability of matter, two different proofs, one ala Lieb-Thirring and one ala Fefferman: 1983.

2. Matter is made up of atoms and molecules: 1983. And for an appropriate temperature and pressure, matter is a gas of those atoms and not just a plasma of electrons and protons: 1985, modulo #3 below. This begins as a refinement of Lieb-Thirring, and then has a life of its own.

3. Better estimates of the stability of matter with a good constant, E_*: 1986, with de la Llave and Trotter. Fefferman needs the better estimates to show that there is a gaseous state of atoms. But, so far, he has been able to develop the estimates only for large atoms. (The Thomas-Fermi model is exact in the $Z=\infty$ limit, and is not as good as we need for low Z, although even there it is remarkably

---

*Recall that the Lieb-Thirring conjecture is for the constant, L, for the sum of the negative eigenvalues in terms of L times an integral of the potential to the 5/2 power—an intermediary in the proof of the stability of matter and crucial for lowering the estimate of E_*, the proportionality constant in the stability of matter.

good.) In effect, he is trying to cut Lieb-Thirring's original estimate for the proportionality constant by a factor of perhaps fifteen, much as Lieb cut Dyson-Lenard's by 10^{13}, so that he can account for the gaseous state. (Lieb and his collaborators, and others, have their own program to do some of this as well, for which see note 68.)

4. Rigorous estimates of the ground state energy of a "large" atom: 1990–1997 with Seco; and, with Córdoba and Seco, rigorous estimates of shell effects.

Fefferman's work comes out his research in partial differential equations, concerning "microlocal analysis" and "pseudodifferential operators." Microlocal analysis is a way of splitting apart an operator or breaking up a space into smaller tractable parts. It is easy to solve the equation on each part, and then one puts the parts back together. Here, phase space is divided into small cubes, effectively quantization. Pseudodifferential operators are a way of dealing with singular potentials (e.g. $1/r$, Coulomb[83]) and so not so nicely behaved differential equations by integrating over the singularity (employing the mathematical device of distributions and test functions).

***Again, what we shall be showing is how a complex and very difficult problem, and an arduous and extensive task, becomes understandable through sequencing and staging. That sequencing and staging is substantive. It is about the world and how it is structured, not just about the proof or the derivation. And of course, it is rhetorical, about making a convincing argument, as well. (Historically, we have been tempted to see such sequencing and staging, considered as a structure, as being mirrored by a structure in the world's organization: the world as God as a human body, or the world as a hierarchy or tree-like structure. Perhaps we are still tempted to do that.) For my purposes here, what is crucial is that the structure is about the world, where by "about" I mean that it reveals features of the physical world that are important and significant. And the devices and techniques are about the world, as well. As for mathematics, there is a mathematical world, these devices and techniques are about that world.

***The dedicated skipper is encouraged to go to "The Lessons, So Far" when the following sections get too heavy.

1. THE STABILITY OF MATTER

In a paper entitled "The Uncertainty Principle" (1983), Fefferman shows how one is to break up phase space into the right size and shape of cubes so that a differential equation is solved or diagonalized on each cube separately, modulo small errors: namely, $Hf=af$, where H is the differential operator on the cube, f is a solution, and a is the eigenvalue.[84] (Recall the notion of bending cubes in phase space.) Fefferman enunciates a principle that in counting eigenvalues as one per cube of phase space, one has to figure out how many whole cubes fit into the space (rather than just dividing the space's volume by the volume of a single cube). A spiky space might not accommodate any whole cubes. Or, depending on its shape, an infinite volume of phase space may only hold a finite number of whole cubes or physical states, and so the number of states and the eigenvalue sum will be finite, not infinite. (For, *pace* infinitesimal eigenvalues, an infinity of cubes would give an infinite sum.)

Fefferman calls this fitting-in rule an "uncertainty principle," in honor of Heisenberg's principle that each quantum state occupies at least one unit of phase space. But now that unit has to be an integral whole. It is "a sharper form of the [Heisenberg] uncertainty principle."[85] Inverting the usual one-liner (which is about combinatorial models of probability distributions), here our task is to understand how to put boxes into or onto balls.

The Lieb-Thirring proof depended on estimating the number of bound states in a system, the number of negative eigenvalues of a differential operator (the Schrödinger equation), in order to estimate and bound the kinetic energy of the electrons in matter. (Again, fermions fill up the levels one by one—actually, two by one, given the electron's spin. The bound on the total kinetic energy of the multielectron atom is, approximately, proportional to the sum of the single-particle energies, the bound states or negative eigenvalues of a stylized single-particle hamiltonian.) Hence, Fefferman's uncertainty principle should allow for a more precise bound by means of better "volume counting," to get the number of states for a given energy and volume of actual phase space. This turns out to be the next order corrections to the optical or WKB approximation.

In the middle of this lengthy survey paper, Fefferman first redoes the Lieb-Thirring proof of the stability of matter (his Theorem 8): namely, fermions and volume counting, Thomas-Fermi no binding, and the electrostatic inequality. Still, Fefferman will eventually need a sharper estimate of the proportionality constant, E_*, to justify that not only is matter stable and "that matter does not merely take up a definite volume, but is actually made of atoms of a fixed size,"[86] his endeavor in this paper—but that those atoms form a gas under a suitable range of temperature and pressure (or chemical potential), what becomes the 1985 paper. And he wants to show that atoms bind into molecules.[87]

In order to prove matter is made up of atoms, Fefferman has to specify what he means mathematically and precisely by an atom: In effect, the electron density is

167

restricted to a cube; the density is greater than some constant for a fixed portion of the cube's volume; there are no serious singularities of charge density in that cube, or that the kinetic energy is bounded; and, at least one nucleus, and at most a fixed number of nuclei belongs in a cube. Having given this formally, he says: "Although lacking in fine detail, these properties capture the idea of an atom as a ball of definite size, carrying a nucleus and some electrons, and possibly sharing its electron cloud with its neighbors."[88]

FIGURE 4.12: STEPS IN FEFFERMAN, 1983, THEOREM 9
PROOF OF THE ATOMIC NATURE OF MATTER

—geometry of cubes;
—estimates of potential energy;
—fermions, exclusion principle, microlocal means one state per cube;
—uncertainty principle, to connect the gradient term or kinetic energy to the bilinear term or potential energy, and hence volume counting;
—then, a brief excursion to provide yet another proof of the stability of matter; and,
—then, a proof of the existence of bound states (atoms), and finally the existence of chemical binding (molecules).
[—Theorem 11: The Thomas-Fermi approximation for the kinetic energy works well when we have in effect pairwise disjoint cubes—hence, no binding.]

NOTE: The problem that remains is the existence of hydrogen as a (di)atomic gas, a problem in quantum statistical mechanics.

Fefferman then goes to work to prove the atomic nature of matter (his Theorem 9) with some steps that are by now quite familiar, and others making use of comparatively proprietary mathematical technology. Fefferman's second (1983) proof of the stability of matter, the brief excursion I refer to in the figure, makes essential use of the fact that the hamiltonian is operating on antisymmetric wavefunctions, and he divides up the hamiltonian much as Federbush did in order to bound its various parts.

Again, these proofs are a small diversion in a much larger paper on general techniques for solving partial differential equations. There are two interesting conclusions, perhaps not unexpected once we have experience with the stability of matter problem. First, "apriori estimates" of solutions to partial differential equations (in effect connecting a solution's growth to its derivatives, the latter given by the equation) and approximate solutions to partial differential equations (valid over small regions) are rather more related than we might have believed. Sharp-enough estimates allow us to solve or diagonalize the equation by restricting our consideration to small enough regions, and hence lead to an approximate solution. Construction of a solution depends on apriori estimates of local problems, so we know how many terms we need to include.[89] So, if we can give a good bound on the stability of matter, then we might give an account of periodic solutions, and vice versa (recalling one of Dyson's observations).

Second, we have connected balls or cubes, and ways of fitting them into the phase space through volume-preserving (or, canonical) transformations that suitably deform them, with estimates of the solution of the equation on or within each cube. The diagonalization or solution is by means canonical transformations, "fourier integral operators," "bending" the operator or the cube so that it is diagonalized there. And so we convert differential operators into multiplication by a scalar, as in fourier analysis more generally, the derivatives bringing down multiples of the exponent of exp ipx. In effect, a differential operator, such as the Schrödinger hamiltonian, is replaced by a weighted sum of its local averages over the phase space.[90] The difficult problem is to get the main terms to dominate over the errors, in effect to choose the right size and shape of bent or distorted cubes or balls, and to match solutions sufficiently well at the boundaries—so leftover parts are not so significant in toto.

Technically, along the way one needs to cut up the space and the operator into small enough pieces so that the operator is diagonalized on each piece (in effect it is constant there); or, one needs to bend the cubes or the microlocal operators so that they best fit into the phase space. There are three stages of cutting and bending:

Level I: Cutting all operators at once, in big pieces, modulo lower-order errors. [Microlocal analysis]

Level II: Cutting a single operator into smaller pieces modulo a lower-order error. [Taking into account the size of derivatives at different places, Calderón-Zygmund decomposition]

Level III: Cutting a single operator into small enough pieces modulo a one-percent error. [Unit cubes in phase space under a canonical transformation][91]

The technology of microlocal analysis and pseudodifferential operators allows for the partitioning, while bending is performed by canonically transforming phase space or the microlocal operators, so preserving phase space volumes. The latter process of canonical transformation is performed using fourier integral operators

169

($\int e(y,\xi)\ \hat{\psi}(\xi)$ exp $iS(y,\xi)\ d\xi$, e is a weight, $\hat{\psi}$ is a fourier transform) which in effect make the canonical transformation be represented by an optical phase shift (iS, S being the Hamilton-Jacobi principal function) leading to a WKB approximation, the trick being that in such an oscillatory integral only the stationary-phase parts add up while the rest cancel out.

***Fefferman's language and mathematical technology would seem to be driven more by pure mathematics and its generic techniques than by the physics. Physicist Thirring wonders if all the work on pseudodifferential operators and the like will really pay off.[92] For in Thirring's work with Lieb, what paid off was more physics, and then the mathematics made it possible to redeem those physical insights. (And in Bach's, and in Graf and Solovej's improvements on Fefferman-Seco, I sense that the physics and the specific techniques of mathematical physics again provided crucial guidance.[93]) And Lieb and Seiringer eventually offer a quite mathematical proof, as we mentioned earlier.

The other distinctive feature here is Fefferman's particular account of fitting boxes into/onto balls. Problems that involve infinite volume limits almost always demand that we figure out a way of dividing up all of space into balls or cubes or whatever so that in the limit, there is a vanishingly small contribution to the quantity of interest due to the interstices between the balls or cubes.

There is in this case, another sort of dividing-up problem. In order to solve a partial differential equation, one wants to solve it on small enough regions of phase space, where in a good approximation the equation is comparatively simple and the solution is slowly changing. Or, in order to figure out the number of eigen-solutions to an equation, one has to find out how actually commodious is that phase space, how to account for its volume in the right way.

2. MATTER AS A GAS

Given the stability of matter, we have yet to prove that matter might be gaseous for suitable temperatures and pressures. This is a problem in statistical mechanics. The technical problem here is making sure that the mathematical limits that would define the thermodynamic state actually exist.[94] Practically and technically, the major requirement is that the free energy per particle be more or less monotone decreasing as the volume goes to infinity, and that changes in the free energy are small for small changes in the volume. Then the suitable bounds will be reached rather than danced-around.[95]

The "gruesome details of the actual proof"[96] of the atomic and molecular nature of matter, say as a gas of atoms (ideally, as a gas of hydrogen molecules) are provided in a 1985 paper.[97] Within suitable limits, "there exist a temperature and density such that on a sufficiently large box the Gibbs measure [the canonical ensemble of statistical mechanics, a system at a fixed temperature and fixed number of particles] describes a gas of hydrogen atoms."[98] This is an extension of the stability of matter, since it requires: (1) that each ball if not empty have just about one nucleus and one electron in it—namely, the vast majority of electrons pair up with protons to form putative hydrogen atoms, the 1983 problem; and, (2) that we have a gas of atoms.[99]

The details are extensive, with suitable deltas and epsilons. In particular, the boundary conditions have to be fixed up after having sliced the world into balls or cubes, so that the boundary contributions are in the end comparatively small or negligible. One starts out with the statistical mechanics partition function for a single ball, makes some Coulomb estimates, and figures out how to divide space up into various-sized balls containing balls, in effect a Swiss cheese as in Lebowitz and Lieb's work. For bounds, one develops a suitable comparison with an "exploded" system in which there is no interaction among the balls. To then get at the inter-ball interactions, one uses an ensemble of Swiss cheeses, in effect to average over all possible configurations of balls, so drowning out peculiar and rare configurations. Such a "brief summary is inaccurate and oversimplified," the actual technical details being rich and delicate.[100]

Again, Fefferman's achievement is modulo a suitable technical definition, that of a gas of atoms, one that proves mathematically tractable and physically realistic.

I think it is fair to say that any probability measure $d\mu$ [of the state of the statistical system] satisfying [technical criteria] (A), (B), (C) for suitable ε, R agrees with our intuitive notion of a gas of hydrogen atoms. In particular, (A) says that atoms form, (B) says that the atoms behave like hydrogen atoms in their ground state, and (C) says that distinct atoms behave nearly independently, as in a gas.[101]

So, criterion (A) is:

Except for a rare event with probability less than ε, we find that all but at most $\varepsilon \times N$ of the particles belong to R-atoms [namely, the electron and proton are within R of each other].[102]

Of course, such an abstract and precise definition is arranged so that what we can actually prove using it is in accord with what we want to show, *and* it reflects our everyday physical intuitions as well.

Still, as indicated earlier, Fefferman's atomic gas proof is also modulo a very good estimate of the proportionality constant for the stability of matter, one we know empirically but for which we do not so far have a rigorous mathematical proof. This estimate is crucial for the very first part of the atomic-gas proof: that a single ball will most likely, with probability $1-\varepsilon$, have no atoms, and of those balls which are occupied they will most likely, $1-\varepsilon$, have just one atom, for a particular range of temperature and density or chemical potential. Is there a value of the chemical potential consistent with the lower bound for the binding energy per hydrogen atom in bulk, E_*? The answer is, Yes, if $|E_*|$ is small enough. In effect, one wants the binding energy per hydrogen atom (in bulk) to be such that the Boltzmann factor, $e^{-\beta E}$, is dominated by no occupation of a ball, and the rest of the terms be dominated by one-atom occupation.[103] This turns out to require that a rigorously proven $|E_*|$ be less than about ½, in atomic units. (1 Rydberg, the first approximation to the (absolute value of the) binding energy of a hydrogen atom, equals ¼ in these units.) The binding energy of an isolated hydrogen molecule is about -0.3/atom and does not change much for hydrogen in bulk.[104] Recall Dyson and Lenard's -10^{14} and Lieb and Thirring's original -6 to -8. So the stability of matter with a good proportionality constant and the atomic/gaseous nature of matter are intimately connected. It would seem that the most recent proofs (given in Lieb and Seiringer) suggest that the best $|E_*|$ they find is 1.4, still about three times too large for Fefferman's purposes.

Again, in general the absolute value of the binding energy per atom in bulk is only slightly larger than the binding energy of the atom itself. We have excellent numerical estimates for the binding energy of an atom from Hartree-Fock calculations (although the errors are not rigorously proven), and even the asymptotic form proved rigorously by Fefferman-Seco is quite good.

3. STABILITY OF MATTER WITH A GOOD CONSTANT

a. *Fefferman 1986*: The basic idea of Fefferman's 1986 derivation of "the stability of matter with a good constant," his Theorem 1, is to note that:

> There is a remarkably delicate balance between the positive and negative terms in the Coulomb potential. If we would replace the negative terms $-1/r$ in the Coulomb potential by, say, $-1.000001/r$ and leave the positive

terms alone, then the next mutual attraction of far-away particles would dominate the interaction. Ordinary matter would then implode.[105]

Second, antisymmetry is necessary if there is to be such a bound:

> . . . the assumption of antisymmetry of our wavefunction ψ. Therefore the kinetic energy grows large if many of the coordinates are close together. So kinetic energy keeps particles apart if the wavefunction is antisymmetric. This saves the system from collapsing. We need to write down the above intuition in a precise, quantitative form.[106]

So far, none of this is news. The task is to find a way of getting a much better value of E_*.

One of Fefferman's collaborators on the numerical computation aspects of the derivation, Hale Trotter, provides a summary that is the kind of gloss that is given in a talk, but not often found written down:

> [T]he idea is to express the total energy as an integral over all balls (of all sizes) in $\mathbb{R}^3$, and then for each ball that contains nuclei to assign its energy in equal shares to all nuclei in it, and for each ball that contains no nuclei to assign its energy to the nearest nucleus. The lowest energy assigned to any nucleus is obviously a lower bound for the average energy per nucleus. . . . [A] lower bound for the energy contributed by [the nearest balls to a nucleus] . . . can be obtained in terms of the sum of the negative eigenvalues of a quadratic form [essentially an approximate hamiltonian] . . . which is essentially the part contributed . . . to the energy for a single electron in the field of a stationary nucleus of charge Z.[107]

Of course, such an account does not tell you, even schematically, what to do mathematically, or just how the physics will make its appearance in the proof.

Technically, one first computes the kinetic and potential energies in a ball that has many atoms, which is comparatively easy, and one integrates over all possible balls to determine the energy. Certain equalities are invoked, which "come as no surprise to the fourier analyst, and an elementary proof comes from the fourier transform," much as the one for $1/|x-y|$ used for the potential energy, and they are eventually the sources of factors like $72\pi/35 = 2.06\pi$ and $18.75/2\pi$.[108] (Namely, for the kinetic energy, $-\Delta$, the expectation value $(-\Delta\psi, \psi) = (18.75/2\pi) \times \int \|\psi - L_\psi\|^2/R^6 \, dz \, dR$, where L_ψ is a linear approximation to ψ on a ball, z being the location in space of a ball of radius R.) The equality may be seen to say that the actual kinetic energy is well approximated by the kinetic energy defined for a ball. One then allocates a ball's energy equally among its nuclei. Then one invokes a potential energy inequality, reminiscent of Onsager's insight; and a kinetic energy inequality, essentially the exclusion principle. (The kinetic energy inequality demands that the number of electrons in a ball be greater than 8 if the kinetic energy is to be large enough, which eventually leads to the estimate for E_* being for $Z \geq 40$.) Not surprisingly, the distance of a nucleus to its nearest neighbor appears, and that will allow one to sum N single-particle energies rather than $N \times N$ interaction energies. Finally, an equation for a one-nucleus many-electron atom, eventually further simplified to noninteracting electrons, and so a one-electron problem, is solved as an eigenvalue problem using provably exact numerical methods, to get a sum of the negative eigenvalues, as usual what is needed here for a bound.

Again, what is notable is that, in the end, some rigorously correct numerical approximations must be computed, namely with provable errors, and here Fefferman, de la Llave, and Trotter have developed techniques for proving that such approximations are rigorous, employing symbolic manipulation programs, and using interval arithmetic to account for roundoff errors (where a number is known by its range of values, say from a to b, and $[a,b] \times [c,d] = [ac$ or ad, bd or $bc]$).

Mean field methods (Thomas-Fermi, Hartree-Fock), per se, which summarize the effects of the electrons on each other by an averaged charge density they all experience, do not play a role here. The big problem is choosing the right balls, allocating the kinetic and potential energies, and averaging in the correct fashion. In the end, here the provably exact or correct bounds for the binding energy per atom are a bit more than two-times too large (in absolute value), compared to the computed values for the binding energy for an isolated atom from approximate Hartree-Fock calculations (which are quite close to the actual binding energies of an atom, and to what we expect to be the binding energy of an atom in bulk). The Hartree-Fock calculations are not rigorously bounded, at least in Fefferman's terms. The provable bounds need to be a bit less than two-times too large (assuming that the binding energy per atom in bulk is only slightly larger in absolute value than the binding energy of an atom) if they are to do the needed work on the gaseous nature of matter problem. And they need to be extended from atomic number $Z=40$–86, to $Z=1$.

b. *Fefferman and Seco, 1990–1997*: In 1990 Fefferman and Seco announced a rigorous proof of the asymptotic formula for the ground state energy of a large *isolated* atom.[109] (For this proof there are no numerical solutions as there were in the 1986 E_* calculation for the lower bound of the binding energy per atom *in bulk*, except to check to be sure that one has a good approximation for the sum of the negative eigenvalues, the "aperiodicity condition," about which more shortly.) This is a rigorous derivation of the Thomas-Fermi/Lieb-Simon/Scott+. . . /Dirac-Schwinger *formula* for the ground state energy, that series in Z ($Z^{7/3}$, $Z^{6/3}$, $Z^{5/3}$, $O(Z^{5/3-1/a})$) we have seen earlier.

FIGURE 4.13: OUTLINE OF THE FEFFERMAN-SECO ARGUMENT
("ON THE DIRAC AND SCHWINGER CORRECTIONS")

1. Upper bound provided by "Hartree-Fock" wavefunctions:

$$E_{hf} = \langle H_{TRUE}\Psi_{hf}, \Psi_{hf}\rangle \geq E_{TRUE}(Z),$$

(Equation 22 of Fefferman-Seco, "On the Dirac and Schwinger Corrections").

2. Lower bound: Improved Lieb pointwise inequality (Eq. 32), leading to a proposed lower bound hamiltonian, H_{LB} (Eq. 39), in which interactions between particles are a small perturbation.

3. (41, the H_{LB} inequality) $\rightarrow$ {$E(Z) = \langle H_{TRUE}\Psi, \Psi\rangle \geq E_{LB}$ (Eq. 39)}

Now, to prove (41):

4. Assume, for the moment,
        {(42,Thomas-Fermi) $\rightarrow$ (43,TF/Ideal-gas-correlations)},
namely, (42)$\rightarrow$(43). Then:

$$\{(42) \rightarrow (43)\} \rightarrow (41)$$
$$\& \text{ NOT}(42) \rightarrow (41)$$

5. Now, prove (42) $\rightarrow$ (43), using free particles in a box.

Again, subsequently Bach, and Graf and Solovej, find the same series rather more directly with better bounds, but they do not compute the coefficient of the crucial $Z^{5/3}$ term. Dirac, in part, and Schwinger had provided more physical derivations. Schwinger had derived the coefficient and the order of the error, but not rigorously.[110] The rigorous derivation will require the bulk of Fefferman and Seco's endeavors, the improved WKB-asymptotics. What Bach, and Graf and Solovej do find is that the mean field solution given by Hartree-Fock or Thomas-Fermi models is very close to the full quantum mechanics for a large Coulomb system.[111]

Fefferman and Seco derive the Dirac-Schwinger term ($Z^{5/3}$) and find the exact coefficient for that 5/3 term as well as a value for a.[112] It can be shown that the $Z^{5/3}$ term represents the electron-electron correlations (Dirac) and the semi-classical

175

one-body asymptotics (for the strongly bound inner electrons, Schwinger). Fefferman-Seco provide a refined computation of the eigenvalue sum that recurrently appears in this problem. One reduces the partial differential equation to an ordinary differential equation employing spherical symmetry and separation of variables (a strategy not available for molecules), and then using quite refined fourier transform methods, also called semiclassical or WKB asymptotics, or geometrical optics with corrections, one estimates and sums the eigenvalues. Much of the proof involves generic ways of improving the WKB estimates. The main structure is roughly as follows: "Our idea is to compare the actual hamiltonian with a weakly interacting one, and then control [or set bounds on] the weakly interacting hamiltonian by using our theorem on free particles [in a box, the periodic boundary conditions to be removed later]."[113] Recall that Lieb-Thirring compare the actual hamiltonian with a stylized non-interacting one.

What do Fefferman and Seco do? They use lots more physics than in Fefferman's earlier work, namely the Thomas-Fermi (ala Lieb and collaborators) and Hartree-Fock methods. As indicated, they develop more accurate approximations to the eigenfunctions and the eigenvalues; and they are now being guided by a physically motivated formula for $E(Z)$. In the end, they use approximate wavefunctions with an exact hamiltonian to estimate the energies.

As for the aperiodicity condition: Again, the number of negative eigenvalues or bound states is, roughly, proportional to the phase space volume enclosed by the boundary.[114] However, this is not a good approximation if there is an infinity of eigenvalues (or periodic solutions) near or at the boundary (the "zero energy orbits").[115]

This problem may be seen to be equivalent to counting the number of lattice points enclosed by an arbitrary plane figure (say we are in two dimensions). So, for example, if we have N fermions in a box, the lattice points of the laplacian's eigenvalues (the kinetic energy) are arrayed rectangularly, and the problem becomes one of estimating the number of such points (and hence the largest eigenvalue) inside a curved plane figure. There will be incremental jumps as one enlarges the surface, but if the points are sparsely located in space, small changes in area will lead to small changes in the number of enclosed points. If however the shape is rectangular (so there is "zero curvature"), then small changes in volume can have very large changes in the number of enclosed points, for you might then include a whole new row of lattice points. The aperiodicity condition avoids just this fact by making sure that the eigenvalues are not so nicely distributed with respect to the enclosing surface ("non-zero curvature").

c. *Córdoba, Fefferman, and Seco, 1994-1996*: It is expected on more general grounds that the $Z^{-1/3}$ series has to break down.[116] Fefferman-Seco have a conjecture for the next term in the asymptotic series, due to atomic energy oscillations or shell effects. Córdoba-Fefferman-Seco prove that conjecture. It leads to a term proportional to $Z^{3/2}$ modulated by a Z-dependent oscillating term. They have to show that this term is large enough to dominate the $O(Z^{5/3-1/a})$ term, and that requires that $a<6$ (that is, $5/3 - 1/a < 3/2$), and that the lower bound of the $Z^{3/2}$ term be large enough. They also need to show that the residual error is $O(Z^{3/2-\varepsilon})$.

***THE LESSONS, SO FAR

I should note that in pure mathematics, there is surely no physical meaning that is probative. But I do not believe my description is restricted to mathematical physics. In pure mathematics there is always a range of examples and cases, relatively concrete, and related theories, for which your mathematical work must account and incorporate.

Structures and sequences of proving are in fact not only instrumental but also meaningful, physically and mathematically. Moreover, some details and particulars determine whether a derivation goes through or not, and many details are crucial to understanding the strategy or meaning of the proof. And, within a single paper or a series of papers the actual presented sequences of proving, at the various levels of organization or of structure, all matter enormously. For an exposition must lead the reader (and the writer, too) to conviction, that what is being proved is actually proven. And earlier work, by the author and by others, allows later work to be more focused and effective.

Of course, parts of a complicated derivation may well be improved in time, in no particular order. So it may turn out that we have svelte sections and others that can be charitably described as portly. A hotch potch, reflecting the cumulation of efforts by a variety of workers, will, hopefully, be eventually presented in a rather more balanced and slim version. And perhaps a new approach will simplify and clarity the proof. This is our hope, and in time it may be fulfilled. But even the most sophisticated technology, no matter how svelte or portly, may not yet yield the required result.[117]

Again, we might ask: Why all the detail, when there is surely not enough detail here to follow the 800 page argument in detail? I do hope to give a feeling for the structure of the argument. And I want to suggest that as one gets closer to the argument, the decipherment of this cathedral continues to prove of interest, both architecturally and theologically. Ideally, the derivation's structure and its meaning are everpresent. We are not just moving around stone blocks or symbols.

MORE ON FEFFERMAN-SECO

First, why are we here? Recall, that Fefferman needs stability of matter with a better constant to show that atoms form ("the atomic nature of matter"), a problem in classical statistical mechanics: Lieb and Thirring have shown that a reasonably sharp lower bound is provided by a functional form that turns out to be the binding energy given by the Thomas-Fermi model of atoms with modified proportionality constants. And, since Thomas-Fermi atoms do not bind with each other, we have the proportionality to N we need. Still, to achieve a proof that we have atoms, we need a sharper proportionality constant for the binding energy per particle. The earlier effort to determine E_* directly was not entirely successful. So we might return to the strategy of doing a better job on rigorous estimates of binding energies based on the Thomas-Fermi model, but now enhanced with better estimates of electron-electron interactions. (The statistical mechanics of the atomic state of matter has already been taken care of by Fefferman's 1983–1986 papers.) On the way to a better E_*, for bulk matter, we might examine an isolated atom.

Technically (yet here still heuristically), Fefferman-Seco evaluate the exact hamiltonian with reasonable wavefunctions so as to get an upper bound and so set a limit on the error term.[118] They use what they call "Hartree-Fock" wavefunctions, that is, an antisymmetrized product of individual-particle wavefunctions (so when you exchange electrons the sign of the wavefunction changes, as we expect for fermions). The latter individual-particle wavefunctions are for an averaged potential (the Thomas-Fermi model of an atom).

Now to get a lower bound, so that the estimate can be fixed to $O(Z^{5/3-1/a})$, previous rigorous calculations by their limitations necessarily missing the $Z^{5/3}$ term. Employing the Thomas-Fermi density, Fefferman-Seco estimate the sum of the negative Thomas-Fermi eigenvalues for single particles in order to compute an improved estimate for the charge density (much as you can hear the shape of a drum) and the electron-electron correlations. Then they determine the Dirac-Schwinger coefficient in the formula. One uses the already-improved sum of the negative eigenvalues to get a better estimate of the Hartree-Fock energies, and the improved density to relate the Hartree-Fock energy to the true energy. More rigorously, one has to take into account the interactions of N particles.

Toward the very end of this long paper (pp. 163–164 of 185 pages), the capstone of all the preliminary papers, they are ready to finally compute the ground-state energy. They make suitable substitutions of the various results from this paper, and WKB results from their earlier papers whose proofs are "long and complicated." They then "recognize the [already known] Scott, Schwinger, and Dirac corrections to the Thomas-Fermi energy."[119] Now, at the end of this rigorous calculation, each of the physically meaningful terms appears as such.

But such talk as I have provided here is surely not enough. What follows is a bit more detailed version of just some of the calculation, demonstrating the importance of staging and organization, finding provable parts, and breaking down a problem into tractable subproblems. And having done so, there are sub-subproblems, new techniques, and necessary due care along the way to control the various errors.

***Again, in order to rigorously obtain $E(Z)$ to within $O(Z^{5/3-1/a})$, there would appear to be no simple pathway, at least yet. The quantum mechanics of electron correlations is hard. Estimates have to be systematically controlled, problems broken down into more tractable subproblems, techniques from mathematical analysis need to be developed, and devices galore are invoked as part of the legerdemain. At the same time, the staging structure hides technicalities, which then appear at a greater level of detail, so that we can focus on well-defined matters at each level of detail. That we can do so, that the problem can be regionalized, so to speak, is notable. For particle physicists, this often becomes an account the hierarchy of forces in Nature, effective field theories, perturbation expansions, and the like. But here the hierarchy is as much a matter of the logic, the features needed at each level of the argument. Such staging, sequencing, and hierarchizing is a remarkable achievement.

THE FEFFERMAN-SECO ARGUMENT: A TECHNICAL INTERLUDE

Here, I summarize and paraphrase some of the introduction to Fefferman-Seco, 1994 (reprinted in Appendix B), where they provide an account and overview of their derivation.[120] Even if the reader does not choose to follow the argument in detail, what is striking I believe is the strategic and structural complexity and ingenuity, not to speak of the technical and tactical moves. Motivations and strategy are sometimes physical, sometimes mathematical, and sometimes both. Step 4 makes use of the logical fact that a statement (41) is true if it is derived from both another statement (42) and that statement's negation (NOT (42)). Moreover, delicate comparison systems are needed to achieve the inequalities.

Following Figure 4.13, we have:

179

1. The Hartree-Fock (HF, or hf) energy, namely the true hamiltonian evaluated for the HF wavefunctions, derived from an antisymmetric product of Thomas-Fermi individual-particle wavefunctions, gives an upper bound to the exact energy. Namely,

$$E_{hf} = \langle H_{TRUE}\Psi_{hf}, \Psi_{hf}\rangle \geq E_{TRUE}(Z)$$

(Equation 22, Fefferman-Seco).

2. In order to fix the true value of the energy to within $O(Z^{5/3-1/a})$, one needs a good lower bound as well. Lieb's "pointwise inequality," converting $N \times N$ to N, needs to be improved, taking into account short-range electron-electron interactions as a perturbation (long-range interactions are screened). The pointwise lower bound is Eq. 32:

Coulomb Energies $\geq$

Σ Individual-particle Energies $- E_0$ [Thomas-Fermi self energy]

$+ \Sigma K(x_i, x_j)$ [short range interactions] $+ O(Z^{1/a})$.

This is what is crucial to obtaining a better lower bound. One now defines a lower bound hamiltonian, H_{LB}, in which the right hand side of (32) replaces the Coulomb energies in the true hamiltonian, so that:

$$H_{TRUE} \geq H_{LB} \text{ (Eq. 39).}$$

What we shall want is that the Hartree-Fock energies differ only slightly from the lower-bound energies $(E_{hf} - E_{LB} \leq O(Z^{5/3-1/a}))$, and hence we have the lower bound we need. The crucial step is to prove a rigorous inequality for the lower-bound hamiltonian: $H_{LB} \geq \ldots + C Z^{5/3-1/a}$ (Eq. 41).

3. In sum, if a rigorous inequality (Equation 41) for the lower-bound hamiltonian is true—namely for the hamiltonian based on (32), evaluated for any antisymmetric wavefunctions—recalling that a hamiltonian is lowest for the true wavefunctions, its own eigenfunctions—then we can have the lower bound we need for the true hamiltonian and the true wavefunctions, with a suitable error. That is, given, $\langle H_{LB}\Psi_{antisym}, \Psi_{antisym}\rangle$, then

$$(41, \text{ the } H_{LB}\text{ineq}) \rightarrow \{E(Z) = \langle H_{TRUE}\Psi, \Psi\rangle \geq E_{LB}\}.$$

To show that (41) is true, we "study wave functions for which the [single-particle-Thomas-Fermi-hamiltonian energy] is nearly as low as possible."[121] We shall employ inequality (42), which expresses an upper bound on the sum of the individual-particle, approximate Thomas-Fermi energies. (41) follows from (42) and its denial; that is, (41) follows from NOT (42), as well: and so (41) is true. The task becomes:

$$\{(42) \to (43)\} \to (41); \text{ and, NOT}(42) \to (41).$$

4. Now, assuming $\{(42) \to (43)\}$, we prove

$$\{(42 \text{ or Thomas-Fermi}) \to (43, \text{TF} + \text{Ideal-gas-correlations})\} \to (41).$$

Again, (42) is an inequality for the Thomas-Fermi energy. (43) bounds the electron-electron interactions in terms of the Thomas-Fermi correlation and the correlation of an ideal gas. So (41) and the needed lower bound for $E(Z)$ follows once we prove (42) $\to$ (43).

5. And (42) $\to$ (43) is proven by that comparison of the actual hamiltonian with a weakly interacting one which is then controlled by an "N free particles in a box" hamiltonian. The comparison is still quite delicate; one might just add in some particles bunched together and reverse the inequality. This observation recalls Lieb's remarks about the Heisenberg Uncertainty Principle's limits in this vein, where again suitable bunching up in *two* places destroys the folk intuition concerning $\Delta x \Delta p \approx 1$; hence, the need for Sobolev's measures of the variance of the density to avoid these pathological cases. As Fefferman says, "computing the energy modulo errors smaller than $Z^{5/3}$ is closely related to finding the variance of the number of particles in a ball located at $Z^{-1/3}$ with radius $Z^{-2/3}$ for an atom in its ground state," namely, discarding or not a potential energy due to that variance.[122] (Most of the electrons are at $Z^{-1/3}$, with just a few outer ones at a radius of order 1, and those outer ones are what gives the atom its size.)

Still, there is a lot more delicate analysis needed to control the error, even given this superstructure.

As for steps #2 ff., the computation of H_{LB} will require improved optical (or WKB) estimates of the wavefunction, as well as employing the model of free particles in a box. As for the WKB methods, one first computes the eigenvalues and eigenfunctions using suitably refined asymptotics, then the sum of their energies or eigenvalues using stationary-phase methods, and then from that sum one computes the charge density—for a one-dimensional problem. Using spherical symmetry and separation of variables to reduce the actual problem to ordinary differential equations, one then does the same for the corresponding three-dimensional mathematical problem, now with an added centrifugal potential, which goes as $\ell(\ell+1)/r^2$, for angular-momentum, ℓ. The approximate wave functions are antisymmetric products of plane waves (with a basis or "support" on a ball in Fermi-momentum space).

One wants to combine: (1) one's knowledge of the exact wavefunctions near the origin and so also avoid the singularity problem there, namely one's knowledge of the well-known low angular momentum states; with (2), the WKB estimates when the radius, r, is larger and the angular momentum is large (where an exact solution is no longer available), and so do the higher-ℓ sums of eigenvalues or energies in terms of stationary-phase integrals.

Since one is in the volume counting business, it is perhaps not too surprising that in the end one has to estimate the number of cubes inside a region such as a sphere, to take into account multiplicities of atomic states or angular momentum degeneracies, what turns out to be a conventional problem in analytic number theory. The formula for the number of enclosed lattice points (from G.H. Hardy) depends on the non-zero curvature of the surface ($\pi R^2 + O(R^{2/3})$, rather than $O(R)$), and that turns out to depend on the "aperiodicity of the hamiltonian flow."

Fefferman-Seco approximate the solution of the crucial differential equation locally, in one region of phase space, in terms of a distorted plane wave ($A(x,p)\times$ exp $-iS(x,p)/\hbar$) as an eigenfunction in that region, the optical model now to be expanded into a power series in $\hbar$ (by first expanding S as a Taylor or power series). And one has to match or to put together all the local solutions to get a global solution, with good error bounds. In effect, one has done a fourier transform of the differential operator, converting it into a sum of evolution operators of the form exp $-iH_{tf}/\hbar$, H_{tf} being the one-particle Thomas-Fermi energy, each operator acting on a small cube in phase space.[123] Fefferman and collaborators call this a Tauberian technique, for connecting the asymptotics of the differential operator to the asymptotics of the function itself.

This level of detail still leaves out much. It does not say how to actually do what it describes. What it does show is the care and ingenuity that is involved in marshalling techniques and in realizing mathematical and physical intuitions. And of course we still need a better value of E_*, the binding energy per atom, the proportionality constant in the stability of matter, if we want to prove that matter exists as a gas.

V

WHY ARE THESE MATHEMATICAL OBJECTS USEFUL IN DESCRIBING THIS SYSTEM?

If you work out a mathematical description of the world, you may end up with mathematical objects which are comparatively esoteric, and for which you have no ready answer to the question that heads this section, Why are these mathematical objects useful in describing this system?--other than that it worked out that way. The system might be a physical one. But it also might be a mathematical object, so we might wonder why parts of algebra are so useful in the topological realm.

We will examine one such case, involving a long and complex calculation of the asymptotic spin-spin correlation function for the Ising model, where strategy and structure demand as well monumental technique and tactics: the 1976 Wu, McCoy, Tracy, and Barouch [=WMTB] paper. In order to begin to answer the above question, and more generally to say what is going on in such a paper, it helps to have alternative derivations (usually done sometime afterward), in which some of these peculiar mathematical objects make another appearance.[124] It helps to examine a particular technique, if it appears recurrently, to see if some physics or some mathematics is embodied within it. What would be nice would be a simple derivation, all the technique and tactics subsumed to more general methods, the meaning and the structure of what is really going on standing out. But that may or may not be available, at least for now.

Our earlier discussion of Yang's calculation of the spontaneous magnetization for a lattice of infinite size presented similar issues. But in that case, there were no obviously peculiar mathematical objects, and the paper was somewhat shorter and less computationally intense.

***Our description of the stability of matter problem emphasized the strategy and structure in each of the various calculations. It suggested how subsequent calculations employ many of the strategic insights of their predecessors, either adopting them or working against them. The physics plays a central role, even for the most hairy of calculations.

Still, some of the time, this semblance of rationality and order may not be available. Imagine exploring in darkest Mathematica or climbing Mount Epsilon. One packs well, knows a great deal about trekking and climbing, brings along native guides, and is prepared for a long journey. Presumably, in the end, the voyage allows access to the source of wealth or for an olympian view of breathtaking proportions. Now, say you actually find that bejeweled source of riches or you achieve those rarified heights, rather than merely getting lost and stranded in Article Minor. Exhausted, you may have little idea of just how you got here, or why. (Here, again, we might recall Yang's testimony—and he had reached Article Major.) Maps and Global Positioning System devices do not tell why a path is good or perspicuous. You want to have some of that fabled olympian strategy, some sense of structure. And is there a more comfortable and straightforward route?

Others follow in your path, or better put, others try to achieve your goal, finding shortcuts or new paths, now that they know that the endpoint is reachable. As a consequence, when you review how you slogged your way to the center or to the top, your account is not so much in terms of what you knew then day-to-day, but what you now know about the flora, the fauna, and the terrain. You now better understand why you followed a path then, even if it now looks circuitous and dangerous. And you appreciate how risky and all but mistaken were the moves you had made.

Scientific and mathematical work is often of this exploratory sort, with much work reaching comparatively dead ends, yet there being subsequent other scientists who find new opportunities for moving forward (or not), and many retrospective recountings of the original voyage. Or, making the best of a dead end, a mathematician can make a minor advance, sometimes of substantial value in itself and which leads into new productive directions. Even given a fruitful outcome, one may not be at all clear why one ended up here (other than to recount the voyage once again). Eventually, it feels rational and strategic, but for the longest time it appears to be tactical and opportunistic, the initial strategic account not at all so satisfying, the outcome not at all inevitable.

***I have been describing what might be called heroic achievements demanding imagination, very hard work, persistence, and a soupçon of mathematical blessing. Big ideas are not enough, and they may not be available. Little ideas are needed again and again. And one must calculate, manipulating long complex expressions.

Kenneth Wilson's algorithmic physical theory as a computational realization of the renormalization group demanded devotion and persistence as hundreds of hamiltonians had to be managed. Lieb, Wu, and Dyson and Lenard, and their collaborators, had to push forward, rearranging and manipulating mathematical expressions, even when the mathematics and the physics were not too helpful. Early calculations in modern quantum electrodynamics demanded scrupulous care and persistence if all the relevant diagrams were to be included and properly evaluated.

Rodney Baxter has been working on lattice models since the late 1960s, when he worked with Lieb at MIT. Baxter is renowned for his inventiveness in finding exact solutions, and the legerdemain in his working them out: in the algebraic manipulations, in the summary objects he invents, in his using symmetries and elliptic substitutions. There would seem to be a magic in his methods, albeit we never see the pages of calculations that do not work out, or the talents and intuitions he has developed over the years. It helps that he is dogged and that he is willing to make assumptions (such as analyticity) on his way. He says that he produces exact but not necessarily rigorous solutions. He reminds me of Onsager, albeit Onsager had a taste for the vatic.

Rather than working combinatorially, Baxter employs symmetries to develop functional equations for the quantities of interest. Moreover, he provides multiple solutions for many of the models, in part to see if those methods will prove useful in less tractable models. And he develops series expansions for the solutions to get an idea of what the exact solution will look like, as did Onsager.

In the background, are analogies and insights from quantum field theory, whether they are realized initially or only later. What is missing from his work (and from Onsager's) is more general theorizing or speculation about why his methods do work, and something about how they are to be seen as systematic methods rather than wonderful jury-rigged devices.

Many pages of calculation may turn out to be byways, crumpled up and sent to the waste basket. Technical details would appear to be inevitable. And sometimes what results is a long paper that is read by few, for in time a more perspicuous derivation is discovered.

Curious mathematical objects, curious in that they have not made much of an appearance in earlier work: Onsager's loop algebra, Painlevé transcendents, Toeplitz matrices, Glaisher's constant, the Barnes function. Eventually, we'd like to understand just why they appear, what is it about the system that makes them all but inevitable. But at first they are unexpected guests at a party. The mathematical objects at one party begin to appear at other parties in different guises. Lenard worked earlier (than his stability of matter work with Dyson) on one-dimensional boson gases. Lenard's ground state encounters those Toeplitz matrices, the same sort of matrix that leads to the correlations in the two-dimensional Ising model.[125] And Dyson worked on the boson gas problem as well.

Ugly but correct initial proofs may not be illuminating. But they show what is possible, and so others are encouraged to find more illuminating proofs.

The discovery at the end of the voyage which I shall now be describing were the Painlevé transcendents, actually Painlevé III, the third of the six transcendents, a function not unlike sines or cosines. Painlevé transcendents are the "next step up" from circular and elliptic functions: sines, and sn or Jacobi elliptic functions, respectively. Just as a Bessel function is asymptotically $z^{-1/2}\sin z$, so a Painlevé transcendent is sometimes asymptotically a Bessel function (if K_0 is a Bessel function, then Painlevé $\approx 1-2/\pi\,K_0(2z)$) or a Weierstrass elliptic function, $\wp(z^{5/4})$. Originally (1900–1906), the Painlevé transcendents were imagined as solutions of specific sorts of nonlinear ordinary differential equations, their actual use for science being moot. But here on this voyage, without seeking them or another particular fountain of youth, WMTB found these functions as the natural solutions. But at first and for some time, one had no idea why they ended up at this point, or the meaning of this treasure. In effect, the problem becomes, Why Painlevé? And in attempting to answer this question, we begin to see how the tactical and the persevering actually succeeds, in detail, and how meaning slowly accumulates as further work is done.

> The correlation functions of integrable statistical mechanical models are characterized as the solutions of classically integrable equations (be they differential, integral, or difference). (B.M. McCoy[126])

While Barry McCoy indicates that he "immediately realized" the connection between correlation functions and solutions of integrable equations when he and his collaborators found Painlevé transcendents in their work, it has taken some time (at least a quarter of a century) and it will take some more time to understand what is really going on. Here is Palmer and Tracy's (1981) account of "Why Painlevé?":

> . . . [Sato, Miwa, and Jimbo (late 1970s)] were aware that Painlevé transcendents occur naturally in the integration of Schlesinger's equations for monodromy-preserving deformations of linear differential equations. . . . they developed new techniques in the theory of Clifford algebras [the algebras employed by Onsager], . . . showed that the scaled n-point [correlation] functions were the coefficients in the local expansion of a basis of multivalued solutions to the Euclidean Dirac equation [recall Kaufman's spinors, now in a continuum theory], and finally used this to demonstrate that the scaled n-point functions satisfy a nonlinear Pfaffian system of differential equations . . . In the case of the two-point function, the Pfaffian system is integrable in terms of the particular Painlevé transcendent appearing in [the paper of WMTB] . . .[127]

Conceptual clarification and calculational simplification may come from the work that builds in the crucial scaling assumption from the start. More generic work provides another sort of support; namely, that the objects initially encountered in this problem are quite natural when seen in that more general context. Some of the rubrics here include solitons (infinite number of conserved quantities, namely, an integrable nonlinear problem, and an associated linear differential equation), Wiener-Hopf and Toeplitz operators and Fredholm determinants, the quantum inverse scattering method, the Bethe Ansatz or nondiffractive scattering (hence many-body equals the sum of two-body), quantum groups, and random matrices. The asymptotics here lead to the Tracy-Widom distributions (much as the central limit theorem leads to the Gaussian), and Tracy-Widom distributions are defined in terms of Painlevé transcendents.

This answer to Why Painlevé?, may require bringing in a completely new technology, some of which is nowadays seen to be well motivated by the physics of the problem. But when Kadanoff and Kohmoto (1980) put the Sato, Miwa, and Jimbo (1977) analysis in more familiar terms (continuum spinor field, operator product expansion, order-disorder operators), they pull back from that new technology into territory perhaps more familiar for many readers (reminding one of the Schultz, Mattis and Lieb derivation of the Onsager result). Yet, again, that new technology proves to be closely associated with other contemporary developments in mathematical physics, so that it becomes somewhat less strange than it might have been.

It is not uncommon that mathematical tactics, formalism, tricks, and devices are ascendant over physical or mathematical insight, so that we are working or performing without the net of a deeper understanding of the physics or the mathematics. Of course, one hopes and expects to eventually achieve a deeper understanding, to get at what is really going on. Techniques and machinery will become meaningful; they are not only mere devices. To see how both tactics and understanding play out, we want to examine some formidable mathematical physics and one of the techniques (Wiener-Hopf factorization) it employs.

TACTICS ASCENDANT: UNAVOIDABLE ALGEBRAIC MANIPULATION AND LENGTHY COMPUTATIONAL PROOFS

One way calculation can be long and elaborate is if there needs to be enormous amounts of creative algebraic manipulation (rather than the epsilons and deltas and the inequalities characteristic of analysis). If the problem can be expressed nicely, say, "factor this polynomial," then symbol-manipulation computer programs can be helpful. So you might have to "use brute force algebra: . . . [W]e had to factorize complex polynomials that were 80 screens long" in SMP (a symbolic-manipulation algebra program).[128]

But some manipulation is not at all amenable to algorithms, at least initially. Here the problem is one of meaning, namely, which expressions or formulas are the right units: for purposes of convenient computation; for purposes of interpretation of what is going on; and for replacement by a portmanteau symbol, a long string of symbols becoming a convenient F or f or phi (φ), which F becomes something we talk about as if it is familiar and understandable. That is, the art lies in defining suitable quantities, re-expressing what we know in a lucid or at least comprehensible and manageable form. And in these cases, it is not at all clear initially that one might avoid a long, computationally intensive calculation by further refinement and structuring, and by new conceptual developments, so distributing the work of the proof to more general theories

FIGURE 4.14: PAPERS IN THE WU, MCCOY, TRACY AND COLLABORATORS' SERIES ON THE ISING MODEL

1966–1968: Theory of Toeplitz determinants and the spin correlations of the two-dimensional Ising model, I–V (III with Cheng).
1968–1970: Theory of a two-dimensional Ising model with random impurities, I–IV.
1973: *The Two-Dimensional Ising Model*, a book.
1976: Spin-spin correlation functions for the two-dimensional Ising model: exact theory in the scaling region (= WMTB).
1977: Painlevé equations of the third kind.
1981: Two-dimensional Ising correlation functions: convergence to the scaling limit (with Palmer).

T.T. Wu, B.M. McCoy, C.A. Tracy, and E. Barouch's formula-dense, "monumental"[129] article, "Spin-spin correlation functions for the two-dimensional Ising model: Exact theory in the scaling [or t] region"(1976), is one of the more lengthy and complicated calculations involving sophisticated manipulations I am aware of, at least in mathematical physics.

First, the title: *Correlation functions* measure the probability that changing something, here atomic spins, at one place will affect something at someplace else, distant in space or in time. Again, the Ising model is a grid with spins at each vertex, classically-interacting magnetically with neighboring spins, so they have a tendency to align, but they are as well shaken up by random thermal interaction.

(See the appendix of chapter 3 for an overview.) Formally, the Ising model in two dimensions can as well be seen to be a model quantum mechanical system in one dimension (a spin chain in an external field) or a two-dimensional quantum field theory. Going down by one dimension, a classical one-dimensional spin chain, with long-range interactions, corresponds to a zero-dimensional quantum system, the Kondo effect of a magnetic impurity in a metal. As in the Ising model, a wide range of energies (here of electrons near the Fermi level in the conduction band) make it difficult to solve. The renormalization group allows us to solve the problem one slice of Fermi momentum at a time, and the Bethe Ansatz leads to an exact solution, again as in the Ising model.

Scaling is a particular generic form for the behavior of a quantity as a function of size or scale, in effect whether it looks the same at all scales. Here the archetypal example is the Gaussian distribution of chapter 2, which scales as $\sqrt{N}$, and other such stable distributions which scale as $N^{1/\alpha}$. Here we are concerned with an asymptotic form in t, for example, $t^{\alpha}F(t)$, α being the scaling exponent, F being the form factor. t is the distance to the critical temperature of the lattice, when it might become permanently magnetizable or not, times the linear size of the lattice, in the limit of the temperature-distance going to zero and the areal size going to infinity, scaling being when t is a good-enough variable to capture the functional form. t is also the ratio of linear size to the correlation length. We are concerned not merely with a point limit, but with an asymptotic form, and hence the inapplicability of the conventional (Szegő) limit theorems employed in deriving the spontaneous magnetization for an infinite lattice.

The 1976 WMTB paper is perhaps the high point of Wu, McCoy, and Tracy and their collaborators' extensive work on the Ising model. They have provided a series of exact calculations of various quantities for various situations, such as the boundary specific heat for a semi-infinite lattice (say, the upper half-plane). WMTB's results also test out in an exact fashion ideas about quantum field theories, since the scaling limit corresponds to a continuum quantum field.

Of course, how one is to count work as a single calculation plays a role in a judgment of length and complication. There is always earlier work that sets the stage for an endeavor. Wu's, and Cheng and Wu's earlier papers—"Theory of Toeplitz Determinants and the Spin Correlations for the Two-Dimensional Ising Model," I (1966) and III (1967), respectively—set the formalism for the WMTB paper: the combinatorial method using Pfaffians, the manner of using analyticity (Wiener-Hopf techniques) to compute the ubiquitous, highly symmetric (Toeplitz) determinant; perturbation expansions, in order to get asymptotic estimates in the scaled temperature, t, rather than in linear distance, N (as in Cheng and Wu's paper), and rather than infinite-distance limits as in Montroll, Potts, and Ward (1963).

Cheng and Wu noted that their own asymptotic results for the correlation function for atoms on a diagonal, at the critical temperature for a square lattice, were already known to Onsager. Kaufman and Onsager's published paper provides a quantitative estimate. Yet, Cheng and Wu's own asymptotics in M and N (where an atom is located M horizontal units and N vertical units from another atom) for the correlation function ($\langle \sigma_{0,0}\sigma_{M,N} \rangle$ or $\langle \sigma\sigma_{M,N} \rangle$) as a function of T, above and below the critical point, do not work well at the critical temperature since the perturbation methods converge too slowly. So Cheng and Wu go back to the methods of Montroll, Potts, and Ward. Their innovation, say over Onsager, or over Montroll, Potts, and Ward, here lies in the method of calculation for the asymptotics and its generalizability.[130] Thirty-seven years after WMTB, and sixty-four years after Kaufman and Onsager's papers, Deift, Its, and Krasovsky (2013) set the project in a larger context. Studies of properties of scaled correlation functions is now a flourishing industry.[131]

Still, I believe WMTB does earn pride of place for a lengthy, formula/algebraically-rich, computationally intensive calculation (artfully presented and structured, to be sure), in its initial resistance to conceptual theorizing, its advance over previous efforts, its length and complexity, its structure and staging, and the curiousness of the objects employed in its solution.

THE WU-MCCOY-TRACY-BAROUCH PAPER

FIGURE 4.15: SECTION HEADINGS OF WU-MCCOY-TRACY-BAROUCH (1976)

I: Introduction

II: Summary of Results and Notation

III: Perturbation expansion for $\langle \sigma\sigma_{M,N} \rangle$ below the critical temperature (T_c) and for large size

    Perturbation expansion, recursion relations

IV: Perturbation expansion for $\langle \sigma\sigma_{M,N} \rangle$ above the critical temperature and for large size

V: Perturbation Expansion for $\langle \sigma\sigma_{N,N} \rangle$ for large N, and small $N \times |T-T_c|$ (the scaling regime)

VI: Scaling Functions in Terms of Painlevé Functions of the Third Kind

VII: Susceptibility

Appendices: Numerical Work; Integral Equations and Painlevé Functions; Various Identities for the Ising Model.

190

4.100 $\quad x(z) = \sum x^{(k)} z^k$

4.105 $\quad x'\,\bar{x} = \bar{x}'x - \bar{x}\,x\,G'(z)$

4.114 $\quad x^{(l)} = \iiint ... d\varphi_i \prod E(\varphi) \prod \sum T(\varphi)$

4.118 $\quad \prod E - \prod T = T \prod E + T \prod E + ...$

4.122 $\quad G = \iiint ... d\varphi_i \prod E \prod T (\prod T \prod E)$

4.129 $\quad S_j = \sum\sum S^{(j)}{}_{lmn}$

4.130 $\quad \sum\prod A_j\,(B_n - A_n) \sum \prod A_j \prod B_j = \sum(\prod B_j \prod A_j) - \prod A_j$

4.145 $\quad \sum \bar{x}\, Gx = \sum(\,\bar{x}\,\bar{x} - x\,\bar{x}\,) = \sum \bar{x}\,x$

4.113 $\quad \sum x\bar{x} = -\sum x\sum G\bar{x} = -\sum x\sum \bar{x}\,G$

FIGURE 4.16: Schematics of formulae from WMTB, Section IV-E2. E = various different exponential functions of the angle variables; T = various trigonometric functions of the angle variables. I have left out most subscripts and coefficients, as well as the limits of the sums and products.

After an introduction to the problem, and an extensive summary of results, WMTB present their staged assault on calculating the correlation functions in the scaling regime. First, they use perturbation expansions to get the correlation function for large values of the scaling variable, below and above the critical point. Then they do the same for small values of the scaling variable. And then they apply a very different approach, wherein they discover that the solution may be expressed in terms of comparatively esoteric transcendental functions, those Painlevé transcendents. (See also, Figure 4.18)

If we examine some of the more computationally-intensive sections of the WMTB calculation (schematically illustrated in Figure 4.16), one can see how much effort must have been required to get anyplace at all, how many less than fruitful paths must have been pursued in search of just the right—or even just better, sufficiently practical—algebraic substitutions and moves. Let me be perhaps too schematic for a moment: Section IV-E2 for example begins with defining generating functions (their equation 4.100) and some differential equations that go along with them (4.105). Then several new objects are defined, each of substantial size typographically (4.114). Then a substitution is made to get a generating function into a better form, and cyclic permutations are invoked (4.118, 4.119). Another product is defined (4.122), and all is reexpressed nicely (4.129). A certain combinatorial result is then needed, involving many summation and product signs (4.130ff). Eventually (4.145), the identity (4.113) is proven.

Given Wu and McCoy's earlier work, perhaps it would not be surprising to have to perform lengthy and complicated calculations in this realm. But, I suspect that only a small part of what was going on in the paper could be imagined by WMTB ahead of time, that is, before WMTB actually did the calculation that they performed. That one might end up with a fairly simple outcome at the end of a very complex calculation, with hard to imagine twists and turns, should be amazing—as amazed as was Yang when his spontaneous magnetization calculation yielded a remarkably simple form (even if he were to already know Onsager's claimed result). Of course, having developed confidence in one's capacity to calculate, and having appreciated the virtue and payoff of perseverance in the face of discouraging dense jungles of algebraic manipulation is an enormous advantage, possessed by few indeed. Even if you are skillful in making perspicuous groupings of terms and the like, the outcome might not be considered simple.

Following the paper line-by-line, or even through an outline, is only a beginning to understanding such a paper. And at first, even that following may in fact be quite discouraging. (Rather, you might try to rederive the result, given your own technical skills, your newfound knowledge of the answer and of some of the devices used to derive it, and your knowing that the problem has been solved and so that it should be tractable.) It may remain difficult to know what is really going on, albeit the technical moves and substitutions start becoming familiar—although surely not so familiar as they were to the senior authors of the paper given their previous work and experience. Still, those moves may begin to be seen as strategic rather than opportunistic.

Eventually, one might hope to say, "In this section they were trying to do X [much as I began to do with Yang's paper], which X we now see may be avoided, or more effectively computed, or be understood naturally as . . ." The latter is one of the messages in the subsequent work, as I have indicated already, connecting Painlevé III to other mathematical objects, such as: Fredholm determinants, $\det(1+\lambda K)$, which arise in solving integral equations[132]; random matrices, whose elements are random numbers; or, "isomonodromic transformations," which I take to be transformations of the differential equations which preserve the general structure of charges, dipole layers, and other such singularities. Moreover, these methods may lead more directly to Painlevé, as we might hope for from subsequent derivations. (See Figure 4.18.)

So we might begin to understand what is going on, albeit schematically.

Before we become further concerned with even more technical matters, it may be helpful to make a number of more general points, so that we might appreciate their appearance in the sequel.

(1)The physics of the problem, namely the lattice's symmetries, would seem to lead naturally to the (2) Toeplitz matrix. There are several ways of ending up with that Toeplitz matrix: combinatorics and Pfaffians; contraction of fermion operators using Wick's theorem; merely as a consequences of calculation (as in Kaufman and Onsager's work). And there are ways of calculating that do not explicitly encounter the Toeplitz matrix.

(Parenthetically, if the lattice is infinite and one is not interested in asymptotic results, again there are limit theorems (originally due to Szegő) that give the correct result for the spontaneous magnetization in terms of the determinant of the Toeplitz matrix.[133] In any case, there is a range of mathematics involved in solving for the spontaneous magnetization of a lattice (such as modular forms) which is apparently different from the range of mathematics involved in solving for asymptotic correlation functions in the scaling regime (those integrable nonlinear differential equations)—although in the limit, the latter should give the former.)

(3) In going to the asymptotic or scaling limit one finds that the Toeplitz determinant is now in Fredholm form ($\det(1+K)$, K a suitably symmetric operator).

(4) And, mathematically, associated (with the correlation functions and) with Fredholm determinants are "solutions to total systems of differential equations."

(5) "The Painlevé equations arise as the simplest examples of these total systems."[134]

Namely: (1) Physical symmetries lead to (2) Toeplitz, which (3) in the asymptotic limit leads to Fredholm, (4) which leads to total systems of differential equations, (5) which leads to Painlevé. And each of these "leads to" depends on the physical or mathematical particulars of our problem. Again, the physics of the problem need not lead to the Toeplitz matrix, even if the problem is exactly solvable by another, seemingly related means.[135]

More generally, integrable nonlinear problems (and that is what the Painlevé equations represent) are integrable, it would appear, because they are associated with the solution to a specific linear problem.

So we might have a formal mathematical account of Why Painlevé?, not yet an account that is physically perspicuous. The latter begins to be achieved when we have many more models solved in this fashion, many more connections among the mathematical objects and the physical processes. As mentioned earlier, Painlevé leads to the natural asymptotic limits for strongly dependent random variables (much as is the Gaussian for independent random variables), and much as the Wigner semi-circle law for random matrices. (See the note on p. 430.)

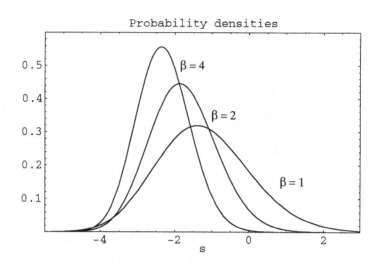

FIGURE 4.17: The Tracy-Widom distributions, $F(s)$, for different values of the weighting variable β.

For example, the Tracy-Widom distributions characterize extreme values of strongly dependent random variables. Typically, the probability is given by:

$$P\{(l_N - 2\sqrt{N}) / N^{1/6} \leq s\} = F(s),$$

where l_N might be the length of a longest increasing subsequence of a random permutation, and F is a Tracy-Widom distribution. The generating function of a related distribution is the determinant of a Toeplitz matrix, and a suitable limit of that determinant leads to F. It can also be shown that F can be expressed in terms of a solution to a Painlevé differential equation.

In this context it is not too difficult to discern a web of connections among a variety of objects, mathematical and physical, which might be called the *Painlevé complex.*[*] The list in the footnote is mere name-dropping, the exact

---

[*]These might include: random matrices, the Gaudin-Mehta-Dyson distribution, the Tracy-Widom distribution, scaling, Young tableaux and permutations, the zeros of the zeta function, the density of eigenvalues and the largest eigenvalue of random matrices, the two-dimensional Coulomb gas; Painlevé, Fredholm determinants, Toeplitz matrices and Wiener-Hopf operators, tau functions (and their analogy to theta functions); inverse scattering methods and quantum groups and Hopf algebras, Lax equations and Lax pairs,

connections being particular facts in each case. Charts are readily made linking these various notions, and links often return to a notion that appeared elsewhere in the net of interconnections.[136] But such a chart is merely the beginning of being able to ask a question about what is really going on in the mathematics and in the world.

Deift, Its, and Krasovsky (2013) suggest:

In [M. Jimbo and T. Miwa, "Studies on holonomic quantum fields XVII," *Proceedings of the Japanese Academy, Series A Mathematics Science* 56 (1980): 405-410.] the authors used a remarkable relations (which they themselves discovered) between the diagonal Ising correlations and isomonodromy deformations of a certain 2×2 Fuchsian system. This discovery laid the foundation for the Riemann-Hilbert method in the theory of correlation functions, random matrices, and Toeplitz and Hankel determinants. (p. 1422)

In preparing this second edition, reading the Deift paper was illuminating. But I am still at a loss to tell a single coherent story.[137] However, reading other papers, by Deift and by Barry Simon one begins to see a single story, for which see the next section.

Why Painlevé?, is then to be answered in the context of this enlarged web of connections. In the next chapter, one of our themes will be that functions such as trigonometric or elliptic functions can have arithmetic-combinatorial properties and analytic-topological ones. And self-similarity and scaling are pervasive phenomena in this realm. Here, too, Painlevé would seem to connect both to combinatorial properties of tiling a surface and to topological properties of covering it.[138] Such observations perhaps begin to make the Painlevé complex more natural. Still, it would be nice to connect Painlevé to something, much as the trigonometric functions connect to harmonic analysis.

To recall and preview: The spontaneous magnetization is derived in a variety of ways, including the infinite separation limit of the correlation function and employing a notional magnetic field to compute the partition function and so its

---

Kortweg-deVries systems, solitons, solvable models and integrable systems, isomonodromic deformations and Riemann-Hilbert problems, holonomic quantum fields, sinh-Gordan; and, of course, scaled Ising-model correlations. An interesting problem is how the Bethe Ansatz is related to scaling (as in as well the Kondo Model), although we know it is related to Yang-Baxter and so transfer matrices (and that leads to k's significance), and quantum inverse scattering. See also notes 8-11of the Preface.

logarithmic derivative. WMTB compute the correlation, away from the critical point in the scaling regime using an n-particle form-factor expansion that eventually leads to P_{III}. The integrals in the form-factor expansion factorize in terms of complete elliptic integrals and exhibit an interesting algebraic structure ("Russian doll," referring to matriyoshka dolls). The magnetic susceptibility, the effect of an external magnetic field on the system, is defined as $\Sigma_{ij} (< >_{ij} - M^2)$, where $(< >_{ij}$ is the correlation between a spin at the origin with one at site (i,j), and M is the spontaneous magnetization. It proves to have many singularities other than at T_c, and there would seem to be a natural boundary, the Nickel boundary, in the complex-T plane, discovered at first by computation.[139]

With all of this retrospective insight, the paper itself, in its detail, still remains difficult and demanding. I should note that, as with Yang's paper, careful multiple readings should make the paper more familiar, the recurrent moves, the use of simple cases before the general one, etc., contributing to that sense of understanding. And knowledge of subsequent work makes a very big difference. (Still, I do not claim that I have followed more than a small fraction of the argument, except architectonically and strategically.)

The clues we might have noted when first reading the paper, and which we now see more clearly after much subsequent work, are appreciated as genuine insights (or not) only later, when the mathematics and the physics have filled out through subsequent work by many others. We still need to understand better how the mathematics and the physics illuminate each other. And, again, that takes place through alternative strategies of proof pursued by others.

Figure 4.18 displays a variety of subsequently found ways of at least partially achieving the WMTB result (although not all of these ways). Immediately, and by now not so surprisingly, and as for the spontaneous magnetization, we might discern something about what is going on through the recurrent appearance in several guises of a formal mathematical device that we might hope is associated with some important physics or mathematics. It would appear that the cruxal point is the (Toeplitz) determinant, det A, expressing notionally the contributions to the correlation function by each of the various paths or linkages between two spins. But now it must do somewhat different work than for the spontaneous magnetization.

In the scaling regime we have $A = A_0(1+\Delta)$, Δ small, in effect a linearized form, which leads as well to a Fredholm determinant $\det(1+K)$ ($= \det A/\det A_0$, WMTB Equation 5.11), where K is the propagator or kernel of an integral equation. ($\int K(x,y) f(x) \, dx = f(y)$). So one subsequent rederivation begins:

FIGURE 4.18: ILLUMINATING WU-MCCOY-TRACY-BAROUCH THROUGH ALTERNATIVE STRATEGIES, PARALLELS, AND RESONANCES

$\langle\ \rangle$ as fermions
--continuum Dirac field, operator product expansion (Kadanoff/Kohmoto) $\rightarrow$
--continuum Dirac field, Clifford algebra (Sato/Miwa/Jimbo) $\rightarrow$ P_{III}
--sinh Gordon (Itzykson/Zuber) $\rightarrow$
--Majorana field (Bander/Itzykson) $\rightarrow$ $|1+K|$

$\langle\ \rangle$ as spinors (Kaufman/Onsager)   Onsager$\rightarrow$long range order
                                   $\searrow$ $\nearrow$
                         TOEPLITZ DETERMINANT
McCoy/Wu (VIII/3) $\nearrow$ $\searrow$ Cheng/Wu
                         $\rightarrow$$t$-small
                     $\downarrow$ $\exp \sum \mathrm{tr}(\Delta A_0^{-1})^n$$\rightarrow$series$\rightarrow$*****
                                $\uparrow$
                     WMTB V$\rightarrow$$|1+\Delta A_0^{-1}|$=$|1+K|$$\rightarrow$Tracy/Widom $\rightarrow$ P_{III}
                                $\downarrow$
                         $\downarrow$ Wilkinson/field theory
                     $\rightarrow$WMTB VI$\rightarrow$integral-equation$\rightarrow$Myers P_{III}

$\langle\ \rangle$ as Pfaffian (Montroll)$\rightarrow$Wu$\rightarrow$McCoy/Wu (VIII/3) + Stephenson
                 $\searrow$Cheng/Wu$\rightarrow$WMTB III/IV$\rightarrow$$\sum \int^n$$\rightarrow$series$\rightarrow$$t$-large$\rightarrow$*****
                 [Toeplitz, asymptotics]

FIGURE 4.18: Methods of solution for: Correlations in the Ising Model, $\langle\ \rangle$, leading to the Scaling Regime, P_{III} or ***** (series—form factor or n-particle). (Montroll = Montroll, Potts, and Ward; McCoy/Wu VIII = *The Two-Dimensional Ising Model*, chapter VIII; WMTB V = section V of WMTB).[140]

Our purpose is to evaluate the Fredholm determinant $F(t)=\det(1+K)$. Although the details will be completely different, our method is strongly motivated by the original derivation of WMTB, in particular we shall relate $F(t)$ to the solutions . . . to the following integral equations of the first kind . . . Once this has been done the problem is essentially solved. Since from the work of Myers [on P_{III}] . . .[141]

There are other ways to get to that Fredholm point (which then leads to the Painlevé functions), from quantum field theory for example, using fermion operators; or, in noting that there are various different determinants or Pfaffians that express the correlation function, as was noted by Montroll, Potts, and Ward, and one of them is a Fredholm determinant. But these various ways of mathematical connection are not so independent of each other, for mathematical, quantitative, and qualitative reasons.[142]

Technically, and more particularly, one can show that the sum of a series of multiple integrals is "the same" as an integral equation and its Fredholm determinant, which is the same as Painlevé III. The $1+K$ or $1+\Delta$ form can be seen to reflect a perturbation expansion where K or Δ is the perturbation (again, Equation 5.11 of WMTB). The form factor, $F(t)$, t being the scaling variable, just what modifies the dominant term in the t-asymptotics, would appear to be a WKB integral, although what this might mean is not so clear to me.[143]

Quantitatively, as I have already indicated, previous results for the spontaneous magnetization, Yang's, must match these asymptotic scaling results as they approach the critical temperature. (Put differently, does a field theory with zero mass ($t=0$) match with a field theory with nonzero mass then evaluated at zero mass?) Moreover, WMTB's perturbation series expansion results (sections III-V) and the Painlevé results (VI) must agree. There is as well a delicious agreement, apparently not readily provable, but to at least 13 digits checkable, concerning an integral of a Painlevé transcendent and a sum of a few transcendental quantities (6.103–6.106) which confirms the goodness of the asymptotic approximation for small and larger differences from the critical point.[144]

Perhaps WMTB, like Yang, were for long stretches working without the net of physics, without what might be called the physical and to some extent even the mathematical meaning of what they were doing. They were working a difficult, arduous, technical problem, guided by their experience or intuitions in solving other such problems. Whatever safety net they had might be called algorithmic or computational (again, much as for Yang), as well as their perseverance and careful checking. The interpretations of WMTB were developed after WMTB, or at least they appeared relevant to the problem only then, through other derivations for other purposes (for example, through developments in mathematical physics such as quantum inverse scattering and solitons). Solving a problem is a prelude to understanding what is going on. To find a solution is vital, but understanding and insight are to be hard won, indeed, through much subsequent work.

WHY WIENER-HOPF AND TOEPLITZ?

At a crucial point in this derivation as well as in the earlier derivation of the spontaneous magnetization, techniques come into play for solving integral equations when a fourier-like transform can only go over the positive real line (so precluding the usual fourier-transform techniques). We might just treat those Wiener-Hopf techniques as formal machinery that does the requisite work. However, I think it is interesting to ask, Why Wiener-Hopf operators and the Wiener-Hopf method, or their discrete analogue, Toeplitz operators and Toeplitz matrices? And, Why Toeplitz determinants?

Mathematically, following from "the celebrated theorem of Noether for dynamical systems," and work on inverse scatting problems (given the outcome, what is the potential or interaction), it was realized that "all $1 + 1$ dimensional integrable systems [the 2-d Ising model corresponds to a $1 + 1$ quantum mechanical system] could be solved by a RHP [Riemann-Hilbert Problem]." RHP are inverse methods, and Wiener-Hopf methods are archetypal. The Toeplitz determinant is the scattering data to which one applies the inverse method. More generally, mathematicians systematically connect Toeplitz matrices or operators with Wiener-Hopf.[145] Namely, Wiener and Hopf showed how to solve an integral equation, whose discrete analog may be expressed in terms of the the Toeplitz matrix and its determinant. Both are about convolution operators. The example Wiener and Hopf use is the Milne equation from astrophysics, and in fact Onsager says that he recognized his Toeplitz-determinant problem as one of solving the Milne equation, and so went to Wiener-Hopf methods.

To preview our physical observations: Toeplitz matrices (let us all them S) apparently build in the translational and k symmetries. Moreover, that Toeplitz matrix as a convolution operator, operates on the vector of a row of spins, much as does a filter in electrical engineering, propagating the effect of a row on adjacent rows, namely S^N. The Toeplitz determinant looks much like a sum (of probability amplitudes) over paths, presumably accounting for the interactions of the spins.

We might ask, What is the meaning of the Toeplitz determinant? The point of view from operator theory suggests that the logarithm of the Toeplitz determinant is like a variance. (I have gone into some of the technicalities in the note.[146]) The Toeplitz matrix is a sort of averaging operator. The logarithm of the Toeplitz determinant is a "variance;" it measures how aligned are those amplitudes. The Wiener-Hopf method allows one to build in the possibility of a phase transition in a robust fashion, for it demands an important modification above the critical temperature, vs. that of below. Also, it links complementary ways of thinking of the crystal at a particular temperature, and both ways should give the same result.

Again, Onsager testifies (in 1979) he used the Wiener-Hopf method in his initial yet unpublished work (in about 1947–1950) on the spontaneous magnetization.[147] He computed the eigenvalues and then multiplied them to get the determinant. The phase factor $\exp i\delta'$ is referred to as a generating function [or symbol] for the Toeplitz matrix whose determinant is the two-point correlation function, which, in the limit of infinite separation, is the square of the spontaneous magnetization.[148] The appropriate Toeplitz determinant for defining the spontaneous magnetization would be equal to "the geometric mean of the generating function [for that Toeplitz matrix]," and,

> After a while, however I realized that the associated difference equation [to the Toeplitz determinant] was just the analog of a Milne integral equation, and I tried Wiener-Hopf technique. (Onsager[149])

The fourier transform of the generating function or symbol, a, provides the terms of the Toeplitz matrix: $S_{ij}(a) = a_{i-j=n} = a_n = \int \exp in\omega\, a(\exp i\omega)\, d\omega$. And the geometric mean of the generating function—since in this case $a(\exp i\omega) = \exp(i\delta'(\omega))$—is $\exp \int \log a(\exp i\omega)\, d\omega = \exp \int i\, \delta'(\omega)\, d\omega$.

But this is a tactical account of, Why Wiener-Hopf? We might just say that the paper's answer to, Why Wiener-Hopf?, is, It works for this particular problem, or perhaps the mathematical answer I mentioned earlier. (Recall that Yang implicitly needed the Wiener-Hopf technique to make his method work.) But we also might want to know what are the features of such a tool that make it effective in this situation.

More formally, the Wiener-Hopf method links two related problem into one. So, for example, in probability, Wiener-Hopf techniques appear when one is describing a random walk and asks both about its behavior (its peak values or "ladder heights") when it stays to the right of the zero, and when it first enters negative values, to the left. Analogously, there is the Kramers Problem of a particle random walking in a potential well, and its escape or not.[150] In general, one sees the two problems as distinguishing, say, $n<0$ and $n>0$, and $n=0$ is shared between them. So if we see Wiener-Hopf, we might ask: What are the two related problems?

Wu used the fact that the Toeplitz determinant, S, that appears in these problems leads to a set of simultaneous equations in which the sums that appear are over $n\geq0$).[151] It would appear that it was a nice device, for by using Cramer's rule on the artificial Wiener-Hopf sum equations, he is able to connect the determinant for an n-by-n matrix, S_n, with the determinant for $n+1$-by-$n+1$, S_{n+1}, as a ratio--for by cutting out the topmost row and the leftmost column, one goes from S_{n+1} to S_n. (The same device allows one to prove the Szegő limit theorem of 1952. In Szegő's paper of 1915, he makes use of this fact about these determinants. Moreover, in discretizing some integral equations, one ends up with the same apparatus.) He speaks of "the similarity between a Toeplitz determinant and the corresponding Wiener-Hopf sum equation." These (Wiener-Hopf sum) equations are solved by Wiener-Hopf methods. One uses an artificial or partial fourier transform (just from zero to infinity), which makes inseparable two distinct regions: the original $n \geq 0$ problem; and, an artificial extension to $n<0$. What we might like to understand are the two actual problems that are intimately related.

Bringing together our various discussions, we might give a physical account of why these mathematical objects may be the means for solving this problem—keeping in mind that we already know that they are such means. At least here, the account will be heuristic and speculative. These are mere sightings of some formal analogies.[152]

(1) The "symbol" or the generating function of the Toeplitz operator—that is, the symbol's fourier transform generates the matrix elements, $a_{ij}=a_{i-j}$—is just Onsager's $\exp i\delta'(\omega)$ ($= [(k - \exp(i\omega))/(k - \exp(-i\omega))]^{1/2}$), where k is the usual temperature-like variable or the modulus of the elliptic functions. Namely, we have $a_n = \int \exp in\omega \times \exp i\delta'(\omega)\ d\omega$. (Note that $\delta'(\omega)$ is real for real k and ω, and that the absolute value of $[(k -\exp i\omega)/(k -\exp -i\omega)]^{1/2}$ is just 1. These observations are crucial for (2) and (6) below.)

In Schultz, Mattis, and Lieb's account of the Ising lattice in terms of a fermion field theory, ω is the momentum of the quasiparticles that make up the lattice, and so the fourier components of $\exp i\delta'(\omega)$ [actually, in their case, it is $\exp i\delta^*(\omega)$] might be said to be the probability amplitude for the contribution to correlation by the quasiparticles of momentum ω.

Also, again, Onsager's $\delta'(\omega)$ measures how much disorder at distance $\sim 1/\omega$ is introduced into an orderly lattice. As Onsager puts it, disorders for ω "small are the most effective in introducing pairs of alternations far apart, which is just what it takes to convert order into disorder."[153] In this particular case, the Toeplitz determinant measures whether amplitudes (the symbol is the phase factor, $\exp i\delta'$) add up coherently or they systematically cancel.

If we look at Figure N.3, we can find out about the behavior of δ^*.[*] We can

---

[*] Although, here we are actually interested in its complement δ' ($\omega+\delta'+\delta^*<\pi$). So we shall have to substitute low for high, large for small, when we substitute δ' for δ^*—and vice versa. More precisely, $\exp i\delta^* = ((1- k \exp -i\omega)/ (1- k \exp i\omega))^{1/2}$. If we replace k by $1/k$, going from low to high temperature, or high to low temperature, then we get the formula for $\exp -i\delta'$. See also, notes 6(d)(i), 6(e), 7, 151, 152.

see that at low temperatures, the fourier components of the phase factor, exp $i\delta^*$, a_n, involve small angles and add coherently: starting at zero and returning to zero as ω goes from zero to π. Yet, since δ^* remains small the amount of disorder introduced (by Onsager's spin-flip operators, $Z(\omega)$, that is $\delta^*(\omega) \times Z(\omega)$) is small. At very high temperatures the fourier components of the phase factor are almost equally distributed in the semi-circle, rotating from π to zero, and so their sum will be small. And here δ^* can be quite large, and so lots of disorder (or spin-flips, especially at large distances) is introduced into the lattice. (Note as well, when ω goes from $-\pi$ to $+\pi$, for high temperatures δ^* winds around the origin once, for low temperatures it does not wind around at all. And, again, we need to reverse high to low for δ'.)

(2) The "index" of a Fredholm operator (the Toeplitz matrix will turn out to be Fredholm here) is the winding number of the symbol. That winding number is either zero or nonzero depending on whether we are below or above the critical point.[154] In this sense, Toeplitz/Fredholm/Wiener-Hopf may be said to build in the possibility of a phase transition.[155] In the Onsager algebraic and spectral derivation, below the critical point there is a twofold asymptotic degeneracy of the maximal eigenvalue. It would be interesting to link the index-zero and degeneracy conditions.

The Wiener-Hopf operator or Toeplitz matrix is Fredholm (and so is invertible) if the symbol is continuous (or has index 0) and does not vanish on the unit circle, both of which are the case here if T is less than T_c, the critical temperature. The phase angle is always real for real k. (And we might ask what does this mean, physically? See (6).) If the Wiener-Hopf technique is to be applied when T is greater T_c the symbol needs to be modified, so that its index is again zero. That formal modification accounts for the different behavior below and above the critical point.

(3) And since the index is a topological invariant, it might well be quite robust to smooth changes in lattice configuration, as we might expect for the physical system's collective behavior. Only a "discontinuous" phase transition will alter the index.

(4) The Toeplitz/Wiener-Hopf form builds-in k-invariance, that what is crucial is k and not the particular coupling constants, through the definition of the symbol. It appears that, the translational invariance of the lattice is built into the fact that a_{ij} depends on $i-j$.

(5) By taking a determinant, one is automatically doing the combinatorics for adding up the contributions of the various "paths" between atoms that contribute to the magnetization.[156] A Toeplitz determinant measures the size (the "variance") of the resulting contribution.

(6) As for two intimately related problems that are linked in the Wiener-Hopf method, I might offer two accounts (that I believe are mutually consistent). One possibility is duality, the formal link of a low-temperature lattice to a high-temperature lattice, order to disorder, and the automorphy of their partition functions. Second, the lattice at a particular temperature might be built up from below, introducing disorder (heat, as spin-flips) into the totally ordered lattice; or, from above, introducing order into the totally disordered lattice (by removing heat, by aligning spins). Presumably, and this is what we mean by a thermodynamic state, in either case the lattice so built is the same as its twin.

If these insights about the two related problems were correct, each should be reflected in the formalism (much as the first and second insights are reflected, separately, in Onsager's formalism for the partition function).

Technically, the Wiener-Hopf factorization separates the solution into two regions of analyticity (here it would seem that $|\exp i\omega| \leq, \geq 1$, where ω is complex (say $v + i\Omega$, v is real, and $\Omega \geq 0$), and apparently $|\exp i\omega n| \leq, \geq 1$ depending on whether n, the index of a series, is greater than or less than zero, with a shared limiting (physical) value on the boundary (and there, for nonzero n, $|\exp i\omega n| = 1$). But in the Wiener-Hopf factorization, separating the analyticity for inside and outside the unit circle, each factor only includes cases when the effect of $i\Omega$ is to lead to an $\exp -\Omega|n|$, that is there is no blowup at infinity or the origin. Notionally, the Ising (or Landau) quasiparticles (see (1) above), in effect spin waves, are given what physicists would call virtual momenta (ω is complex away from the unit circle).[157] I take this to mean that the virtual quasiparticles are localized in space (n is like an x). We would need to show how these virtual quasiparticles build up the lattice or its dual, or build up the lattice from the extremes of perfect order or perfect disorder, within the context of the Wiener-Hopf formalism.

***The Mathematics and the Physics.* These observations are more or less speculative. Yes, it would appear that Toeplitz operators and determinants, and Wiener-Hopf techniques, build in some of the physics naturally. But, at least in this case, such building-in is noted only after the techniques and mathematical apparatus have been employed successfully in solving a range of problems. (Pure mathematics developed out of this problem and related ones, and then it begins to explain the connection of Toeplitz and Wiener-Hopf method.) It could be that were we to make such observations about a technique ahead of time, even as numerous and as detailed as we have made above, that technique might not in the end prove to be appropriate for the problems at hand. Or, perhaps these observations are not the reason why the method works. So when we ask, Why Wiener-Hopf and Toeplitz?, the only good answer we might provide is a very deep account of the essential physics and a demonstration that the mathematical technique fully embodies or models those essentials (and no others we do not want)—much too great a demand in general, at least without a good deal of subsequent work.[158] And the developments in pure mathematics are not enough, at least for my purposes.

Similarly, when we ask, Why are there the ubiquitous elliptic functions in this set of problems?, a correct but still limited answer is that they allow for translational symmetry, and scaling and inversion symmetries, they are in fact of relevant combinatorial significance, and they build in the Yang-Baxter relation (a triple product rule, not unlike that for theta functions). While none of these observations are decisive, they are I believe comforting, much as those I made about Wiener-Hopf. And that comfort is called understanding if it is ramified and holds up to further inquiries. If we are J.-M. Maillard, we say that "the 2-d Ising model is nothing but the theory of elliptic curves" and devote a great deal of energy in showing how this is the case from many aspects and for many lattice models.

When mathematical objects are entrenched in a subject (so that they do not appear opportunistic), as is the Gaussian in statistics and are Lie groups in particle physics, then those objects are pervasively identified with the world and its nature. Such identifications are achievements of ramified understanding.

VI

FIGURING OUT MORE OF WHAT IS REALLY GOING ON

***The mathematical story I have been telling connecting the various features of the Painlevé complex is I believe still something of a folktale, not even a folk theorem or a heuristic. One regularly encounters new papers that make further connections among the various mathematical objects and provide new objects. But so far, I have yet to see an overview where the connections are systematic rather than still being comparatively occasional. Eventually, all the features might be tightly linked, formally, mathematically, and physically. Or, perhaps they will form groups of notions, each member of a group tightly linked, but the groups themselves are quite distinct. So, for example, Wiener-Hopf, Toeplitz, and Fredholm may be connected to phase transitions, while Painlevé will be connected to the Gaudin-Mehta-Dyson and the Tracy-Widom asymptotic distributions.

Again, for all we once knew, many of the connections could have turned out otherwise. Prospectively, the connections are often neither apriori necessary nor quite sufficient.[159] So, Poincaré is said to have compared the Painlevé's discovery (the transcendents) to an island, isolated from the continent of mathematics. But, it would now appear that it is not so much an island as a new world linked broadly to the old continent, a world filled with promise and mathematical fruit.[160] And the connections are not only casual but would appear to be systematic.

Formal analogies are sometimes sufficient for appreciating a connection, at least prospectively, but usually we need detailed and particular examples or cases to guide us in what to analogize and how to do so—as we shall see in the next chapter. When we have found the solution to a concrete problem, we can begin to tell a tentative story of connections among disparate concepts. Working on various alternative derivations, it then becomes a bit clearer why we ended up where we did end up, much as successive assaults on Mt. Epsilon make it clearer why a particular circuitous path actually worked out. Yet, still, a perspicuous conceptual derivation may elude us. (But, again in this particular arena, conceptual, physical, and formal analogies with quantum field theory are suggestive and illuminating, and they provide another set of useful suggestions.) Here is another general characterization by a mathematician:

> When asked what it was like to set about proving something, the mathematician likened proving a theorem to seeing the peak of a mountain and trying to climb to the top. One establishes a base camp and scaling the mountain's sheer face, encountering obstacles at every turn, often retracing one's steps and struggling every foot of the journey. Finally, when the top is reached, one stands examining the peak, taking in the view of the surrounding country side and then noting the automobile road up the other side! (Robert J. Kleinhenz[161])

One would still hope to be able to do this sort of calculation ever more efficiently, and perhaps with greater physical insight. So, work on special cases, or on part of a derivation, or on mathematical connections between disparate objects that makes clearer what is going on are significant contributions. Rigorous solutions are complemented by series expansions and quantitative studies, pointing to phenomena otherwise not seen.

Recall that some parts of the Fefferman-Seco proof were simplified and improved by Bach and Graf-Solovej, while other parts (the lengthiest ones) remain untouched. And, complementarily, the Lieb-Thirring constant has been materially sharpened over the years, yet it is still tantalizingly too large.

It would appear that much of WMTB's work will not remain unavoidably complex.[162] The perturbation series expansion can be understood in terms of its singularities and in terms of dispersion series and fermions. There are various subsequent derivations that lead to the Painlevé transcendents (Figure 4.18), sometimes building in the scaling or continuum regime from the beginning. These are perhaps calculationally much briefer, but may involve more technical apparatus than WMTB. (The next few paragraphs are filled with concept- and name-dropping, and they repeat some of what I have mentioned already. My purpose is to indicate a set of only partly realized connections, a set that entrances and tantalizes many researchers.) So the work of Sato, Miwa, and Jimbo provides a more natural setting for this calculation.[163] And it is then translated into a perhaps more familiar and simpler form of Green's functions by Kadanoff and Kohmoto, and it is made more rigorous by Palmer and Tracy. The quantum inverse scattering method and the Bethe Ansatz are another approach that can build in the asymptotic or scaling symmetry, and they have been connected to fermions (through the Jordan-Wigner transformation) and to the Yang-Baxter relation.[164] Bander and Itzykson, and Itzykson and Zuber, use quantum field theory of fermions; Wilkinson provides a continuum derivation. At the critical point, conformal field theory discrete complex analysis provide a natural formulation, and various symmetries of Lie algebras (E_8) characterize the system and its massive particles. The harder problem is to be able to say something away from the critical point, or with an applied magnetic field.[165]

The connection to Painlevé can be elucidated and made more natural and generic, as in Tracy and Widom's work on Wiener-Hopf, Fredholm determinants, Painlevé, and random matrices. Or, again, we might say that to integrable nonlinear problems there is an associated linear problem. And the set of connections between the various objects can be made generic, our saying something like: It is well known by now that the various objects in the Painlevé complex are part of a single story, although the connections still require a good deal of calculation and manipulation to be shown and proven, and it is still not clear just how pervasive are these connections. A similar constellation of concepts appears in combinatorial theory, number theory, in random matrices, and integrable systems more generally. The asymptotic distribution functions (and Painlevé transcendents) appear ubiquitously, reflecting symmetries of the problems. And it would seem that when they are not exactly the case, they are the natural asymptotic limits in many of these situations.

Such a set of deeper connections is the generic situation for much work that was originally lengthy and complicated, if we are fortunate in our subsequent work. (So fifty or more years after Onsager's 1944 exact solution for the two-dimensional Ising lattice's partition function, the various aspects and kinds of solutions seem rather more natural and physical.[166]) But knowing what the answer is, and that it is computable, is a very great advantage, as Hermann Weyl suggests in the quote that begins this chapter. Whether we shall eventually have a one-page derivation may be a matter of how much the required techniques become part of the common knowledge of the scientist, so that they can be presumed in writing up that one-page derivation.

CLUES ALONG THE WAY: IDENTIFYING THE FORMAL AND THE ABSTRACT WITH SUBSTANTIVE MEANING

Hints of connections and analogies become hints rather than chance coincidence or peculiarities when echoes and resonances lead us to insight and proof. Often, the hints are technical and formal (formula A looks like formula B) rather than conceptual or about ideas. The various connections that are suggested are at least first discovered practically, by deliberately going through a detailed calculation. We could not have imagined the connection until we had it staring us in the face. So it may be useful to give a bit more detailed account of some of WMTB's calculation, in part to appreciate its structure, in part to note where the hints appear. Yet this is a retrospective retelling, identifying the formal and the abstract with the substantive and the concrete and the meaningful, much of which was discovered latterly.[167] (See Figure 4.15, page 190; Figure 4.18, page 197.)

1. Technically, the WMTB paper begins with perturbation series expansions for the correlation function above and below the critical temperature for finite but large N, and then one goes to the scaling limit: Sections III and IV, for large-t, and Section V for small-t. Namely, they use Montroll, Potts, and Ward's Pfaffian method, as adapted by Wu, and by Cheng and Wu, there to achieve asymptotic estimates in N for fixed T. WMTB eventually obtain a perturbation series expansion for the correlation function in terms of multiple integrals, again below and above the critical temperature, multiplicity n for the nth term. Then they go to the scaling regime, first for large t (which builds in large N). However, these expansions are not nicely convergent for small-t.[168]

2. In the small-t case, in Section V, WMTB start out with a determinant of an antisymmetric matrix, a Pfaffian, and another determinant (with recurrent entries), taken from Cheng and Wu's exact closed-form expression for the diagonal correlation function in N. Then they go to large N and small distance from the critical temperature, the scaling regime expressed through t.

At the beginning of Section V they do obtain a Fredholm determinant form for the correlation function, $\langle\ \rangle = \det \hat{A} = \det (A_0 + \Delta) = \det A_0 \times \det (1 + \Delta A_0^{-1})$. ($A_0$ is the value at the critical temperature.) This form is reminiscent of Montroll, Potts, and Ward's $\det(1 + A^{-1}\delta)$ (their Equation 19, δ here being Montroll, Potts, and Ward's variable, a skew-symmetric matrix, not the phase angles, δ^* or δ', referred to earlier), and Potts and Ward's earlier $\det(1 - P'PQ)$ (Potts and Ward's Equations 20,

25, and 27). [Again, a Fredholm determinant is of the form $\det(1+\lambda K)$, K being the kernel of an integral equation, the determinant being nonzero for eigen-solutions.] But, they leave Fredholm behind, and they then follow the route of a perturbation series expansion as in Sections III and IV, but now for the small-t asymptotics for Δ.

    3. In Section VI, they take a different tack on the small-t case, and convert the Toeplitz determinant series expansion of the correlation function of Section V, in the scaling limit, into an integral equation good for all t. Then they find as solutions "scaling functions [or form factors] in terms of Painlevé functions of the third kind [P_{III}]." Or, as described elsewhere,

> [they] start from the diagonal correlation function $S_N(T)$ given by the Toeplitz determinant and define a quantity . . . that is the ratio of two determinants, they then used Cramer's rule to express it in terms of the solution to a certain set of linear simultaneous equations [Wu's Wiener-Hopf sum equations]. Finally, in the scaling limit they were able to relate the solution of these simultaneous equations to those of the integral equations . . . [169]

    3a. More technically, one goes from an expression for the correlation or two-point function, a Pfaffian, to an integral equation in the scaling regime. That integral equation—say, $\int K_0(|x-y|)\, \varphi(y)\, dy = \psi(x)+\kappa\varphi(x)$, K_0 being a Bessel function which arises in scaling—appears in scattering of electromagnetic waves from a thin strip, and earlier had been shown (by Myers) to lead to a Painlevé III equation. Namely, in the scaling regime: $\langle\sigma\sigma\rangle \rightarrow$ MontrollPottsWard $\rightarrow$ ChengWu $\rightarrow$ [|Fredholm| $\rightarrow$] $\int K_0\varphi=\psi+\kappa\varphi \rightarrow$ Painlevé. As indicated earlier, in Section V they show that the determinant is a Fredholm determinant, although that fact is not used here. Hence, I have bracketed the connection.

    3b. WMTB has been converted to the language of perturbative quantum field theory and the language of S-matrix theory, the natural languages of many theoretical physicists. A bit more technically, and interpolating some observations: The various intermediate scatterings or diagrams that contribute to the correlation function or two-point function are enumerated by a Pfaffian, following the 1963 work of Montroll, Potts, and Ward, and then Wu and collaborators: namely, it turns out, a (Fredholm) determinant derived from an integral equation. That integral equation is equivalent to a total system of ordinary differential equations and hence the Fredholm form, reflecting a series of multiple integrals: $a +b\int \ldots +c\iint \ldots + \ldots$, which can be expressed as $\exp \int\eta$. That exponential is expressible as a solution to a Painlevé differential equation.[170] In effect, the Painlevé transcendent "sums the dispersion integral representation of the two-point function," the contributions of the various connection paths between particles (the paths expressed in fourier transform space). Namely, we have "an infinite series of multiple integrals . . . [taken as] the coordinate space analog of the dispersion integral representation of the two-point function."[171]

Eventually, we might hope for a pellucid derivation, where all the hard work is built into apparatus we have mastered already, the essential points now standing out clearly. Strategy and structure, once so necessary yet obscure in their deeper motivations, become embedded into techniques, so that all one need do is apply technique in a straightforward fashion. For, in the end, whatever else, a superbly crafted many-talented watch, called a "complication" in French, does tell the time, tick-tock tick-tock. The resonances with other problems and techniques, once appreciated, eventually become part of the theory, now necessary, and they are obvious as well. And the tactical manipulations have been substantially reduced. We see what is going on, and strategy, structure and sequence, and tactics and techniques, are no longer at issue. Of course, while this is going on, new problems appear whose solutions are now complicated and resonant and lengthy.

It may be useful think in terms of a continuum: from formal and sometimes algorithmic calculation and manipulation, filled with tactical moves, to strategic and conceptual proof, filled with ideas. For some of the time complications cannot be avoided, and at others sophisticated machinery and models or deep insights about symmetry can provide a "399th solution" which is brief and even perspicuous.[172] And still it may be hard to know why it does the work. The Wu, McCoy, Tracy, and Barouch article lies perhaps toward one end of the continuum, with its various substitutions and manipulations and comparatively little physical motivation to be given, or so it would seem. Farther along is Yang's paper, and parts of Fefferman's articles. But in retrospect, in both cases there is lots of physics and mathematics (perhaps unavoidably computational) to be seen, as there will be for WMTB, as we are beginning to see. And Fefferman-Seco is filled with quite powerful mathematical strategy. For the path to their achievement will not come by simply hacking through a forest, no matter how talented the hatchet bearer. Then comes Onsager (1944), although perhaps my intimate acquaintance with it allows me to give it more conceptual clarity than it has been conventionally credited.[173] And Dyson and Lenard's proof is in fact filled with physics and guidance along the way, despite the talk of hacking through a forest of inequalities, but again this may be my retrospective interpretation. Surely, the Lieb-Thirring proof is strategic and conceptual, the physics and the mathematics quite manifest and meaningful once we appreciate the significance of the Thomas-Fermi model. And, again, I believe that there are many places within Fefferman's articles where this is the case as well.

Of course, and this is in the nature of scientific work, the formal and computational are often eventually made conceptual. And as importantly, the conceptual and strategic and deep are converted into algorithms and tricks—as in the calculus—which enable ordinary folks to do the work without thinking.

5

Analogy

A Syzygy Between a Research Program in Mathematics and a Research Program in Physics

Analogy in Mathematical Work; The Mathematical Analogy, Historically; The Analogy in Practice. *A Syzygy Between Fields of Mathematics, Expressed in Various Ways*; The Onsager Program in Langlands Program Terms; The Onsager Program and the Langlands Program. *The Import of the Analogies and the Syzygy*; Prime Factorizations and Elementary Excitations; The Analogy Made Concrete; How These Ideas Came to Be Seen as an Interrelated Complex. *André Weil's Rosetta Stone*, Learning Riemannian and Translating. *Sygyzy and Functoriality*; The Programs in Practice.

This chapter provides a foundation for the Larry and Bernie conversation of chapter 1.

I

ANALOGY IN MATHEMATICAL WORK

Analogy has proven to be a powerful strategy in mathematical research. The analogy may be a picture or an intuition linking different phenomena, one that allows us to solve a particular problem. The analogy may be formal and axiomatic, connecting two different fields. So, for example, there develops a dictionary-like correspondence between polynomials and their properties, considered as algebraic objects, and the graphs of those polynomials and their properties, considered as geometric objects.[1] Or the analogy might be tentative and inexact, at first allowing for a speculative correspondence between disparate fields and endeavors, perhaps eventually leading to a more precise understanding of how the fields are more or less the same.[2]

The analogy may be called a philosophy, as in Harish-Chandra's "philosophy of cusp forms," understanding automorphic forms from the point of view of representation theory. The philosophy of cusp forms is a predecessor of the Langlands Program which we shall discuss below.

Analogy may be a program for mathematical work, setting forth how problems and solutions in one area are suggestive for those in another. But just how the analogy is substantively fulfilled is not known ahead of time and is only dimly guessed; but once found and worked out, the substantive content of the analogy is now comparatively straightforward. And those contents go from being the fulfillment of an analogy to being just facts about the world which "happen to" or necessarily fulfill the analogy—depending on how well we understand the analogy. (In Appendix D I have provided a translation of an essay by the mathematician André Weil in which many of these points are developed. We shall turn to that essay later in the chapter.)

Typically, at first the analogy is not at all so clear, there are mismatches that do not disappear readily, and even if there would appear to be an analogy, it may not be clear just what is analogous to what in each particular case. Such an incomplete chart of potential correspondences demands that you fill in the blanks to set the analogy into motion. But for some of the places you haven't more than a clue about what should be there. Of course, in the end, the analogy is just that. It is not a proof, not a fact. It is speculative. (If you can prove the analogy, that is spectacular. But this is a rare moment indeed, it would seem.) Even with a good idea of how the analogy should be fulfilled, enormous effort may be required to make the analogy an honest one. Finding provable or factually true things to fill in the blanks is quite nontrivial.

Here I want to take on one such analogy, which will turn out to be an analogy between analogies, and explore its life within mathematical work. For the mathematicians, the analogy has been central and programmatic over the last two centuries, and is associated with Riemann, Dedekind and Weber, Hilbert, Hasse, Hecke, Artin, Weil, and Langlands. I will call it the *Dedekind-Langlands Program*, since I shall be combining some of Dedekind's analogies (ca. 1880, as later presented by Klein and by Weil) with some of Langlands' (known as the Langlands Program[3], 1967). In the background are conjectures (the Taniyama-

Shimura-Weil conjecture, ca. 1955ff, and the Hasse-Weil conjecture, ca. 1936ff) connecting geometry, spectrum, and partition functions (that is, elliptic curves, automorphic forms, and L-functions), the foundation for the proof of Fermat's Last Theorem.

The same analogy would seem to be behind a good deal of detailed work by physicists, both in exact solutions and in series approximations, on particular models in statistical mechanics (and thermodynamics) over the last seventy years. In honor of Lars Onsager's germinal contribution of 1944, I call it the *Onsager Program*, although it has never been so announced. But, for the physicists, the analogy has been only occasionally appreciated as such, the detailed calculations dominating their concern.

The analogy between these two analogies, the mathematicians' and the physicists', has been noted of late. (But it is not so clear what to do with that observation, in the mathematics or in the physics. Frenkel and Witten speak of duality and the geometric Langlands program.) At the least, as we shall see, the mathematicians' analogy allows us to systematically organize many disparate and seemingly occasional physical facts and calculations. And the physicists' analogy gives concreteness and particularity to the variety of linked mathematical ideas.

FIGURE 5.1: THE ANALOGY AS SEEN IN THE TRIGONOMETRIC
FUNCTIONS (VERSION Z)

Column 1	*Column 2*	*Column 3*
Analysis/Geometry	Algebra	Arithmetic, Combinatorial
$\sin(x+2\pi) = \sin x$ / Circle	$\sin Nz = f(\sin z)$	t_ψ "+" $t_\varphi = t_{\psi+\varphi}$ for a circle / Factorials

Version Z[*]: The analogy we shall be describing (at least the Dedekind part) is exemplified by the various ways of thinking about the trigonometric functions: (1) Analytically and geometrically, the trigonometric functions are the periodic functions known by their zeros or their periods. They can be used to label the points on a circle: $t_\varphi=(x,y)_\varphi=(\cos \varphi, \sin \varphi)$, each point on the circle known by the rotation angle, φ. (And by definition, here $x^2+y^2=1$). (2) There are algebraic relationships connecting functions like $\sin Nz$, where N is an integer, to $\sin z$.

---

[*]See p. 222 for a list of the versions of the analogy in this chapter.

Namely, transformations of the independent variable of the function, or transformations of the function itself, as in $d(\sin z)/dz$, lead back to functions of the same sort (what is sometimes called automorphy). (3) Arithmetically, the points on the circle have an addition, t_ψ "+" $t_\varphi = t_{\psi+\varphi}$. And, combinatorially, the series expansion of the sine function "packages" the factorials of the odd integers:

$$\sin x = \Sigma \, (-1)^n \, x^{2n+1}/(2n+1)!$$

Version Y: As we shall see more generally, mathematicians often associate what are called generating or partition functions to objects, the generating functions packaging properties of those objects into a single function. And relationships among those functions connect arithmetic properties of objects to analysis of functions. The number of ways of composing each integer into a sum of smaller integers of a certain sort might be packaged into a function whose properties are expressed without direct reference to those combinatorial numbers. But those properties do depend on the regularity of those combinatorial numbers. The decomposition of integers into prime factors is packaged into the Riemann zeta-function, $\zeta(s)$, which has its own functional properties. And in harmonic analysis, geometric features of objects are packaged into a spectrum or a spectral function, so the shape of a drum determines its sound.

Zeros of those partition or generating functions may sometimes be connected to the eigenvalues and symmetries of an operator or a matrix, one that often is in effect a representation of a group. So another spectrum is connected to symmetry. Such operator symmetries may characterize dynamical processes, say the symmetries of a physical system. One might hope to find a physical system whose eigenvalues (that is, the eigenvalues of a matrix associated to the system) are the prime numbers and so study the zeta function, or attach a zeta function to a dynamical system and use number theory to study the system's dynamics. For example, the statistical mechanics partition function for a particular one-dimensional system with long-range interactions is a ratio of zeta functions: $\zeta(s-1)/\zeta(s)$.[4]

Or, the analysis of those partition functions may lead to functional equations, epitomizing the partition functions' properties or the asymptotic properties of the objects being studied. So the Riemann zeta-function fulfils a simple functional equation, and tells about the asymptotic frequency of the prime numbers.

In sum, combinatorial properties of objects are connected to functions, arithmetic to analysis. These are functions whose zeros may be connected to geometry or the symmetries of a matrix, and functions that may have nice functional equations and clean asymptotic properties.

THE MATHEMATICAL ANALOGY, HISTORICALLY

> . . . symmetries create redundance, and this disorderly redundance has to be reduced to an orderly, perceptible uniqueness. (Robert Langlands[5])

Version A: The analogy begins with Riemann's (1851) geometrical and topological account of an algebraic function, $w(z)$, a smooth function defined by a polynomial equation, $P(w,z)=0$, with complex coefficients. The topology of a Riemann surface associated with $w(z)$ embodied the analytic and the algebraic facts. (See Figure P.1) The analogy here is between readily visualized pictures and algebraic properties of functions. Subsequently, Dedekind and Weber (1882) push that account in an "arithmetical" direction, the analogy now between properties of those functions and properties of number systems. Then these analogies are extended by Hilbert (1897 and later); Emil Artin (1927, "Artin reciprocity"); André Weil (1940); and, Robert Langlands (1967), among others. The system of analogies is sometimes called a yoga ("a system of exercises for attaining bodily or mental control and well-being,"[6] says my dictionary). A current version of the analogy is called the Langlands Program and Langlands reciprocity.

(Here "reciprocity" means a connection between the divisibility or factoring properties of a number or a polynomial and properties of another object, here the analytic properties of a function, an "automorphic form." What would be difficult to discover by actually searching for factors, can be discovered by studying another object that proves to be rather more tractable. Analogously, were we deaf, we might discover how a drum sounded by measuring its area and periphery.)

Dedekind and Weber wanted to use algebraic and functional means (what André Weil will call the middle column, here Column 2, see Figures 5.1 and 5.2) to prove as much as was possible of Riemann's "transcendental" (or topological or analytic) results (Column 1), so extending the analogy between algebraic functions (Column 2) and geometry and topology (Column 1). They wanted to use algebraic means to obtain algebraic results, avoiding assumptions or geometric intuitions about continuity and about curves and surfaces. To do this, they employed what was already known about algebraic numbers, solutions to algebraic equations with rational coefficients (Column 3), to develop a theory of algebraic functions (Column 2), namely, they developed an analogy of Column 3 to Column 2, in order to work out the analogy of Column 2 to Column 1. More technically, defining a field of algebraic functions in analogy to a field of algebraic numbers, they then work out a theory of primes (ideal numbers and prime ideals) in the realm of functions. And the prime ideals in that realm are shown to provide the algebraic analogue to a point on a (Riemann) surface, so avoiding geometric intuition.[7]

Subsequently, Hilbert, in his great survey of number theory, what is now called the *Zahlbericht* (1897), refers in his preface to the "close connection of number-theoretic questions and algebraic problems." And in the next paragraph, he then connects both algebraic function theory and Riemannian function theory to number theory, "arithmetizing" both subjects. He adds that "even in function theory a fact can count as proved only when in the last resort it is reduced to relations between rational integers."[8] (In time, Hilbert (1901) and Weyl (1913) redo Riemann's theory so that the prejudice against the geometric intuition, and the transcendental and the analytic, is no longer tenable.) Weil later pursues the analogy in a profound way.

Of course, one has to understand what the analogy means, what that close connection is. So, as Weil describes it, in Artin's thesis (1921), he "elaborately carries over the whole classical theory of quadratic number-fields to quadratic extensions of the field of rational functions over a prime field of odd characteristic, with a never ending delight at finding the correct analogy for every classical concept."[9] Langlands pursues the insight of Artin's later work, which employed one-dimensional group representations of Galois groups to epitomize the regularities within number theory, now for higher dimensions.

Now, in the last seventy-five or so years, and essentially independent of the mathematicians, physicists have been solving the problem of deriving the thermodynamic properties (the statistical mechanical partition function and the spin-spin correlation function) of the Ising model of ferromagnetism for a two-dimensional crystal endowed with a spin at each lattice. This is a problem in classical statistical mechanics, albeit it has formal analogy with a one-dimensional quantum mechanical system. The physicists have solved the problem in an astounding variety of ways—one solution, using a functional equation, refers to itself as the "399th solution."[10] What is remarkable is that each of many of these various ways may be seen to exhibit one moment or another of the mathematicians' analogy, in a rarely available level of detail and specificity, and so gives some more flesh to the schematic analogy. The various mathematical objects employed are just those one might expect from the mathematicians' work (*Versions Y, A, B*).

The Analogy in Practice

There is some sort of combinatorial object, *O*, combinatorial in that we might count the number of ways of arranging its parts distinctly, or the number of solutions to an equation in a particular number system, and so allows us to define a partition function (=*PF*) that packages all the combinatorial information about it in a convenient form. In such counting, as in putting together a puzzle, it does not matter how we group and subgroup the parts, much as in arithmetic that $a+(b+c)=(a+b)+c$, namely, the associativity of addition.[11] I will refer to the third or arithmetic column as having the property of "associativity" (admittedly, in part to alliterate with the first column's analyticity and the second column's automorphy).

Figure 5.2: The Analogy Connected to Some Matrix Systems
(Version *B*)

Partition Function (PF)

Geometry/Topology Analyticity/Continuity	Algebraic Automorphy	Arithmetic Associativity
Curve/equation, *E*, as surface within a $\leftrightarrow$ complex space (e.g. Elliptic curve)	Object, *f*, taken as scale-symmetric, $\leftrightarrow$	Object, *O*, taken as combinatorial or arithmetic
$\downarrow$		$\downarrow$
Matrix System (M_E) PF =trace M_E	*f* itself is a group representation	Matrix System (M_c) PF =trace M_c

Version B: The basic ideas and moments in the analogy are indicated in Figure 5.2, recalling Dedekind, Artin, and Langlands.

Now it may be that the partition function might actually be calculated. But, rather than forming the partition function by counting things up or doing the combinatorics for each situation or configuration, one attaches to the object, O, a nice function, f, which is perhaps scale symmetric: say, $f(Nx)=\sqrt{N}f(x)$, as is the case for the Central Limit Theorem of statistics. (In the case of the Central Limit Theorem, where we are adding up N random variables, the scaling constant is $\sqrt{N}$.) It turns out that by employing f, one can calculate the partition function rather efficiently. The problem that remains is to figure out how to find f for a particular O. In some sense, O's symmetries should be reflected through f.

More generally, f is said to be automorphic: a transformation, γ of x ($x \rightarrow \gamma(x)$), leads to another function of the same sort as f, as in $f(\gamma(x))=Cf(x)$, where C is a suitable constant or "modulus."[12] Automorphy is an algebraic property (here, of an analytic object, f), namely that functions transform into functions of the same sort. Again, for example, sin $2x$ is expressible in terms of the self-same trigonometric functions ($= 2$ sin x cos x).

The classical example of going from O to f in mathematics is the Riemann zeta function, $\zeta(s)$, a combinatorial partition function enumerating the prime numbers, whose properties may be understood through those of the theta function, $\theta(\tau)$, a function that exhibits automorphy. The combinatorial partition function, zeta, is transformed into the modular function, theta, by the fourier-like Mellin transform. Theta's automorphy provides for a good definition of zeta (its analytic continuation) and leads to a functional equation for zeta.

One might also associate a curve or surface, E, to O, much as we might associate a graph to a complicated mechanical system, that graph displaying the performance of the system as a whole. And geometric or topological features of that curve or surface may be illuminating for understanding the partition function. A crucial feature of these curves is their smoothness or continuity or analyticity, as well as their cuts, holes, and singularities and their topology.

During the late eighteenth and early nineteenth centuries, mathematicians learned how to attach symmetry groups to objects, such as the Galois group (of permutations) to an equation, reflecting the symmetries when exchanging its roots. One of the discoveries of the early part of the twentieth century was that for many difficult mathematical problems, an often convenient side-path is to associate a linear object—a matrix or a matrix system, what is called a group representation—to the objects one is studying, here to O or f or E.[13] One has to figure out how to make such an association, just how and when a group may be associated with O or f or E, and what sort of representation one needs. In the particular case of O or f or E for the Langlands Program, the trace of such an appropriate matrix (its group character) will provide the combinatorial information as well, and it leads naturally to the partition function.[14] Note that here f itself is a (automorphic) group representation.

Echoing our initial presentation of the analogy in terms of trigonometric functions such as sines and cosines, the leitmotif throughout is provided by the elliptic functions, the doubly-periodic cousins of the singly-periodic circular or trigonometric functions. Corresponding to the sine function, sin u, there is the Jacobian elliptic function sn (k,u); corresponding to $1/\sin^2 z$ is the Weierstrass elliptic function, $\wp(\omega_1,\omega_2,z)$; and corresponding to $q^{1/4}\cos z$ are the elliptic theta functions, $\theta(q,z)$—where k, $\omega_{1,2}$, or q account for the two periods of these functions, and the arguments z and u are simply related to each other. As do the trigonometric functions, they have arithmetic (and in this case, combinatorial) properties, they are automorphic and so have function-theoretic properties, and they have geometric or topological properties—all of which are intimately connected to each other. Those interrelationships are a model for the analogy that appears in the number theoretic and in the statistical mechanical realms. Elliptic functions, or their cousins the automorphic functions and modular forms, appear pervasively in each realm, initially in surprising places.

Historically, in computing the arc length along a circle, one finds the integral $\int_0^{\sin z} dx/\sqrt{(1-x^2)}$, $z \leq \pi/2$. Now in computing the arc length along an ellipse, one integrates $dx/\sqrt{P(x)}$, where $P(x)$ is a third or fourth degree polynomial. Appropriately, these are called elliptic integrals. (The elliptic functions are their inverses.) If $P(x)$ is a nice cubic, and say as well that $P(x)=y^2$, then $y^2 = 4x^3 - ax - b$ is called an elliptic curve. Elliptic curves are a recurrent example of the triplet. They, too, possess arithmetic and combinatorial properties, function theoretic properties, and geometric properties—again, all of which are intimately connected.[15]

Of course, a leitmotif is not a proof. Rather it is a temptation to show that what might be true is in some particular sense actually true.

Each of the various paths to the partition function is distinct. Usually the analogy is not so clear, and the paths we might follow, or things we might actually compute, are rather more restricted. So it is striking that one can sometimes compute the partition function directly through O, or through an f, or through an E. Moreover, the matrix-system paths are also distinct, and are only sometimes available. Whatever the approach, we expect that the partition function we end up with is the same in each applicable case. But, each of the various paths builds into its formulation features that may well be very difficult to prove if we follow another path. For example, automorphy can be built into one derivation, but be a conclusion in another; analytic continuation may be provable if we follow one path, but not at all apparent if we follow another. So if we can show that there are two roads to the same partition function, we can then infer distinct features of that function from each of those paths. Multiple derivations of a result are not at all repetitions; rather, they are aspectival variations allowing us to see the result in various lights.

The deepest problem, put baldly, is to understand how and why this threefold analogy works in different realms, how and why the various computations agree, how and why the matrix representations do the same work, and how and why the physics analogy and the mathematics analogy could be analogous to each other. And knowing how the analogy works may not tell us enough about why it works. For that we may need an overarching or deeper theory or many more examples seen as aspectival variations. Moreover, parts of the analogy might prove to be coincidental and idiosyncratic, and so be put aside. And the analogy may not be so formally precise, but just highly suggestive.

In some realms these associations are more or less worked out; or, the associations were facts found along the way in a calculation, and the task is to show that the connections are more than just occasional. In other realms, the associations form a set of conjectures for which we have only the most hazy prospects for verification. An ancillary but formidable problem, and in fact where almost all the work has been focused, is to understand how this analogy works out in particular cases, in detail, in the mathematical and in the physical realms, and why it works out in each case. As the analogy becomes better understood and more precisely stated, language and models from one realm may influence another realm, fruitfully or not.

II

A Syzygy Between Fields of Mathematics, Expressed in Various Ways

As we shall see again and again, deep analytic/geometric facts, algebraic facts, and arithmetic facts have turned out to mirror each other in mathematical statistical mechanics and in number theory. And that mirroring within each field is then mirrored in the other. There is an analogy *between* fields of the analogy *within* each field. Yet, historically, the analogy within each field had fairly independent courses of development.

Whether it be solving a functional equation, given assumptions about the solutions (their smoothness or analyticity, and the location of poles and branch cuts); or, working with a set of matrices (potentially a group representation), matrices that embody the crucial arithmetic and analytic facts; or, counting up the number of configurations or combinations of objects, under particular constraints, say in terms of counting by scales or a cluster decomposition—all mirror each other. And properties of a cubic curve, or one of higher degree, would seem to mirror all these sets of facts, again in each field. Still, the details matter enormously. But having enunciated the analogy, just what is mirrored and how it is mirrored is not a given. Only in its provable and factual correspondences will we find the analogy truly informative rather than merely speculatively suggestive (which suggestiveness is no mere achievement in itself).

We have here what might be called a "syzygy," to use J.J. Sylvester's (1850) term of art for an analogy between analogies. A physicist who came upon number theory, or a mathematician who came upon statistical mechanics of the Ising model of ferromagnetism, might surely be struck by this analogy of analogies, as were Larry and Bernie. And they have been so struck. Each threefold-analogy is itself striking. In 1940, the mathematician André Weil called one of them a Rosetta stone.[16] In retrospect, considering that number theory and statistical mechanics are counting enterprises, perhaps the syzygy is not so surprising, at least generically. Still, in its detail, the analogy retains its striking and remarkable quality.

Deciphering such an inscribed Rosetta stone is sometimes the work of mathematicians. In the background, there is always a guiding philosophy, a sense of how the world must go, a sense of which analogies are worth taking seriously. A mathematician might say about such analogies:

[I have been] describing things as they may be and, as seems to me at present, are likely to be. They could be otherwise. Nonetheless it is useful to have a conception of the whole to which one can refer during the daily, close work with technical difficulties, provided one does not become too attached to it, but takes pains to ensure that it continues to conform to the facts, and is prepared to abandon it when that is called for. (R. Langlands[17])

Again, the purpose of this chapter is to show one way analogy is employed by mathematicians and by mathematical physicists in doing their work. But it is worth repeating that that work, in the end, has to stand on its own, more or less independently of the analogy that suggested it—unless the analogy itself is proven.

We already have a schematic history (*Version A*) and a chart (*Version B*) that summarizes some of the analogies we shall encounter. Now, to the several Rosetta stones. Since such decipherment benefits from various partially redundant texts in different languages, there is somewhat more repetition in this chapter than would otherwise be needed. For it is the repetitions that provide the hints, the keys, and the analogies themselves.

By my count, I provide twelve versions of the analogy within a realm or of the analogy between realms:

Part I:

Version Z: Properties of the Trigonometric Functions

Y: Objects and their Generating Functions

A: Historical, On the Mathematical Side.

B: Basic Three-Column Scheme, Connected to Matrix Systems

Part II:

C: The Onsager Program in Dedekind-Langlands-Program terms.

D: The Statistical Mechanics (*D1*) and the Number Theory (*D2*) stories, so presented that there are apparent parallels between them, on account of the partition functions, the symmetries and functional equations and elliptic curves, and the matrix systems or group representations.

Part III:

E: Motivations: Prime Factorizations and Elementary Excitations.

F: Rubber Sheet Geometry, Fractals, and Jigsaw Puzzles (Analyticity, Automorphy, and Associativity).

G: The Mathematical Objects.

Part IV:

H: André Weil's Three Columns, as he presents it historically (*H2*), as it might be arranged in terms of correspondences (*H3*), and its predecessor in Dedekind-Weber as presented by F. Klein (*H1*).

Part V:

I: Shibboleths: Langlands functoriality and the Yang-Baxter relation.

J: Summary scheme: An Object and Its Partition Functions

As we have already seen in *Versions Z*, *Y*, and *A*, respectively, some versions concern the threefold analogy (*Z*, *B*, *F*, *H*), some describe how partition functions are connected to objects (*Y*, *I*, *J*), and some are stylized histories and comparisons of the two fields (*A*, *C*, *D*, *E*, *G*). To say there is one analogy here is surely speculative on my part, in part because of my very different levels of understanding of the different fields. But in my reading of Langlands, Onsager, Weil, and modern mathematicians and physicists, the parallels I draw are not too far from what they themselves say.[18]

I now want to describe the syzygy in more concrete terms, and the distinct programs for research in statistical mechanics and in number theory that it represents, the Onsager Program and the Dedekind-Langlands Program, respectively. Part III goes further into what it means to follow a research program, and how analogy functions within such a program, how the analogy is in practice almost always quite particular and restricted even if we believe it applies more generally. André Weil's presentation of the analogy for number theory follows in Part IV. I then say a bit more about how such analogies are useful, and speculate on the prospects for these research programs.

We want to show how a significant analogy is understood in actual practice, how ideas from one moment of an analogy may lead to ideas for the other moments, and how an analogy in one field might give practical support, justification, and moral encouragement to an analogy in another field. At first, analogies may be technically incorrect, but still they are suggestive of a true connection or a true fact which in the end can be proven. The analogy between different research programs may still need lots of work for it to be made technically correct, but the encouragement it provides along the way may be crucial to progress in each field. On the other hand, the just-rightness of an analogy is an important demand. Close is not enough, because the additional requirements that make an analogy just right can be illuminating about what is really going on. In the end, theorems, derivations, and facts are what counts. But, until we fully understand an analogy, the analogy proves to be indispensible, as Weil suggests. It leads us to new insights about what might be true. I am not sure we ever fully appreciate the most profound of the analogies, for they seem to be recurrently productive.

THE ONSAGER PROGRAM IN DEDEKIND-LANGLANDS-PROGRAM TERMS

The physicists employed statistical mechanics in the context of classical mechanics to work out the thermodynamic properties of a model of a magnetic system (the Ising model in two dimensions), developing whatever techniques they needed along the way. They constructed formal mathematical models; they

FIGURE 5.3: THE ONSAGER PROGRAM IN DEDEKIND-LANGLANDS-PROGRAM TERMS

—The partition function for the Ising model, Q, is the *trace of a matrix* (tr V) a matrix that is drawn from a *group representation* associated with the braid group (Yang-Baxter relation), or with an *elliptic curve*, or with an *automorphic form* (actually here the automorphic form is the partition function) or another group representation associated with a Lie algebra. One can solve for the partition function *analytically, algebraically,* or *arithmetically* (and *combinatorially*).

—Q has a natural *combinatorial interpretation* and a derivation based on counting graphs. Q may be computed as a series expansion or by renormalization group methods, as well as by the determinant or Pfaffian of a matrix.

—The symmetries of the lattice system lead to a *functional equation* for Q, connecting physical and unphysical regions (within the complex plane). That equation is solvable, using elliptic functions, if we assume *analytic or meromorphic continuation.* That we can *extend the partition function to the complex plane* would seem to reflect *regularities in the partition numbers.*[19]

—What is difficult to discern in one derivation of the partition function, is straightforward or even deliberately built-in in another.

—All *the various partition functions are the same* (as we might hope). Notably, they appear to reflect the *Riemann/Dedekind-Weber/Hilbert/Artin/Weil analogy* of transcendental algebraic functions (understood in terms of continuity and analysis), algebraic functions understood algebraically, and algebraic numbers.

—For a square lattice (equal horizontal and vertical couplings), Q exhibits *self-duality.* By observation of the matrix elements, or direct calculation, or as a consequence of the *Poisson sum formula* of harmonic analysis, Q is a *modular function.*

—In simple cases, Q's *zeroes lie on a line* in the complex plane, and in general they are well behaved for exactly-solvable models. We might say that Q's *zeroes are the duals of the combinatorial numbers* encoded in Q. Since Q is a sum of strictly positive terms, Q's zeroes are off the real axis but pinch it in the infinite volume limit.

—Q may be expressed as a *product* of the partition functions of elementary (fermion) components of the lattice.

calculated, more or less exactly, more or less rigorously. The mathematicians, on the other hand, had a longstanding, generic problem of understanding prime factorization (what came to be called "reciprocity laws"). The physicists were able to quite concretely instantiate the generic program of the mathematicians, without being much aware of or concerned with that program. And the mathematicians can suggest more generally why the physicists might well end up employing certain mathematical objects or encounter connections between such objects. Still, the concrete particulars of the physicists' problems may exceed the generic capacities of the mathematics; the generic structures of the mathematics may not apply directly to the particular calculations of the physicists' model.

Version C: We might retell much of the story of the Ising model in mathematical statistical mechanics in light of the work in analytic number theory and representation theory, namely, we might tell the story of the Onsager Program in mathematical statistical mechanics in terms of the Dedekind-Langlands Program of number theory and representation theory—realizing of course that this is a retrospective reconstruction, not an historically chronological account. The generic Holy Grails of the mathematicians, the italicized terms, objects, and relations in Figure 5.3, some of which are not the terms usually employed in discussing the Ising model) are the ordinary spectacular achievements of the physicists, achieved on the way to solving their problem. Each of the panoply of solutions to the Ising model is not just an advance, or an improvement on a predecessor solution, or a simplification. It turns out to be as well a representative of one moment or another of the mathematicians' analogy and schema. So we might sketch the story of the Ising model and the Onsager Program in Dedekind-Langlands Program terms (here, mostly Langlands Program terms). (I realize that for many readers the resonances won't be readily available, given the disparate nature of the two different programs. But I believe I have been fair and I have not pressed the parallels too strongly.)

I have restated the range of results and methods of solution of the Ising model, employing the mathematicians' notions and language (perhaps sometimes with a bit of a speculative stretch). That restatement itself is of some interest if it adequately covers the range and methods employed for solving the model: it provides an organization for the disparate methods. Again, the various different methods of solution are not necessarily superseded by new methods. For each method of solution illuminates a different aspect of the model, here organized by a pattern from an overarching mathematical account. The solutions bear a larger analogy to each other. The methods themselves are linked, not only by their leading to the same result, but by their being representatives of this overarching analogy.

THE ONSAGER PROGRAM AND THE DEDEKIND-LANGLANDS PROGRAM

In order to appreciate the keying of the Onsager Program to Dedekind-Langlands Program terms, it will be useful to have an account of each program in its own terms, while arraying in parallel the features of each program. I can imagine that some readers will want to skim the technical details of the next two sections, on the statistical mechanics and the number theory, and go immediately to part III, where I summarize the argument and draw some lessons. I hope that for other readers, the details presented here will be just the flesh they need to appreciate the schematic presentation so far.

*Version D*1 (Statistical Mechanics): In the 1940s and the early 1950s it was at least implicitly noted if not always proved, that solutions to the two-dimensional Ising model of ferromagnetism, as a lattice of atoms interacting with their neighbors through their spins, possessed the following features:[20]

(1) Multiplicative Partition Function and its Zeros. As usual, there is a counting function, $Q(K_1,K_2)$, called the partition function, which counted the number of ways of arranging, say, an Avogadro's Number of molecules so that macroscopically, as in a gas or solid, we could not tell one particular microscopic arrangement of atoms from another. K_1 and K_2 are measures of the interaction among adjacent molecules, horizontally and vertically for a two-dimensional rectangular grid, and they implicitly include the temperature (large K_i means low temperature). Hence $Q=Q(T)$, where T is the temperature. Away from the critical temperature for a phase transition, T_c, Q can be expressed as a power series in its macroscopic variables.

Kramers and Wannier (1941) appreciated there was a temperature-like variable kappa, nicely related to the K_is, and at $\kappa=1$ there was likely to be a phase transition in this Ising model. Onsager (1944) employed another temperature-like variable, k, nicely related to κ (by what is called a Gauss-Landen transformation), which was as well the modulus of an elliptic function. So $Q=Q(K_i)=Q(\kappa)=Q(k)$, although the functional forms in each case are of course different.

Notably, for two independent non-interacting systems at the same temperature and pressure, now considered as a whole, the partition function is the product of their partition functions. Namely, log Q was proportional to the thermodynamic free energy, F, and F was an intensive function of temperature and an extensive function of volume. The thermodynamic prescription is that one simply adds the free energies of noninteracting system at the same temperature and pressure to obtain the free energy of the combined system.

Formally, in the canonical ensemble of statistical mechanics, the model of a system at a fixed temperature, $Q=\Sigma e^{-\beta E}$, where β is the inverse temperature and E is the energy of each of the arrangements, and the sum is over all the possible energy states of a system of molecules. Phase transitions take place when $Q=0$ and β is non-negative and real (the range of values one would expect for an inverse temperature). But as is apparent from its definition, Q would seem to be strictly positive, a sum of positive terms. However, that is not the case if β is a complex number, and we might imagine, as we approach the real line in the complex β plane, that the $e^{-\beta E}$, now complex, might cancel out in the sum much as we get destructive interference of light waves. (Note that it is crucial that Q be well defined outside of its original area of definition, the positive real β line—if we are to analytically continue Q so as to study its zeros as β approaches the real line.) In some nice models, we know that the zeros can and do pinch the real line, at the critical temperature, in the limit of an infinite number of particles at a fixed finite density, the "infinite volume limit." For simple models, the zeros of the partition function lie on a line in the complex plane.[21]

It was also realized that the partition function corresponded to a graph-counting combinatorial problem, leading to series solutions of the problem and so the possibility of exploring the zeros of the partition function and its behavior when the magnetic field is nonzero. Eventually, it was shown by graph-theoretical means how the known exact solutions (which were originally derived algebraically based on the symmetries of the system) did the combinatorial work. The partition function could be calculated quite effectively, and exactly for many situations, by combinatorial means, by setting up a matrix, A, whose determinant (actually, a "Pfaffian") did the counting work. The method has a direct analogy to calculating scattering processes in quantum field theory, in effect a sum histories ala Feynman, and to the Wick product.

(2) Symmetry, Functional Equations, and Elliptic Curves. $Q(k)$ exhibited automorphy: an inversion (or "duality") symmetry and a scaling symmetry. More precisely, $Q(K_1, K_2)$ exhibits a symmetry connecting Q for values of the K_is that are low, with Q for values of the K_is which are high, originally called duality, referring to a geometric dual of the lattice (points exchanged with lines), effectively connecting a high temperature partition function with a low temperature partition function: $Q(\text{high } T)=(k_{\text{low}})^{N/2}Q(\text{low } T)$, where N is the number of particles. This would seem to be an automorphy. It was realized that

employing the normalized set of variables, k and u—one a measure of temperature, the other a measure of the relative strengths of horizontal and vertical forces in the lattice—would make this symmetry transparent as an inversion symmetry since the temperature-like variable went from k to $1/k$ in a duality transformation, while u becomes u's complement. So $Q(k)$ and $Q(1/k)$ are related, much as are the elliptic functions. (For Onsager (1944), already, these variables (k, u) or close cousins emerged out of their role in the elliptic functions and elliptic integrals needed to solve the problem.) Again, in particular, $k=1$ is a special point (as is the corresponding $\kappa=1$), and in fact a phase transition does take place at this point.

Q also exhibits a scaling symmetry (very roughly, scaling up differentially by a power of k). Again, k, in $Q(k)$, behaved like the modulus of an elliptic function. Also, one could use scaling and an inversion symmetry to write a functional equation for Q.[22] And, given "reasonable" assumptions about analyticity, one could solve that equation. Moreover, Q's zeros are often located on a line or within a region of symmetry of a functional equation for the partition function.

k's significance is expressed in terms of a polynomial equation (here an elliptic curve) connecting k with the K_is, and that one can write nice functional equations for $Q(k,u)$.[23]

(3) Group Representations (for free). One could find a matrix system, V, such that the trace of the matrix was exactly the partition function of interest, namely, taking the trace computes the partition function, summing up the relevant Boltzmann weights ($\exp - \{\beta \times [\text{Energy of a configuration}]\}$, β being the inverse temperature). Its algebra (or commutation rules) might be seen to represent symmetries of the system such as thermal equilibrium (that is, k is a temperature-like variable). In fact, there were at the least two such matrix systems: one that led to the elementary particles of the thermodynamic system—essentially patterned rows of spins that might be said to make up the interacting lattice of spins (say, V_r, where r stands for row) and, another which exhibited automorphy or k-symmetry most explicitly (say, V_d, where d stand for diagonal).[24] Matrices characterized by different (K_1,K_2) but the same k commuted, and that symmetry was encoded by the polynomial equation or elliptic curve relating the K_is to k. (Properly, we would want to be explicit about which groups are being represented; why group characters (those traces) and those conjugacy classes are relevant (say, the same thermodynamic state); and just how the representation is automorphic (say, the same way as in the Langlands Program).)

There is as well a third counting matrix, V_c (c standing for corner), expressed in terms of half-rows and half-columns, accounting for the interactions in a quadrant of a lattice, and again taking the trace computes the partition function. And recall, there is another matrix, A_{Pf}, that does the graph-counting or combinatorial work more directly, the links being weighted by Boltzmann weights, and its Pfaffian gives the partition function. Originally, Kac and Ward developed a jury-built counting matrix V_{KW}, based on Kauffman's spinor method, whose determinant was the partition function. (I do not believe that A_{Pf} or V_{KW} are group representations.)

Remarkably, the matrix system, V_d which expressed the automorphy most directly, had an algebra (the Yang-Baxter equation) that was like the rules for the algebra of braids that make up a knot. The combinatorial matrix, A_{Pf}, effectively counting either closed polygons on a lattice or colorings of a map or graph, led to the same partition function as the group representations V_r and V_d, as we might expect. Again, examining the matrices in detail, we encounter the recurrent connection between scaling or automorphy and combinatorial problems.

More generally, V_r and V_d were analogous to evolution operators in quantum mechanics (that is $V \approx \exp-"H"\tau$, where τ is goes from row to row, say, and "H" is a pseudo-hamiltonian), the V_c analogous to a Lorentz boost in special relativity (along the u axis), the matrix A_{Pf} analogous to calculation of a sum of amplitudes for different paths between beginning and final states in quantum mechanics and field theory.[25]

Originally, as the subject actually developed, $Q(K_1,K_2)$ was expressed as the trace of a matrix, $V_r(K_1,K_2)$—just because and only because the trace of that matrix in effect added up the contributions of the atomic configurations' Boltzmann weights to the partition function. It was an ingenious device, not a group representation.

More technically, the partition function's sum of terms, $Q=\Sigma e^{-\beta E}$, each of which was a product of the contributions from each of the atoms, so that the exponents or energies added up to the total energy of a configuration of spins, turned out to be expressible as the trace of a product of matrices. Those matrices, at first, were just convenient devices to keep track of the sums of products— albeit they allowed one to think in terms of the matrix mechanics techniques physicists were used to employing for quantum mechanical problems. Earlier, the formal symmetries of the matrix's elements suggested the high-temperature/low-temperature symmetry (expressed by Onsager as "$B \leftrightarrow A$").

Even when Onsager treated the matrix system as a group representation, figuring out the symmetries and so diagonalizing the matrices, what it was a representation of (other than spatial symmetry and duality) was not a main issue. Onsager found that $V_r(K_1,K_2)$ could be diagonalized by means of a Lie algebra of what we now would call energy operators of the putative particles of the system. Onsager also found that if the rectangular $V_r(K_1,K_2)$, which added up interactions a horizontal row or a vertical column at a time, is replaced by a diagonal-to-diagonal $V_d(k,u)$, which added up interactions along the diagonal, then V_d-matrices with the same k but different (K_1,K_2) commuted. Later, it was found that products of V_d-matrices with the same k, might be said to conserve momentum or rapidity, u—as if we have particles (*not* the particles referred to as rows of spins) with different u colliding, and out come particles that conserve total u. Again, the algebra here is reminiscent of the braid group.[26] In effect we have two-body elastic scattering (what eventually becomes the Bethe Ansatz.)

In sum, $Q(T)$=trace $V_r(K_1,K_2)$=trace $V_d(k,u)$.

In all this work, it was not so clear why one ended up with elliptic functions, or the particular algebras or commutation relations, or why one would have matrices appear in this sort of problem, matrices which would appear to a mathematician as possibly a group representation—except that they did the work. Initially, it just turned out that way. (Onsager's 1944 paper is filled with casual but cryptic remarks, which may in retrospect be seen to insightfully presage much of what was to come.) For the physicists, it was eventually seen that the elliptic functions reflected the k-invariance of the lattice, or in a quantum field theory analogy, Lorentz invariance; Onsager's energy operators could be seen to belong to particular Lie groups that in effect generated the lattice; again, the transfer matrix or group representation might be shown to work in the same way as does the evolution operator in quantum mechanics (although this is a classical problem); and, the algebraic commutation properties of the Ising V_d's could be seen to reflect features of an analogous account of particle scattering.

However, for the mathematicians, in their own work, the elliptic functions or the group representations would have been taken to be the most natural of objects to be found in this realm, whatever their physical significance.

Version D2 (Number Theory): Over the last two-hundred-plus years, mathematicians have noticed:

(1) Multiplicative Partition Function (Harmonic Series) and its Zeros, and Elliptic Curves. There is a counting function, nowadays called the Riemann zeta function or, more generally, the Dirichlet L-function, which can be used to enumerate or encode the prime numbers in an arithmetic progression: say, 1, 5, 9, *13*, *17*, 21, 25, *29*, 33, ... Or, we might enumerate the number of rational

solutions to a diophantine equation, such as the cubic $y^2=4x^3-g_2x-g_3$ (an elliptic curve), modulo a prime, p. L is a harmonic series: $L=\Sigma\chi(n)/n^s$; $\chi(n)$ depending on some arithmetic property of n, for example, that n is in an arithmetic series and is or is not prime, or how many solutions there are to a diophantine equation modulo n. Notably, the $\chi(n)$ may be multiplicative, namely that $\chi(mn)=\chi(m)\chi(n)$. Then L may be expressed multiplicatively as an "Euler product," for all primes p, of terms like $\chi(p)/(1-p^{-s})$. (Technically, when the denominator is expanded in a power series, and the terms are multiplied, one obtains each of the $\chi(n)/n^s$ terms, by unique factorization of the integers.) If all the $\chi=1$, then L is the zeta (ζ) function of Riemann. (Note that L recalls the partition function defined in statistical mechanics, $\Sigma \exp -\beta E$: $L=\Sigma \exp\{-s \log n + \log \chi(n)\}$.)

If $s=1$, the L-function will usually be infinite, unless $\chi(n)=0$ for $n>N$, for some finite N. More generally, the sum that defines the L-function will be convergent rather than infinite, for real s greater than s_0, where s_0 depends on χ. In any case, L is defined only for real s greater than s_0, say 1 or perhaps 3/2. Yet important number theoretic information, such as the prime number theorem about the frequency of primes, depends on knowledge of L's behavior, such as its zeros, for values of s outside of L's range of definition (in this case, for $s=1/2$ and, in others, in the neighborhood of $s=1$).

We need another way of defining the same L-function, one that builds in the analytic continuation and its definition for these crucial places.

(2) Symmetry, Functional Equations, and Elliptic Curves. There is another counting function, called the theta function, $\theta(z,\tau)$, which is also an elliptic function, defined for all values of z (it is an "entire" function).[27] (In general, these are the theta-like automorphic forms.) Theta and its rth-powers counted the ways of forming a number from the squares of r numbers. The fourier coefficients of $\theta(0,\tau)$ are just these numbers. Obviously, for the theta function itself ($r=1$), this is only the case for n being a square itself (referring to the nth term in the fourier series expansion), and then the number of ways is just 1. If $n=m^2$, then theta is a power series consisting of terms: $\exp i\pi zm^2 \times q^{m^2}$, where $q=\exp i\pi\tau$, so it is a fourier series in τ. Again, there may be a canonical way, much like fourier transforms, of associating to zeta, and more generally to many an L-function, a theta-like function (the automorphic form). Making use of the theta function's good behavior and well-definedness, and its symmetries, one could define (or analytically continue) the L-function in areas of interest; and, one could write a functional equation for $L(s)$, so relating $L(s)$ to $L(1-s)$, for example.

Theta also exhibited an automorphy, an inversion symmetry, much as we described for Q, what is called the Jacobi imaginary transformation.[28] Theta has two almost-periods, namely when it is shifted by one of the periods, it is merely multiplied by a number (a modulus).

Again, we have reason to believe that the complex zeroes of the L-function lie in a strip of symmetry suggested by the functional equation. The Riemann hypothesis says that for the zeta function, the zeros lie on a vertical line in the complex plane, with the real part equal to 1/2. Notionally, we might say that the regularity of the combinatorial numbers allows for theta's good behavior: its pseudo-periodicity, its functional equation, its analyticity. Put differently, the regularity of the combinatorial numbers packaged in an L-function allows for the automorphy of its associated theta function, and vice versa.

(3) Group Representations. And it turns out that, sometimes, one can find a mathematical group that reflects the symmetries of the object (say, the Galois group of the diophantine equation). That group may be represented by the group of $n \times n$ matrices, n usually being 1 or 2, at least to start. In addition, one might also associate an automorphic form to another representation, although how one is to do this is very far from obvious. The trace of the matrices (the characters of these group representations) then encode the combinatorial numbers of interest: whether or not the equation has a solution, modulo p, a prime integer; or, the number of solutions to the equation modulo p. And those combinatorial numbers define an L-function. Remarkably, those numbers are as well the coefficients in that automorphic form's power series expansion in q, or alternatively, its fourier series expansion in τ. For an equation of an elliptic curve, the appropriate automorphic form parametrizes the curve (that is, the form leads to functions $x(t)$ and $y(t)$, such that $y^2 = P(x)$, as required).

If the L-function is generated directly from an equation's combinatorial properties, or if the L-function is generated from the group representation of the Galois symmetry group of the equation, we know that the L-function is about that object. Alternatively, if the group representation is generated through an automorphic form, we can often prove that the L-function has nice properties, a good functional equation, and analytic continuation. Again, we might be able to predict the automorphic form from the equation, but in general this is very difficult. More speculatively, we might compare an L-function generated automorphically to one generated combinatorially or through the Galois group, show they are the same, and then readily infer the good properties of an equation's L-function from those of its putative twin, the L-function generated automorphically. The model is the zeta function's inheritance of good properties from the theta function, although the right twin here is somewhat different.[29] The regularity of the combinatorial numbers is explained by the powerful internal constraints (for example, its scaling symmetry) on the automorphic form associated to the equation.

More technically, the Galois group (of the symmetries of the coefficients or roots of the diophantine equation) leads to one group representation. On the automorphic side, one is seeking a group representation that reflects as well the symmetries of an automorphic form, F. If g, an element of the group we want to represent, is given by a matrix $[\begin{smallmatrix} a & b \\ c & d \end{smallmatrix}]$, define $\varphi_F(g) = (ci+d)^{-k} F((ai+b)/(ci+d))$, and for another element h similarly. Then the representation of h $\varphi_F(g)$ is $\varphi_F(gh)$.[30] In this way the automorphic form leads to a group representation, φ_F. In effect, the automorphic form and the group representation are one and the same. The partition functions are defined by the (automorphic) group representation, the automorphic form, or the actual combinatorial numbers. (Presumably, in the Ising-model analogy, we have something like $Q_k(u)$, the automorphic form, and $\varphi_Q(V_d)$, Q and $V_d(k,u)$ being the partition function and transfer matrix, respectively, and k and u index the representations and their algebra. Products of transfer matrices build up the lattice, atom by atom, or row by row, or plane by plane. Recall Figure 5.3.)

(An analogy with spherical harmonics, $Y_{lm}(\theta,\psi)$, the natural functions on the surface of a sphere, might prove helpful for some readers. Here, θ is the azimuthal angle, and ψ is rotation around the z-axis. In analogy, a spherical representation would be one in which the action of g is a rotation (θ',ψ'), φ is a $\varphi_Y(g)$ and gh is a product of two rotations. The subscripts of Y_{lm} point to the irreducible representations of the rotation group (angular momentum states), and the Y_{lm} commute with the angular laplacian.)

For the simplest cases, reciprocity laws (see *Version E* below) lead to the automorphic object: the theta functions, as carrying the combinatorial numbers, is a transform of the zeta function. (More correctly as an analogy in the Langlands Program, an Eisenstein series, $\sum_{m,n} \frac{1}{2}(mz+n)^{-k}$, which is an automorphic form, corresponds to a product of zeta functions. See note 42.) For the next more complicated step (due to Emil Artin), there is an explicit group representation, which reduces in simple cases to the automorphic form (and hence an "automorphic representation"). The Langlands Program is a generalization of this pattern.

III

THE IMPORT OF THE ANALOGIES AND THE SYZYGY

***The Onsager Program can be described in Dedekind-Langlands Program terms (*Version C*), and the programs can be described in parallel series of observations (*Version D*). The parallel series (*Versions D*1 and *D*2) have not been arranged to fit the three-fold columnar division (*Versions Z* and *B*). But it is apparent that in each series of observations, there are aspects that are analysis-transcendental (continuous manifolds and surfaces), others that are algebraic-automorphic, and others that are arithmetic-combinatorial. There are as well partition functions, variously defined and computed. And, there are at least two sorts of matrix systems (group representations) that lead to the partition functions: one expressing the automorphy; the other expressing symmetries of an equation or a system and the associated combinatorics. Each of these aspects is a moment in the threefold analogy, within either statistical mechanics or number theory. And, there is a parallel or analogy between the realms, each of which is understood in terms of the three columns, without claiming that every feature of one series of observations has a corresponding feature in the other series, although for the most part they do so correspond.

The critical move is to claim that each series of observations is in some sense necessary. And that necessity is expressed, as well as formally organized, by the threefold columnar structure. In part, I have so indicated by describing the Onsager Program in Langlands-Program terms (*Version C*). Namely, I have deliberately projected the dogma of the Dedekind-Langlands Program onto the found materials of the Onsager Program.[31]

The distinctive feature of the mathematicians' program (the Dedekind-Langlands Program) is its insistence on the various objects and their interrelationship: group representations, automorphic forms, and functional equations and the analyticity of partition functions, as well as the parallel geometric/analytic, algebraic, and arithmetic objects—with each moment of the analogy to be in detailed (or perhaps still only speculative) correspondence with each of the two other moments. The crucial feature of the physicist's program (the Onsager Program) is the actual calculations, which by the way find these various objects and their interrelationships. What the physicists find, the mathematicians would expect. But that expectation does not make the mathematicians more capable of solving the physicists' problems, for the specificity of the physicists' problems are in general beyond the capacity of the mathematicians' apparatus. It is not clear how to apply the mathematicians' theory to the physical problem in any direct way. The mathematicians' expectations are just that, expectations or programmatic statements, enormously valuable for organizing what is known and for suggesting where to look for more insight.

On the other hand, what the mathematicians seek, the physicists take for granted from their hard won calculations. The physicists already have a jury-rigged (or perhaps it is jerry-built) version of the Dedekind-Langlands Program, achieved through more than eighty years of specific calculations—much as, for physicists, fluid flow concretely instantiated and anticipated the abstractions of Riemann's account of functions of a complex variable (as in Arnold Sommerfeld's anecdote, quoted at the beginning of chapter 6, in which the physicists find Riemann's ideas intuitive and clear, and the mathematicians have a much harder time of it). On the other hand, the mathematicians already have a generic account of the Onsager Program, achieved through two hundred years of conceptual development and proofs.

In practice, as far as I can discern, the mathematicians' analogy and program played little if any role in the physicists' work, at least until the 1990s. And, again until recently, the physicists' program, in their concrete calculations, has not much emboldened or encouraged the mathematicians—except that the developments in physics provided models of the theta functions: namely, Fourier's analytical theory of heat.[32] There have been recurrent observations by physicists and mathematicians of the analogy between the fields, and a hope and desire to learn from the other field.[33]

Within each realm, the analogies, whether explicitly presented or not, played valuable roles. So when the physicists were doing their various calculations they were learning about the physics of the Ising lattice by examining it from different perspectives in various presumably equivalent ways, equivalent in their end points at least. They were not deliberately instantiating the Dedekind-Langlands Program and its predecessors.

Thinking like a physicist, and the work that ends up building an identity in a manifold presentation of profiles, the various quite different calculations of the partition function for the Ising model are not just improvements or clarifications. They are discoveries of aspects of the natural world, in terms of a model system. But, whether intended or not, they are as well explorations of different aspects of the analogy that founds the more generic mathematics within the Dedekind-Langlands Program. Each calculation might be attached to one of the moments of the analogy: the analytic, the algebraic, and the arithmetic (or the geometric, the function theoretic and scale symmetric, and the combinatorial), or the group representations associated with them (*Version A*, and appendix to chapter 3).

The exact labels on these columns is less important than whether the alignments are appropriate, the relations among columns leads to a justification of the variety of solutions to the Ising model, and to fruitful suggestions for what to prove or define in one column given what is the case in another. And then we shall say that the geometric, the automorphic, and the arithmetic mutually imply each other or are analogous to each other, at least notionally.

Notably, some of these connections are provably and factually true, within each realm. The claim in each case is that there should exist such connections more generally. That is the big leap. For one is going from some scattered if systematic observations to a claim about a more systemic connection, a connection that usually cannot be proven so generically. The problem is to figure out under which constraints such a connection is true—hopefully, the constraints not completely vitiating the analogy.

PRIME FACTORIZATIONS AND ELEMENTARY EXCITATIONS

In practice, an analogy is not abstract. It exists as a program for research, as tantalizing facts, and as concrete and particular sets of objects and calculations. Here the analogy also represents the deepest, "elementary" motivations of mathematics or of physics: What are the natural, simple objects (the "elements") that make up a system, and how do they compose it?

Version E: Langlands' problem is to extend the correspondence I have already mentioned between counting or zeta- or L-functions and automorphic forms such as theta functions. This turns out to be a restatement of the very old problem of determining when numbers are prime, with no factors or divisors other than themselves and 1, and when numbers have (unique) factorizations—now generalized. To know theta is to know zeta, and thus to know the primes.

Historically (Euler, Legendre, Gauss), the Langlands Program is a continuation of the phenomenon of (quadratic) reciprocity observed in diophantine equations or congruences: the number of distinct sets of q mod p integer solutions of a quadratic equation ($x^2=q$ mod p, that is, $x^2-q=Np$, where N is an integer) is simply related to the number of p mod q solutions. It is called "reciprocity" since the relevant counting numbers are noted as (^q_p) and (^p_q). The counting function or partition function that "encodes" or combines all the (^p_q) for a particular equation, a version of the Riemann zeta function, is nicely connected to a well-behaved analytic function, a sort of theta function, an automorphic form. The automorphic form's fourier coefficients become the zeta function's harmonic series coefficients, which are just the (^P_q)s. So we might study a function not unlike the sines or cosines or the elliptic functions in order to learn about congruences.

We study congruences, such as the one for quadratic reciprocity, because we want to solve algebraic equations. We want to find out if the equations have solutions in some particular number system. But solving equations is equivalent to factoring a polynomial expression: Polynomial $=(x-a)(x-b)$ . . . , which says that a, b, . . . are solutions to the equation. If the polynomial does not have solutions in a particular number system, it is a prime. Geometrically, factoring a polynomial tells where its graph will intersect the x-axis. So we are learning geometric information as well.

Onsager's problem is a generic one: to exactly compute the properties of models of bulk matter (such as the Ising model and its relatives) using classical statistical mechanics. In his 1944 paper he links: (1) the analyticity properties of a function of a complex variable, that it is determined by its poles and zeros (namely, see his detailed appendix on elliptic functions, meant ostensibly to fix the notation); (2) to a symmetry between high and low temperature situations ("duality"), and another symmetry between lattices of different shapes (the "star-triangle relation"), both to be expressed quite naturally through the well known properties and identities of elliptic functions; (3) to the algebra of ("transfer") matrices whose trace is the statistical mechanics partition function; (4) to symmetries of what appears to be a group representation (the algebra of the matrices), and to what I have also called the energy operators. [34] (Some of this is only hinted at in his paper, although it seems that he had already done a good deal of the necessary calculation.)

Twenty years later, Schultz, Mattis, and Lieb showed how the Ising lattice might be considered as a medium populated by (Cooper-) pairs of thermally excited fermion particles, what are called elementary excitations—essentially, patterned rows of spins. (See the appendix to chapter 3 for some further details.) Traditionally, physicists have searched for the collective vibrations or modes of an elastic medium (the fourier components, the harmonic tones), Hooke's law generalized, as a first approximation to the medium's complexity. Corresponding to the arithmetic problem of reciprocity or factorization would seem to be the physical problem of finding the right degrees of freedom or the right particles (the collective modes or the elementary excitations or the normal modes) that characterize the system, the natural independent objects into which it might be decomposed. [35] (These are not the original spins for the Ising lattice, since in their mutual interaction they lose their naturalness as the elementary particles, much as the resonant frequencies of a complex interacting system are not usually the resonant frequencies of its original components.)

We compute the partition function in order to search for those natural components, recognized because the partition function is a product of the partition functions of those components. (Recall our discussion of the free energy.) And we study the elementary excitations because we want to find out how the medium responds to impacts and influences. How does a metal bar sound when you hit it? How does an Ising lattice respond when you heat it?

THE ANALOGY MADE CONCRETE

> The existence of analogies between the central features of various theories implies the existence of a general theory which underlies the particular theories and unifies them with respect to their central features. (E.H. Moore, 1908[36])

Analogy is destiny in myth and in mathematics. If I may be allowed a fairy tale, the witch in "Hansel and Gretel" looks very much like the stepmother, at least in some interpretations. This is no coincidence. The "general theory" here is that of psychological projection, that the bad guys literally embody our fears, here that mother will turn on or abandon us. The analogy is of course deadly serious, with enormous consequences for Hansel and for Gretel.

Much the same is true for mathematics as it is practiced. Analogy suggests that what is true in one field may well be true in another, although, since mathematics is deadly serious about such a claim, effective techniques, theory, and actual proof are still required. For just what is analogous to what, in which way, has to be found and shown. Yet, as Langlands notes in a related but different context, " . . . a frontal mathematical attack without any clear notion of the possible conclusions has little chance of success. We are dealing with a domain in which the techniques need to be developed."[37] And looking analogous is not enough:

> [These ideas] arise in a physical context rich in experience and inspiration whose sources of insight are unfamiliar to the mathematician, and of difficult access, so that, intimidated and sometimes at sea, he hesitates to apply his usual criteria. . . the authors have not at all persuaded themselves that they fully comprehend to what extent the arguments are formal, inspired by the physical and historical

connotations of the symbols, and to what extent they involve precisely defined mathematical entities. (Langlands[38])

For the physicists, usually what is precisely defined is the system one is studying, at least as an abstraction. Mathematical precision and rigor may well be of great benefit. But it is only rarely a primary concern. For most theoretical physicists, an analogy must lead to a calculation.

[I]f you work as a theoretical physicist in the United States, and wish to publish in *The Physical Review*, you had better *calculate* something concrete and interesting with your new theory pretty soon! (Michael Fisher[39])

And a calculation must be correct or at least a good approximation, even if it is not quite rigorous. Hopefully, it is a good description of the world. Ideally, one might like to have mathematical elegance go along with computational exactness and efficiency. Some of the time that is the case—but only some of the time.

An analogy, in creating a complex of related ideas, will make some otherwise disparate connections more natural. The Dedekind-Langlands Program's connection between automorphic forms and elliptic curves and diophantine equations becomes, through the Taniyama-Shimura conjecture, a foundation for the proof of Fermat's Last Theorem. Recall that the Theorem says that the Pythagorean Theorem has solutions for integer sides of a right triangle (3, 4, 5, for example), but if the exponent were larger than two there are no such integer or rational solutions. The proof of the Fermat Theorem has two main parts: Say there were such a solution for $n>2$: $a^n + b^n = c^n$. Form an elliptic curve from that solution: $y^2 = x(x+a^n)(x-b^n)$, where a and b are integers. (I am leaving out some nontrivial points here.) One can show that such a particular curve cannot be parametrized ($x=x(t)$, $y=y(t)$) by elliptic functions or automorphic forms, $x(t)$ and $y(t)$. Second, show that almost all elliptic curves, including those like the one of interest, can in fact be so parametrized. (Langlands' Program plays a crucial role here.) We have a contradiction, and so Fermat was right.[40]

The strategy of proof employs the connection between: counting the number of solutions of diophantine equations such as Fermat's, in various number systems; the ability to parametrize certain curves by automorphic forms; and matrices that represent the symmetries in that counting. In the background is Riemann's geometric way of thinking about analytic functions.

Version F: The analogies I have discussed are abstract, but in return they bring to bear lots of technical machinery. We might make the analogy more concrete and intuitive, although perhaps too cute by a half. What we give up for the moment is the machinery. What we gain is another sense of the disparate areas that are linked. Rubber sheet geometry, fractals, and jigsaw puzzles instantiate moments of the analogy I have been discussing:

—Rubber Sheet Geometry (Analyticity): The general shape of the world is determined by its singularities and boundaries. The tent poles that stretch a rubber sheet, the pegs that hold it down, and the 2-by-4s laid along the edges, all determine its shape. And it turns out that we may have, in effect, an arithmetic or combinatorics of these singularities and boundary objects.

—Fractals and Self-Similar Patterns (Automorphy): The world is scale symmetric, it looks the same at all scales.

—Doing Jigsaw Puzzles (Associativity): It does not matter much, at least in this realm, whether you put a puzzle together by first putting the figures together and then the background, or first the borders, then the background, etc., or even if you just put down parts willy-nilly in the right places. In the end, it does not matter in which order, in the sense of grouping, you put something together out of its parts.

The remarkable claim is that these disparate phenomena are intimately linked, perhaps through another object.

***At this point, it is fair to ask once more, Why all these versions of the syzygy, when much is repeated among the versions? If we fully appreciated the connections, presumably we might tell the story once and for all, precisely and rigorously and efficiently, with little repetition or backtracking. But that is not a luxury we might afford with analogies that are not at all fully understood. Each version, like each manuscript of an ancient text that we have only in pieces, contributes to a more general understanding. Some versions will work much better for some readers than others. Overlap would seem to highlight the significance of a variety of facts and observations, suggesting some overarching scheme perceived through many layers, an overarching scheme subject to revision as we learn more. Again, we glimpsing an identity in a manifold presentations of profiles.

HOW THESE IDEAS CAME TO BE SEEN AS AN INTERRELATED COMPLEX

Version G: Over the last two hundred years, many of the ideas and objects I have been discussing came to be seen by mathematicians as interrelated, and not merely independent or accidentally connected. Analysis and geometry, algebra, and arithmetic, and group representations (matrix systems) spoke to each other in ramified ways. As expected, the lesson will be: Nice smooth functions with lots of symmetries may be connected to partition functions that package lots of combinatorial information. A matrix system may be used to express the symmetries, and the traces or group characters are the combinatorial numbers of interest.

The history might be schematized as follows:

1. Fourier as a Model: In solving many a differential equation, if we formally decompose the solution, $f(x)$, into its fourier components (so that $f(x)=\int dq \, g(q) \exp iqx$), then the differential equation is reduced to an algebraic equation for each of the fourier components. We might think of $f(x)$ as packaging the information in the $g(q)$, for all q.

2. The Heat Equation: Fourier's solution to the heat equation, using fourier components, leads to functions with remarkable symmetries, functions which as well have combinatorial significance: the theta functions.[41] In general, solutions to the heat equation are smooth; namely, the temperature at a point is an average of the temperatures in its neighborhood (as long as we are not at a sharp boundary or at a source or sink of heat). Also, there is here something like the relationship of resonant tones and the shape of a drum discussed in the introduction, a relationship between spectrum and geometry, now between the time-rate of cooling off and the shape of the object.

Since thetas are smooth (as solutions to the heat equation) and are combinatorial, we might say interesting combinatorial things by studying their asymptotics or the geometry of an object.

3. Partition Functions as Analytic Objects: It is hard to say much about a partition function, as such, when it is defined only in terms of packaging lots of data, as a series. Might you associate a nice smooth function to that partition function, much as the theta is associated to itself as a partition function? When Riemann connected his zeta function, concerning primes, to the theta function, he made such an association. He needed to know about the zeta function's values in places where the original combinatorial definition does not seem to work. Theta's nice analytic properties translate into a good definition for zeta, even where zeta would appear to be badly defined (that is, it is apparently infinite).

Curiously, we are connecting a function that packages the prime numbers with a function that packages the ways of building up a number out of smaller squares. Presumably, what grounds theta's gift to zeta is the fact that there are regularities in the numbers zeta is packaging.[42] Or, theta's fourier coefficients are not in fact arbitrary, but have internal relations among themselves (they are nonzero only for the n^2-terms), and that regularity is theta's gift to zeta.

4. Symmetries and Group Representations: Early in the twentieth century, it was realized that one might study the orderliness or symmetry of an object by studying a matrix system (a group representation) which in a precise sense embodied that symmetry—much as fourier analysis studies the symmetry of an object. And it was also realized that the trace of such matrices (the group character) would be a powerful way of studying those systems.

Symmetries and their group representations have proved to be the correct way to study elementary particles, atoms, molecules, and crystals. That strategy also led to the correct generalization of the nineteenth-century reciprocity laws. Namely, rather than computing the combinatorial numbers of a system directly, and by-hand packaging that information into a partition function, pay attention to the symmetries of the system. They may be automorphies, as for the theta functions, or symmetries in an algebraic equation (the Galois group). Form a suitable group representation of those symmetries—a nontrivial task—and traces of the matrices will turn out to be the combinatorial numbers of interest.

IV

An analogy signifies a program for research; it is a tantalizing, uncanny set of connections; and it is discovered at first in quite concrete and particular cases, not at all apparently generalizable. For the mathematician André Weil (1906–1998), such an analogy is a powerful magnet drawing him into its field of influences expressed in three languages, which we have to master and learn to translate between: Riemannian, Galoisian-Dedekindian (Italian algebraic geometry), and Gaussian-Dedekindian (German). Mathematicians speak one or more of these languages, more or less selfconsciously. The physicists have been speaking all three and translating between them, albeit without being so aware of their linguistic competences. They were just talking.

242

Arithmetisch	*Funktionentheoretisch*
Ausgangspunkt eine ganzzahlige irreduzible Gleichung $f(x) = 0$.	Ausgangspunkt eine irreduzible Gleichung $f(\zeta, z) = 0$, welche z rational enthält (deren Koeffizienten also nach Heraufmultiplizieren mit dem Generalnenner ganze rationale Funktionen von z sind, mit irgendwelchen Koeffizienten, die hier nicht interessieren).
Körper aller $R(x)$.	Körper aller $R(\zeta, z)$, d. h. aller algebraischen Funktionen, die auf der Riemannschen Fläche eindeutig sind.
Herausheben der ganzen algebraischen Zahlen des Körpers.	Herausheben der ganzen algebraischen Funktionen des Körpers, d. h. derjenigen Funktionen $G(\zeta, z)$, die nur für $z = \infty$ unendlich werden.
Zerlegung in reale und ideale Primfaktoren bzw. Einheiten.	Ideelle Zerlegung der Funktionen $G(\zeta, z)$ in solche Faktoren, deren jeder nur in einem Punkte der Riemannschen Fläche verschwindet bzw. in Bestandteile, die nirgends verschwinden.

FIGURE 5.4: Felix Klein's account of the Dedekind-Weber (1882) "bilingual" text, the analogy between arithmetic and function theory (or Column 3 and Column 2).[43] For a translation, see the note.

*Version H*1: Felix Klein provides an account of the languages in 1926. Referring to Kronecker's and Dedekind and Weber's work, he notes, "[T]here is a wide-ranging analogy between number theoretic (the theory of integers of a number field) and function theoretic thought (the theory of algebraic functions on a Riemann surface over the z-plane)"[44] The analogy is between theories of algebraic numbers and theories of algebraic functions, between the third column and the first and second columns. Figure 5.4 is from Klein's history of 19th century mathematics.

Dedekind and Weber (1882) wanted to provide a surer foundation to Riemann's geometrical insights (for Weierstrass seemed to have invalidated Riemann's methods) by applying the model of the theory of algebraic numbers to a theory of algebraic functions.[45] Klein indicates how significant such analogies are, in that they prevent the death of science through the inability of scientists to understand one another—in his case the Italian and the German algebraic geometers, who as in the Tower of Babel came to speak different languages, the various speakers eventually no longer understanding one another.[46]

ANDRÉ WEIL'S ROSETTA STONE, LEARNING RIEMANNIAN AND TRANSLATING

> [T]here is nothing more to do about algebraic functions of one variable, because Riemann discovered just about all that we know about them. . . Riemann's memoir of 1851 is one of the greatest pieces of mathematics that has ever been written; there is not a single word in it that is not of consequence. (A. Weil[47])

> But the experts in the arithmetic theory and in the Galoisian theory, no longer knew how to read Riemannian . . . (A. Weil[48])

In a 1940 letter to his sister Simone (see Appendix D for a translation), André Weil describes the role of analogy in mathematics. He discusses an unusually productive analogy, which he calls a trilingual text and which he arranges in three columns—the analogy which has been the subject of this chapter. Elsewhere he refers to the Rosetta stone, recalling Champollion's (ca. 1800) decipherment of Egyptian hieroglyphics through the Rosetta Stone's three columns and three languages.[49]

> I have some ideas about each of the three languages [or three columns]: but I know as well there are great differences in meaning from one column to another, for which nothing has prepared me in advance. In the several years I have worked on it, I have had little pieces of the dictionary. Sometime I work on one of the columns, sometime under another. My large study that appeared in the *Journal of pure and applied mathematics* made nice advances in the "Riemannian" column; unhappily, a large part of the deciphered text surely does not have a translation in the other two languages.[50]

Weil is describing his ongoing mathematical work, in particular how a specific analogy among areas of mathematics suggests for him problems and results for one field from those in another. Traditionally such analogies might have been called a "metaphysics," following usage of the 18th century: the metaphysics of the calculus or the metaphysics of the theory of equations, referring to vaguely appreciated analogies.[51] Where once Lagrange saw analogies, we nowadays see theorems. Once one understands the exact results, the metaphysics and the analogies will then disappear.

Version H2: Weil's *first column*'s label is Riemann's transcendental (or continuous) and geometric (or topological) theory of algebraic functions, emblematically given by the Riemann surface. Algebraic functions, such as the circular (trigonometric) and elliptic functions, are functions that are solutions of polynomial functional equations, equations whose coefficients are polynomials over the complex numbers. Characteristic of Riemann, one tries to relate global configurations to local properties. In the electromagnetic (or hydrodynamic) analogy of Felix Klein, one relates the singularities or sources such as charges and charge sheets to the generic electric field configuration, in effect the spectrum to the field.[52] *Analyticity* and smoothness, and their lacunae, are the main themes.

The *third column* is arithmetic algebraic number theory, the story of number fields (having the four operations: addition, subtraction, multiplication, and division) and their prime numbers. Here, arithmetic or combinatorial properties are encoded in partition functions, such as the Riemann zeta function and Dirichlet L-functions. In honor of the fact that such combinatorics would seem to be independent of the precise grouping when breaking something apart, I called this feature *associativity*.

The *second column* is algebraic functions in one variable, with coefficients from a finite field rather than the complex numbers. The integers of the third column will correspond to the polynomials in x in the second column. Note as well that the zeros and poles of the rational functions inhabiting the first column might be seen to correspond to features of rational numbers in the third (taken as "p-adic" numbers).

In order to determine crucial properties of third-column partition- or L-functions, one needs to determine their values at points for which they would seem to be undefined or infinite. What saves the day is the *automorphy* of the second or middle column—namely, that there is a nice relationship between $f(x)$ and, say, $f((ax+b)/(cx+d))$.[53]

Weil provides a notional history of the subjects that concern him, and so discusses the sources and evidence for the analogy.[54] He sets the stage with a discussion of reciprocity laws, as Gauss's ways of analyzing what divides what, namely, prime numbers and factorization. And since factoring algebraic expressions is equivalent to solving equations in particular number systems, this is a very large realm indeed. After some discussion of Riemann, as further stage setting, the movement of decipherment and translation goes as follows: There is a subject, under the auspices of one of the columns (here referring as well to the columns of text on the Rosetta stone). It would appear that there is some sort of

FIGURE 5.5: WEIL'S HISTORY OF THE ANALOGY

Column One	*Column Two*	*Column Three*
Riemann	Galois, Dedekind	Gauss, Dedekind
(A) Transcendental, algebraic functions over complex numbers		
Transcendental over complex numbers →	(B) Function theory over a field (Dedekind-Weber)	←Ideals over the field of algebraic numbers (Dedekind)
	Algebraic functions	Algebraic numbers
	Polynomials in x	Integers
	Rational fractions of polynomials	Rational numbers
	?	Decomposition into prime ideals
(C) Riemann-Roch → (Hilbert)		The different ideal
	(D) Abelian functions	Abelian extensions = Class field theory
	Classes of divisors	Classes of ideals
(E) Point at infinity Birational invariance→	Point at infinity Birational invariance→	←the Different ?
(F) Point at infinity→	Point at infinity→	Valuations p-adic fields
Point at infinity	Point at infinity	←Prime ideal at infinity
	(G) Power series in the neighborhood of a point; local field	←Prime ideals (?abelian extensions, class field theory)
Calculus, analytic functions	←(H) Algebraic functions over a Finite field→	Zeta functions, Artin functions

analogue under another column, or at least we believe it should. An advance is made that then confirms that suspicion. But that advance then sets up expectations among the other columns. And so forth.

Version H3: Here is Weil's account of the back and forth flow of influence and analogy. (Figure 5.5, next page. See also pp. 329–333 of our translation.):

A: (Column 1) Riemann made the great advance with his transcendental account (that is, using continuity) of algebraic functions of one variable with coefficients in the complex numbers. Riemann seems never to have taken advantage of his friend and colleague Richard Dedekind's lectures on abstract group theory and its application to algebraic equations.

B: $(1 \rightarrow 2 \leftarrow 3)$ In 1882, Dedekind and Weber show how to get some of Riemann's apparently transcendental results (involving geometric intuitions) for algebraic functions, now algebraically and for functions whose coefficients are in any field, exploiting an analogy between Dedekind's ideal numbers and algebraic number fields (Column 3) and algebraic functions (Column 2).

C: $(1 \rightarrow 3)$ Hilbert then shows how one of Dedekind's notions of a prime factor or ideal (the different) corresponds to the Riemann-Roch theorem, a geometric and arithmetic fact concerning the topology of Riemann's surfaces.[55]

D: $(2 \leftrightarrow 3)$ A standard set of correspondences between algebraic numbers and algebraic functions is established. But it is not enough to achieve deeper results.

E: $(3 \rightarrow 1,2)$ In order to implement the different in the Riemannian and function realms, one has to add in a "point at infinity." Along the way it is realized that a crucial feature of both Riemann's and Dedekind-Weber's account of algebraic functions is the invariance of their objects to birational trans-formations, essentially that no point on a curve or no number is "special." How was one to make the point at infinity not be special?

F: $(1 \rightarrow 3)$ One discovered that the correct correspondent to the point at infinity is a "prime ideal at infinity," under a regime of something like absolute values (called valuations, and p-adic numbers).

G: $(3 \rightarrow 2)$ And those prime ideals may be seen to be like the power series expansions of algebraic functions at a point.

H: $(1 \leftarrow 2 \rightarrow 3)$ Still the distance between the transcendental and the algebraic formulations, between Column One's great power and Columns Two and Three is quite far indeed. A bridge, actually a rotating bridge ("un plaque tournante"), what Weil describes as "Et voila justement que Dieu l'emporte sur le diable: ce pont existe,"[56] turns out to be provided by the German school of algebraic number theorists (Hasse's school). One considers algebraic functions of one variable where the coefficients are not the real or complex numbers but come from a finite set of numbers (a finite field).

So Weil constructs the history of the decipherment of this now trilingual text. In about 1948, Weil employs techniques from Column One to prove in Column Two an analogy of a conjecture in Column Three (the Riemann Hypothesis, still not yet proved for Column Three). Weil is self-consciously aware of his especial skill in discerning and employing analogies. Notably, not all or many mathematicians work in this way so systematically.

Weil's epistolary rhetoric is remarkable for its metaphors, which include those from the military and artillery and logistics, epigraphy and lexicography, religion, and marriage and sexuality. So, for example, Weil deliberately speaks of the interplay of mathematical fields in terms of incest and intimacy and adultery among the columns.[57] Weil's rhetoric is technically correct anthropologically. Such mixtures of what is ordinarily taken as not to be mixed do produce offspring that are surprising and radically unexpected.

Weil's three columns parallel the three moments of the analogy I have been describing, what I have called analyticity, automorphy, and associativity. In so far as it applies to different areas, it allows for a speculative epitome of the syzygy. And it provides an analysis of both the Dedekind-Langlands Program and the Onsager Program.

V

SYZYGY AND FUNCTORIALITY

These ramified relations of relations are practically useful. The metaphysics here lies in the actual doing, in the fact that a theory or a program or an analogy is a tool to do work. Learning to read and speak Riemannian demands knowledge of other languages, which then resonate with Riemannian, and vice versa. Apparent systematic (or "functorial"[58]) relationships, those syzygies, are promissory notes that need to be paid back in the coin of detailed definition and proof.

To intertwine these columns, exploring the relationship between analyticity, automorphy, and associativity, between the continuous, the algebraic, and the arithmetic is to be in the realm of the Langlands Program for research in number theory and representation theory and, it turns out, the program in exactly-solved models in statistical mechanics exemplified by Onsager's (and E. Lieb's and R.J. Baxter's) work.

FIGURE 5.6: SOME OF THE ANALOGIES THAT MAKE UP THE SYZYGY

Transcendental/ Riemann	*Algebraic/Galois*	*Arithmetic/Gauss*
Geometry Column One	Function Theory Column Two	Arithmetic Column Three
Rubber Sheet Geometry	Fractals	Jigsaw Puzzle, Random Walk
Analyticity	Automorphy	Associativity (Arithmetic, Combinatorial)
Zeros of Partition Function or ζ-Function	Scale Symmetry of Partition Function	Regularities in Counting, Partitions, Arithmetic
Lee-Yang Circle Theorem, $Q(\beta)=0$	Riemann Hypothesis	Riemann Hypothesis
Algebraic functions over complex numbers	Algebraic functions over a finite field	Arithmetic algebraic number theory
Yang-Baxter as elliptic curve	Yang-Baxter as symmetries in group representation	Yang-Baxter as Star-Triangle Relation
$L(s,E)$, from elliptic curve	$L(s,f)$,$L(s,\pi_f)$, from an automorphic form or representation	$L(s,\chi)$ or $\zeta(s)$, $L(s,\pi)$: from an arithmetic series, or from a representation of a Galois group
Analytic continuation, functional equation	Uniformization by automorphic form	Group of rational solutions of elliptic curve

Such syzygetic relations are recurrently noted by mathematicians and physicists, sometimes even between the Langlands and the Onsager Programs.[59] At the level of a syzygy between research programs in different realms, these connections are still to be worked out, in principle and in their technical detail. From our experience so far we have some confidence that what we are talking about here is quite real, but just what that reality is, whether it is substantive or only suggestive, is still to be shown. And, once we understand it precisely, the analogy may fade away as an analogy or syzygy and become just an obvious fact.[60]

Version I: Two shibboleths are recurrent in these realms: "Langlands functoriality," and the "Yang-Baxter relation."

As for Langlands functoriality: In effect I believe this is much like the physicists' universality. Suitably similar groups will lead to the same L-functions. Moreover, operations on partition functions (L-functions) correspond to operations on automorphic forms, which are as well group representations.[61] Just which operations and how they correspond has to be specified, to be sure. But the central theme is that the combinatorial corresponds to the scale-symmetric and group representational, the associative to the automorphic, all in the service of analytic continuation and a functional equation. (Technically, corresponding to the Galois group of an equation and its representation, there is an automorphic form and a group representation, and the L-functions associated to each are the same.)

Yang developed an S-matrix for one-dimensional particle scattering and its reduction to a series two-body processes, so that a three-body scattering would be a consequence of two-body scatterings, expressed as an algebraic statement.[62] Baxter's elaboration of Onsager's star-triangle relation concerning symmetries of the "lattice Boltzmann weights" ($\exp -\beta E$), led to an algebraic statement as well. Those statements are formally the same.

The Yang-Baxter relation might look like $ABA=BAB$, or:

$$U_{i+1}(u_1)U_i(u_1+u_3)U_{i+1}(u_3)=U_i(u_3)U_{i+1}(u_1+u_3)U_i(u_1),$$

where A, B, and U_i are transfer or scattering matrices, and $u_1+u_3+u_2=I'$ (u_2 is the third particle's momentum or rapidity, I' the imaginary half-period of the elliptic function), in effect conservation of momentum.[63] It can be shown to expresses the two-body nature of a scattering matrix or the commutativity of a group representation (the transfer matrices). And, it is the basic equation for the braid group. Elliptic functions often model, encode, and operationalize these facts, the theta functions having analogous identities ("Riemann's quartic theta identity"[64]). Weierstrass's theorem to the effect that a polynomial equation in $f(u)$, $f(v)$, and $f(u+v)$, has a solution in terms of elliptic and rational functions is to the point. It has been suggested that the Yang-Baxter equation will become as generic as the Jacobi identity that defines Lie algebras: $[[x,y],z]+ [[y,z],x]+ [[z,x],y]=0$.

These shibboleths carry the import of the analogy and the syzygy: the combinatorial is connected to the automorphic through a group representation mediated by a partition function; the arithmetic, the algebraic, and the analytic are in analogy. In each case, there is still the task of working out the analogy, the task of translation and encoding.[65]

Object————————————————$\Rightarrow$Combinatorial Encoding, χ, $L(s,\chi)$

$\Downarrow$ $\searrow$ $\Updownarrow$

Group Representations, π, $L(s,\pi)$$\Leftrightarrow$Automorphic Encoding, f, $L(s,f)$

FIGURE 5.7: The analogy (J) as an object and its partition functions, $L(s, \ )$.

Version J: Generically, we have some sort of object that admits of a combinatorial or number-theoretic interpretation: the statistical mechanics of an Ising lattice; the number of modulo-p rational solutions of an elliptic curve with integer coefficients; whether or not a diophantine equation has a solution, modulo p. We can encode that information in a partition function. We are especially interested in the properties of these functions for particular values of their argument (s or β), often away from regions where the functions are readily defined.

We also note that there are symmetries of the object of interest, such as those of the Ising lattice or of an equation, which may be expressed as a group, and eventually in terms of matrices or linear operators, a group representation. Traces of an appropriate group representation are measures of the combinatorics. And then we may encode or package that information in a partition function.

Finally, we also note that the object may possess a functional symmetry, such as a scaling symmetry (for example, that a random walk looks the same at all scales). The object may be parametrized by a modular function. And again we might generate a partition function, derived from the fourier coefficients of that automorphic form, or from a group representation mirroring that automorphy.

When the same object is considered in these various ways, the ways of computing its partition function all result in the same function, and for good reason. As significantly, features of the partition function built-in by one method may be very difficult to prove if the partition function is derived by another method. Hence, the various modes of derivation, the analogy, is of practical import as well.

On his own, Langlands sketched a vast program which aimed to establish the connection between Eulerian products and group representations. . . . The connection between his work and mine was not apparent to me until much later. . . . It is a sort of generalized reciprocity. (Weil[66])

Langlands gave a number of conjectures relating Galois groups [the permutation group of roots of an algebraic equation] with "automorphic forms" . . .
    The study of non-abelian Galois groups occurs via their linear "representations." For instance, let l be a prime number. We ask whether $GL_n(F_l)$, or $GL_2(F_l)$, or $PGL_2(F_l)$ occurs as a Galois group over $\mathbb{Q}$, and "how." The problem is to find natural objects on which a Galois group operates as a linear map, such that we get in a natural way an isomorphism of this Galois group with one of the above linear groups.
    . . . Of course the product and series [for the automorphic form] have [here] been pulled out of nowhere. To explain the somewhere which makes such a product and series natural would take another book. (S. Lang[67])

The Onsager Program has been wide-ranging and productive of results that would fit under each of Weil's columns, with demonstrations, typically in appendices to the papers, that the partition functions in the various approaches are in fact the same. Why all these various approaches work remains more or less obscure (unless, perhaps, we keep the Dedekind-Langlands Program in mind). The Langlands Program has also been productive, but is apparently rather more difficult and recalcitrant. As Langlands puts it:

The aesthetic tension between the immediate appeal of concrete facts and problems on the one hand, and, on the other, their function as the vehicle to express and reveal not so much universal laws as an entity of a different kind, of which these laws are the very mode of being, is perhaps more widely acknowledged in physics . . . than in mathematics, where oddly enough, especially among number theorists, conceptual novelty has frequently been deprecated as a reluctance to face the concrete and a flight from it . . .

It may be that we are hampered by the absence of a central unresolved difficulty and by the extremely large number of currently inaccessible conjectures, at whose extent we have hardly hinted. Some are thoroughly tested; others are in doubt, but they form a coherent whole. What we do in the face of them, whether we search for specific or general theorems, will be determined by our temperament and mood.[68]

It is perhaps useful to have further concrete examples that are more tractable by whatever means we have available to us than is the more general and abstract problem. So it makes sense that Langlands and his collaborators have studied Ising-like models and scaling symmetry (the "renormalization group") numerically using computer simulations to search for both the expected conformal invariance in two-dimensions and the right variables (the fixed points or critical points) to describe the system, and what that might correspond to mathematically and rigorously.[69]

Inadvertently, the mathematicians' work provides a framework for organizing the physicists' results. And the physicists' work on the Ising model provides hints about what must be the case more generally in the mathematicians' program. Or, better put, the physicists' work contributes evidence for the correctness of the mathematicians' program. A more general theme is the relationship of spectrum to geometry, of sources to fields, of discrete properties to continuous ones, of randomness to asymptotic scaling and automorphy.[70]

To have a partially deciphered trilingual text is not to have a dictionary. Enormous imaginative effort is required to infer what might be written down in the blank places in the chart of correspondences. All we have is what we know of Weil's three columns, Dirichlet L-functions and various forms of reciprocity (Artin, Langlands), and the classical statistical mechanics of the Ising model of a two-dimensional ferromagnet and closely related models. In a sense, our only source of these languages is given in the writings we have already.[71] But, we might hope to surmise what goes in a blank space and then test that surmise by proof and calculation. We might work on simplified toy models or particular cases to get further hints about what goes with what. We might even hope to develop more general theories that make of this potential analogy an elementary instance. So we had better learn to read Riemannian, as well as Italian and German.

Syzygies are relations among relations, analogies of analogies, originally the lining up of pairs of heavenly bodies, nowadays taken as simultaneous equations or polynomials, so yoked together much as are planets. We began with a threefold relation in mathematics and its apparent analogy with a threefold relation in physics. The metaphysics here is technical and sometimes provable.

Whatever we call our observations—aspects, subrealms, moments—the mathematical and physical work is phenomenological, as the philosophers would call it: taking the world in every case as it presents itself to us, yet at the same time trying to get at a conception of what is there that could so present itself to us in so many aspects or moments, understandable in so many realms—an identity in a manifold presentation of profiles. For to say what an object is, specifying it fairly precisely, as for example the integers or an Ising model, does not tell us yet how it presents itself to us under various auspices. And to say how it presents itself to us does not tell us why it is capable of its variety of presentations. We are called to understand that fact.

Philosophy in any case always comes on the scene too late to give instruction on how the world ought to be. As the thought of the world or the theoretical enterprise, philosophy appears only when actuality is already there cut and dried after actuality's process of formation has been completed. The history of science or the teaching of the concept, which is also history's inescapable lesson, is that it is only when actuality is mature that the ideal or the theoretical first appears over against the real. The ideal or the theoretical apprehends this same real world in its substance and builds it up for itself into the shape of an intellectual realm, a theory. When philosophy or theorizing has put it all together, when it paints its gray in gray, then has a shape of life grown old. By philosophy's gray in gray actuality and the shape of life cannot be rejuvenated but only understood. The owl of Minerva spreads its wings only with the falling of the dusk. (Hegel, *Philosophy of Law*, 1821[72])

Minerva is goddess of wisdom, her wise owl its retrospective perspective. In *The Phenomenology of Mind* (1807), Hegel systematically organized all of our world and its history and phenomena in an ascending sequence of formal analogies, a massive syzygy, a ramified syzygy of syzygies. Formally, they were threefold: thesis, antithesis, and synthesis. They were dialectical, to use the term of art from rhetoric. Substantively, they included art, nature, science, and history. Yet Hegel was no determinist with respect to his syzygies. He could not tell ahead of time how they would be realized. His formalism was a way of organizing and explaining what there was, not a way of anticipating it. For how analogies work out is a genuine surprise. And this is true for the syzygy I have been describing. At the same time, one might well wield this system as a guide to action, the analogies suggesting what is true and what we ought make true if at all possible. Again, of course, only in being worked out do we discover if our supposed truth is actually true, or whether we have misjudged and misread the ways of the world.

6

In Concreto

The City of Mathematics

Riemann and Maxwell as an Epitome; "What Is Really Going On" in the Mathematics; Philosophies of Mathematics; The Library of Mathematics; Library Circulation During 1900–1930. *Mathematics and the City in the Nineteenth Century. Order and Structure*; Center and Periphery; Process, Pattern, and Palimpsest. *Employing Your Body to Prove a Theorem*: The Snake Lemma of Algebraic Topology; Train Tracks and Pairs of Pants. *God's Transcendence as Analogized to Mathematical Notions*; Ascent to the Absolute; Sets as Models of God's Infinitude. *The City of Mathematics*.

I

RIEMANN AND MAXWELL AS AN EPITOME

In effect, we have been examining the legacies of Gauss and Faraday, and of Riemann and Maxwell: the Maxwell-Boltzmann statistical distribution of molecular velocities or the bell-shaped curve or the gaussian; the Riemann surface of multiple-valued functions, such as the square root or an indefinite integral or the electromagnetic potentials (and their gauge invariance); and, Maxwell's equations for electromagnetism, which may be readily expressed in algebraic topological terms. And, notably, the legacies have been shaped for our time through Hermann Weyl's work on the eigenvalue problem, topology, and group representations.

Maxwell (1873) is aware of the topological aspects of his formulation, and is enthusiastic about the early topologist J.B. Listing's (1847) as well as Riemann's (1851, 1857) work.[1] Maxwell's equations, and much of Faraday's original insights that founded them, are essentially topological, connecting the local to the global. Nonzero line integrals of the electric fields ($\oint E \, dl \neq 0$) say that the field is a "closed but not exact" vector field. There are currents present, so there are nonvanishing cycles or "cyclosis" (to use Listing's term). The intersection of lines of force and currents leads to electromagnetic induction, so there are intersection integrals (ala Gauss). Kirchoff's analysis of electrical circuits (1845, 1847) is topological. The set of connections that form a circuit are paths in a network subject to local constraints.[2]

Continuity and connectivity assumptions link the local to the global. Local conservation of a source or charge is linked to global boundary conditions of a flow or a field. Gauss's law links volume integrals of charge to surface integrals of fields. There is an arithmetic of the dimension of a space or subspace, the number and kind of sources (charges, dipoles, . . ., twists); the number of holes in that space, such as the doughnut's hole; and the number of possible inequivalent field configurations or initial conditions. The kinds of flows or fields possible in a space are intimately related to its shape and to its sources.[3]

Riemann and Maxwell were seen as being close in their styles of thinking. Riemann's ideas were congenial to physicists:

> Riemann's dissertation at first was foreign to his contemporary mathematicians who reviewed it as if it were a book published for the cognoscenti [*sie wurde wohl gelegentlich das Buch mit den sieben Siegeln genannt*—referring to the seven seals in the Book of Revelation]. . . . it was closer in its way of reasoning to physics than to mathematics . . . Once Riemann spent his vacation together with Helmholtz and Weierstrass. Weierstrass took Riemann's dissertation along on holiday in order to deal with what he felt was a complex work in quiet circumstances. Helmholtz did not understand what complications mathematical specialists could find in Riemann's work; for him Riemann's exposition was exceptionally clear. (Arnold Sommerfeld[4])

Maxwell loved diagrams, curves generated by rotating figures, stereoscopic images, and images of lines of force and their duals.[5] Pictures and geometric intuition inform Riemann's work.

For a new generation, Weyl (1913) reformulates Riemann's theory of algebraic functions and their integrals in more modern and what becomes topological terms. Emmy Noether (1926) develops modern abstract algebra to comprehend a variety of examples and brings its technology to Riemannian and algebraic geometry and to algebraic topology, under the auspices of what we now call commutative algebra. Moreover, Weyl (1911) describes the asymptotic distribution of the eigenvalues of a partial differential equation, the shape and sound of the drum: a story of variances, topology, uncertainty, and the analogy between geometry and analysis—themes of the last four chapters.

So might go an epitome of our discussion, internal to the mathematics. Now let us venture outside a bit.

"What Is Really Going On" In the Mathematics

It is a truism that mathematics, much like other human endeavors, is done in specific places by actual persons, situated in intellectual and material culture. And those actual persons who do the mathematics employ intuitions and ideas drawn from their imaginative and kinesthetic experience. Historically, mathematics has been part of sacred and theological discourse and part of practical and scientific discourse—as mathematics. So the Babylonian theory of the planets is about ominous phenomena, and it is an exact mathematical science.[6] The connections drawn by Georg Cantor (1845–1918) between the finite, the transfinite, and the Absolute are the subject of mathematical investigation. Whatever their ancillary theological import for Cantor, they are to be studied mathematically. Mathematics is not cultural theory, not anatomy, not theology. And mathematical results are objective, independent of place and person.[7] But, in historical fact the unsaid assumptions in mathematical work may depend on place and on person, only to be revealed when new examples show that the original clear presentation was not quite so clear or overarching. Mathematics is realized "through our human endeavors in our historical existence, but forming an indissoluble whole transcending any particular science," to recall the quote from Hermann Weyl that began the Introduction.[8]

Permutations may be the "metaphysics" of equations, as Lagrange (1771) suggested,[9] and "even theology has its merits," as Paul Gordan (ca. 1890) realized concerning abstract methods.[10] But, again, such metaphysics and theology is recognizably mathematics. In contrast, the three tales I shall relate in this chapter, about the City, the Body, and God, are just that, meditative and suggestive linkages between technical mathematics as it is practiced and the world in which it is actually done. The tales are not mathematics, they are not scholarly history. (On the other hand, the examples I explore in the earlier chapters are mathematics, fully fledged.) What the tales do show is the variety of concrete resonances and linkages of the mathematics and the world.

Much of this chapter is speculative. My hope is that it is interesting enough speculation, from which might be isolated rather more narrow claims, perhaps each worthy of a monograph.

Doing mathematics is craft employing tools and artful performances, with motivations that guide the work and which are occasional rather than eternal. The production of objective mathematical knowledge is much like the production of other objects through craftwork. So in describing what mathematics is, we might list tools and methods and masterpieces.

But I do not know what mathematics is, as such. As may be apparent, I have had little to say that is generically about mathematics. "Convention" or "analogy" are topics which then allow me to indicate significant features of the detailed cases of doing mathematics. The features are surely more general than those examples, but I do not know how to show they are more general other than by doing further case studies.

The epitome I offered of the book's themes, in terms of the legacies of Riemann and Maxwell, is convenient and perhaps charming. Surely, other epitomes are possible, in particular an account emphasizing algebraic themes. Still, an epitome does not exhaust the content of a paper or a book. The overviews in research papers, in the prefaces to those papers, in textbooks, and in histories allow us to see the work as a whole. A good introductory page or two to a highly technical research paper can make sense of the moves in the argument or proof, provide motivations, and point out places worthy of a second glance, showing more of what is really going on in the mathematics.[11] But there is always more that is going on.

"What is really going on" in the mathematics is the strategy—the motives, the basic ideas, the deeper facts—that allow for the technical apparatus and structure in proofs and derivations. But knowing the motives, ideas, and deeper facts does not make a proof go through. Motives, ideas, and facts need to be employed effectively. What is really going on is really going on only when it works, in detail and in full rigor. And what is really going on is often revealed through comparisons among different proofs or derivations of what is apparently the same result, that identity in a manifold presentation of profiles.

What is really going on lies also in our more general ideas. Weyl (1927, 1949) describes what he takes to be the essential features of mathematics and its applications to natural science, again understanding his own mildly technical and philosophically sophisticated description within an "historico-philosophical" context.

And yet science would perish without a supporting transcendental faith in truth and reality, and without the continuous interplay between its facts and constructions on the one hand and the imagery of ideas on the other.[12]

(Edmund Husserl's phenomenology (ca. 1910) informs Weyl's notion of construction; Leibniz and Kant are the great figures for the imagery of ideas.[13])

Earlier, in *The Concept of a Riemann Surface* (1913), Weyl shows some of what is really going on in a field of mathematics (algebraic topology and functions of a complex variable), by the way codifying and making sense of it through even deeper and more rigorous mathematical ideas. He develops a theory that encompasses Riemann's ideas and Felix Klein's concrete interpretation in *On Riemann's Theory of Algebraic Functions and Their Integrals* (1882). But, Weyl's account of what is really going on in Riemann is an interpretation in light of fifty years of mathematical development subsequent to Riemann.[14]

Ideally, for any field, one would want as pithy and deep a comment as, again, Lagrange's "permutations are the metaphysics of equations," reflecting his work on the invariants of a polynomial equation, such as its coefficients, when you permute its roots according to particular rules, just what eventually becomes Galois theory.[15] We might say that algebraic topology is concerned with the algebraic image of topology: spaces transformed or imaged into groups, maps into group homomorphisms. Or, homology theory does for continuous maps what matrices and their irreducible decomposition do for linear algebra.[16] Or, homology linearizes topology by attaching additive linear invariants to topological objects (much as Hilbert's invariant theory "linearized" polynomials in terms of generators and relations,[17] and the calculus is the linearization of analysis). Each of these sayings promises in the matter of course a mathematical explication. And in part one wants that mathematical explication to say why you can get away with these metaphysical claims—and in part that will be a matter of understanding the historical developments that pushed mathematics in these mutual directions.

I take from Weyl the warrant for a philosophy of mathematics that is focused on "what is really going on," and on the connections with the historico-philosophical context.[18] But, whatever else, such philosophizing is done through mathematical means, examining and explaining actual definitions, constructions, theorems, proofs, derivations, and examples.[19] Of course, one may prescribe how things ought to go, following an analogy or a program. But the facts of the matter—which definitions turn out to be most fruitful, what you can actually prove—are in the end what counts.[20] And "fruitful," "interesting," and "mathematics" are not given a priori. They depend on the actual practice of mathematics, mostly by professional mathematicians.[21]

The concrete, the particular, and the exceptional carry along the abstract, the arbitrary, and the norm. Concrete models such as the Gaussian distribution are made to carry along with them abstract values and tags, such as statistical independence and plenitude (whatever is possible will happen). The Central Limit Theorem is taken to justify variances as the right statistics. Infinite variance distributions, such as the Cauchy or gunshot distribution, are often employed contrastively to highlight the power of the finite-variance Gaussian.

We recognize what we are doing, as such, when we make strange what we take as obvious, perhaps through historical, comparative, or phenomenological analysis (what the Russian critic Viktor Shklovski (1916) called defamiliarization). Yet, to claim that topology is a continuation of Riemann's complex analysis, does not deny that in time topological notions transcend that origin, nor that Riemann's understanding of his work was different than our understanding of his work. Conversely, we may "familiarize" what is foreign, addressing questions such as: How much of what we now know did they appreciate then? Can we rewrite their work in our terminology and notation?[22] What do we lose by doing so?

The mathematics itself, that ramified collection of concepts, proofs, methods, motivations, and examples, is a philosophy of mathematics saying what is really going on in the mathematics. It says what is really going on in the formalism so that the formalism is taken to be obvious or at least a provision of certain truth: You want to understand a special function such as the gamma function? Then here it is, or at least here is one way it is, as a story of distribution-and-occupancy, a combinatorial story of balls-into-boxes.[23] You want to understand eigenvalues of partial differential equations? Then here it is, or at least here is one way it is, as a story of how boxes fit into balls.[24] But there are multiple ways, and it pays to appreciate them all.

Felix Klein, in his history of 19th century mathematics, returns again and again to the social context of mathematics and mathematical research, for "the great social displacements, such as the French Revolution and its consequences, had an influence on scientific life."[25] Even if one believed that mathematics, as we understand it, formulate it, and come to know it in human history, is a transcendent gift, as revealed as are the Hebrew Scriptures, we might well take mathematics (and the Scriptures) to be as embedded in society and history as much as any other activity, practical or ideational. This does not mean that mathematical truth is relative.[26] But the issues that come up, how they are worked on, what we consider an understanding, and the analogies, models, and rubrics we employ, do reflect other problems in a society and the practical tasks of commerce, manufacture, and exploration.[27]

Philosophies of Mathematics

> I want to make clear that I have nothing to say about the ontology of mathematics: about whether the subject matter of mathematics is platonic objects or whether it is other things, such as structures, concepts, or proofs. I also have nothing to say about the semantics of mathematical discourse. Nor do I have anything to say about whether evidence for accepting mathematical assertions is evidence for taking them as realistically true or is merely evidence for adopting them in some conventional or methodological sense. In short, I have nothing to say about the content of mathematical propositions. . . .
>
> The first and most important reason for my not discussing ontology and semantics is that I have no good ideas or even opinions about these questions. I confess that they mystify me. On the one hand, it is hard to avoid—in any account of mathematics that is at all realistic—an appeal to some kind of mathematical objects. On the other hand, there is a strong intuition that objects do not really matter. . . .
>
> Another, and perhaps more acceptable, reason for studying epistemology while ignoring ontology and content is that we seem to have more intuitions—in the intersubjective sense, at least—more reliable intuitions about mathematical truths and evidence for them [for example, axioms] than we do about the other questions.[28] (Tony Martin, a mathematician and professor of philosophy.)

There is a sub-field called philosophy of mathematics. For the most part it has employed mathematics and mathematical objects, notions, and practices as ways of working on traditional philosophical problems in ontology, metaphysics, semantics, and epistemology. The mathematics that is the subject of these studies usually has been drawn from mathematical logic, set theory, and sometimes category theory, all very respectable parts of mathematics. Rarely are the subjects drawn from algebra, or analysis, or number theory, or geometry (other than Euclid, and non-Euclidean)—say as taught to undergraduate mathematics majors. Except for the above fields, there has been little detailed discussion of advanced work in more conventional areas of mathematics. Of late there has been an interest in the philosophy of mathematical practice, but again traditional issues in philosophy are the main themes.[29] There is as well sociology of mathematical practice, which again tries to get hold of how mathematics is actually done. But, again, traditional sociological questions are usually the motive and the audience is defined by those concerned by these questions. I am told it is hard for a mathematician to recognize their own practice in those descriptions.

I would like to think that the work in this book may be of interest to the professional philosophers and other scholars. But I am not sure I have anything useful to say directly about these scholarly fields as they are currently constituted.

A mathematician (Frank Quinn) has argued that rigor is a distinctive characteristic of mathematics as it has been practiced for the last century.[30] Rigor means that arguments and proofs may be followed step by step by ordinary mathematicians, their finding no serious gaps or if gaps are found they either have to be repaired or the work itself is not mathematics (yet). In contrast, in earlier eras, intuition and appeals to pictures or examples might be used to make mathematical arguments—and eventually, often well after they are accepted as reliable, every once in a while it was found that such arguments did have gaping holes. To do mathematics before, say, 1900 one needed great intuition; to do mathematics since, one needed to be scrupulous and to learn the methods of reliable proof. Rigor forces one to recognize peculiar cases or unrecognized assumptions, and those cases and assumptions are not merely technicalities but actually tell us something about the world. As Quinn puts it, rigor is "error displaying"; I think of it as detail revealing.

A sociologist (Eric Livingston) studies mathematics "at the blackboard" and shows how mathematicians convince each other of the credibility (and so truth) of their arguments (he focuses on logic and set theory, and basic geometry). The mathematicians are trying to show others how the argument must go, and supply support for passages that would not seem to be obvious. That their audience is well-trained means that some part of the argument need not be very detailed, but in the audience there may well be someone who will ask how you get from one point to another, and the speaker is expected to explain that path in sufficient detail that the questioner sees the road ahead. Some of the time, the questioner will say that there is an (counter-)example that suggests that the argument at a particular point could not be true, and again the author must repair the argument or show that the counterexample is not to the point. These are fabulous moments in mathematics.

Set theorists, as mathematicians, have arguments that might be of philosophical interest. Which axioms are essential, which are not needed for much of mathematics?[31] They may construct mathematics without using certain axioms or weakened forms of certain axioms. In any case, the mathematics itself is evaluated by mathematicians in the usual manner, even if they are dubious about whether such an endeavor is worthwhile. In the case of large cardinals, where one needs to add axioms that are stronger than the conventional ones, they may have opinions on whether such axioms are valid or necessary, but in the end the mathematics that is done is evaluated conventionally. They are much like the number theorist who might say, "*If* the Riemann Hypothesis is the case, *then....*," Presumably one has confidence that eventually one will be able to prove the Riemann Hypothesis, or if not, enough of the Riemann Hypothesis will survive and that will be enough for one's if-then.

When mathematicians argue about these questions of stronger axioms, they adduce the kind of evidence others would bring to bear in their own specialties. Their speculations are informed and grounded, even if they were to end up being false. You can give good reasons to believe that the Continuum Hypothesis is true or it is false, but what matters in the end is the mathematics you produce that allows you make a well-founded claim. (The Continuum Hypothesis says that the size of the real line, the number of its elements, in effect the number of infinite decimal numbers, is equal to the next larger cardinal number after the size (the number of) of ordinary integers.) In the case of the Continuum Hypothesis, one big issue is what kinds of axioms are needed to make such a decision or adduce useful evidence, and the believability of those axioms (although their believability may well depend on what they allow you to prove).

A curious feature of arguments in set theory is the recurrent reference to Kurt Gödel's (1906–1978) writing *about* mathematics and set theory. It is striking that one would refer to authority (Gödel as Bible, needing interpretation, and allowing for several), even if it is merely to set forth one of the conventional positions.

In this book, the analytic descriptions of doing mathematics do not qualify as philosophy or sociology or anthropology or mathematics per se. My goal is to describe the mathematical work in a way that a mathematician would recognize it as familiar, and lay persons could see that work as not so far from what they do in other realms. Remarkably, there is comparatively little analytical description of doing mathematics that is faithful to how the mathematics was or is done. Mathematicians do not much care, and it would seem others are unwilling to get into the detail of current mathematics. (There is a great deal written to explain mathematics to the lay reader, some of which is quite good.) It is as if someone were writing about burlesque dancing: the dancers do not much care, and the writer would rather stay away from the burlesque theater.

THE LIBRARY OF MATHEMATICS

Mathematics as a discipline systematically reviews its literature, indexes that literature by detailed subject categories, and produces surveys, often in books, that organize what is known and develops it further. Moreover, progress is often denoted by mathematical theories that are more general than predecessors, and incorporate much of what is already known. (However, it turns out to be useful to survey earlier work, since in the process of advancing and synthesizing, insights and particular methods may well be left behind and forgotten.) Such a more general synthesis is a substantial achievement, often demanding deep insight into the research literature. Of course, there are textbooks, but advanced expository textbooks are as likely to be read by research mathematicians as by students. And scientists, who are not mathematicians, who make use of that literature may well advance it in the process of doing their own research.

LIBRARY CIRCULATION DURING 1900–1930

—What we call the Dirac delta-function was employed by P. A. M. Dirac in his account of quantum mechanics (§15 of his textbook). The function is zero everywhere on the real line, except at zero where it is infinite. The integral of the function from minus infinity to infinity is 1, and what it does is pick out the value of an associated function at $x=0$. Namely:

$$\int_{-\infty}^{\infty} f(x)\delta(x)\ \mathrm{d}x = f(0)$$

Dirac's delta function is the continuous analogue to the Kronecker delta, δ_{ij}. Poisson, Fourier, and Cauchy employed similar functions, and Heaviside developed a related "calculus" for engineering purposes in 1893. But in no conventional sense is the function continuous, and its peculiar properties might well disturb proper mathematicians. Schwartz (1945) eventually developed a theory of these generalized functions or "distributions" making the physicist's use of them rigorous. Moreover, that theory had much wider implications for mathematics. In this case, a physicist re-invented a mathematical object to do certain work, and the mathematicians eventually made that object proper and well-behaved and fruitful for pure mathematics.

—When Schroedinger was reviewing de Broglie's work on the wave nature of matter, he was encouraged by his boss Pieter Debye to find a partial differential equation to make de Broglie's account rigorous—for their teacher Sommerfeld demanded a differential equation if something is to become mathematical physics rather than hand waving. Schroedinger came back a few weeks later and had found a partial differential equation ("I have found an equation."), whose physical meaning took a while to be understood. But it worked.

—When Dirac was trying to provide a quantum mechanical account of the electron, a particle that had "spin" and a magnetic moment, and often was energetic enough, going fast enough, to be in the realm of special relativity, he conceived of such a partial differential equation that was first order (unlike Schroedinger's, which was second order in space and first order in time, that is, it had second and first derivatives). Here, the space and time derivatives had to be symmetrical. Curiously, a solution would be look like a 4-vector, if the solution was to be consistent, rather than a scalar algebraic expression. And eventually its physical meaning was discerned (spin led to two dimensions, matter and antimatter led to two more). It was a new equation, not imagined before by the mathematicians, one for which they previously had no need.

—When Heisenberg was developing his account of quantum mechanics, one that avoided Schroedinger's wave function, and was expressed only in terms of observable quantities, he found that he had to multiply two physical quantities not as $a \times b$, but as $\Sigma_j a_{ij} b_{jk}$. That multiplication rule made good physical sense, since it represented atomic transition between one energy level and another (i to k) through all the relevant intermediate levels (j), but was quite peculiar. Max Born recognized that Heisenberg's multiplication rule was that for matrices, and hence we have "matrix mechanics." (About twenty years later, in solutions of the Ising model, Montroll, and Kramers and Wannier discovered that a nice way of expressing all the possible interactions of a lattice of spins with their neighbors, is in terms of matrix multiplication. Eventually it is seen that such formalism expresses an analogy between classical statistical mechanics (the Ising model) in two dimensions, a plane, and quantum mechanics in one spatial dimension (a line, a one-dimensional spin chain) and one temporal dimension. It was not by chance that matrix multiplication appeared in this classical realm. But you had to do the work, the actual calculation, in detail, to discover that what in retrospect might be seen as manifest.

—And when physicists tried to classify the atomic energy levels of molecules and eventually the elementary particles into natural families, they discovered (Wigner, 1926; von Neumann) that the natural families were defined by mathematical (Lie) groups that reflected the symmetries or properties of each of the families of levels or particles.

—When Einstein (1905) tries to account for the diffusion of a gas as a molecular phenomenon, he treated the molecules as randomly bumping into each other, and so he eventually connected the temperature of the gas with the average distance a molecule moves between collisions, and so the rate of diffusion of the gas (think of a perfumed lady's fragrance wafting across the room). Diffusion is a random walk.

Ten years later, Einstein wanted to express his intuition that gravitational forces, say that we fall down, are just another way of describing the geometry of the physical world. In effect, mass and energy globally determine the way we fall down at a particular point: Mass-Energy is much the same as Geometry. The problem was to figure out how to express this sameness, with the requisite generality. While the physicists had some idea how to express mass-energy (the stress-energy tensor), they did not readily have the most general geometrical description—but the mathematicians had been working on this problem. Eventually—actually over a short period Einstein (with the aid of Marcel Grossmann) had to try three or four mathematical expressions—he was able to find an expression of geometry that matched his expression of mass-energy, in effect searching that library of differential geometry that had developed over the previous three decades.

What happens again and again is that those who use formalism to describe the world find that they are speaking mathematics (as Galileo said), and at the same time they are inventing practical notions and language, slang and neologisms, that enlarges mathematics, often in significant ways. And that mathematics is not only descriptive, it is also informative, extending intuitions and revealing new aspects. (Mathematicians often go to their own lending library, as well.) It seems that we have a Library of Mathematics, where those who borrow books sometimes deface them in useful ways, and others, usually mathematicians, keep adding to the library, with new books and rebinding old ones. What is remarkable is that some of those defacings and graffiti are outlandish yet they work and are eventually made into proper Library volumes, the graffiti becoming text. Paul Lévy's (1948) account of a random process, in effect expressing the fact that variances are the core notion, was to talk not about dx but about $d\sqrt{x}$, saying that in the random world change takes place in the square-root of time or space. And that peculiar notation was made to work.

The big problem is to know what is in the Library. A well-trained scientist learns some of the basic outlines. What is wonderful is that there are bibliographic experts on each of the catalog numbers, and if you are lucky you have one nearby, even a friend, who can tell you where you might look more specifically. Max Born and Heisenberg, Marcel Grossmann and Einstein. You might well discover the relevant volume on your own, and scanning the shelves is surely worthwhile. What makes all of this work is that there is such a Library of Mathematics classification system, and even if that classification system changes over many decades, usually there are guides that translate one generic catalog number to the new classification.

Some volumes gather dust, are lost or forgotten, or the parchment is consumed by vermin, or the paper crumbles. And sometimes, there are no volumes in the Library useful for your purposes, and when you try to invent your own, you may come up empty-handed. If you are lucky, you'll find a volume that is a reasonably helpful substitute for what you want, but not always. Or you'll change what you want to suit what is available.

There is a large literature discussing the nature of this Library. In part, some of mathematics was developed in conjunction with the needs of natural science. In part, mathematicians are often drawn by their own work to explore areas that are interesting and productive of more mathematical insight. But there is nothing mysterious here, I believe, since there is always a large number of volumes in the Library that never seem to circulate, and there are patrons who leave without finding the book they needed. Many patrons are mathematicians, and as I mentioned earlier, what is remarkable is the unreasonable effectiveness of mathematics in the mathematical realm.

II

MATHEMATICS AND THE CITY IN THE NINETEENTH CENTURY

> With each simple thought-act something actual and lasting enters into our minds. These somethings appear to us as a unity, apparently contained by an inner manifold. I call them mental objects [Geistesmasse].
>
> These imagined mental objects fuse and are tied to each other to a certain degree, parts under other parts, parts with earlier mental objects. (Riemann, "Zur Psychologie und Metaphysik"[32])

> Riemann surfaces are not merely a device for visualizing the many-valuedness of analytic functions, . . . [it is] their native land, the only soil in which the functions grow and thrive. (Hermann Weyl[33])

Mathematical conceptions and ideas have an autonomy within mathematics. Yet, by the words we use and the pictures we employ, these ideas resonate, deliberately or not, with other societal notions. Often the resonance is suggestive, sometimes it is playful, sometimes it is misleading. In any case, it is hazardous to bring extramathematical conceptions to mathematical notions without carefully examining their import. The mathematics itself delimits the meaning and value of a metaphor. Conversely, mathematical conceptions and ideas may be employed to describe nature and society. And nature and society delimit the meaning and the value of the metaphor.

It would appear that there is a lovely analogy between Riemann's concept of a surface on which lies the values of a function—the surface suitably disentangled so that even for many-valued functions the function has only one value at each point of that surface—with notions of the 19th century industrializing city drawn from modern historiography. This is surely fantastic speculation, as I have already indicated. Still, even if the mathematical notions do have resonances with contemporary phenomena, there need be no causal link. The mathematicians and the society's thinkers draw from a common culture. But we only know of that cultural richness from the withdrawals made from it.

Bernhard Riemann (1826–1866) was known as "the mathematician from Göttingen." (Of course, so were Gauss and Hilbert.) He provided one answer to the question, Where do functions live?, in a time in which the more general question was, Where are people to live and grow and thrive? Here, think of the First Industrial Revolution of water and steam power, the factory, and the rise of urbanization; the novels of Charles Dickens or Émile Zola; the reconstruction of Paris under Napoleon III and Haussmann; and the soon-to-come Second Industrial Revolution of electricity, chemistry, and the further transformation of agriculture. Consequently, in the city all sorts of things were now mixed together: classes, values, roles. Those things and statuses might be separated out into a simpler less mixed-together world. Technically, such a mixture of what is not to be mixed together is called pollution. And the mathematicians and the city planners aimed to purify and make sense where there was once pollution and disorder.[34]

For people and society, the presumably nicely ordered landed estates are being replaced by complex and alienating and exciting urbanizing places, mixed and polluted. So, in *The Marriage of Figaro* (1786, out of Beaumarchais' plays of the 1770s), Mozart and his librettist da Ponte show how much work would be needed to restore the crumbling and now artificial order of those estates, the multitude of cultural layers having collapsed upon each other. Young Figaro almost has to marry a woman his mother's age, who then turns out to be his long-lost mother. When must *droit de seigneur* yield to the differentiation, the new order, and the romanticism of urban alienation? And when will the Baron Haussmanns arrive to reorder the city by eviscerating it?

The seventeenth- and eighteenth-century answers to, Where do functions live?, once expressed as polynomials or in terms of "Cartesian" square-ruled paper (the graph of a function), are inadequate to the subtleties of functions of a complex variable. The graph is to be replaced by sets of pairs of real or complex numbers, now coordinates, $(z, f(z))$, but each of these is a two-dimensional number (real and imaginary part): the absolute value of $f(z)$, to be displayed by those three-dimensional plaster models of functions found in glass cases outside mathematics departments (and nowadays by computer graphics displays of functions, the quadrant in which the argument of the complex number $f(z)$ resides indicated by a color). Or, we have an articulated surface on which $f(z)$ is single valued, the value being attached to each point on the surface (the Riemann surface). (For the latter, see the engraving, on our frontispiece, drawn from C. Neumann's 1884 [1865] book on Riemann surfaces.[35])

There are a number of themes that permeate conceptions of nineteenth-century and later cities *and* geometry using algebraic and topological methods.

First, *the palpable and visible is in tension with the formal and conceptual.* The once manifest and particularistic visual and spatial structures of class or guild or profession become more fluid and less obvious, and relations are mediated by the market and its financial abstractions, eventually to be frozen out into a new ordering of places, at least for a while. One may give up, at least provisionally, the details and specifics of concrete, everyday life for the formal structures of interrelationship and systematic repetitions of those structures in economy, bureaucracy, and managerial hierarchy. So a modern mathematician, Pierre Cartier, reflecting on the formal, algebraic, anti-pictorial character of much of modern mathematics, puts it in a societal context:

> The Bourbaki [group of mid-twentieth century mostly-French mathematicians] were Puritans, and Puritans are strongly opposed to pictorial representations of their faith. The number of Protestants and Jews in the Bourbaki group was overwhelming. And you know that the French Protestants especially are very close to Jews in spirit. I have some Jewish background and I was raised as a Huguenot. We are people of the Bible, of the Old Testament, and many Huguenots in France are more enamoured of the Old Testament than of the New Testament. We worship Jaweh more than Jesus sometimes. (Pierre Cartier[36])

A counter-theme is to *simplify and smooth*: to rezone or to resegregate incompatible activities, to freeze what is fluid and to solidify distinctions. In reconceptualizing a problematic object, in effect reconstructing it, one might start with a nice smooth surface in three dimensions, but when sliced through by a plane, in a certain fashion, the curve that results is not smooth and even has disconnected parts. (Here, think of a conic section such as a parabola, as that curve that results from slicing through a cone; other such slices through other surfaces produce unsmoothnesses and disconnectedness.) So a problematic object is shown to be a consequence of fairly innocent simple moves. If there is polysemy or ramification or unsmoothness, go to a more complex, roomy, higher-dimensional space or environment in which ramification and discontinuity are no longer apparent or problematic. (There is in effect a suburbanization of society, although actual suburbanization is quite an intermixed world.)

Or, if there is a multiplicity of meanings or values associated with a place, there is a consequent strategy reminiscent of multi-storey buildings and urban utilities. Attach to each place a layered structure (say, the utilities below ground), with each of the properties or meanings associated with that place having its own layer. Or, as in an electric power distribution system for a building, attach to each place a structure that can hold all the information for that place (the distribution box on the side of the building), that structure at one place intimately connected to the structure at a nearby place.

As a consequence of these efforts to simplify, smooth, and segregate, *order and structure* come to encompass meaning and reference, syntax dominates semantics, formalism dominates detail and idiosyncrasy. Calculus and geometry become algebraic and topological, urban life becomes urban structure and stratification. And urban planning comes further to the fore as a profession and as a means of formally redefining the city.

Fourth, we understand the complexity of places in terms of their *centers and their peripheries*, or their poles or downtowns, and our locations and activities with respect to those centers. And if we know about the spatial distribution of activities, we can infer where the central places are. If we know about the electrical charges, we can infer the electrical fields from the inverse square law, and vice versa. Similarly if we know the zeroes of a polynomial, we know how it factors, and where its graph crosses the x-axis (if $P(x_i)=0$, then we have $P(x)=(x-x_1)(x-x_2)\ldots$).[37]

Moreover, matters of division and overlap among places and centers, or matters of foreignness and propinquity, are to be understood in terms of distances from the center and costs of access, a spatial economy variously construed, rather than in terms of the particulars of those places and their activities, a spatial ethnography.

And so, fifth, there develops a notion that *processes and patterns* determine each other. The cumulation of the individual market processes of buying and selling of property produces the regular and hierarchical patterns of city form. Random walks produce spatial patterns that look the same at all scales. Random molecular scattering produces the nice thermodynamic properties possessed by the temperature, pressure, and entropy of a gas. Computation and algorithms define what something is, they are what is called an effective construction, as much as do that something's properties.

In sum, we might say that cities became systemic, disentangled, alienated, centralized, and algorithmic. And we might understand what we mean by these notions through mathematical models of structures of interrelationship and of formal decomposition, algebraic systems, homology, and computation. Yet, these analogies are never probative. Rather they are suggestive and heuristic, their value attested by their ability to describe actual places and phenomena.

III

We might put a bit more flesh on the analogy. One of the lessons mathematicians began to learn in the nineteenth century was how to make use of local knowledge, of contiguity of neighborhoods, of what is nearby to what, at least if there was some sort of continuity. (Or, as Stone put it much later, "One must always topologize."[38]) It was realized, surely by Riemann if not much earlier, that we could learn some of the global properties of a space or an object from the kinds of localized survey information that the space would accommodate, the kinds of functions that could reside within it. Out of the local detail emerge the invariant features, invariant features of the underlying space or global object that allow for the displayed local properties.[39] Gravimeters can tell us about the shape of the earth, once we know Newton's law of gravitation. If we think statistically, local census information can tell us about the condition of a city.

Conversely, we might study the kinds of localized idiosyncrasies allowed by a known global object, its creases and cuts and discontinuities, and identify them with observed singular phenomena. Henry Adams (1909) used such a method, employing J. Willard Gibbs' phase rule as his model, to develop a historiography of discontinuous historical change.[40] Generic changes in the shape of the "space" of history, or in how we slice through it, may lead to localized discontinuities as historical transitions, much as there is the possibility of a change in physical phase in the material realm, say from ice to water. And of course, besides sharp breaks or creases, there might be multiple layers as in phyllo dough.

Parenthetically, I should note that the description that I have been providing is meant to be true to what we understand about urban life. But it is also meant to be true to those mathematical objects—in particular, Riemann surfaces and the mid-twentieth century notion of sheafed spaces—that allow *at each basepoint* for a collection of properties (say a vector, as in an electric field) or a local space or algebraic object, those properties or structures to vary smoothly from point to point—again with the proviso that there are lines of discontinuity where those variations need not be smooth.

The nineteenth-century industrializing city in both its scale and its scope became the stuff of contemporary novels and painting and opera: Hugo, Zola, Dickens; Monet and Caillebotte; or *Carmen's* cigarette factory. This was a city that was gridded and multilayered, literally: horizontally, the Paris sewers with their auxiliary pipes of potable water and, eventually, pneumatic tubes and telephone and electrical lines; and vertically, in the growth in multistorey buildings and the elevators that enabled them, not to speak of the underground layers, different classes residing at different heights. And figuratively, the city was gridded and multilayered as well: in the variety of social relations taking place, relations that were graded by race, gender, class, criminality, and sexual preference.

Walter Benjamin (ca. 1930) pointed out that Paris stands over a system of caves, grottos, and catacombs, which have developed since the Middle Ages. And one can take a tour of these wonders. But in the Middle Ages these were not seen as tourist attractions, they were an underworld.[41] Benjamin then provides a wonderful panoply of this underworld and its revolutionary, subversive capacities—no mere words when Paris is the topic.

In these nineteenth century cities, there were enormous flows of people and goods, and what happened in one place surely influenced its neighborhood. Sources and sinks, of fluids, sewerage, and of goods and money, and dynamic relations between city and hinterland were crucial to urban development and urban life. Much as in the Freudian analysis of libido, here too there was a hydrodynamics in which flows were conserved, and what disappeared here had to appear elsewhere.

Again, places were ramified, garrets over apartments over stores over cellars, the occasion for the display of a multiplicity of values and statuses, many of them in conflict. Properties are no longer so readily and simply associated with real or even movable property. The meaning of a fetish might be multiple and context-dependent. There was a many-to-many mapping of geography and society. If a city might have been for a while divisible into sectors or *arrondissements*, now it would appear that the complex horizontal, vertical, and social structures would vary from place to place, somehow producing the appearance of a unitary city, those structures somehow coherently linked, perhaps by markets, perhaps by culture, law, and custom.[42]

At the same time, within a city there were vast disparities of wealth and class, cheek by jowl, separated by borders that it would appear were virtually unbreachable. Yet, one of the themes of the novel and of opera was exactly how such disparate regions, tangent to each other, were penetrated and breached. One had the sense as well that Adam Smith's Invisible Hand (1776) worked not only in benign ways but also in diabolical ones, unseen and unpredictable, the fate of the likely-to-be-bankrupt entrepreneur or of the working class now up for grabs.

Enormous effort went into articulating the proper and smooth linkages among the various levels and regions in order to create an apparent unity and continuity while maintaining segregation. Urban etiquette or decorum was the knitting that held together the various layers or sheets and surfaces of society. Again, the structure of a particular place had to interdigitate with the structures of nearby places, so that it all fit together. Much as a thatched roof provides shelter from the rain, doing more than a collection of leaves might do on their own, here too the city, as such, provided an overlay which was as significant as the activity in each of its places.

In sum, it would appear that the nineteenth century city is a highly ramified or multilayered surface, at each spatial point inhabited by quite complex structures, structures which are in their own ways linked to each other from place to place.

For the analyst of urban life, the task was to figure out how things tied together, so that, figuratively, when you pulled both ends of the string the knot did not just disappear. So Gauss (1777–1855), Riemann's doctoral examiner, tried to understand what made a knot a knot. How can the self-intersections and intertwinings of places produce a tapestry rather than a pile of skeins? For Michael Faraday (1792–1867), the great electrician, the dynamic intersection of the electromagnetic lines of force generates electricity through the dynamo effect called induction.

ORDER AND STRUCTURE

More generally, over each point in actual planar space there is a bundle of information about activities and exchanges: by location, by sector, by interaction with other places. We want to encode that information so that the information is disentangled, so that it is suitably structured, and so that we connect the information at one point with the information at another, actually the activities and exchanges at one point with those at another point, assuming perhaps some sort of smoothness or continuity while allowing for natural breaks and mismatches such as a river. We might layer the bundles of information, each layer concerned with a particular sort of information—much as has been done in cartography. Cutting through the layers are connections among the information at a point. Then we might imagine the city as a sheaf of stalks of grain, each stalk at one point, nearby stalks being related to each other. Ideally, we might hope that the layers and stalks are compatible with each other.

(Since in actual life, exchanges take place between what are apparently distant points, we may need to change our notion of contiguity (our topology) or the shape of space, or allow for action-at-a-distance, or imagine intermediate exchanges, each with adjacent points, that mediate the original exchange.)

The layers might be a thatching of sublayers or overlapping layers, or a skeleton (Brouwer's simplicial approximation), or a crystalline ordering, or a map projection, all of which can allow for hiatus, mismatch, and fissure. Cauchy (1789–1857) called these breaks *lignes d'arrêt* (branch cuts), delineators of where we cannot cross continuously: we have to pay a toll or absorb a jump, or add what Riemann called moduli to the function.

Whatever the layering, the information is in effect checked for consistency by constraints. There is usually some sort of conservation law that connects the facts at one point with those at its neighbors. Whether it be in hydrodynamics or electricity or thermodynamics, stuff *is* conserved. And if it appears not to be conserved, we can patch up our notions and spaces so that stuff is *made* to be conserved. What disappears here must reappear elsewhere: usually right nearby—unless there is a source or sink of that stuff at that point, or an underground pipe between the two points. So, for example, it was discovered that economic activity destroyed in one place by ongoing industrialization and capitalism reappeared someplace else, often *not* nearby, often further enhanced, the source of that enhanced vitality the great problem of political economy. (The vital factor-input is technology and information and organization, as Smith argued. The other crucial input was movable rather than real property, capital and labor rather than land and its aristocracy.)

In order to mathematically and formally guarantee that stuff is conserved, there developed the notion of unseen potentials—as in potential energy, gauge fields, or electrical and thermodynamic potentials such as voltage or Gibbs free energy.[43] Maxwell presented to Gibbs, in honor of his achievement of a richer potential theory in thermodynamics, a plaster model (a "thermodynamic mountain") of the mathematical surface representing those potentials for water.

Another constraint is provided by invariant features of a city or a space, such as the river that divides a city or a hole in a doughnut. The local information has to acknowledge these hiatuses. And another constraint are the local rules that connects the various points on a single stalk. If the layers and stalks have to be compatible, further constraints are introduced.

Riemann's surfaces, and the more general idea of a sheaf of stalks, can accommodate all these features and incorporate others as well.[44] So if there is sufficient smoothness, the value of a function at a place might be an average of its value in neighboring places (and hence, perhaps, the remark, "location,

location, location," in discussing the contributors to real estate value). The real and imaginary parts of the function are "harmonic functions," described by the partial differential equation for a drum or a rubber membrane.[45]

Maxwell's fluidic electromagnetism (1873) epitomizes this ordering and structuring of space by fields and potentials.[46] Eventually it leads, through Hertz and Heaviside and Edison, and many others, to the electrification and the electrical lighting of the city, and to electrical engineering.[47] Electrification's implications for urban life have been enormous and pervasive, both technically and culturally. Concretely, electricity allows for elevators, centralized power generating plants, and dispersed factory locations, allowing for the city's disentanglement into an interconnected and alienated world.[48]

Culturally, the implications are rather more varied. So, for example, the poet Francis Ponge was commissioned by the French electrical utility (in 1954) to make sure that architects did not forget the implications of electrification. He wrote a paean to electricity, in part to make sure builders provided for adequate electrical power lines and outlets in their designs. It is a beautiful text, with speculation on whether the Biblical instructions for constructing the Tabernacle, in effect a capacitor, included "charging points" in its design. He noted that Moses and Aaron both wore long priestly robes made of gold threads, which touched the ground, itself a fine conductor, so providing for an alternative path for discharging the capacitors, so that Moses and Aaron are not electrocuted.[49]

The nineteenth century's holistic energetics became for some thinkers a source of animism, and an analogous inner life of plants and the constellations (Fechner, 1801–1887), or of the earth taken as a body (A. Ritter, 1826–1909).[50] If you are like Riemann and are to some extent influenced by a selective reading of Romantic *Naturphilosophie*, here Herbart (1776–1841) and Fechner most directly, you might want to see the diversity and multiplicity of emergent city life as aspects of a whole, as a body composed of systems, as a unity in multiplicity, and search for universal principles and a unified field in the physical sciences.[51] Moreover, Riemann adopts a principle of energy minimization that goes along with potential theory, what he called the Dirichlet Principle (since he learned it from Dirichlet). It is also known as Thompson's principle (after Lord Kelvin), which has been called the foundation for "energy and empire."[52]

Sigmund Freud's *Interpretation of Dreams* (1900) was a scientific embodiment of this romanticism. Dreams *meant* something, they indicated our place in the world. Dreams were proleptic because we would try to actually fulfill them, albeit as perversely as the dreams are fulfillments of our waking lives. So, Leopold Kronecker could write in 1880 about the central problem in his research, that it was the most beloved dream of his youth (hence nowadays called "Kronecker's *Jugendtraum*").[53] For Kronecker is proving something, the essential point of which he "knew" to be true in his youth twenty years earlier.

Or, Andrew Wiles, who in 1994–1995 provided the last crucial link in the proof of Fermat's Last Theorem, says in 1996:

> There's no other problem [Fermat's Last Theorem] that will mean the same to me. I had the very rare privilege to pursue in my adult life what had been my childhood dream. . . . If one can do this. . . . If one can really tackle something in adult life that means that much to you it is more rewarding than anything I could imagine.[54]

Dreaming here is a form of magic, to dream is to make something happen, albeit through the means of inspiring the dreamer to work sufficiently hard at a problem that it might be solved.

The magic here is to find unity and meaning in multiplicity and variety, in urban life and in mathematics and in a career. One finds out about an object through the inferred generic features displayed in its particular instances. Again, an identity in a manifold presentation of profiles. One finds out about a place by figuring out generic features of the activities that can take place there. As the mathematician Alexandre Grothendieck (1928–2014) recurrently says: strip a concept of its immediate and visual intuition, and find what would appear to be its core; any statement about a concrete object can be viewed also as a statement concerning a more generic abstract object; and it appears that many such statements "make sense and are true."[55] And we can then recover a new sort of concrete or geometric intuition in this more abstract realm.

CENTER AND PERIPHERY

Another feature of the industrializing nineteenth-century city, at least as an ideal type, was a sharp sense of its center versus the periphery: the central business and manufacturing districts vs. the city's edge; the city vs. the country; the urban vs. the heartlands—in politics, in everyday life, in real estate development. Centers and peripheries became mutually dependent, in the flows of goods and people between them.[56] Reflecting this nineteenth-century phenomenon, Los Angeles has recurrent anxiety about its not having a real center, just being merely suburbs. So we rebuild and re-inhabit a once-central downtown in Los Angeles, as in Babel: "Come, let us build ourselves a city, and a tower with its top in the heavens, and let us make a name for ourselves; otherwise, we shall be scattered abroad upon the face of the whole earth."(Genesis 11:4)

Analogously, we possess a geometrical and topological description of a smooth ("analytic") function of a complex variable, a function that possesses well-defined slopes or derivatives. We attend to its zeros and its infinities (its singularities), and its boundary and its values there, or to just how and where it is multiply valued (the topology of its Riemann surface).[57]

So we might epitomize a city's structure by a smoothed envelope, and that envelope is epitomized by its singularities and boundaries or its shape. David Hilbert's 21st and 22nd mathematical problems for the twentieth century (1902) are restatements of the center/periphery theme.

> [Problem 21] To show that there always exists a linear differential equation of the Fuchsian class, with given singular points and monodromic group. [Namely, finding a function which has prescribed poles and discontinuities]
>
> [Problem 22] Uniformization of Analytic Relations by Means of Automorphic Functions. . . . That is, if any algebraic equation in two variables be given [for example, $x^2 + y^2 = 1$], there can always be found for these variables two such single valued automorphic functions of a single variable [here, $x=\sin t$, $y=\cos t$] that their substitution renders the given algebraic equation an identity [$\sin^2 t + \cos^2 t = 1$].[58]

So spectrum, geometry, and mapping are linked. As I have indicated, these are recurrent associations. One hundred years after Riemann, Grothendieck studied rather more general objects in terms of their zeros, their "spectrum" in the most general sense.[59]

PROCESS, PATTERN, AND PALIMPSEST

In nineteenth-century Paris we have a center that is deliberately layered, with both a tower and subterranean passageways, a city deliberately triangulated by Baron Haussmann's rebuilding. A chthonic Paris: the catacombs, the old stone quarries, the cave at Châtelet, the underground passages beneath the forts, the sewers, and eventually the Métro.[60] So, eventually, New York and London develop the emblematic Subway and Underground, and Chicago has its Elevated and its Loop.

The manifest surface is a realization of the palimpsest, and that palimpsest is the product of a process. So, in nineteenth century geology, decoding the history of the earth became a matter of correctly reading the meaning of those strata of fossil remains revealed through geological uplift and erosion.[61]

If we were to believe in a unity of the world's manifold variety, then the process of its formation, the patterns it exhibits, and the multiplicity of those patterns, might be one. In effect, the partial differential equations of mathematical physics, a great legacy of the nineteenth century, exemplify this Hegelian unity—especially if those equations' solutions are understood as Riemann understood them: in terms of their poles and boundaries, and their jumps (or single-valuedness or "monodromy" properties), or the surface over which the solutions lie.

Could a surface epitomize the layers? Not quite, and not completely. Newton appreciated that for a spherically symmetric body the gravitational field experienced outside of that body might be epitomized by the body's total mass, now located at its center. Different spherically symmetric mass distributions will give the same gravitational field outside the distribution. And Gauss and Poincaré showed more generally that whatever the complexity of the mass distribution inside a body, perhaps not spherically symmetric, whatever the layers and inhomogeneities, one might distribute that mass onto the surface, in effect sweeping it out, and still achieve an identical external gravitational field.

Finally, recall that Newton, using a prism, showed that white light was composed of a multiplicity of colors. He also proposed (1672) a three-color theory of color vision. Two hundred years later, Maxwell (ca. 1855–1861) employed a color wheel (a "Chromatic Teetotum") to indicate how the visible colors might be formed additively from primaries (Vermillion, Ultramarine, Emerald Green, and Chrome Yellow for example), and demonstrated the possibility of color photography using lantern slides and several projectors.[62] (Maxwell eventually settles on Scarlet, Blue, and Green, plus White and Black, for hue and brightness.) Subsequently, a variety of attempts were made to develop full color photographic images. There might be two or three projectors, as nowadays in projection television receivers, each with one of the primaries; or, perhaps a lenticular screen, as in modern displays or televisions or in Polaroid instant color slide film, with alternating dots or lines of the three primaries. Or, we might "subtract" color from white light, so achieving red by taking out the blue and the green (that is, the cyan), just what a red filter does. Technicolor (ca. 1925–1935) was perhaps the most remarkable realization of this subtractive process, the original image recorded simultaneously through three primary color filters on three black-and-white emulsions (shared between two rolls of film). The reproduced image was formed by combining each of those suitably dyed or pigmented color-separation images (actually, gel release positives), in register on top of each other, by means a diffusion process not unlike lithography ("dye imbibition"). The dye was in the complementary color. The world could be recreated in glorious Technicolor—named, by the way, for the Massachusetts Institute of Technology, the inventors' alma mater—if all the layers were properly prepared and registered. Subsequently, in Kodachrome film (1935), all three color-sensitive layers were coated onto one surface. And, after reversal to positives, in chemical development appropriate complementary color was added to each layer separately. There have been other such processes, some of which were two-color rather than three.[63]

So pattern, process, and palimpsest provided for a unified manifold that incorporated the variety of phenomena. In each case, the task is to figure out the right primaries or their complementaries, and to make sure that the various images registered so well that the fact that the world is multilayered is not at all apparent from what is seen.

***A multilayered surface, whether it be sheaves of stalks over a field, a Riemann surface, or an urban palimpsest, allows for the dream of a unity in multiplicity: an analogy of order and structure, center and periphery, and process and pattern. Still, the syzygy I have discerned here between mathematical themes and societal ones is at best notional and fantastic. In a rigorous historical sense, little follows from the kind of account I have presented. We know that everyday images influence mathematicians' concepts. And we know that notions of geometry and form have been important for city design, and that planners have been at times entranced by mathematical formalism as a solution to their problems. A resonance between ideas about urban development and those in mathematics is not yet an analogy. Still, as in Kronecker's and Wiles' dreams of their youth, that resonance encourages me to believe there is something more familiar in the most esoteric of mathematics, and something more systemic in the most varied of urban phenomena.

IV

I have no trouble believing that mathematics can be formalized and abstracted, so that the mathematician and her experience are apparently absent. There would be just definition, theorem, proof, and remarks, presumably with a canonical way of reading that material. But, in the actual doing of mathematics, the mathematician's store of examples, everyday images and analogies (ones that would as well be appreciated by her listeners or readers), the fact that she has a body and that she has a particular range of experiences and everyday ideas, play a significant if occasional role (perhaps eventually to be dismissed). Different groups of researchers may employ very different concrete analogies: sometimes pictures and images; at others diagrams and charts, or choreographed moves, or symbolic systems. What is taken to be abstract almost always has within it canonical meaning, and concrete objects or symbols which draw from everyday life or from a variety of mathematical objects. So, conversely, Emmy Noether (1926) comprehended the combinatorial topology of Alexandroff in terms of abstract groups and their properties.[64]

I take it mathematics is an activity within a larger culture, from which it borrows ideas and images, and toward which it contributes more than its share of notions and conceptions. And this is surely the case among mathematical subfields. None of this denies the objectivity of mathematical truth.

EMPLOYING YOUR BODY TO PROVE A THEOREM: THE SNAKE LEMMA OF ALGEBRAIC TOPOLOGY

In the first minutes of the film *It's My Turn* (1980, directed by Claudia Weill), the actress Jill Clayburgh, playing a professor of mathematics, choreographically proves the Snake or Serpent Lemma of algebraic topology. Her hands, arms, and upper body trace the snakelike path of inference:

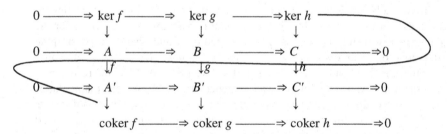

FIGURE 6.1: The diagram employed in proving the Snake Lemma. The "snake" is the object one constructs in the lemma. (Given a commutative diagram of abelian groups and homomorphisms between them, and the rows are exact sequences; then the kernels and cokernels of the vertical homomorphisms form an exact sequence.)

And the snake is, in the mathematics, in effect biting its own tail. (By the way, and continuing in this eponymous vein, in algebraic topology there is also the Horseshoe Lemma, the Braid Lemma, the Five Lemma, the Baby Ham Sandwich Theorem, and the Zig-Zag Lemma, to draw from the index of one text. The diagram of each notionally resembles the object designated by its name.)

The Snake Lemma and its brethren are part of the basic apparatus of algebraic topology. That apparatus enables the mathematician to decompose an object—implicitly a topological space, one now known through an algebraic representation—into simpler objects, and to keep track of what is lost at each stage.[65]

The proof is a series of backward inferences (if Z then Y, if Y then X, . . .) along the serpentine path that provides the needed mappings. One can lay out the argument linearly, leaving out the mathematician, her body, and the chart or diagram. But then one loses the mnemonic and formal devices that encourage the ready understanding and application of the ideas:

> Of course, it is easy to describe the snake diagram sequentially in a suitable more or less formal, linear language. However, such a procedure requires an artificial and not uniquely determined breaking up of a clearly two-dimensional picture (as in scanning a television image). Moreover, without having the overall image in mind, it becomes harder to recognize the analogous situation in other contexts and to bring the information together into a single block. (Yu. Manin[66])

Proverbially, each textbook in algebraic topology or homological algebra announces that the proof of the Snake Lemma is best left as an exercise for the student, that working it out on paper is tedious, and that it is best done narratively—on your own, or by watching the movie. (By the way, it would appear that the film's advisors made sure that even the side-blackboard material was authentic. And Clayburgh was coached by a Princeton mathematics professor; she claimed to actually have more or less understood the proof and its meaning.[67]) In this film, we have a chance to watch mathematics as it is proved and as it is instructed, albeit by Jill Clayburgh enacting that professorial role.[68]

The Snake Lemma is itself an object or at least a diagram, and to appreciate its narrative helps one to employ it in other situations. The proof—rather, going through the proof yourself—is a training ground for future use of the object. And the proof itself when enacted is a bodily and spatial process, as much as it is an intellectual one.

TRAIN TRACKS AND PAIRS OF PANTS

The range of our own experience is the constitution for the analogies and examples and ways of working we employ. So William Thurston (b. 1946) says:

Informally, in Teichmüller space, we pay attention not just to what metric a surface is wearing, but also to how it is worn. In moduli space, all surfaces wearing the same metric are equivalent. The importance of the distinction will be clear to anybody who, after putting a pajama suit on an infant, has found one leg to be twisted.[69]

. . . As long as you have ever spent time dressing an infant, that is. For Riemann or Weierstrass and their contemporary readers that is not so likely, while for Thurston and his readers dressing your infant presumably went along with being a mother or a father. It appears he knows of what he speaks:

But, remember, to determine a point in Teichmüller space we need to consider how many times the leg of the pajama suit is twisted before it fits onto the baby's foot.[70]

Thurston is a modern innovator in two- and three-dimensional geometry and topology, based in part on his developed intuitions of a noneuclidean hyperbolic geometry. ("I gradually built up over a number of years a certain intuition for hyperbolic three-manifolds, with a repertoire of constructions, examples and proofs. This process actually started when I was an undergraduate . . ."[71])

Thurston's agenda or program is to find decompositions and classifications of the various manifolds or spaces in these lower dimensions. Pasting, cutting, gluing, and plumbing fixtures are the technologies of choice. They are here encoded: (1) in diagrams of train tracks, switches, and branches, forming a layout that has no dead ends and does not have isolated circles; (2) in pairs of decorated short pants, what is in effect a sphere with three holes (or a disk with two—put your feet through the holes, and pull the disk's perimeter up to your waist), the surfaces marked with geodesics and flows and field lines, as well as train tracks in a layout; and, (3) in "pants-decompositions" of an object into pants and annuluses.[72]

Continuing the tailoring analogy, Thurston describes his task:

Given a large supply of some sort of fabric, what kinds of manifolds can be made from it in a way that the patterns match up along the seams? . . . For open manifolds, Gromov's theorem gives a good answer for a wide variety of fabrics. . . .

A foliation is a manifold made out of a striped fabric—with infinitely thin stripes, having no space between them. The complete stripes, or *"leaves,"* of the foliation are submanifolds.[73]

Earlier, Möbius (1863) showed how a decomposition leads to caps (*a*), annuluses (*ab*), and pants (*abc*), where the letters *a*, *b*, and *c* label the borders of the basic object (zero, one, and two holes, respectively). He developed a canonical mode of describing an object in terms of "words" composed of these letters.

It is only too cute by a half to ask, Who's wearing the pants and playing with trains? Someone who crosses the once-conventional gender roles, doing mathematics and being pretty, doing mathematics and taking care of baby (and broadcasting taking care of baby).[74] In 1979 Charles Fefferman described the way that he liked to work:

> I like to lie down on the sofa for hours at a stretch thinking intently about shapes, relationships and change - rarely about numbers as such. I explore idea after idea in my mind, discarding most. When a concept finally seems promising, I'm ready to try it out on paper. But first I get up and change the baby's diaper. . . . New ideas are not easy to find. If you are lucky enough to be working on an idea which is actually right, it can take a long time before you know that it's right. Conversely, if you are going up a blind alley, it can also take a long time before you find out. You can end up saying 'Oops, I've been working for years on something wrong.' A good mathematician must have the courage to take a lot of work and throw it away.[75]

None of the specific metaphors or examples or analogies is necessary, in that one might present the mathematics without them. But perspicuous presentations rely on particular examples, analogies, and metaphors. And those examples, analogies, and metaphors are likely to be drawn from and embedded in the ordinary everyday life of the people employing them. What is remarkable is how such contingent circumstances enable us to produce mathematical facts that are more generally reliable (or, for that matter, art that appears transcendent, or governments that endure). On the other hand, things don't always work out and are forgotten.

V

Finally, we turn to theology, and to how mathematics is theological, *as mathematics*. I mean by this that mathematics may be taken to help answer theological questions about the transcendent, that mathematics has been so taken for much of history, and that mathematics provides models of theological concepts, especially in its notions of the infinite.[76] Still, the facts of the mathematics insist on themselves. They are not likely to be compliant to any particular theological agenda. And they provide a variety of very different models of the infinite, the inaccessible, and other theological notions, so they do not prescribe the theology, either.

But, in the matter of course, the mathematics may be stripped of these theological vestments so that they are not even vestiges. The major loss would a sense of the meaning and purpose the mathematical facts possess for their creators and interpreters at various times.

My examples could begin with the Babylonians and their exact sciences, where ominous phenomena and mathematical technology of a high order went hand in hand. Or, we might recall the Pythagoreans and their sacred forms. Here we shall examine how the scientific revolution interacts with philosophic theology in the seventeenth century, and how those concerns then are replayed in the last century or so.

GOD'S TRANSCENDENCE AS ANALOGIZED TO MATHEMATICAL NOTIONS

For Descartes (1640), God is a source of infinitude, mathematically understood. God as an infinity is analogized to the natural numbers, there being no largest one, $N+1$, such that it predecessor, N, is finite:

> . . . when I count, I cannot reach a largest number, and hence I recognize that there is something in the process of counting which exceeds my powers. . . . I have the power of conceiving that there is a thinkable number which is larger than any number that I can ever think of, . . . something which I have received not from myself but from some other being which is more perfect than I am.[77]

In Descartes' analysis of God's transcendence, transcendence is modeled by the infinite:

> . . . there is more reality in an infinite substance than in a finite one, and hence that my perception of the infinite, that is God, is in some way prior to my perception of the finite, that is myself.[78]

And theological notions are modeled by mathematics:

... it is quite evident that existence can no more be separated from the essence of God than the fact that its three angles equal two right angles can be separated from the essence of a triangle, or than the idea of a mountain can be separated from the idea of a valley.[79]

And so we might imagine the richness of a Cartesian theology were Descartes to have access to Cantorian set theory and noneuclidean geometry.

Spinoza (ca. 1660–1675) takes on arguments of the following sort concerning God's corporeality and substance:

If corporeal substance is infinite, they say, let us conceive it to be divided in two parts. Each part will be either finite or infinite. If the former, then an infinite is composed of two finite parts, which is absurd. If the latter, then there is one infinite twice as large as another, which is also absurd. . . . [80]

He draws from intuitions of finite numbers, or objects, or geometrical figures:

That thing is said to be finite in its own kind that can be limited by another of the same nature. . . . By God I understand a being absolutely infinite, . . . I say absolutely infinite, not infinite in its own kind; for if something is only infinite in its own kind, we can deny infinite attributes of it . . . A substance which is absolutely infinite is indivisible.[81]

We might imagine Spinoza wielding the technical apparatus of modern set theory or nonstandard analysis (in which the algebra and analysis of infinitesimals is developed).

In Leibniz (1686), the orderliness of the world, its composition and structure, is modeled by a smooth algebraic curve that can be drawn linking a set of points (what is now called an algebraic variety):

. . . someone jots down a number of points at random on a piece of paper, as do those who practice the ridiculous art of geomancy. I maintain that it is possible to find a geometric line whose notion [or rule] is constant and uniform, following a certain rule, such that this line passes through all the points in the same order in which the hand jotted them down.[82]

Modern algebraic geometry would provide a scheme and a spectrum of points vastly more general, enabling Leibniz to say even more about the world's orderliness.

For Descartes, Spinoza, and Leibniz, improved mathematics would allow for a richer set of analogies and models in their theological analyses. The world would not become "disenchanted" by modern science (*Entzauberung*, to use Schiller's notion); its enchantment would be even more complex and ramified.[83] As the historian Amos Funkenstein put it concerning this era, "Theological concerns were expressed in terms of secular knowledge, and scientific concerns were expressed in theological terms."[84] So, for example, Newton was both a theologian and a natural philosopher, by his own intention, at the same time.

The theological vocation concerned with God's unboundedness, infinity, and indivisibility, often taken as matters of God's omnipresence and God's inaccessibility to us, has a curious resonance with mathematics. Earlier, for Augustine, in both the *Confessions* (390 CE) and *The City of God* (420 CE), allegory and analogy are the only ways we can know of, or at least speak of, God's infinite nature. Time, as we know it, begins with the Creation, much as it does with the modern notion of the Big Bang in cosmology. So, for Augustine, the transcendent God is outside of time (outside of or "before" time). Insofar as we have any access to God, God is accessible to us only through mundane if sacred events and through mundane language. There is an infinity to which we have no direct access at all.

A more modern model of such inaccessibility-made-commensurable or at least speakable is provided by extensions of number systems: Much as $\sqrt{2}$ and $\sqrt{-1}$ are extensions of the rational numbers, whose combinations with the rational numbers generate what is called an ideal, so God might be thought of as an ideal element. It is incommensurable, but may be combined with the commensurable in a formal way: $1 + 2i$, $\frac{1}{2} + \frac{2}{3}\sqrt{2}$, *man + God*. Thereby, one restores the unity of the universe, albeit formally, much as ideal elements in mathematics restore "the validity of simple laws," such as unique prime decomposition of numbers.[85]

It is worth keeping in mind that the $\sqrt{-1}$, the imaginary i, was once a fantastic idea, made commonplace and visual by $a+b\sqrt{-1}$ taken as a vector in the x-y plane (the Argand diagram).

Cohen proposed that we might add what he called a generic real to our world, and so doing he could prove the independence of the Continuum Hypothesis. Analogously we might define a binary number between zero and one as $0.0010011...$, the value at each place determined randomly. We could not be sure if the number were rational (repeating sequence of digits) or not because we cannot know its yet to be determined places.

Set theory, which becomes the study of the infinite by finite means (much as is theology), speaks of the "reflection" of transcendent properties in more ordinary albeit infinite objects, or of the similarity of an object to itself.[86] As is exemplified in Dante's *The Divine Comedy* (1308–1321), any image or idea we have of another world, Heaven or Hell or whatever, must be reflected into the ordinary terms of our everyday world. How else could we have an image of it, except by reflection and transformation into the everyday and mundane?

(Technically, the mathematician's argument from reflection goes something like the following: Generally, if we can conceive of some sort of ordinal number or set characterized by a property that is about its Absolute unavailability to us (much as is God), that conception itself suggests there is a set with that property in a more modest world than the Absolute one or the class of all sets. Properties we take as characteristic of the transcendent Absolute are said to be reflected into or instantiated by the more everyday mathematical world. Otherwise, the Absolute would be distinctively characterized; yet its defining feature is its inconceivability. There must be, to use Cantor's term, transfinite but not Absolute objects that already possess the property.)

We are told that Cantor's set theory (ca. 1870) became for Cantor of theological interest, about the Absolute infinite and the more accessible transfinite (the models of which are the infinite numbers). Point-set topology allows Cantor to connect those theological intuitions about the relationship of the Absolute to the transfinite, to problems in the analysis of functions and their properties, about the "sets of uniqueness" of a fourier-series representation of a function.[87] Modern descriptive set theory, an account of the "definable" sets of the reals, might be said to take the continuum, say the real line, as an Absolute and then make it more commensurable in terms of a hierarchy of nice subsets of that continuum.

ASCENT TO THE ABSOLUTE

Cantor's transfinite numbers were meant to mediate between our finite realm and God's Absolute and inaccessible realm. The systematic articulation of a hierarchy of these numbers creates a realm neither inaccessible nor Absolute. The transfinite numbers were the transfinite ordinals, such as ω, the first infinity "following" the natural numbers $(1, 2, 3, \ldots, \omega)$ and in effect ω is inaccessible from below, and the transfinite cardinals, such as the $\aleph$s. $\aleph_0$ is the number of integers or natural numbers, what we often mean by an infinite number of discrete objects. $\aleph_1$ is the next cardinal. Cantor writes that the transfinite numbers he discovers are justified as being numbers (like the other numbers) by God's perfection and divinity, and by their scientific usefulness:

> One proof [of an infinite creation] proceeds from the concept of God and concludes first from the highest perfection of God's essence the possibility of creating a *transfinitum ordinatum*. It then goes on to conclude from His divinity and glory the necessity of the actual successful creation of the *transfinitum*. Another proof shows a posteriori that the assumption of a *transfinitum* in *natura naturata* [a passive God or Nature] yields a better (because more complete) explanation than the opposite hypothesis, of the phenomena especially of the organisms and the psychical facts.[88]

Hilbert (1925) emphasizes the mathematical usefulness of ideal elements, and so, by analogy, the usefulness and validity of adding-in ideal propositions. Here the infinite is a Kantian idea, one that transcends experience but allows us to make sense of that experience.

> No one [who finds contradictions] shall be able to drive us from the Paradise that Cantor created for us. [Aus dem Paradies, das Cantor uns geschaffen, soll uns niemand vertreiben können.] . . . Let us remember that we are mathematicians and as such have already found ourselves in a similar predicament, and let us recall how the method of ideal elements [for example, $i=\sqrt{-1}$], that creation of genius, then allowed us to find an escape. . . . so we must here adjoin the ideal propositions [about the existence of infinite sets] to the finitary ones in order to maintain the formally simple rules of ordinary Aristotelian logic. . . . The role that remains to the infinite is, rather, merely that of an idea— if, in accordance with Kant's words, we understand by an idea a concept of reason that transcends all experience and through which the concrete is completed so as to form a totality . . .[89]

For both Cantor and Hilbert and their contemporaries, one had not only to discover these objects but also to place them within the realm of other mathematical objects. Articulating that mediating realm—mediating the finite and the Absolute, mediating the empirical and the rational—into an inclusive hierarchy of larger and larger cardinals, each one constructed "from below" out of smaller ones by formal rules (of replacement, succession, and recombination), provides a ladder toward the Absolute.

It would appear that ". . . the universe goes on through as many transfinite stages as it can. Various large large-cardinals are defended on grounds arising from the idea that the hierarchy of stages is so complex and rich that it must contain stages that resemble one another in certain ways."[90] So we have an orderly and systematic articulation of the large cardinals.[91] But there are limitations on size. Some proposed sets of large size or cardinality are inconsistent with the rest of set theory. And when we unrestrictedly speak of the set of all sets, we may get into logical trouble because we do not acknowledge our own conceptual limitations.

$0=1$ (inconsistent!)
Huge$\downarrow$
Supercompact$\downarrow$
Woodin$\downarrow$
Strong$\downarrow$
0^+exists$\downarrow$
Measurable$\downarrow$
Ramsey$\downarrow$
Jónsson$\downarrow$
$0^\#$ exists$\downarrow$
Weakly compact$\downarrow$
Mahlo$\downarrow$
Inaccessible$\downarrow$
Weakly inaccessible

FIGURE 6.2: Chart of large cardinals, drawn from A. Kanamori, *The Higher Infinite* (edn. 2, p. 472), arrows indicating direct implication or relative consistency or both. Weakly inaccessible is already larger than $\aleph_0$, and inaccessible is by definition larger than two-to-the-aleph-nought. Ordinary set theory axioms do not determine the size of the continuum (or real line, or two-to-the-aleph-nought).

Some of these large cardinal numbers are initially defined by particular properties. Although they are not at first obviously built up from below, it turns out they can be, albeit by highly technical devices for combining smaller sets, such as "mice" and "trees." It is not a given ahead of time that large cardinals might form an inclusive hierarchy of sets or numbers which include smaller sets or numbers, insofar as that is the case. The inclusive hierarchy is a discovery, another curious feature of these models of the infinite. What is even more remarkable is that the natural extensions of set theory are indexed, so to speak, by the hierarchy of large cardinals.

(Technically, the hierarchy of sets of reals developed by Borel and Lusin, what is now called the Borel hierarchy, based on countable $(1, 2, 3, \ldots)$ unions, complementation, and projection-by-continuous-functions of the open sets, has a systematic formal analogy provided by Kleene in terms of the complexity of "primitive recursive functions" of the natural numbers, with a host of further correspondences, such as consistency strength.[92])

The continuum remains deeply problematic. Those large cardinals are not adequate for measuring the size of the continuum.[93] The continuum, as the real numbers or the real line, is given to us by recombination: namely, the continuum is the set of all subsets of the natural numbers (say the binary or decimal numbers, each one being a set), namely $2^{\aleph_0}$ ("two-to-the aleph-nought") or $10^{\aleph_0}$ (they are equal), the power-set construction. But the continuum may not be accessible by any process that builds up from below by "replacement."[94] Hugh Woodin has suggested how one might settle the Continuum Hypothesis, of the size of the continuum, by strengthening the logic we use (much as an appended Axiom of Projective Determinacy settles traditional questions, such as measurability, about many sets).[95] He suggests as well that the Continuum Hypothesis is false, that in fact the continuum is larger than the first cardinal number following $\aleph_0$: $2^{\aleph_0} > \aleph_1$, and in fact it equals $\aleph_2$.

Transcendence, that $> \aleph_1$, is given a technical meaning, intimately connected to our capacities to think and reason. And those capacities are also given a technical meaning in terms of a strengthened logic. So we articulate the ascent to the Absolute.

The various infinities and the proposed large-cardinal axioms (if they are consistent) provide a language so that we may talk about and understand what was thought to be inconceivable. They provide mathematics with greater explanatory power, and we can say just which kinds of statements they allow for. I think of the large cardinals as the neutrinos of set theory. Recall, neutrinos were proposed to preserve conservation of energy in beta radioactivity, and they allowed for a much richer theory as well. Their experimental detection came much later. In effect, one proposed an axiom or an object to preserve what one takes to be fundamental about the world, and then sees if they prove to be systematically useful.

So, for example, to understand the nature of the continuum, one has to develop a technology that avoids the impact of Cohen's invention ("forcing" and generic reals). Woodin not only needed to strengthen logic, he also needed an infinitude of "Woodin cardinals" to defeat the forcing's capacity to convert the $\aleph$s into countable objects.

IN CONCRETO

SETS AS MODELS OF GOD'S INFINITUDE

The realm of the transfinite numbers was a discovery within mathematics, driven as far as I can tell by internal mathematical problems of nineteenth century analysis. Surprises and structures abound in the ascent to the Absolute. No theological intuitions, or other intuitions for that matter, could have prepared one for those transfinite numbers—even if those transfinite numbers might be fit more or less well within a metaphorical and theological account once they were discovered (as Cantor did so fit some of them).

So, within the history of analysis, it at first seemed as if the ordinary operations and functions would create sets and functions that were of the same sort as we started out with. For example, Henri Lebesgue claimed to have proved that the projection onto an axis $((x,y) \rightarrow x)$ of a "good" set (that is, a Borel measurable set) would also be good. This turns out not to be the case, as Lebesgue describes in his preface to Lusin's 1930 book on set theory. It would seem that:

> Analysis supposedly had within itself a boundary or limit principle. . . .
> The proof [by Lebesgue] was simple, short, but false. [Souslin and Lusin provided a counterexample in about 1918.] . . . Thus, analysis does not have within itself a limit principle. The extent of the family of Baire functions is so large as to give one vertigo; the field of analysis is even vaster. . . . The field of analysis is perhaps bounded, in any case it embraces the infinity of the types of functions.[96]

The richness of sets of the natural numbers, the real numbers, and of sets of the reals, is vertiginous. These collections of sets are sometimes called baroque or botanical or pathetic, referring to the complexity and variety and apparent idiosyncrasy of their constructions. But, not only are there systematic organizing principles for this richness, such as hierarchical inclusion, there is, again, an analogy with another hierarchy of functions of the natural numbers that makes the principles apparently less arbitrary. And these sets do tell us something about the nature of the continuum. Notionally, if we think of these sets or numbers as models of God's infinity, then we find that God is metaphorically "unknowable," "playful," "very big," and so forth—to use the everyday names that are sometimes attached to quite technical features of these sets' constructions.[97]

***Mathematical notions of the infinite, in models and analogies and objects, may be employed to express some of the traditional concerns about God's infinitude and uniqueness. And they have been so employed, sometimes with great technical seriousness as in Descartes, Spinoza, and Leibniz. Problems of theology, re-expressed and modeled as best we can as mathematics, do not then disappear nor do they avoid subtlety and complication. And the mathematics may develop in ways unsuited to the theology, or even to our previous mathematical expectations. The infinite continues to provide an enormous challenge to our conceptual resources, mathematically and theologically. The ascent to the Absolute is, it seems, more than a technical climb.

Moreover, those large cardinals help mathematicians better understand more mundane problems in algebra, analysis, and number theory.

VI

THE CITY OF MATHEMATICS

> "A city, according to St. Augustine, is a group of people joined together by their love of the same object. Ultimately, however, there can be only two objects of human love: God or the self. . . . It follows that there are only two cities: the City of God, where all love Him to the exclusion of self, and the City of Man, where self-interest makes every sinner an enemy to every other. The bonds of charity form a community of the faithful, while sin disperses them and leaves only a crowd."[98]

Augustine's account of the presence of God in this world was a matter of urban culture. Augustine asked, How can transcendent Jerusalem be present in finite, mundane Babylon? Nine hundred years later, Dante's *Inferno* immortalizes Florence as Babylon in Jerusalem. The metaphorical and the actual city has succeeded the Israelites' desert as the (transcendental) condition for the appearance of order and transcendence among us. God is present in the mundane city much as i $(=\sqrt{-1})$ is present in the real numbers, as something other but familiar, modeled by a palpable everyday notion, which itself is admittedly incommensurable. And the sociologist Émile Durkheim (1912), sounding much like Hilbert's reflections on Kantian themes, suggests that within a community's mundaneness and society, there arises the possibility of the sacred: the sacred as something apparently incommensurable, as other and different than the community's everyday productions, a "collective representation" we could not imagine our being able to provide.[99]

Mathematics, too, is a product of the city as understood here by Augustine, "a group of people joined together by their love of the same object." So the City of Mathematics would be a City of God. And it is perhaps not too surprising that conceptions of God and of infinity address similar problems.

Mathematical facts and truths appear within actual history. The facts and truths are discovered, proved, reproved, fixed up so that they are just right, codified, generalized, reformulated in new languages, abandoned or at least summarily dismissed—by actual historical persons. If we try to read a scientific paper that is even a generation or two in the past, much that is taken for granted, or much about the style and mode of argument, may well be foreign to us. Not always, but often enough. Our perhaps more sophisticated understanding and the advancements in the field may make us less capable of appreciating what it might mean to have, say, a more concrete and example-based understanding or a more computational sense of what the mathematical task is taken to be. Surely, techniques and specific examples, once abandoned, can be rescued for future use. But if they have not been so rescued they are likely to feel strange and uncanny, until we have accustomed ourselves to the literature of a particular era. Historians of mathematics argue about the meaning of Euclid in a way that indicates just how foreign Greek mathematics is to us.[100] But they argue as well about what Newton was doing, and they have to do careful reconstructions if we are to appreciate the Italian school of algebraic geometry of the first part of the twentieth century.[101]

Mathematical facts and truths appear within a cultural matrix that provides the everyday intuitions we employ more generally. I have suggested how highly technical mathematics shares models with city design, everyday human life, and theology. But, that mathematics is embedded in society and material culture *as mathematics*. Following a proof with your body and with chalk; choosing apt names and diagrams; employing a philosophy of mathematics that is a generic idea of what is mathematically interesting and why, what is sometimes called a research program; finding important ideas embedded in particular functions or other mathematical objects; working out transcendent questions through technical mathematical means; and, finding again and again that one's mathematical specifications of objects allow-in other objects that were not anticipated, and hence the productivity of the world compared to our imaginations—all these *are* mathematics as it is actually practiced.

Mathematics is a craft skill. And such skills are local and particular, not always readily conveyed to outsiders. For example, topology would seem to be a regionalized branch of mathematics. We might account for this regionalization in terms of specific historical contingencies, for example, the deliberate choice in Poland (after World War I) to develop a nascent strength in set-theoretic topology as a foundation for national eminence in mathematics, taking advantage of the already close connection with the French constructivists such as Borel and Lebesgue.[102] There is much to the craft knowledge of topology and other fields of mathematics, conveyed least well by scholarly papers, that advantages local specialization and face-to-face-to-blackboard interaction. But, to have a local dialect of examples and procedures does not mean that that dialect is not to be translated, or to be mastered by others even if they are not native speakers. To be located is to be specific and actual and ongoing in the work, even if in writing up papers we might generalize, abstract, and erase that locatedness as best we can.

In so far as mathematics is cultural, it is so by being mathematics. The diagrams we employ have to do mathematical work. The metaphors and illustrative examples must help us understand the mathematics better and suggest useful directions in which to go. Mathematics may be rhetorical and generic and philosophical and theological and historical. But it is only especially interestingly so as mathematics.

Each field of human intellectual endeavor might insist on its autonomy. Surely, its conceptual structures are a reflection of its capacity to mirror the world. And it is conceded that human activity is embedded in human culture. What is striking to those who do mathematics is the productivity of their conceptual structures, the structures' objectivity, and their naturalness.

> When I invented (I say invented, and not discovered) uniform spaces, I did not have the impression of working with resistant material, but rather the impression that a professional sculptor must have when he plays with a snowman. (A. Weil, See Appendix D.)

To think of mathematics as something else (rhetorical, etc.) is to not represent the experience of actually doing and understanding mathematics. *As mathematics* is an insistence on the integrity of that experience of doing mathematics.

Epilog

Saharon Shelah is a phenomenal mathematician. . . . In set theory Shelah is initially stimulated by specific problems. He typically makes a direct, frontal attack, bringing to bear extraordinary powers of concentration, a remarkable ability for sustained effort, an enormous arsenal of accumulated techniques, and a fine, quick memory. When he is successful on the larger problems, it is often as if a resilient, broad-based edifice has been erected, the traditional serial constraints loosened in favor of a wide, fluid flow of ideas and the final result almost incidental to the larger structure. . . . In the end, however, we are still left to marvel, like Caliban peering at Prospero, at how all of this could have been accomplished, and, unable to take it all in, to gaze silent. (A. Kanamori[1])

The results on chains stated here are proved by computing the connection in the special case of a boundary in normal form. Details are formidable. (Charles Fefferman[2])

PLACING BOXES INTO BALLS AND TILING THE BATHROOM FLOOR

Let us recall some of the mathematics we have employed in our survey of ways of doing mathematics. Initially, I suggested that the mathematics I would be discussing might be arrayed around the sound of a drum, the relationship of spectrum to geometry. The latter theme is the story of fourier or harmonic analysis, symmetry, and the uncertainty principle.[3] One counts up the boxes that fit into balls of phase space, or one nicely tiles the bathroom floor. (Lebesgue employed a similar picture in attempting to prove the "invariance of dimension" for topology, a *Pflastersatz* or paving principle: In an n-dimensional domain that is covered by small enough closed sets, there will be $n+1$ non-empty intersections.[4]) It is as well a story of a collection of spinning tops. These are much like the orbits of Kepler's planets, acknowledging that the planets (and the tops) influence each other. The system of tops provides those boxes of *frequency×angular-momentum* that are to fill up phase space.[5]

Harmonic analysis began its life as Fourier's way of solving the heat equation. The analyst finds nice parts that may be combined arithmetically (or linearly), nice because the parts reflect the symmetries of the situation. Being harmonic, the energies go as the squares of the displacements, namely variances. Harmonic analysis is comparatively insensitive to sparsely located errant points, so there are "sets of uniqueness," a foundation of modern set theory and topology. An uncertainty principle connects the fineness of our spatial and momentum (or frequency and angular-momentum) analyses. And, again, the spectrum is a reflection of geometry.

In the descriptions I provide of convention, subject, calculation, and analogy, there is a curious interplay of a transcending truth or facticity with our attempts to grasp the larger meanings through concrete mathematical fantasies and metaphors. As mathematical achievements, these results have not come easily. Often, details *are* formidable, even with modern proofs. Fefferman and Shelah find the simple truths only after enormous efforts at construction and computation. Perhaps Brouwer's complicated constructions are archetypal rather than exceptional. The image of in-your-face truth, forcing you to slap your forehead when it is shown to you, such obviousness or "evidence," is an endpoint of many years of reformulation and reproving. And early on, there is enough preparatory calculation so that when you reach the solution you are as much relieved as surprised.

My point in this all too neat account is to justify as canonical what is of course to some extent a writer's particular choice of examples and themes: how a subject matter becomes canonical, a field is defined, arduous computations are simplified, and analogies become destinies—*as mathematics*, and what you can actually do.

PHYSICAL REVIEW VOLUME 85, NUMBER 5 MARCH 1, 1952

The Spontaneous Magnetization of a Two-Dimensional Ising Model

C. N. YANG
Institute for Advanced Study, Princeton, New Jersey
(Received September 18, 1951)

The spontaneous magnetization of a two-dimensional Ising model is calculated exactly. The result also gives the long-range order in the lattice.

IT is the purpose of the present paper to calculate the spontaneous magnetization (i.e., the intensity of magnetization at zero external field) of a two-dimensional Ising model of a ferromagnet. Van der Waerden[1] and Ashkin and Lamb[2] had obtained a series expansion of the spontaneous magnetization that converges very rapidly at low temperatures. Near the critical temperature, however, their series expansion cannot be used. We shall here obtain a closed expression for the spontaneous magnetization by the matrix method which was introduced into the problem of the statistics of a two-dimensional Ising model by Montroll[3] and Kramers and Wannier.[4] Onsager gave in 1944 a complete solution[5] of the matrix problem. His method was subsequently greatly simplified by Kaufman,[6] and the result has been used to calculate the short-range order in the crystal lattice.[7]

The Onsager-Kaufman solution of the matrix problem will be used in the present paper to calculate the spontaneous magnetization. In Sec. I we define the specific magnetization I and express it as an off diagonal element in the matrix problem. By introducing an artificial limiting process this calculation is reduced to an eigenvalue problem in Sec. II. This is solved in the next three sections and the final result given in Sec. VI. The relation between I and the usual long-range order is discussed in Sec. I.

It will be seen that the final expression for the spontaneous magnetization is surprisingly simple, although the intermediate steps are very complicated. Attempts to find a simpler way to arrive at the same result have, however, failed.

I. SPONTANEOUS MAGNETIZATION

Using Kaufman's notation[6] we have for the two-dimensional square lattice the following expression for the partition function:

$$Z = (2 \sinh 2H)^{n/2} \text{ trace}(V_2 V_1)^m, \qquad (1)$$

where

$$V_1 = \exp\{H^* \sum_1^n C_r\}, \qquad (2)$$

and

$$V_2 = \exp\{H \sum_1^n s_r s_{r+1}\}. \qquad (3)$$

H^* and H are given by

$$e^{-2H} = \tanh H^* = \exp[-(1/kT)\{V_{\uparrow\downarrow} - V_{\uparrow\uparrow}\}]. \qquad (4)$$

The following abbreviation will be useful:

$$x = e^{-2H}. \qquad (5)$$

If a weak magnetic field is introduced the partition function becomes

$$Z_{\mathfrak{IC}} = (2 \sinh 2H)^{n/2} \text{ trace}(V_3 V_2 V_1)^m, \qquad (6)$$

where

$$V_3 = \exp\{\mathfrak{IC} \sum_1^n s_r\}. \qquad (7)$$

For a large crystal only the eigenvector of $V = V_3 V_2 V_1$ with the largest eigenvalue is important. We shall be interested in the limiting form of this eigenvector as $\mathfrak{IC} \to 0$.

It has been shown by Onsager[5] that below the critical temperature, i.e., for

$$x < \sqrt{2} - 1,$$

the largest eigenvalue of $V_2 V_1$ is doubly degenerate. This is evidently also true of the symmetrized matrix $V_1^{\frac{1}{2}} V_2 V_1^{\frac{1}{2}}$. Let ψ_+ and ψ_- be the even and odd eigenvectors corresponding to the largest eigenvalue λ.

$$V_1^{\frac{1}{2}} V_2 V_1^{\frac{1}{2}} \psi_+ = \lambda \psi_+, \quad V_1^{\frac{1}{2}} V_2 V_1^{\frac{1}{2}} \psi_- = \lambda \psi_-. \qquad (8)$$

The even eigenvector remains unchanged when the spins of all atoms are reversed while the odd eigenvector changes sign. Introducing the operator

$$U = C_1 C_2 \cdots C_n,$$

that reverses the spins of all atoms we have

$$U\psi_+ = \psi_+, \quad U\psi_- = -\psi_-. \qquad (9)$$

With the introduction of the magnetic field $\mathfrak{IC}$ the degeneracy is removed. Since we are only interested in the limit as $\mathfrak{IC} \to 0$, we may perform a perturbation cal-

[1] B. L. van der Waerden, Z. Physik **118**, 473 (1941).
[2] J. Ashkin and W. E. Lamb, Jr., Phys. Rev. **64**, 159 (1943).
[3] E. Montroll, J. Chem. Phys. **9**, 706 (1941).
[4] H. A. Kramers and G. H. Wanner, Phys. Rev. **60**, 252, 263 (1941).
[5] L. Onsager, Phys. Rev. **65**, 117 (1944).
[6] B. Kaufman, Phys. Rev. **76**, 1232 (1949).
[7] B. Kaufman and L. Onsager, Phys. Rev. **76**, 1244 (1949).

culation and consider the largest eigenvalue of

$$V_1{}^{\frac12}VV_1{}^{-\frac12}=V_1{}^{\frac12}V_3V_2V_1{}^{\frac12}$$

$$=V_1{}^{\frac12}V_2V_1{}^{\frac12}+\mathfrak{IC}V_1{}^{\frac12}(\sum_1^n \mathbf{s}_r)V_2V_1{}^{\frac12}. \quad (10)$$

The last term is a matrix that anticommutes with U. It has, therefore, no diagonal matrix element with respect to either ψ_+ or ψ_-. It is, besides, a real symmetrical matrix. Ordinary perturbation theory shows immediately that the eigenvector of (10) with the largest eigenvalue approaches, as $\mathfrak{IC}\to0$

$$\psi_{\max}=(1/\sqrt2)(\psi_++\psi_-), \quad (11)$$

if the phases of ψ_+ and ψ_- are so chosen that they are real and that[8]

$$\psi_+{}'V_1{}^{\frac12}(\sum_1^n \mathbf{s}_r)V_2V_1{}^{\frac12}\psi_-\gtrless0. \quad (12)$$

The average magnetization per atom is, from the general definition of the matrix method,

$$I=\frac{1}{mn}\frac{m\ \mathrm{trace}(V_3V_2V_1)^m \sum_1^n \mathbf{s}_r}{\mathrm{trace}(V_3V_2V_1)^m}$$

$$=\frac{1}{n}\frac{\mathrm{trace}(V_1{}^{\frac12}V_3V_2V_1{}^{\frac12})^m(V_1{}^{\frac12}\sum_1^n \mathbf{s}_rV_1{}^{-\frac12})}{\mathrm{trace}(V_1{}^{\frac12}V_3V_2V_1{}^{\frac12})^m}$$

$$=\frac{1}{n}\psi_{\max}{}'V_1{}^{\frac12}(\sum_1^n \mathbf{s}_r)V_1{}^{-\frac12}\psi_{\max}.$$

As $\mathfrak{IC}\to0$ this becomes by (11)

$$I=\frac{1}{2n}(\psi_+{}'+\psi_-{}')V_1{}^{\frac12}(\sum_1^n \mathbf{s}_r)V_1{}^{-\frac12}(\psi_++\psi_-).$$

But $V_1{}^{\frac12}(\sum \mathbf{s}_r)V_1{}^{-\frac12}$ anticommutes with U, and therefore has no diagonal matrix element with respect to either ψ_+ or ψ_-. Besides, by the use of (8), one shows easily that

$$\psi_-{}'V_1{}^{\frac12}(\sum_1^n \mathbf{s}_r)V_1{}^{-\frac12}\psi_+=\frac{1}{\lambda}\psi_-{}'V_1{}^{\frac12}(\sum_1^n \mathbf{s}_r)V_2V_1{}^{\frac12}\psi_+,$$

$$\psi_+{}'V_1{}^{\frac12}(\sum_1^n \mathbf{s}_r)V_1{}^{-\frac12}\psi_-=\frac{1}{\lambda}\psi_+{}'V_1{}^{\frac12}(\sum_1^n \mathbf{s}_r)V_2V_1{}^{\frac12}\psi_-, \quad (13)$$

which are obviously equal. Hence at zero magnetic field the spontaneous magnetization is

$$I=\frac{1}{n}\psi_-{}'V_1{}^{\frac12}(\sum_1^n \mathbf{s}_r)V_1{}^{-\frac12}\psi_+, \quad (14)$$

which is always positive by (13) and (12).

[8] We use the notation $A'\equiv A$ transposed.

Intuitively one would infer that the summation $\sum \mathbf{s}_r$ in (14) can be replaced by $n\mathbf{s}_1$ so that

$$I=\psi_-{}'V_1{}^{\frac12}\mathbf{s}_1V_1{}^{-\frac12}\psi_+. \quad (15)$$

This can also be shown in detail by introducing the orthogonal operator L that is equivalent to the cyclic permutation of the n spins:

$$L\sigma_iL^{-1}=\sigma_{i+1}, \quad L\sigma_nL^{-1}=\sigma_1.$$

Evidently L commutes with V_1, V_2, and U. Therefore $L\psi_+$ is also an even eigenvector of V_2V_1 with eigenvalue λ. Hence

$$L\psi_+=a\psi_+.$$

L and ψ_+ are real. Therefore a is real. Since further $L^n=1$, we have $a=1$, and

$$L\psi_+=\psi_+. \quad (16)$$

Similarly

$$L\psi_-=\psi_-.$$

Now

$$\mathbf{s}_r=L^{(r-1)}\mathbf{s}_1L^{-(r-1)}.$$

Substituting this into (14) and using (16) we obtain (15).

The spontaneous magnetization I per atom is exactly the usual long-range order parameter s which may be defined as the average of the absolute value of the total spin of the lattice divided by the number of atoms. That I is equal to s is easily seen from the fact that the introduction of a vanishingly weak positive magnetic field merely cuts out all states of the lattice for which the total spin is negative.

One may ask, as Zernike[9] did, what is the average value of the total spin of the lattice if it is known that at a given lattice point the spin is $+1$. We can show that the answer is NI^2 in the following way: The total spin is either $+NI$ or $-NI$. If a given lattice point has a spin $+1$, it assumes the former value more frequently than the latter in the ratio of $\frac12(1+I):\frac12(1-I)$. Hence the average total spin is

$$NI(1+I)/2-NI(1-I)/2=NI^2.$$

The long-distance order can also be investigated as the limit of the short-distance order which has been studied by Kaufman and Onsager. Onsager[10] has done this and obtained the correlation of the spins of two atoms in one row at an infinite distance from each other. It can be shown that the long-distance order can be obtained from this, and the result agrees with the findings of this paper.

II. REDUCTION TO EIGENVALUE PROBLEM

A.

To calculate the spontaneous magnetization as given by (15) we notice that it is the off-diagonal ele-

[9] F. Zernike, Physica 7, 565 (1938).
[10] L. Onsager, unpublished; see also Nuovo cimento 6, Suppl. p. 261 (1949). The author wishes to thank Bruria Kaufman for showing him her notes on Onsager's work.

ment of the matrix $V_1^{\frac{1}{2}}s_1V_1^{-\frac{1}{2}}$ between the vectors ψ_+ and ψ_-. Onsager and Kaufman[7] have shown how to calculate diagonal elements by reducing the $2^n \times 2^n$ matrix problem to one of $2n \times 2n$. Their method, however, does not apply to off-diagonal elements. To resolve this difficulty we shall in the present section introduce an artificial limiting process and reduce the problem to an eigenvalue problem of an $n \times n$ matrix.

From Kaufman's[6] Eq. (60) we have, except for a multiplicative phase factor:

$$\psi_- = S^{-1}(T_-)\tau, \qquad (17)$$

where

$$\tau = \mathbf{g}\begin{bmatrix} 1 \\ 0 \\ 0 \\ \vdots \\ 0 \end{bmatrix} = 2^{-n/2}\begin{bmatrix} 1 \\ 1 \\ \vdots \\ 1 \end{bmatrix}. \qquad (18)$$

Similarly

$$\psi_+ = S^{-1}(T_+)\tau.$$

Now since T_- is real it follows that $S(T_-)$ is unitary. Hence taking the complex conjugate transposed of Eq. (17) we obtain

$$\psi_-' = \tau' S(T_-).$$

The reality condition of ψ_- has been used. Eq. (15) therefore assumes the form

$$I = \tau' S(T_-)V_1^{\frac{1}{2}}s_1V_1^{-\frac{1}{2}}S^{-1}(T_+)\tau. \qquad (19)$$

As we have just mentioned, if the expression were of the form

$$\tau' S(T_-)\cdots S^{-1}(T_-)\tau,$$

it would have been easy to reduce because $S(T_-)\cdots \times S^{-1}(T_-)$ induces a rotation in the $2n$ dimensional space formed by the Γ's.[11] We could, however, in the present case still utilize this reduction by first writing

$$I = \text{trace}\,V_1^{\frac{1}{2}}s_1V_1^{-\frac{1}{2}}S^{-1}(T_+)\tau\tau' S(T_-). \qquad (20)$$

Now

$$\tau\tau' = (1/2^n)(1+C_1)(1+C_2)\cdots(1+C_n), \qquad (21)$$

does not induce a rotation. But we notice that

$$1+C_1 = \operatorname*{Lim}_{a\to i\infty}(\cos a)^{-1}(\cos a - iC_1 \sin a)$$
$$= \operatorname*{Lim}_{a\to i\infty}(\cos a)^{-1}\exp(-iaC_1), \qquad (22)$$

and $\exp(-iaC_1)$ does induce a rotation. Write

$$M = \begin{bmatrix} \cos 2a & \sin 2a & & & 0 \\ -\sin 2a & \cos 2a & & & \\ & & \cos 2a & \sin 2a & \\ & & -\sin 2a & \cos 2a & \\ 0 & & & & \ddots \end{bmatrix} \qquad (23)$$

so that

$$\exp(-ia\sum_1^n C_r)\Gamma_\alpha \exp(ia\sum_1^n C_r) = \sum_\beta M_{\beta\alpha}\Gamma_\beta. \qquad (24)$$

[11] The Γ's are defined in Kaufman's paper (see reference 6). There is a mistake of sign in her Eq. (11) which should read
$$\Gamma_{2r} = -C \times C \times \cdots \times isC \times 1 \times 1 \times \cdots = Q_r.$$

We have from (21) and (22)

$$\tau\tau' = \operatorname*{Lim}_{a\to i\infty}(2\cos a)^{-n}\exp(-ia\sum_1^n C_r)$$
$$= \operatorname*{Lim}_{a\to i\infty}(2\cos a)^{-n}S(M).$$

Substitution back into (20) gives

$$I = \operatorname*{Lim}_{a\to i\infty}(2\cos a)^{-n}\,\text{trace}\,V_1^{\frac{1}{2}}s_1V_1^{-\frac{1}{2}}S(T_+{}^{-1}MT_-). \qquad (25)$$

B.

This can easily be calculated if we know the eigenvalues and eigenvectors of the $2n$-dimensional rotation $T_+{}^{-1}MT_-$. The rotations T_+ and M have determinants equal to 1 while T_- has a determinant equal to -1. Thus $T_+{}^{-1}MT_-$ is an improper rotation and must have eigenvalues 1, -1, $e^{\pm i\theta_2}$, $e^{\pm i\theta_3}$, $\cdots e^{\pm i\theta_n}$. Let ζ be an orthogonal matrix that transforms $T_+{}^{-1}MT_-$ into the canonical form

$$\zeta T_+{}^{-1}MT_-\zeta^{-1}$$

$$= \begin{bmatrix} 1 & & & & & 0 \\ & -1 & & & & \\ & & \cos\theta_2 & \sin\theta_2 & & \\ & & -\sin\theta_2 & \cos\theta_2 & & \\ & & & & \cos\theta_3 & \sin\theta_3 \\ & & & & -\sin\theta_3 & \cos\theta_3 \\ 0 & & & & & \ddots \end{bmatrix} = W. \qquad (26)$$

W is evidently orthogonal. We shall compute, first, instead of (25), the more general expression

$$\text{trace}\,\Gamma_j S(T_+{}^{-1}MT_-), \qquad (27)$$

where Γ_j is as defined in Kaufman's paper.[11] By (26)

$$\text{trace}\,\Gamma_j S(T_+{}^{-1}MT_-) = \text{trace}\,\Gamma_j S(\zeta^{-1})S(W)S(\zeta)$$
$$= \text{trace}\,S(\zeta)\Gamma_j S(\zeta^{-1})S(W). \qquad (28)$$

Now

$$S(\zeta)\Gamma_j S(\zeta^{-1}) = \sum \zeta_{\alpha j}\Gamma_\alpha,$$

where $\zeta_{\alpha j}$ are the matrix elements of ζ. Moreover, the explicit form of $S(W)$ is known:

$$S(W) = iP_1(P_2Q_2)(P_3Q_3)\cdots(P_nQ_n)\exp(\tfrac{1}{2}\sum_2^n \theta_\beta P_\beta Q_\beta).$$

(28) therefore reduces to

$$\text{trace}\,\Gamma_j S(T_+{}^{-1}MT_-)$$
$$= i\,\text{trace}(\sum \zeta_{\alpha j}\Gamma_\alpha)P_1(\prod_2^n P_\alpha Q_\alpha)\exp(\tfrac{1}{2}\sum \theta_\beta P_\beta Q_\beta)$$
$$= i\zeta_{1j}\,\text{trace}\prod_2^n P_\alpha Q_\alpha \exp(\tfrac{1}{2}\theta_\alpha P_\alpha Q_\alpha)$$
$$= i(-1)^{n-1}2^n\zeta_{1j}\prod_2^n \sin(\theta_\alpha/2). \qquad (29)$$

Returning to (25) we notice that

$$V_1^{\frac{1}{2}}s_1V_1^{-\frac{1}{2}} = P_1\cosh H^* - iQ_1\sinh H^*. \qquad (30)$$

(25), (29), and (30) give

$$I=(\prod_2^n \lambda_\alpha)i(\xi_{11}\cosh H^*-i\xi_{12}\sinh H^*),\qquad(31)$$

where

$$\lambda_\alpha=\operatorname*{Lim}_{a\to i\infty}(-\cos a)^{-1}\sin(\theta_\alpha/2),\qquad(32)$$

and

$$\xi_{\alpha\beta}=\operatorname*{Lim}_{a\to i\infty}(\cos a)^{-1}\zeta_{\alpha\beta}.\qquad(33)$$

C.

In this subsection we shall derive a formula for λ_α as the eigenvalue of an $n\times n$ matrix.

The matrices $\mathbf{T}_+$ and $\mathbf{T}_-$ are real, so that we can write

$$\mathbf{T}_+^{-1}\mathbf{MT}_-=\tfrac12\mathbf{G}\exp(-2ia)+\tfrac12\mathbf{G}^*\exp(2ia),\qquad(34)$$

where $*$ means complex conjugate, and $\mathbf{G}$ is *independent* of a and is given by

$$\mathbf{G}=\mathbf{T}_+^{-1}\begin{bmatrix}1 & i & & & 0\\ -i & 1 & & & \\ & & 1 & i & \\ & & -i & 1 & \\ 0 & & & & \ddots\end{bmatrix}\mathbf{T}_-.\qquad(35)$$

Now in Eq. (34) the eigenvalues of the left-hand side are 1, -1, $e^{\pm i\theta_2}$, $e^{\pm i\theta_3}$, $\cdots e^{\pm i\theta_n}$. As $a\to i\infty$, the second term of the right-hand side becomes negligible, and we see that

$$\operatorname*{Lim}_{a\to i\infty}2e^{2ia}e^{i\theta}\alpha=l_\alpha,$$

where l_2, l_3, $\cdots l_n$ are the nonvanishing eigenvalues of $\mathbf{G}$. A relation between the l's and the λ's is found by squaring (32):

$$\lambda_\alpha{}^2=\operatorname*{Lim}_{a\to i\infty}(\cos a)^{-2}\sin^2(\theta_\alpha/2)=\operatorname*{Lim}_{a\to i\infty}4e^{2ia}\sin^2(\theta_\alpha/2)$$
$$=-\tfrac12 l_\alpha.\qquad(36)$$

We therefore want to find the eigenvalues of the $2n\times2n$ matrix $\mathbf{G}$ defined by (35). Now explicit matrix elements of $\mathbf{T}_+$ and $\mathbf{T}_-$ have been exhibited by Kaufman.[6] Using these matrix elements and rearranging the rows and columns of all $2n\times2n$ matrices so that the order of the Γ's is changed into $\mathbf{P}_1$, $\mathbf{P}_2\cdots\mathbf{P}_n$, $\mathbf{Q}_1$, $\mathbf{Q}_2\cdots\mathbf{Q}_n$, we arrive at the following expression for $\mathbf{G}$:

$$\mathbf{G}=\begin{bmatrix}\mathbf{D}_+^{-1} & 0\\ 0 & -i\mathbf{D}_+^{-1}\end{bmatrix}\begin{bmatrix}\mathbf{p}_+^{-1}\\ \mathbf{p}_+\end{bmatrix}\begin{bmatrix}\mathbf{p}_- & \mathbf{p}_-^{-1}\end{bmatrix}\begin{bmatrix}\mathbf{D}_- & 0\\ 0 & i\mathbf{D}_-\end{bmatrix},\qquad(37)$$

where

$$\mathbf{D}_-=n^{-\frac12}\begin{bmatrix}\epsilon^2 & \epsilon^4\cdots\epsilon^{2n}\\ \epsilon^4 & \epsilon^8\cdots\epsilon^{4n}\\ \cdots & \cdots\quad\cdots\\ \cdots & \cdots\quad\cdots\\ \epsilon^{2n} & \epsilon^{4n}\cdots\epsilon^{2nn}\end{bmatrix},\quad\epsilon=\exp(\pi i/n),\qquad(38)$$

$$\mathbf{D}_+=\mathbf{D}_-\begin{bmatrix}\epsilon^{-1} & & & 0\\ & \epsilon^{-2} & & \\ & & \ddots & \\ 0 & & & \epsilon^{-n}\end{bmatrix},\qquad(39)$$

$$\mathbf{p}_-=\begin{bmatrix}e^{i\delta_2'/2} & & & 0\\ & e^{i\delta_4'/2} & & \\ & & \ddots & \\ 0 & & & e^{i\delta_{2n}'/2}\end{bmatrix},\qquad(40)$$

and

$$\mathbf{p}_+=\begin{bmatrix}e^{i\delta_1'/2} & & & 0\\ & e^{i\delta_3'/2} & & \\ & & \ddots & \\ 0 & & & e^{i\delta_{2n-1}'/2}\end{bmatrix}.\qquad(41)$$

The quantities δ' are defined in Kaufman's paper. Explicit expressions for them will be given later in Eq. (60). The four matrices $\mathbf{D}_-$, $\mathbf{D}_+$, $\mathbf{p}_-$, and $\mathbf{p}_+$ are all unitary. Writing the eigenvector of $\mathbf{G}$ as $\begin{bmatrix}\phi\\ \eta\end{bmatrix}$ and by the use of (37) one obtains the following eigenvalue problem

$$\mathbf{D}_+^{-1}\mathbf{p}_+^{-1}(\mathbf{p}_-\mathbf{D}_-\phi+i\mathbf{p}_-^{-1}\mathbf{D}_-\eta)=l\phi,\qquad(42)$$
$$-i\mathbf{D}_+^{-1}\mathbf{p}_+(\mathbf{p}_-\mathbf{D}_-\phi+i\mathbf{p}_-^{-1}\mathbf{D}_-\eta)=l\eta.\qquad(43)$$

If $l\neq0$, this shows that

$$\mathbf{p}_+\mathbf{D}_+\phi=i\mathbf{p}_+^{-1}\mathbf{D}_+\eta.$$

With the aid of this, η could be eliminated and the eigenvalue problem is finally reduced to

$$(\mathbf{D}+\mathbf{p}_-^{-2}\mathbf{D}\mathbf{p}_+{}^2)\phi_1=l(\mathbf{p}_-^{-1}\mathbf{p}_+)\phi_1,\qquad(44)$$

where

$$\phi_1=\mathbf{D}_+\phi,$$

and

$$\mathbf{D}=\mathbf{D}_-\mathbf{D}_+^{-1}.\qquad(44a)$$

D.

The calculation of $\xi_{1\beta}$ will be reduced in this subsection to the eigenvector problem of an $n\times n$ matrix.

From the definition of ζ in (26) we see that the column matrix

$$\zeta_1=\begin{bmatrix}\zeta_{11}\\ \zeta_{12}\\ \vdots\end{bmatrix}$$

which is the first column of ζ^{-1} is an eigenvector of $\mathbf{T}_+^{-1}\mathbf{MT}_-$ with the eigenvalue $+1$:

$$(\mathbf{T}_+^{-1}\mathbf{MT}_-)\zeta_1=\zeta_1.\qquad(45)$$

It is easily shown that if a column matrix ξ_1 could be found such that

$$\mathbf{G}\xi_1=0,\qquad(46a)$$

and

$$\mathbf{G}\xi_1^*=2\xi_1,\qquad(46b)$$

then in virtue of (34)

$$\zeta_1=\tfrac12(e^{-ai}\xi_1+e^{ai}\xi_1^*)\qquad(47)$$

does satisfy (45). It is to be emphasized that the ξ_1

APPENDIX A

defined by (46) is *independent* of a so that as $a \rightarrow i\infty$ (47) shows that ζ_1 becomes proportional to ξ_1, and the first column of the matrix $\|\xi_{\alpha\beta}\|$ is exactly ξ_1.

We now tackle Eqs. (46). Equations (37) and (46a) lead to

$$\left[\mathbf{p}_-\mathbf{p}_-^{-1}\right]\begin{bmatrix}\mathbf{D}_- & 0 \\ 0 & i\mathbf{D}_-\end{bmatrix}\xi_1 = 0,$$

showing that there exists an $n \times 1$ column matrix y such that

$$\begin{bmatrix}\mathbf{D}_- & 0 \\ 0 & i\mathbf{D}_-\end{bmatrix}\xi_1 = \begin{bmatrix}\mathbf{p}_-^{-1} \\ -\mathbf{p}_-\end{bmatrix}y, \quad (48)$$

which is both necessary and sufficient for the fulfillment of (46a). Solving (48) for ξ_1 and substituting into (46b) one obtains

$$\mathbf{D}_+^{-1}\mathbf{p}_+^{-1}(\mathbf{p}_-\mathbf{D}_-^2\mathbf{p}_- + \mathbf{p}_-^{-1}\mathbf{D}_-^2\mathbf{p}_-^{-1})y^*$$
$$= 2\mathbf{D}_-^{-1}\mathbf{p}_-^{-1}y, \quad (49)$$

$$-\mathbf{D}_+^{-1}\mathbf{p}_+(\mathbf{p}_-\mathbf{D}_-^2\mathbf{p}_- + \mathbf{p}_-^{-1}\mathbf{D}_-^2\mathbf{p}_-^{-1})y^*$$
$$= 2\mathbf{D}_-^{-1}\mathbf{p}_-y. \quad (50)$$

We shall show that

$$\mathbf{p}_-\mathbf{D}_-^2\mathbf{p}_- = \mathbf{p}_-^{-1}\mathbf{D}_1^2\mathbf{p}_-^{-1}. \quad (51)$$

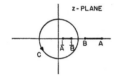

FIG. 1. Cuts in z-plane.

First, from (38)

$$\mathbf{D}_-^2 = \begin{bmatrix} 0 & 1 & 0 \\ & 1 & \\ & \ddots & \\ 1 & & \\ 0 & & 1 \end{bmatrix}. \quad (51a)$$

But from Kaufman's definition of δ',

$$\delta_{2r}' = -\delta_{2n-2r}', \quad (52)$$

and

$$\exp(i\delta_{2n}') = -1, \quad T < T_C. \quad (53)$$

Hence by (40) $\mathbf{D}_-^2\mathbf{p}_-^2\mathbf{D}_-^2 = \mathbf{p}_-^{-2}$; and using $\mathbf{D}_-^4 = 1$ one immediately proves (51). (49) and (50) now simplifies to

$$\mathbf{D}\mathbf{p}_+^{-1}\mathbf{p}_-\mathbf{D}_-^2\mathbf{p}_-y^* = \mathbf{p}_-^{-1}y, \quad (54)$$

$$-\mathbf{D}\mathbf{p}_+\mathbf{p}_-\mathbf{D}_-^2\mathbf{p}_-y^* = \mathbf{p}_-y. \quad (55)$$

Elimination of y^* and simplification leads finally to

$$(\mathbf{D}^{-1} + \mathbf{p}_+^{-2}\mathbf{D}^{-1}\mathbf{p}_-^{-2})(\mathbf{p}_-^{-1}y) = 0. \quad (56)$$

Equations (54) and (56) together determine y, which in turn gives ξ_1 through (48).

The normalization of ξ_1 is determined by substitution of (47) into

$$\zeta_1'\zeta_1 = 1. \quad (57)$$

This results in

$$\xi_1'\xi_1 = 0, \quad (58)$$

and

$$\xi_1'\xi_1^* = 2. \quad (59)$$

(58) is automatically satisfied by virtue of (48), (51), and the fact that $\mathbf{D}_-$ and $\mathbf{p}_-$ are symmetrical matrices.

E.

To summarize the results of this section: The spontaneous magnetization I is given by (31), in which the λ's are related through Eq. (36) to the eigenvalues l of Eq. (44), and in which ξ_{11} and ξ_{12} are the first and the $(n+1)$th element of the column matrix ξ_1 calculated through (48) from the column matrix y which in turn is determined by (54) and (56). ξ_1 is to be normalized according to (59).

III. LIMIT FOR INFINITE CRYSTAL

A.

The procedure just outlined simplifies greatly when we approach the limit of an infinite crystal. To show this let us first introduce the variable

$$z = e^{i\omega} \quad (\omega = r\pi/n, r = 1, 2, \cdots n). \quad (59a)$$

The relationship between δ' and ω is given by Kaufman's Eq. (52). In terms of z this can be reduced to

$$e^{2i\delta'} = \frac{\tanh^2 H^*(z - \coth H \coth H^*)(z - \tanh H \coth H^*)}{(z - \coth H \tanh H^*)(z - \tanh H \tanh H^*)}. \quad (60)$$

From this we obtain $e^{i\delta'}$ and we shall write it as

$$\Theta(z) = e^{i\delta'}$$
$$= (1/AB)^{\frac{1}{2}}[(z-A)(z-B)/(z-A^{-1})(z-B^{-1})]^{\frac{1}{2}} \quad (61)$$

where

$$A = \coth H \coth H^* = [(1+x)/x(1-x)],$$
$$B = \tanh H \coth H^* = [(1-x)/x(1+x)]. \quad (62)$$

For $T < T_C$, $A > B > 1$. $\Theta(z)$ is analytic everywhere except at the points $z = A$, B, $1/A$, or $1/B$ where it has branch points. The square root in (61) is defined to be that branch of the function that takes the value -1 at $z = 1$, in accordance with (53). (See Fig. 1.)

Consider Eq. (44). For a very large crystal

$$\mathbf{p}_- = \mathbf{p}_+ = \mathbf{p},$$

and we have

$$(\mathbf{D} + \mathbf{p}^{-2}\mathbf{D}\mathbf{p}^2)\phi_1 = l\phi_1. \quad (63)$$

By the definition of $\mathbf{D}$, Eq. (44a), the matrix elements of $\mathbf{D}$ are

$$(\mathbf{D})_{rs} = \frac{1}{n}\sum_{t=1}^{n} \epsilon^{2rt}\epsilon^t \epsilon^{-2ts} = -\frac{2}{n}\frac{1}{1 - \epsilon^{2s-2r-1}}.$$

Hence $\mathbf{D}$ operating on any vector ϕ gives

$$(\mathbf{D}\phi)_r = \sum (\mathbf{D})_{rs}\phi_s = -\frac{2}{n}\sum_{s=1}^{n}\phi_s\left[1-\exp\frac{\pi i}{n}(2s-2r-1)\right]^{-1}$$

$$= -\frac{2}{n}\sum_{s=1}^{n}\frac{\phi_s}{1-z_{2s}/z_{2r}\epsilon}, \qquad (64)$$

where z is the variable defined in (59a). For $s = 1, 2, \cdots n$ the values assumed by z_{2s} are the n nth roots of unity. As $n \to \infty$ the summation in (64) therefore becomes an integral around the unit circle:

$$(\mathbf{D}\phi)_r = -2\int_C \frac{1}{2\pi i}\frac{dz}{z}\frac{\phi(z)}{1-(z/z_r)}, \qquad (65)$$

where

$$z = \exp(2\pi i s/n) \quad \text{and} \quad z_r = \exp(2\pi i r/n).$$

The contour C is the unit circle. At the point $z = z_r$ the principle value of the integral is to be taken. This is necessary because of the factor ϵ in the expression (64) which prevents the denominator from assuming the value zero. Alternately, we might make a detour around the point z_r and make up the difference by adding a term to (65):

$$(\mathbf{D}\phi)_t = -\frac{1}{\pi i}\int_{C'}\frac{dz}{z}\frac{\phi(z)}{1-(z/t)}+\phi(t). \qquad (66)$$

We have here used the more convenient notation t for z_r. With this definition it is evident that the point t does not have to be on the unit circle. If, however, t is inside the unit circle, it is more convenient to use the following equivalent of (66):

$$(\mathbf{D}\phi)_t = -\frac{1}{\pi i}\int_{C''}\frac{dz}{z}\frac{\phi(z)}{1-(z/t)}-\phi(t), \qquad (66a)$$

where C'' is as shown in Fig. 2.

The definitions (66) and (66a) for $\mathbf{D}$ are valid when $\mathbf{D}$ operates on any function $\phi(z)$ that is analytic in a region that contains the circumference of the unit circle in its interior. It is important to notice that this region does not have to be singly connected.

We quote a few interesting properties of the operator $\mathbf{D}$:

$$\mathbf{D}z^m = t^m \quad \text{for } m = \text{integer} \geq 1,$$
$$\mathbf{D}z^m = -t^m \quad \text{for } m = \text{integer} \leq 0, \qquad (67)$$

$$\mathbf{D}^2 = 1. \qquad (68)$$

Now return to Eq. (63). Since $\mathbf{p}^2$ is a diagonal matrix with diagonal element $\Theta(z)$ given by (61), it is evident that (63) reduces to

$$2\phi_1(t) - \frac{1}{\pi i}\int_{C'}\frac{dz}{z}\frac{\phi_1(z)}{1-(z/t)}\left(1+\frac{\Theta(z)}{\Theta(t)}\right) = l\phi_1(t). \qquad (69)$$

This integral equation will be solved in the next two sections.

B.

There still remains, according to the results of the last section, the problem of solving (54) and (56). By virtue of (68), (56) reduces to the same form as (63) with $l = 0$. Thus $p_-^{-1}y$ is proportional to that eigenvector $\phi_1 = \Phi$ of (69) belonging to the eigenvalue $l = 0$:

$$2\Phi(t) - \frac{1}{\pi i}\int_{C'}\frac{dz}{z}\frac{\Phi(z)}{1-(z/t)}\left(1+\frac{\Theta(z)}{\Theta(t)}\right) = 0. \qquad (70)$$

For the convenience of normalization we shall write

$$\mathbf{p}_-^{-1}y = n^{-\frac{1}{2}}\Phi. \qquad (71)$$

Then by virtue of (48), Eq. (59) reduces to

$$(1/n)\Phi'\Phi^* = 1,$$

or

$$\frac{1}{2\pi i}\int_C |\Phi(z)|^2\frac{dz}{z} = 1. \qquad (72)$$

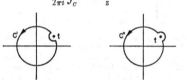

FIG. 2. Contours in z-plane.

The first and $(n+1)$th elements of ξ_1 are, according to (48) and (71):

$$\xi_{11} = \frac{1}{n}\sum_{s=1}^{n}\epsilon^{-2s}\Phi_s = \frac{1}{2\pi i}\int_C \frac{dz}{z^2}\Phi(z),$$

$$\xi_{12} = \frac{i}{n}\sum_{s=1}^{n}\epsilon^{-2s}\Theta(\epsilon^{2s})\Phi_s = \frac{1}{2\pi}\int_C \frac{dz}{z^2}\Theta(z)\Phi(z). \qquad (73)$$

The question of the fulfillment of Eq. (54), which now reduces to

$$\mathbf{DD}_-^2\Phi^* = \Phi,$$

is best discussed with the aid of the introduction of the function $\Phi^\dagger(z)$ defined by

$$\Phi^\dagger(z) = [\Phi(z^*)]^*. \qquad (74)$$

If Φ is analytic in a region containing the circumference of the unit circle in its interior, $\Phi^\dagger$ would be analytic in a similar region. Equation (51a) shows that

$$(\mathbf{D}_-^2\Phi^*)_s = [\Phi(z^*)]^* = \Phi^\dagger(z).$$

Thus (54) is fulfilled if

$$\Phi^\dagger(t) - \frac{1}{\pi i}\int_{C'}\frac{dz}{z}\frac{\Phi^\dagger(z)}{1-z/t} = \Phi(t). \qquad (75)$$

APPENDIX A

C.

The integral equation (70) is easily solved by inspection:

$$\Phi(z) = Fz[(A-z)(B-z)]^{-\frac{1}{2}}, \qquad (76)$$

where F is a normalization factor. F will turn out to be real, so that according to (74)

$$\Phi^\dagger(z) = \Phi(z).$$

It is easy to prove that (75) is satisfied. This completes the verification that (73) does indeed give the correct matrix elements of ξ_1.

FIG. 3. Contour in u-plane.

To find F we substitute (76) into (72) and obtain

$$\frac{F^2}{2\pi i} \int_C \frac{dz}{[(A-z)(B-z)(Az-1)(Bz-1)]^{\frac{1}{2}}} = 1. \quad (76a)$$

In the integrand the sign of the square root is to be so taken that at $z=1$ the integrand is positive. The integral is a complete elliptic integral and can be reduced to the standard form by a projective transformation. The result is

$$F^{-2} = \frac{4}{\pi}\frac{1}{A-B}k_{-1}{}^{\frac{1}{2}}K(k_{-1}),$$

where

$$k_{-1} = \left[\frac{(A^2-1)^{\frac{1}{2}} - (B^2-1)^{\frac{1}{2}}}{A(B^2-1)^{\frac{1}{2}} + B(A^2-1)^{\frac{1}{2}}}\right]^2, \quad (77)$$

and K is the complete elliptic integral of the first kind.[12] It is convenient to change the modulus and define[13]

$$k = 2k_{-1}{}^{\frac{1}{2}}/(1+k_{-1}) = 4x^2/(1-x^2)^2 = \sinh^{-2}2H. \quad (78)$$

Then

$$F^{-2} = 2kK(k)/\pi(A-B). \quad (79)$$

The values of ξ_{11} and ξ_{12} are obtained from (73) and (76):

$$\xi_{11} = F(AB)^{-\frac{1}{2}}, \quad \xi_{12} = 0.$$

Substitution of these into (31), with the use of (36), leads to

$$I^4 = \prod_2^n (l_\alpha{}^2/4) F^4 A^{-2} B^{-2} \cosh^4 H^*.$$

We have here taken the fourth power of I to eliminate the undetermined phase factor that was introduced into the expression for I as early as Eq. (17). When the

[12] E. T. Whitaker and G. N. Watson, *Modern Analysis* (Cambridge University Press, London, 1927), fourth edition.
[13] The modulus k is the same as that used in references 5 and 7.

explicit expressions for A, B, F, and H^* are introduced, the expression for I^4 further simplifies to

$$I^4 = \left(\prod_2^n \frac{l_\alpha{}^2}{4}\right)\frac{\pi^2}{4}\left[\frac{1}{K(k)}\right]^2. \quad (80)$$

IV. ELLIPTIC TRANSFORMATION

It remains to find the eigenvalues l from (69) and substitute into (80). To do this we first introduce an elliptic transformation[12] that was essentially the one used in evaluating the integral in (76a):

$$z = -(\mathrm{cn}u - i[1+k]^{\frac{1}{2}}\mathrm{sn}u)(\mathrm{dn}u - i[k+k^2]^{\frac{1}{2}}\mathrm{sn}u)/ \\ (1+k\,\mathrm{sn}^2 u), \quad (81)$$

the modulus k being given by (78).[13] This is the same transformation as was used by Onsager,[5] and Kaufman and Onsager[7] in their calculations. It serves to eliminate the square root in the function Θ.

$$\Theta = e^{i\delta'} = \mathrm{cn}u + i\,\mathrm{sn}u. \quad (82)$$

It is easy to verify that

$$\frac{1}{z}\frac{dz}{du} = -i\frac{1-k^2}{(1+k)^{\frac{1}{2}}}\frac{1}{\mathrm{dn}u - k^{\frac{1}{2}}\mathrm{cn}u}. \quad (83)$$

We shall need the following properties of the transformation (81): (A) z is doubly periodic in u with periods $4K$ and $4iK'$.

(B) z is everywhere analytic, except at $u = iK'/2$, $3iK'/2$ (mod. $4K$, $4iK'$), where $z = \infty$.

(C) In a unit cell in the complex u-plane, to every value of z there correspond exactly two values of u, except for $z = A$, B, $1/B$ or $1/A$ for which there corresponds only one value of u, namely, $u = +iK'$, $2K+iK'$, $2K-iK'$ or $-iK'$ (mod. $4K$, $4iK'$).

(D) If for a value of z there correspond in a unit cell two values of u, then at those two values Θ assume equal values but have different signs. Thus a unit cell of the u plane corresponds to both sheets of the Riemann surface in the z plane of Fig. 1 with respect to the function $\Theta(z)$.

The substitution, suggested by (76), into (69), of

$$\phi_1(z) = z[(z-A)(z-B)]^{\frac{1}{2}}\phi$$

gives, with the use of (81), (82), and (83)

$$2\phi(u') + \int_0^{4K} J(u', u)\phi(u)du = l\phi(u'), \quad (84)$$

where

$$J = \mathbf{I}\,\mathbf{II}\,\mathbf{III}\,\mathbf{IV}; \quad (85)$$

and

$$\mathbf{I} = [1 - z(u)/z(u')]^{-1}, \quad (86)$$

$$\mathbf{II} = 1 + \Theta(z)/\Theta(z') = 1 + (\mathrm{cn}u + i\,\mathrm{sn}u)/(\mathrm{cn}u' + i\,\mathrm{sn}u'), \quad (87)$$

$$\mathbf{III} = \frac{z(u)}{z(u')}\left[\frac{\{z(u')-A\}\{z(u')-B\}}{\{z(u)-A\}\{z(u)-B\}}\right]^{\frac{1}{2}}, \quad (88)$$

and

$$IV=-\frac{1}{\pi}\frac{1-k^2}{(1+k)^{\frac{1}{2}}\,\mathrm{dn}u-k^{\frac{1}{2}}\,\mathrm{cn}u}.$$

V. SOLUTION OF INTEGRAL EQUATION (84)

We proceed by investigating the analytic behavior of $J(u', u)$ *with respect to the variable* u.

(A) I, II, and IV are all doubly periodic with periods $4K$ and $4iK'$. But III is doubly periodic with periods $4K$ and $8iK'$. It changes sign at periods $4iK'$:

$$III(u+4iK)=-III(u). \qquad (89)$$

(B) III is analytic everywhere except at $z=A$ or $z=B$, i.e., $u=iK'$ or $2K+iK'$ (mod. $4K$, $4iK'$) where III has simple poles.

(C) II is analytic everywhere except at $u=-iK'$ or $2K-iK'$ (mod. $4K$, $4iK'$) where it has simple poles.

(D) IV is analytic everywhere except at

$$u=\pm iK'/2, \quad \pm3iK'/2(\text{mod. }4K, 4iK'),$$

where it has simple poles.

(E) I is analytic everywhere except at $z(u)=z(u')$. According to the last section in each cell there are, in general, two values of u where this exception occurs. At these two points I has simple poles.

However, there is *only one pole* for J in each unit cell ($4K$ by $4iK'$). This is so because of the following considerations:

(F) At $u=\pm iK'$, $\pm iK'+2K$ (mod. $4K$, $4iK'$), IV has simple zeros.

(G) At $u=iK'/2$, $3iK'/2$ (mod. $4K$, $4iK'$), $z(u)=\infty$, so that I has simple zeros.

(H) At $u=-iK'/2$, $-3iK'/2$ (mod. $4K$, $4iK'$), $z(u)=0$, so that III has simple zeros.

(I) According to property D, Sec. IV, in a unit cell ($4K$ by $4iK'$) at one of the solutions of $z(u)=z(u')$, $\Theta(u)=\Theta(u')$ so that II=2. At the other solution, II has a zero.

Thus inside the rectangle in Fig. 3, J has only one pole at $u=u'$ which we assume to be inside of the rectangle. In the neighborhood of this pole II=2, III=1, and I·IV=$i\pi^{-1}(u-u')^{-1}$. Hence the residue of J at $u=u'$ is $2i/\pi$.

The solution of (84) is given by

$$\phi=\exp(im\pi u/2K), \quad m=\pm\text{integer}. \qquad (90)$$

To show that this is indeed a solution we note that ϕ is periodic with period $4K$. Hence calling

$$\mathcal{J}=\int_0^{4K} J(u', u)\phi(u)du,$$

one obtains by performing a contour integration around the rectangle of Fig. 3:

$$2\pi i(2i/\pi)\phi(u')=\mathcal{J}[1+\exp(-2m\pi K'/K)]. \qquad (91)$$

[The integration along the two vertical sides cancel

each other and that along the top reduces to $\mathcal{J}$ multiplied by a factor, in virtue of (89).] This gives

$$\mathcal{J}=-4\phi(u')/(1+q^{2m}), \qquad (92)$$

where

$$q=\exp(-\pi K'/K). \qquad (93)$$

(84) is therefore satisfied with

$$l=2-4/(1+q^{2m})=2(q^{2m}-1)/(q^{2m}+1). \qquad (94)$$

For $m=0$ this gives, as expected, the solution $l=0$ which was already found by inspection in Sec. IIIC.

Knowing all the nonvanishing eigenvalues we can now calculate

$$\prod_2^\infty\frac{l_\alpha{}^2}{4}=\prod_{\substack{m=-\infty\\m\neq0}}^\infty\left(\frac{1-q^{2m}}{1+q^{2m}}\right)^2=\prod_1^\infty\left(\frac{1-q^{2m}}{1+q^{2m}}\right)^4.$$

This infinite product can be[14] expressed in terms of the ϑ functions which are related to K. We get finally

$$\prod_2^\infty\frac{l_\alpha{}^2}{4}=\frac{4}{\pi^2}[K(k)]^2(1-k^2)^{\frac{1}{2}}=\frac{4}{\pi^2}K^2\frac{1+x^2}{(1-x^2)^2}(1-6x^2+x^4)^{\frac{1}{2}}. \qquad (95)$$

VI. FINAL RESULTS

The spontaneous magnetization I is obtained from (95) and (80) as

$$I=\left[\frac{1+x^2}{(1-x^2)^2}(1-6x^2+x^4)^{\frac{1}{2}}\right]^{\frac{1}{4}}. \qquad (96)$$

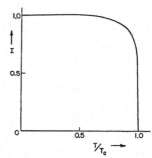

FIG. 4. Spontaneous magnetization.

At low temperatures this gives the same expansion in powers of x as obtained in previous works of Van der Waerden[1] and Ashkin and Lamb:[2]

$$I=1-2x^4-8x^6-34x^8-152x^{10}-714x^{12}-\cdots.$$

This series is convergent all the way up to the critical

[14] See reference 11, especially p. 472.

Appendix A

point, where

$$x = x_C = \sqrt{2} - 1.$$

Near the critical point, I has a branch point:

$$I \cong [4(\sqrt{2} + 2)(x_C - x)]^{1/8}.$$

In Fig. 4, I is plotted against the temperature.

This work was completed in the summer of 1951 while the author was at the University of Illinois. He wishes to take this opportunity to thank the staff of the Department of Physics, University of Illinois for the hospitality extended him during his stay. He also wishes to thank Bruria Kaufman for many stimulating discussions.

ADVANCES IN MATHEMATICS **107**, 1–185 (1994)

On the Dirac and Schwinger Corrections to the Ground-State Energy of an Atom

C. FEFFERMAN* AND L. A. SECO

*Department of Mathematics, Princeton University, Princeton, New Jersey; and
Department of Mathematics, University of Toronto, Toronto, Ontario, Canada*

Contents.

INTRODUCTION

In this article and [FS2–FS7] we prove an asymptotic formula for the ground-state energy of a non-relativistic atom. The ground-state energy $E(N, Z)$ for N electrons and a nucleus of charge Z is defined[1] as the infimum of the spectrum of the Hamiltonian

$$H_{NZ} = \sum_{k=1}^{N} (-\Delta_{x_k} - Z |x_k|^{-1}) + \sum_{1 \le j < k \le N} |x_j - x_k|^{-1}, \tag{1}$$

acting on antisymmetric $\Psi(x_1 \cdots x_N) \in L^2(\mathbb{R}^{3N})$. The ground-state energy of an atom is then defined as

$$E(Z) = \min_{N \ge 1} E(N, Z), \tag{2}$$

and our problem is to compute $E(Z)$ asymptotically for large Z. Building on the previous work of Thomas, Fermi, Dirac, and Scott (see the survey article of Lieb [L2]), Schwinger [S] proposed the refined formula

$$E(Z) \approx -c_0 Z^{7/3} + \tfrac{1}{8} Z^2 - c_1 Z^{5/3}, \tag{3}$$

---

* Partially supported by the NSF.
[1] In the Introduction, we neglect electron spin to simplify notation. When we prove our main results, we take spin into account.

1

for explicit positive constants c_0, c_1. After the early work of Lieb and Simon [LS] on molecules, Hughes and Siedentop and Weikard [H, SW] gave a rigorous proof of the "Scott conjecture," namely

$$E(Z) = -c_0 Z^{7/3} + \tfrac{1}{8} Z^2 + O(Z^\gamma) \qquad \text{with} \quad \gamma < 2. \tag{4}$$

Recently, Ivrii and Sigal [IS] proved the analogue of the Scott conjecture for molecules. Our main result, announced in [FS1], is as follows.

THEOREM. $E(Z) = -c_0 Z^{7/3} + \tfrac{1}{8} Z^2 - c_1 Z^{5/3} + O(Z^{5/3 - \varepsilon_0})$ with $\varepsilon_0 = 1/2835$.

It would be interesting to prove an analogous result for molecules by combining our ideas with those of Ivrii and Sigal. We also want to know more accurate asymptotic formulas for $E(Z)$, containing additional correction terms beyond $Z^{5/3}$. Prediction of chemical phenomena from first principles requires knowledge of $E(Z)$ and its analogue for molecules far beyond our current understanding.

We sketch the physical reasoning that leads to Schwinger's formula (3), and then give the strategy of the proof of the Scott conjecture. Next we give a crude outline of the proof of our theorem. Then we explain further the subset of our proof which appears in this paper. Our Introduction concludes with a conjecture on the next term in the asymptotics of $E(Z)$.

The starting point in discussing atoms is an elementary observation on free particles in a box. For N free particles in a box $Q \subset \mathbb{R}^3$, the minimum possible kinetic energy $KE(N, Q)$ is equal to the lowest eigenvalue of $-\Delta$ acting on antisymmetric $\Psi(x_1, ..., x_N) \in L^2(Q^N)$ with appropriate boundary conditions. One computes $KE(N, Q)$ trivially, by separation of variables. For large N, the answer is

$$KE(N, Q) \approx c_{\text{TF}} \rho^{5/3} |Q|, \tag{5}$$

where $\rho = N/|Q|$ is the density of particles in the box, and c_{TF} is a universal constant.

This suggests a way to approximate the energy $\langle H_{NZ} \Psi, \Psi \rangle$ of a wave function $\Psi(x_1 \cdots x_N)$ in terms of the electron density

$$\rho(x_1) = N \int_{\mathbb{R}^{3(N-1)}} |\Psi(x_1, x_2, ..., x_N)|^2 \, dx_2 \cdots dx_N. \tag{6}$$

In fact, we set

$$\varepsilon_{\text{TF}}(\rho) = c_{\text{TF}} \int_{\mathbb{R}^3} \rho^{5/3}(x) \, dx - \int_{\mathbb{R}^3} \frac{Z}{|x|} \rho(x) \, dx + \frac{1}{2} \iint_{\mathbb{R}^3 \times \mathbb{R}^3} \frac{\rho(x) \, \rho(y)}{|x-y|} \, dx \, dy. \tag{7}$$

Here, the first term on the right is an approximation of the kinetic energy motivated by (5), and the remaining terms on the right are simply the classical electric potential energy for a charge density ρ and a nucleus of charge Z. Thomas and Fermi independently proposed that the ground-state energy $E(Z)$ is approximately equal to the minimum of $\varepsilon_{TF}(\rho)$ over all possible densities $\rho(x)$. Moreover, they proposed the minimizing density ρ_{TF} for (7) as an approximation to the electron density for an atom in its ground state. This is an immense simplification, since the original problem deals with $\Psi(x_1 \cdots x_N)$ for $N \gg 1$, while Thomas–Fermi theory deals merely with a function on $\mathbb{R}^3$. An elementary computation with the Euler–Lagrange equation for (7) leads to an ordinary differential equation for ρ_{TF}, which may therefore be understood in great detail. In particular, Thomas–Fermi theory predicts that

$$E(Z) \approx -c_0 Z^{7/3}, \tag{8}$$

which is correct as far as it goes, but is much too crude.

A more refined prediction for $E(Z)$ comes from the *Hartree–Fock approximation*.[2] The idea is that since the electron density is approximately ρ_{TF}, each electron behaves as if it were moving in a potential

$$V_{TF}(x) = -\frac{Z}{|x|} + \int_{\mathbb{R}^3} \frac{\rho_{TF}(y)\,dy}{|x-y|}. \tag{9}$$

Therefore it is reasonable to approximate the ground state of the true Hamiltonian H_{NZ} by that of the much simpler Hamiltonian

$$H_{hf} = \sum_{k=1}^{N} (-\Delta_{x_k} + V_{TF}(x_k)), \qquad \text{acting on antisymmetric } \Psi(x_1 \cdots x_N). \tag{10}$$

Unlike the original Hamiltonian, (10) can be diagonalized using separation of variables, and the state of lowest energy can be written explicitly in terms of the eigenfunctions of $-\Delta + V_{TF}$. So again, the problem is reduced from $3N$ to 3 variables. In fact, suppose E_k are the (negative) eigenvalues of $-\Delta + V_{TF}$, and let $\varphi_k(x)$ be the corresponding (normalized) eigenfunctions. Then the ground-state wave function for (10), which we call Ψ_{hf}, is an antisymmetrized product of the φ_k. As an approximation to the ground-state energy of an atom, it is natural to use

$$E_{hf}(Z) = \langle H_{NZ} \Psi_{hf}, \Psi_{hf} \rangle. \tag{11}$$

Note that we use the exact Hamiltonian H_{NZ} in (11), even though Ψ_{hf} arose from the simplified Hamiltonian (10). Elementary computation gives the formula

---

[2] This is not exactly the same as the usual Hartree–Fock approximation.

$$E_{hf}(Z) = sneg(-\Delta + V_{TF}) - \frac{1}{2} \iint_{\mathbb{R}^3 \times \mathbb{R}^3} \frac{\rho_{TF}(x)\,\rho_{TF}(y)}{|x-y|}\,dx\,dy$$

$$-\frac{1}{2} \iint_{\mathbb{R}^3 \times \mathbb{R}^3} \frac{|\mathscr{S}(x,y)|^2}{|x-y|}\,dx\,dy + \frac{1}{2} \iint_{\mathbb{R}^3 \times \mathbb{R}^3} [\rho_{hf}(x) - \rho_{TF}(x)]$$

$$\times [\rho_{hf}(y) - \rho_{TF}(y)] \frac{dx\,dy}{|x-y|}, \tag{12}$$

with

$$sneg(-\Delta + V_{TF}) = \sum_k E_k, \tag{13}$$

$$\rho_{hf}(x) = \sum_k |\varphi_k(x)|^2, \tag{14}$$

$$\mathscr{S}(x, y) = \sum_k \varphi_k(x)\, \overline{\varphi_k(y)}. \tag{15}$$

To obtain more explicit information from (12), we approximate $sneg(-\Delta + V_{TF})$, ρ_{hf}, and $\mathscr{S}$. The *semiclassical approximations* for these quantities are as follows.

$$sneg(-\Delta + V_{TF}) \approx -\frac{1}{15\pi^2} \int_{\mathbb{R}^3} |V_{TF}(x)|^{5/2}\,dx, \tag{16}$$

$$\rho_{hf}(x) \approx \frac{1}{6\pi^2} |V_{TF}(x)|^{3/2}, \tag{17}$$

$$\frac{1}{2} \iint_{\mathbb{R}^3 \times \mathbb{R}^3} |\mathscr{S}(x, y)|^2 \frac{dx\,dy}{|x-y|} \approx c_D \int_{\mathbb{R}^3} \rho_{TF}^{4/3}(x)\,dx,$$

$$\text{for a universal constant } c_D. \tag{18}$$

We omit the motivation for (16), (17), and (18) and content outselves with the remark that they are closely related to Weyl's theorem on eigenvalues of the Laplacian. Formula (18) and its application to atoms are due to Dirac.

Putting (16), (17), and (18) into (12), we obtain the semiclassical approximation for $E_{hf}(Z)$. From the first two terms on the right in (12), we recover the Thomas–Fermi energy $-c_0 Z^{7/3}$. The third term on the right of (12) takes the form $-c_1' Z^{5/3}$ for a universal constant c_1'. The final term in (12) vanishes in the semiclassical approximation, by virtue of (9), (17), and the Euler–Lagrange equation for ρ_{TF}. Altogether, we have

$$E_{hf}(Z) \approx -c_0 Z^{7/3} - c_1' Z^{5/3}. \tag{19}$$

The last term in (19) is called the Dirac correction. Comparing (19) with the correct formula (3), we see that the Z^2-term is missing from (19), and the $Z^{5/3}$ coefficient is wrong. The trouble is that (16) is only a crude approximation.

A refined form of (16) was proposed by Scott (see [L2]) and Schwinger [S]. For potentials V with a Coulomb singularity $V(x) \approx -Z|x|^{-1}$ at the origin, their formula is

$$\text{sneg}(-\Delta + V) \approx -\frac{1}{15\pi^2} \int_{V<0} |V|^{5/2} + \frac{1}{8} Z^2 + \frac{1}{48\pi^2} \int_{V<0} |V|^{1/2} \Delta V. \quad (20)$$

Scott guessed the Z^2-term by working out the elementary example

$$V(x) = E_0 - Z|x|^{-1}. \quad (21)$$

Schwinger deduced the last term in (20) from the form of the heat kernel for $e^{-t(-\Delta+V)}$, which in turn he guessed from the known case of the harmonic oscillator. Using (20), (17), and (18) to approximate the right-hand side of (12), we obtain Schwinger's formula (3) for the ground-state energy. This concludes our discussion of heuristic methods.

Next we explain some ideas from the rigorous discussion of atoms. There are two main issues: justifying the approximations (16), (17), (18), and (20) in the calculation of the Hartree–Fock energy; and comparing the Hartree–Fock energy $E_{hf}(Z)$ with the true ground-state energy $E(Z)$. The second issue is deeper, since it forces us to understand an interacting N-particle system. In this Introduction, we concentrate on comparing $E_{hf}(Z)$ with $E(Z)$. Note that

$$E_{hf}(Z) = \langle H_{NZ} \Psi_{hf}, \Psi_{hf} \rangle \geq \inf_{N, \Psi} \langle H_{NZ} \Psi, \Psi \rangle = E(Z), \quad (22)$$

so the problem is to prove a lower bound for $E(Z)$.

The main tool used previously to prove lower bounds for $E(Z)$ is a pointwise inequality of Lieb [L1] for the Coulomb potential. With $V_{\text{Coulomb}}(x_1 \cdots x_N) = -\sum_{k=1}^{N} Z|x_k|^{-1} + \sum_{1 \leq j < k \leq N} |x_j - x_k|^{-1}$, Lieb's inequality is

$$V_{\text{Coulomb}}(x_1 \cdots x_N) \geq \sum_{k=1}^{N} W(x_k) - E_0, \quad (23)$$

with $W(x)$ close to $V_{\text{TF}}(x)$, and E_0 close to $\frac{1}{2} \iint_{\mathbb{R}^3 \times \mathbb{R}^3} \rho_{\text{TF}}(x) \rho_{\text{TF}}(y) \, dx \, dy / |x-y|$. We explain how to use (23), and then explain how to prove it. Assuming (23), we see at once that

$$H_{NZ} = \sum_{k=1}^{N} (-\Delta_{x_k}) + V_{\text{Coulomb}} \geq \sum_{k=1}^{N} (-\Delta_{x_k} + W(x_k)) - E_0,$$

and the right-hand side may be diagonalized by separation of variables. Hence,

$$E(Z) \geqslant \mathrm{sneg}(-\Delta + W) - E_0. \tag{24}$$

We expect the right-hand side of (24) to approximate

$$\mathrm{sneg}(-\Delta + V_{\mathrm{TF}}) - \frac{1}{2} \iint_{\mathbb{R}^3 \times \mathbb{R}^3} \frac{\rho_{\mathrm{TF}}(x)\, \rho_{\mathrm{TF}}(y)}{|x - y|}\, dx\, dy.$$

Comparing this expression with (12), and recalling the semiclassical approximations (17) and (18), we guess that the right-hand side of (24) is $E_{\mathrm{hf}}(Z) + O(Z^{5/3})$. So if we can compute the right-hand sides of (12) and (24), then (22) and (24) will give rigorous upper and lower bounds for $E(Z)$, which we expect to differ by $O(Z^{5/3})$. This reduces the computation of $E(Z)$ from $3N$ dimensions to 3, provided we are willing to tolerate errors $O(Z^{5/3})$. The solution of the Scott conjecture is based on (23).

To prove Lieb's inequality (23), we can start with the elementary identity

$$|x - x'|^{-1} = \frac{1}{\pi} \iint_{\substack{y \in \mathbb{R}^3 \\ R > 0}} \chi_{x,\, x' \in B(y, R)} \frac{dy\, dR}{R^5} \qquad \text{for} \quad x, x' \in \mathbb{R}^3. \tag{25}$$

Except for the value of the constant $1/\pi$, identity (25) is forced by the fact that both sides transform in the same way under translations, rotations, and dilations. Summing (25) over all pairs of particles $x = x_j$, $x' = x_k$, we obtain

$$\sum_{j < k} |x_j - x_k|^{-1} = \frac{1}{\pi} \iint_{\substack{y \in \mathbb{R}^3 \\ R > 0}} \frac{N_{yR}(N_{yR} - 1)}{2} \frac{dy\, dR}{R^5}, \tag{26}$$

with $N_{yR} = [\text{number of } x_j \in B(y, R)] = \sum_{j=1}^{N} \chi_{B(y, R)}(x_j)$. In general, a potential

$$V(x_1 \cdots x_N) = \iint_{\substack{y \in \mathbb{R}^3 \\ R > 0}} F(N_{yR}, y, R)\, dy\, dR \tag{27}$$

has the form $\sum_{k=1}^{N} W(x_k)$ if $F = f(y, R) \cdot N_{yR}$. If instead $F = f(y, R) \cdot (N_{yR}(N_{yR} - 1)/2)$ with $f \geqslant 0$, then (27) has the form $V = \sum_{j < k} K(x_j, x_k)$ for a non-negative symmetric two-body interaction $K(x, y)$. Finally, if F depends only on y and R in (27), then the potential $V(x_1 \cdots x_N)$ is merely a constant.

For an atom in its ground state, we guess that the number N_{yR} of electrons in $B(y, R)$ is approximately $\bar{N}_{yR} \equiv \int_{B(y, R)} \rho_{TF}$. So in (26) it is natural to complete the square and write

$$\tfrac{1}{2}N_{yR}(N_{yR} - 1) = \tfrac{1}{2}(N_{yR} - \bar{N}_{yR})^2 + f(y, R)N_{yR} + g(y, R). \qquad (28)$$

When we put (28) into (26), the term $f(y, R)N_{yR}$ will contribute $\sum_k W(x_k)$, and the term $g(y, R)$ will contribute a constant. The term $\tfrac{1}{2}(N_{yR} - \bar{N}_{yR})^2$ is hard to understand, but its contribution is non-negative. If $\bar{N}_{yR} \gg 1$, then we hope $(N_{yR} - \bar{N}_{yR})^2$ will be negligibly small compared to the other terms in (28), because $N_{yR} \approx \bar{N}_{yR}$ for statistical reasons. Hence we get a rigorous lower bound

$$\tfrac{1}{2}N_{yR}(N_{yR} - 1) \geqslant f(y, R)N_{yR} + g(y, R), \qquad (28a)$$

which is useful when $\bar{N}_{yR} > 1$ (say). When $\bar{N}_{yR} \leqslant 1$, we use instead the trivial lower bound $\tfrac{1}{2}N_{yR}(N_{yR} - 1) \geqslant 0$. Therefore, (26) implies that

$$\sum_{j<k} |x_j - x_k|^{-1} \geqslant \iint_{\{\bar{N}_{yR}>1\}} f(y, R) N_{yR} \frac{dy\,dR}{R^5} + \iint_{\{\bar{N}_{yR}>1\}} g(y, R) \frac{dy\,dR}{R^5},$$

which has the form

$$\sum_{j<k} |x_j - x_k|^{-1} \geqslant \sum_k W(x_k) - E_0. \qquad (29)$$

Adding $-\sum_k (Z/|x_k|)$ to both sides of (29), we obtain Lieb's inequality (23). This proof is different from that of [L1], but is closely related to it.

The above proof shows clearly that inequality (23) inevitably sacrifices an error $O(Z^{5/3})$ in estimating the ground-state energy. In fact, we discarded the term

$$\frac{1}{2\pi} \iint_{\{\bar{N}_{yR}>1\}} (N_{yR} - \bar{N}_{yR})^2 \frac{dy\,dR}{R^5} \qquad (30)$$

in the potential energy.

To keep (30) as small as possible, we tried to pick $\bar{N}_{yR}$ close to the expected number of electrons in $B(y, R)$ for an atom in its ground state. Nevertheless, the expected value of $(N_{yR} - \bar{N}_{yR})^2$ will be at least as large as the variance of the random variable N_{yR}. Hence by applying (23), we discard an expected potential energy at least as large as

$$\frac{1}{2\pi} \iint_{\{\bar{N}_{yR}>1\}} \text{Variance}(N_{yR}) \frac{dy\,dR}{R^5}. \qquad (31)$$

For a Hartree–Fock atom, a plausible semiclassical approximation to (31) has the order of magnitude $Z^{5/3}$. The main contribution to (31) comes from (y, R) with $|y| \sim Z^{-1/3}$, $R \sim Z^{-2/3}$, $\bar{N}_{yR} \sim 1$.

This shows that the use of (23) leads to an error $\sim Z^{5/3}$ in $E(Z)$. It shows also that computing $E(Z)$ modulo $o(Z^{5/3})$ is closely related to finding the variance of N_{yR} for an atom in its ground state, with $|y| \sim Z^{-1/3}$ and $R \sim Z^{-2/3}$.

Next we give a crude outline of our proof of our main theorem on the ground-state energy. There are four main steps:

(A) Lower bound.

$$E(Z) \geqslant \operatorname{sneg}(-\Delta + V_{\mathrm{TF}}) - \frac{1}{2} \iint_{\mathbb{R}^3 \times \mathbb{R}^3} \frac{\rho_{\mathrm{TF}}(x)\,\rho_{\mathrm{TF}}(y)}{|x-y|}\,dx\,dy$$

$$- c_D \int_{\mathbb{R}^3} \rho_{\mathrm{TF}}^{4/3}(x)\,dx - CZ^{5/3-\varepsilon_0}.$$

(B) Density.

$$\iint_{\mathbb{R}^3 \times \mathbb{R}^3} [\rho_{\mathrm{hf}}(x) - \rho_{\mathrm{TF}}(x)][\rho_{\mathrm{hf}}(y) - \rho_{\mathrm{TF}}(y)] \frac{dx\,dy}{|x-y|} < CZ^{5/3-\varepsilon_0}$$

(C) Correlation function.

$$\iint_{\mathbb{R}^3 \times \mathbb{R}^3} |\mathscr{S}(x,y)|^2 \frac{dx\,dy}{|x-y|} > c_D \int_{\mathbb{R}^3} \rho_{\mathrm{TF}}^{4/3}(x)\,dx - CZ^{5/3-\varepsilon_0}.$$

(D) Eigenvalue sum.

$$\operatorname{sneg}(-\Delta + V_{\mathrm{TF}}) = -\frac{1}{15\pi^2} \int_{\mathbb{R}^3} |V_{\mathrm{TF}}(x)|^{5/2}\,dx + \frac{1}{8}Z^2$$

$$+ \frac{1}{48\pi^2} \int_{\mathbb{R}^3} |V_{\mathrm{TF}}(x)|^{1/2}\,\Delta V_{\mathrm{TF}}(x)\,dx + O(Z^{5/3-\varepsilon_0}).$$

Here, (A) is our substitute for Lieb's pointwise inequality (23), while (B), (C), and (D) justify the semiclassical approximations as corrected by Scott and Schwinger.

From (A)–(D) we can easily read off our Main Theory on $E(Z)$. In fact, (12), (22), and (A) provide upper and lower bounds for $E(Z)$ that differ by

$$\frac{1}{2} \iint_{\mathbb{R}^3 \times \mathbb{R}^3} [\rho_{\mathrm{hf}}(x) - \rho_{\mathrm{TF}}(x)][\rho_{\mathrm{hf}}(y) - \rho_{\mathrm{TF}}(y)] \frac{dx\,dy}{|x-y|}$$

$$+ \left\{ c_D \int_{\mathbb{R}^3} \rho_{\mathrm{TF}}^{4/3}(x)\,dx - \frac{1}{2} \iint_{\mathbb{R}^3 \times \mathbb{R}^3} |\mathscr{S}(x,y)|^2 \frac{dx\,dy}{|x-y|} \right\} + CZ^{5/3-\varepsilon_0}.$$

This expression is less than $C'Z^{5/3-\varepsilon_0}$, by (B) and (C). Therefore,

$$E(Z) = \text{sneg}(-\Delta + V_{TF})$$

$$-\frac{1}{2} \iint_{\mathbb{R}^3 \times \mathbb{R}^3} \rho_{TF}(x)\, \rho_{TF}(y) \frac{dx\, dy}{|x-y|} - c_D \int_{\mathbb{R}^3} \rho_{TF}^{4/3}(x)\, dx + O(Z^{5/3-\varepsilon_0}),$$

so that (D) yields Schwinger's formula for $E(Z)$ with an error $O(Z^{5/3-\varepsilon_0})$. Thus, our main theorem is reduced to (A)–(D).

We indicate some of the ideas in the proof of the lower bound (A). The first step is to prove a pointwise lower bound

$$V_{\text{Coulomb}}(x_1 \cdots x_N) \geqslant \sum_{k=1}^{N} W(x_k) - E_0 + \sum_{j<k} K(x_j, x_k), \qquad (32)$$

with $K(x, x')$ a short-range non-negative two-body interaction. This follows by changing slightly the proof of (23). In fact, using (26), we can break up the Coulomb interaction as

$$V_{\text{Coulomb}} = V_{\text{Long-range}} + V_{\text{Short-range}}, \qquad (33)$$

with

$$V_{\text{Long-range}} = \frac{1}{\pi} \iint_{N_{yR} \geqslant Z^{\varepsilon_0}} \frac{N_{yR}(N_{yr}-1)}{2} \frac{dy\, dR}{R^5} \qquad (34)$$

and

$$V_{\text{Short-range}} = \frac{1}{\pi} \iint_{N_{yR} < Z^{\varepsilon_0}} \frac{N_{yR}(N_{yR}-1)}{2} \frac{dy\, dR}{R^5}. \qquad (35)$$

Applying (28a) to the integrand in (34), we get

$$V_{\text{Long-range}} \geqslant \sum_{k=1}^{N} W(x_k) - E_0. \qquad (36)$$

On the other hand,

$$V_{\text{Short-range}} = \sum_{j<k} K(x_j, x_k), \qquad (37)$$

with

$$K(x, x') = \frac{1}{\pi} \iint_{N_{yR} < Z^{\varepsilon_0}} \chi_{x, x' \in B(y, R)} \frac{dy\, dR}{R^5}. \qquad (38)$$

315

APPENDIX B

Moreover, (38) shows that $K(x, x')$ is short-range in the following sense. If the particles $x_1 \cdots x_N$ are independent and distributed according to ρ_{TF}, then each x_j will interact with only $O(Z^{\varepsilon_0})$ of the x_k. Thus we have easily carried out the first step in the proof of (A). Later, we give another proof of (32), closer in spirit to Lieb [L1].

The point of (32) is to bound H_{NZ} from below by a Hamiltonian $H_{\text{lower bound}}$ in which the interaction between particles may be treated as a small perturbation. In fact, (32) yields at once

$$H_{NZ} \geqslant \sum_{k=1}^{N} (-\Delta_{x_k} + W(x_k)) - E_0 + \sum_{j<k} K(x_j, x_k) \equiv H_{\text{lower bound}}. \quad (39)$$

Regarding the term $\sum_{j<k} K(x_j, x_k)$ as a small perturbation in (39), we get the approximate formula

$$\text{sneg}(-\Delta + W) - E_0 + \left\langle \sum_{j<k} K(x_j, x_k) \Psi_0, \Psi_0 \right\rangle \quad (40)$$

for the lowest eigenvalue of $H_{\text{lower bound}}$, where Ψ_0 is the ground-state eigenfunction for $\sum_k (-\Delta_{x_k} + W(x_k))$. Using a semiclassical approximation for the last term in (40), we obtain the heuristic formula

(lowest eigenvalue of $H_{\text{lower bound}}$)

$$\approx \text{sneg}(-\Delta + W) - E_0 + \tfrac{1}{2} \iint_{\mathbb{R}^3 \times \mathbb{R}^3} K(x, y)$$

$$\times \{ \rho_{TF}(x) \rho_{TF}(y) - |\mathscr{S}_{\rho_{TF}(x)}(x-y)|^2 \} \, dx \, dy,$$

where $\mathscr{S}_\rho(y)$ is an elementary function (the correlation for an ideal gas). We prove the rigorous inequality

$$\langle H_{\text{lower bound}} \Psi, \Psi \rangle$$

$$\geqslant \text{sneg}(-\Delta + W) - E_0$$

$$+ \tfrac{1}{2} \iint_{\mathbb{R}^3 \times \mathbb{R}^3} K(x, y) \{ \rho_{TF}(x) \rho_{TF}(y) - |\mathscr{S}_{\rho_{TF}(x)}(x-y)|^2 \} \, dx \, dy$$

$$- CZ^{5/3 - \varepsilon_0} \quad (41)$$

for antisymmetric $\Psi(x_1 \cdots x_N)$ of norm 1.

From (39) and (41) follows immediately

$$E(Z) \geqslant \text{sneg}(-\Delta + W) - E_0$$

$$+ \tfrac{1}{2} \iint_{\mathbb{R}^3 \times \mathbb{R}^3} K(x, y) \{ \rho_{\text{TF}}(x) \, \rho_{\text{TF}}(y) - |\mathscr{S}_{\rho_{\text{TF}}(x)}(x-y)|^2 \} \, dx \, dy$$

$$- C Z^{5/3 - \varepsilon_0},$$

which is equivalent to (A). Thus, (A) is reduced to (41).

The idea in proving (41) is to study wave functions Ψ for which $\langle \sum_k (-\Delta_{x_k} + W(x_k)) \Psi, \Psi \rangle$ is nearly as low as possible. Our results show that

$$\left\langle \sum_{k=1}^{N} (-\Delta_{x_k} + W(x_k)) \Psi, \Psi \right\rangle \leqslant \text{sneg}(-\Delta + W) + Z^{7/3 - \varepsilon_1} \qquad (42)$$

implies

$$\left\langle \sum_{j<k} K(x_j, x_k) \Psi, \Psi \right\rangle$$

$$\geqslant \tfrac{1}{2} \iint_{\mathbb{R}^3 \times \mathbb{R}^3} K(x, y) \{ \rho_{\text{TF}}(x) \, \rho_{\text{TF}}(y) - |\mathscr{S}_{\rho_{\text{TF}}(x)}(x-y)|^2 \} \, dx \, dy$$

$$- C Z^{5/3 - \varepsilon_0}. \qquad (43)$$

To deduce (41), we first suppose that (42) holds. In that case,

$$\langle H_{\text{lower bound}} \Psi, \Psi \rangle$$

$$\geqslant \text{sneg}(-\Delta + W) - E_0 + \left\langle \sum_{j<k} K(x_j, x_k) \Psi, \Psi \right\rangle$$

$$\geqslant \text{sneg}(-\Delta + W) - E_0$$

$$+ \tfrac{1}{2} \iint_{\mathbb{R}^3 \times \mathbb{R}^3} K(x, y) \{ \rho_{\text{TF}}(x) \, \rho_{\text{TF}}(y) - |\mathscr{S}_{\rho_{\text{TF}}(x)}(x-y)|^2 \} \, dx \, dy$$

$$- C Z^{5/3 - \varepsilon_0},$$

by (43) and the definition of $H_{\text{lower bound}}$. So (41) holds whenever Ψ satisfies (42). On the other hand, if (42) fails, then trivially

$$\langle H_{\text{lower bound}} \Psi, \Psi \rangle \geqslant \left\langle \sum_{k=1}^{N} (-\Delta_{x_k} + W(x_k)) \Psi, \Psi \right\rangle - E_0$$

$$\geqslant \text{sneg}(-\Delta + W) + Z^{7/3 - \varepsilon_1} - E_0.$$

This implies (41), since the integral in (41) has order of magnitude $Z^{5/3 + (2/3) \varepsilon_0} \ll Z^{7/3 - \varepsilon_1}$. Hence (41) holds in either case. So the main point in the proof of (A) is that (42) implies (43). Our discussion of (A) concludes with some explanation of this implication.

Any wave function Ψ that satisfies (42) ought to behave much like Ψ_0, the ground state of $\sum_{k=1}^{N} (-\Delta_{x_k} + W(x_k))$. Hence it is plausible that

$$\left\langle \sum_{j<k} K(x_j, x_k) \Psi, \Psi \right\rangle \approx \left\langle \sum_{j<k} K(x_j, x_k) \Psi_0, \Psi_0 \right\rangle,$$

which is close to (43). However, one has to be careful. It is quite simple to produce Ψ satisfying (42), for which $\langle \sum_{j<k} K(x_j, x_k) \Psi, \Psi \rangle$ is much larger than $\langle \sum_{j<k} K(x_j, x_k) \Psi_0, \Psi_0 \rangle$. (We just modify Ψ_0 by adding many particles in a small ball.) So the sign of the inequality is essential in (43).

We prove that (42) implies (43) by first establishing an analogous result in which $-\Delta + W$ is replaced by $-\Delta$. Thus, the heart of the matter is a theorem on N free particles in a box, which we now state. Let $T = \mathbb{R}^3 / L\mathbb{Z}^3$ be a flat torus, and let $\Psi(x_1 \cdots x_N) \in L^2(T^N)$ be antisymmetric with norm 1. Let $K(x, y)$ be a short-range Coulomb interaction on $T \times T$. That is, we suppose $0 \leq K(x, y) = K(y, x) \leq |x - y|^{-1} \chi_{|x-y| < r_{max}}$.

Assuming that Ψ has kinetic energy near the minimum possible, we will control $\langle \sum_{j<k} K(x_j, x_k) \Psi, \Psi \rangle$. Recall that the ground state Ψ_0 for the kinetic energy has

$$\|\nabla \Psi_0\|^2 \approx c_{TF} \rho^{5/3} L^3 \qquad \text{with} \quad \rho = NL^{-3} = \text{density},$$

and

$$\left\langle \sum_{j<k} K(x_j, x_k) \Psi_0, \Psi_0 \right\rangle \approx \tfrac{1}{2} \iint_{T \times T} K(x, y) \{ \rho^2 - |\mathscr{S}_\rho(x - y)|^2 \} \, dx \, dy.$$

THEOREM ON FREE PARTICLES. *Let T, K, and Ψ be as above. Assume that $\|\nabla \Psi\|^2 \leq (1 + \delta) c_{TF} \rho^{5/3} L^3$ and that*

$$\delta < c_0 (\rho r_{max}^3 + 1)^{-m_0}, \qquad N > C_0 (\rho r_{max}^3 + 1)^{m_0}$$

for positive universal constants c_0, C_0, m_0. Then

$$\left\langle \sum_{j<k} K(x_j, x_k) \Psi, \Psi \right\rangle \geq \tfrac{1}{2} \iint_{T \times T} K(x, y) \{ \rho^2 - |\mathscr{S}_\rho(x - y)|^2 \} \, dx \, dy$$

$$- (\delta + N^{-1})^{\varepsilon_0} \rho^{4/3} L^3$$

for a universal constant $\varepsilon_0 > 0$.

Our full result on free particles is slightly stronger than the above, but is more complicated to state.

Let us summarize our discussion of (A). The standard idea for proving lower bounds for $E(Z)$ is to compare the actual Hamiltonian with a non-interacting one. Our idea is to compare the actual Hamiltonian with a weakly interacting one, and then control the weakly interacting Hamiltonian by using our theorem on free particles.

We illustrate what we know about weakly interacting systems by a simple example. Let $T = \mathbb{R}^3/L\mathbb{Z}^3$ as before, and define a Hamiltonian

$$H^{\tau NL} = \sum_{k=1}^{N} (-\Delta_{x_k}) + \tau \sum_{1 \leqslant j < k \leqslant N} \chi_{|x_j - x_k| < 1},$$

acting on antisymmetric $\Psi(x_1 \cdots x_N) \in L^2(T^N)$. Here, τ is a real parameter. For τ real and $\rho > 0$, we pass to the thermodynamic limit by defining

$$E(\tau, \rho) = \lim_{\substack{N, L \to \infty \\ N/L^3 \to \rho}} \left\{ \frac{\text{lowest eigenvalue of } H^{\tau NL}}{N} \right\}. \tag{44}$$

Thus, $E(\tau, \rho)$ is the energy per particle for a zero-temperature gas of interacting fermions, and τ controls the strength of the interaction. It is easy to see that the limit (44) exists when $\tau \geqslant 0$, but not when $\tau < 0$. Our results show that $(\partial/\partial\tau) E(\tau, \rho)$ exists at $\tau = 0$ and is given by an obvious perturbation-theoretic formula. We want to understand more fully how $E(\tau, \rho)$ behaves as $\tau \to 0$. This is probably closely related to atoms.

In our proof of (A), we used a very weak assumption (42). Simpler proofs and sharper theorems might follow if we knew how to exploit a stronger hypothesis.

Our Introduction is nearly complete. In this paper we prove (A) and (C), leaving (B) and (D) to be established in [FS2–FS7]. Our exposition includes many details, to lighten the task of checking the correctness of our results. Curiously, (C) follows as a consequence of our proof of (A). The proofs of (A) and (C) go over easily for molecules, while our discussion of (B) and (D) applies only to atoms, since it is based on ordinary differential equations. From [FS7] it is natural to conjecture a sharp formula for $\text{sneg}(-\Delta + V_{\text{TF}})$, which refines (D). The formula is as follows.

For $l \geqslant 0$ and $r \in (0, \infty)$, define $V_l(r) = l(l+1)/r^2 + V_{\text{TF}}(r)$, and let $l_{\max}$ be the largest integer for which $V_l(r)$ is negative for some r. Define

$$n_l = \int_{\{V_l < 0\}} |V_l(r)|^{-1/2} \, dr \qquad \text{and} \qquad \psi_l = \frac{1}{\pi} \int_{\{V_l < 0\}} |V_l(r)|^{1/2} \, dr$$

for $1 \leqslant l \leqslant l_{\max}$. Finally, set $\beta(t) = -1/12 + \min_{k \in \mathbb{Z}} |t - k|^2$. Then we conjecture that

$$
\begin{aligned}
\mathrm{sneg}(-\Delta + V_{\mathrm{TF}}) \approx {} & -\frac{1}{15\pi^2} \int_{\mathbb{R}^3} |V_{\mathrm{TF}}(x)|^{5/2}\, dx + \frac{Z^2}{8} \\
& + \frac{1}{48\pi^2} \int_{\mathbb{R}^3} |V_{\mathrm{TF}}(x)|^{1/2}\, \Delta V_{\mathrm{TF}}(x)\, dx \\
& + (\text{const.}) \sum_{1 \leqslant l \leqslant l_{\max}} \frac{(2l+1)}{n_l} \beta(\psi_l).
\end{aligned}
\tag{45}
$$

As explained in [F], the last term in (45) is related to classical theorems of analytic number theory.

From (45) we expect that

$$
E(Z) \approx -c_0 Z^{7/3} + \frac{1}{8} Z^2 - c_1 Z^{5/3} + (\text{const.}) \sum_{1 \leqslant l \leqslant l_{\max}} \frac{(2l+1)}{n_l} \beta(\psi_l).
$$

Finally, we refer the reader to the last section of this paper for additional results on quantum states for an atom having nearly the lowest possible energy.

Note added in proof. The reduction of our main result to (B) and (D) above has been simplified and improved by Bach, and again by Graf-Solovej.

FREE PARTICLES IN A BOX

In this section, we study N-particle wave functions Ψ on a box in $\mathbb{R}^3$, with periodic boundary conditions. We compare Ψ with the state Ψ_0 of lowest possible kinetic energy. Our goal is to prove that if Ψ has kinetic energy near enough to the minimum, then Ψ behaves much like Ψ_0.

The precise setup is as follows. We fix $L > 0$ and define that flat torus $T = \mathbb{R}^3/L\mathbb{Z}^3$. Suppose we are given a function spin: $\{1 \cdots N\} \to \{1 \cdots q\}$ and a complex-valued function $\Psi(x_1 \cdots x_N)$ on T^N, with L^2 norm 1. We assume that the wave function satisfies

$$
\Psi(x_{\sigma_1} \cdots x_{\sigma_N}) = (\mathrm{sgn}\ \sigma)\, \Psi(x_1 \cdots x_N)
$$

for permutations σ that preserve spin.   (1)

The kinetic energy of Ψ is defined as $\|\nabla\Psi\|^2_{L^2(T^N)}$. Given the map spin: $\{1 \cdots N\} \to \{1 \cdots q\}$, let Ψ_0 be a wave function of norm 1 and satisfying (1), with lowest possible kinetic energy. The properties of Ψ_0 are elementary

ON THE GROUND-STATE ENERGY OF AN ATOM

and

$$\frac{1}{\text{Vol } A} \int_{x_0 \in A} \left| \text{Var}(x_0, r_0) - q \left\{ \rho_{\text{TF}}(x_0) \cdot \frac{4\pi}{3} r_0^3 \right. \right.$$

$$\left. \left. - \iint_{x, y \in B(x_0, r_0)} |\mathscr{S}_{\rho_{\text{TF}}(x_0)}(x - y)|^2 \, dx \, dy \right\} \right| dx$$

$$\leqslant C'(\eta + Z^{-\gamma_2})^{1/2}.$$

The constant C' depends only on c, C, q.

REFERENCES

[F]   C. FEFFERMAN, Atoms and analytic number theory, *in* "Proc., American Math. Society Centennary Symposium."

[FS1]   C. FEFFERMAN AND L. SECO, On the energy of a large atom, *Bull. Amer. Math. Soc.* **23**, No. 2 (1990), 525–530.

[FS2]   C. FEFFERMAN AND L. SECO, Eigenvalues and eigenfunctions of ordinary differential operators, *Adv. Math.* **95**, No. 2 (1992), 145–305.

[FS3]   C. FEFFERMAN AND L. SECO, The density in a one-dimensional potential, *Adv. Math.*, to appear.

[FS4]   C. FEFFERMAN AND L. SECO, The eigenvalue sum for a one-dimensional potential, *Adv. Math.*, to appear.

[FS5]   C. FEFFERMAN AND L. SECO, The density in a three-dimensional radial potential, *Adv. Math.*, to appear.

[FS6]   C. FEFFERMAN AND L. SECO, The eigenvalue sum for a three-dimensional radial potential, *Adv. Math.*, to appear.

[FS7]   C. FEFFERMAN AND L. SECO, Aperiodicity of the hamiltonian flow in the Thomas–Fermi potential, *Revista Mat. Iberoamericana* **9**, No. 3 (1993), 409–551.

[FS8]   C. FEFFERMAN AND L. SECO, Asymptotic neutrality of large ions, *Comm. Math. Phys.* **128** (1990), 109–130.

[H]   W. HUGHES, An atomic energy lower bound that agrees with Scott's correction, *Adv. Math.* **79** (1990), 213–270.

[IS]   V. IVRII AND I. SIGAL, On the asymptotics of large Coulomb systems, *Ann. Math.*, to appear.

[L1]   E. LIEB, A lower bound for coulomb energies, *Phys. Lett. A* **70** (1979), 444–446.

[L2]   E. LIEB, Thomas–Fermi and related theories of atoms and molecules, *Rev. Modern Phys.* **53**, No. 4 (1981).

[L3]   E. LIEB, Bound on the maximum negative ionization of atoms and molecules, *Phys. Rev. A*

[LS]   E. LIEB AND B. SIMON, Thomas–Fermi theories of atoms, molecules, and solids, *Adv. Math.* **23** (1977), 22–116.

[S]   J. SCHWINGER, Thomas–Fermi model: The second correction, *Phys. Rev. A* **24**, No. 5 (1981), 2353–2361.

[SSS]   L. SECO, I. SIGAL AND J. SOLVEJ, Bound on the ionization energy of large atoms, *Comm. Math. Phys.*

[SW]   H. SIEDENTOP AND R. WEIKARD, On the leading energy correction for the statistical model of the atom: Interacting case, *Comm. Math. Phys.* **112** (1987), 471–490.

Appendix C

Jean Leray was still in a German prisoner of war camp when his article appeared. He had converted himself into a pure mathematician, rather than the applied mathematician who might be recruited into the German war effort. André Weil (Appendix *D*) wrote from prison, in Rouen, having resisted the draft. He was released a few months later and went into the service.

Translated from Jean Leray, "Sur la forme des espaces topologiques et sur les points fixes des représentations," *J. Math. Pures Appl.* 24 (1945): 95–167. Reprinted in Jean Leray, *Selected Papers, Oeuvres Scientifiques*, volume 1 (Berlin: Springer, 1998), pp. 60–64. I have not included references, but I have included relevant notes. The translations aim to be reasonably faithful not only to the meaning but also the syntax of their authors. Page numbers of the original publication are indicated in brackets. Braces indicate footnotes in the original. Braces and brackets are my editorial comments.

[96] Introduction

History—Topology is the branch of mathematics that studies continuity: it does not consist solely in studying those properties of objects that are invariant under topological maps {1-1 and continuous in both directions}: the work of Brouwer, H. Hopf, and Lefschetz has also given it the task of studying maps (1-1 and continuous) and equations. It begins with the definition of topological spaces; these are abstract spaces in which the following notions are meaningful: open and closed sets of points, and their mappings. It continues with the introduction of new algebraic-geometric objects: complexes, groups and rings. Set theoretic topology is the part of topology which employs only the following operations: union, intersection, and closure of sets of points; we will assume that the reader knows the excellent exposition provided in the first two chapters of Alexandroff and Hopf, *Topologie*. Algebraic topology (or combinatorial topology) is the part of topology that employs algebraic-geometric notions; our subject here is algebraic topology, more precisely the theory of homology and its applications to the theory of equations and mappings. {[When Leray uses the term "homology" he means what

322

we nowadays call "cohomology," the dual theory. See A. Borel's introduction to Leray's papers on algebraic topology, p. 4.]}

In recent years the theory of homology has made two advances, the origin of which is the method developed by E. Cartan for determining the Betti numbers of spaces of closed groups.

On the one hand, deRham, by developing this method, has identified the characters of the Betti groups with certain classes of Pfaffian forms; the fact that the Pfaffian forms constitute a ring means that these characters themselves form a ring: the homology ring. Alexander, [97] Kolmogoroff, Čech, and Alexandroff have successfully extended the definition of this homology ring to locally bi-compact spaces, then to normal spaces; while Kolmogoroff and Alexandroff have studied simultaneously the properties of the homology ring and the Betti groups, Alexander has noted in contrast that it is sufficient to study the properties of the homology ring, the properties of the Betti groups resulting from it by duality. The latter point of view will be ours; among other advantages it is congenial with Cartan's geometric conception: we know that the ring of Pfaffian forms plays an almost exclusive role there. {The theory of Pfaffian forms permits Cartan to resolve the problem of equivalence, that is to say to find the *necessary and sufficient* conditions for there to be certain topological mappings, whereas topology, in its current state, only succeeded in establishing *necessary* conditions for two spaces to be homeomorphic.}

On the other hand, the results which furnished the Betti numbers of the four main classes of simple groups suggested to H. Hopf an extremely original study of the homology ring of spaces which have mappings between themselves of a particular sort; Hopf applied his arguments only to spaces which have closed and oriented multiplicities, spaces in which the Betti groups may be identified with the homology ring.

Recall, moreover, the developments in the theory of equations and mappings: The fundamental work is by Brouwer; it is based on the notion of a simplicial approximation and concerns pseudo-multiplicities. Brouwer's essential results were extended to abstract linear spaces by means of passage to the limit.

My initial plan was to develop a theory of equations and mappings that applied directly to topological spaces. I needed to have recourse to new methods, and abandon the classical procedures, and it is impossible for me to develop this theory of equations and mappings without, on the one hand, giving a new definition of the homology ring, and, on the other hand, adapting the arguments of Hopf to more general hypotheses than his.

Summary—I introduce, beside the notion of a covering, which belongs to set theoretic topology, a more flexible notion, [98] that of a "couverture," which belongs to algebraic topology; that notion and that of the intersection of complexes, which I believe is original, allow for a definition of the homology ring which is extremely simple and is useful for the study of mappings. On the other hand, I do not do any subdivision of complexes, I make no hypotheses concerning orientability and I do not employ any simplicial approximations; I never suppose that the space is locally linear. {Nowadays we avoid formulating a priori the conditions that a space is locally linear or homeomorphic to a subspace of a linear space, and on the contrary we want to establish a posteriori that such conditions come from rather more general hypotheses.} When I need to specify a space, I discuss only the properties of mappings of the space into itself; and then follow the arguments of Hopf or closely similar ones. I had initially employed the "continuous homology group" of a space in my articles in the *Comptes Rendus* of 1942, well before my ideas took their current form. I have since succeeded in eliminating this notion; taking the already-cited position of Alexander, that it is superfluous and mistaken to introduce the Betti groups of a space: the pth Betti group is nothing more than the character group of the set of dimension-p homology groups.

Still, the Betti groups of a space play an essential role as do the duality theorems of Pontrjagin in the work that constitutes Chapter II, concerning bi-compact Hausdorff spaces which have a convex covering; I prove that these spaces have the same homology properties as do polyhedrons. I do not know if the procedure I describe for determining their homology ring is similar to a known procedure.

Chapter III, concerning fixed points of mappings, begins my theory of equations. Its brevity and the generality of its theorems justify our interest in the preceding notions, which now find an application therein.

The first part of my course in algebraic topology [this sequence of papers] is restricted to problems in which there is no special role or feature for subspaces of the space being studied, that is to say, to problems concerning the form of the space. On the contrary, what will follow is entitled, *On the situation of a closed set of points in a topological space; their equations and mappings*. Its [99] main interest will be for the theory of equations it sets forth. It will begin with an extension to normal spaces of Alexander's duality theory; such extensions have already been given, in the case of locally bi-compact spaces, by Alexandroff, Pontrjagin, Kolmogoroff and Alexander; moreover, they constitute the sole application that these authors give for their definition of the homology ring.

Appendix D

From André Weil, *Oeuvres Scientifiques, Collected Papers*, volume 1 (New York: Springer, 1979), pp. 244–255. Again, page numbers of the original publication are indicated in brackets. Braces indicate footnotes in the original. Braces and brackets are my editorial comments.[1]

[244] "A [14-page] letter . . . to Simone Weil {[A. Weil's sister]}," written from Bonne-Nouvelle prison.

Rouen, March 26, 1940

Some thoughts I have had of late, concerning my arithmetic-algebraic work, might pass for a response to one of your letters, where you asked me what is of interest to me in my work. So, I decided to write them down, even if for the most part they are incomprehensible to you.

The thoughts that follow are of two sorts. The first concerns the history of the theory of numbers; you may well believe you are able to understand the beginning; you will understand nothing of what follows that. The other concerns the role of analogy in mathematical discovery, examining a particular example, and perhaps you will be able to profit from it. I advise you that all that concerns the history of mathematics in what follows is based on insufficient scholarship, and is derived from an a priori reconstruction, and even if things ought to have happened this way (which is not proven here), I cannot say that they did happen this way. In mathematics, moreover, as much as in any other field, the line of history has many turning points.

With these precautions out of the way, let us start with the history of the theory of numbers. It is dominated by the law of reciprocity. This is Gauss's *theorema aureum* (? I need to refresh my memory of this point: Gauss very much liked names of this sort, he had as well a *theorema egregium*, and I no longer know which is which), published by him in his *Disquisitiones* in 1801, which was only beginning to be read and understood toward 1820 by Abel, Jacobi, and Dirichlet, and which remained as the bible of the number theorist for almost a century. But in order to say what this law is, whose statement was already known to Euler and Legendre [Euler had found it empirically, as did Legendre; Legendre claimed moreover to give a proof in his Arithmetic, which apparently supposed the truth of something which was approximately as difficult as the theorem; but he complained bitterly of the "theft" committed by Gauss, who, without knowing Legendre, found, empirically as well, the statement of the theorem, and gave two very beautiful proofs in his *Disquisitiones*, and later up to 4 or 5 others, all based on different principles.]: it is necessary to backtrack a bit in order to explain the law of reciprocity.

Algebra began with the task of finding, for given equations, solutions within a given domain, which might be the positive numbers, or the reals, or later the complex numbers. One had not yet conceived of the ingenious idea, characteristic of modern algebra, of starting with an equation and *then* constructing ad hoc a domain in which it has a solution (I am not speaking ill of this idea, which has shown itself to be extremely productive; moreover, Poincaré has somewhere or other some beautiful thoughts, a propos of the solution by radicals, on the general processes whereby, after having searched long and [245] in vain to solve such a problem with a foreordained procedure, mathematicians inverted the terms of the question and started from the problem to devise adequate methods). The problem had been solved subsequently for all second-degree equations which had solutions in negative numbers; when the equation had no solution, the usual formula having led to the imaginaries, about which there remained many doubts (and it was thus until Gauss and his contemporaries); just because of the suspicion of these imaginaries, the so-called Cardan and Tartaglia formula for the solution of the equation of the 3rd degree in radicals produced some discomfort. Be that as it may, when Gauss began the *Disquisitiones* with the notion of congruences for building up his systematic exposition, it was also natural to solve congruences of the second degree, after having solved those of the first degree (a congruence is a relationship among integers a, b, m, which is written $a=b$ modulo m, abbreviated $a=b$ (mod m) or $a \equiv b(m)$, meaning that a and b have the same remainder in division by m, or $a-b$ is a multiple of m; a congruence of the first degree is $ax+b \equiv 0(m)$, of the second degree is $ax^2+bx+c \equiv 0(m)$, etc.); the latter lead (by the same procedure through which one reduces an ordinary second degree equation to an extraction of roots) to $x^2 \equiv a$ (mod m); if the latter has a solution, one says that a is a quadratic residue of m, if otherwise, a is a non-residue (1 and −1 are residues of 5, 2 and −2 are non-residues). If these notions were around for some time before Gauss, it is not that the notion of congruence was a point of departure; but the notion itself arose in diophantine problems (solutions of equations in integers or rationals) which were the object of Fermat's most important work; the first degree diophantine equations, $ax+by=c$, are equivalent to first degree congruences $ax \equiv c$ (mod b); the second degree equations of the type studied by Fermat (decompositions in terms of squares, $x^2+y^2=a$, and equations $x^2+ay^2=b$, etc.) are not equivalent to congruences, but congruences and the distinction between residues and non-residues play a large role in his work, in truth they did not appear explicitly in Fermat's work (it is true that we do not possess his proofs, but he seems to have employed other principles about which we can make some approximate inferences), but which, as far as I know (based on second-hand evidence) were already well in evidence in Euler.

The law of reciprocity permits us to know, given two prime numbers p, q, whether q is or is not a (quadratic) residue of p, if one knows already whether, (a) p is or is not a residue of q; (b) if p and q are respectively congruent to 1 or -1 modulo 4 (or for $q=2$, if p is congruent to 1, 2, 5, or 7, modulo 8). For example, $53 \equiv 5 \equiv 1$ (mod 4), and 53 is not a residue of 5, *therefore* 5 is not a residue of 53. Since the problem for non-primes leads naturally to the problem for primes, this law gives an easy means of determining if a is or is not a residue of b as soon as one knows their prime factorization. But this "practical" application is insignificant. What is crucial is there be *laws*. It is obvious that the residues of m form an arithmetic progression of increment m, for if a is a residue, it is the same for all $mx+a$; however it is beautiful and surprising that the prime numbers p *for which* m is a residue are precisely those which belong to certain arithmetic progressions of increment $4m$; for the others m is a non-residue; and what is even more amazing, if one recalls on the other hand that the distribution of prime numbers in any given arithmetic progression $Ax+B$ (which one knows [246] from Dirichlet will have an infinity of primes as long as A and B are relatively prime) does not follow any other known law other than a statistical one (the approximate number of primes which are $\leq T$, which, for a given A, is the same for any B prime to A) and appears, for each concrete case that one examines numerically, to be as "random" as a list of numbers generated by a roulette wheel.

The rest of the *Disquisitiones* contains above all:

1. the definitive theory of quadratic forms in 2 variables, $ax^2+bxy+cy^2$, having *among other* consequences the complete resolution of the problem which gave birth to the theory: to know if $ax^2+bxy+cy^2=m$ has solutions in integers.

2. the study of the n-th roots of unity, and, as we would say, the Galois theory of *the fields given by these roots* and their subfields (all without using imaginaries, nor other functions other than the trigonometric ones, and ending up with the necessary and sufficient conditions for the regular n-gon being constructible by ruler and compass), which appeared as an application of earlier work in the book, as preliminary to the solution of congruences, on the multiplicative group of numbers modulo m. I will not speak of the theory of

quadratic forms of more than two variables since it has had little influence until now on the general progress of the theory of numbers.

Gauss's subsequent research was to study cubic and biquadratic residues (defined by $x^3 \equiv a$ and $x^4 \equiv a$ (mod m)); the latter are a bit simpler; Gauss recognized that there were no simple results to be hoped for by staying within the domain of ordinary integers and it was necessary to employ "complex" integers $a+b\sqrt{-1}$ (a propos of which he invented, at about the same time as Argand, the geometric representation of these numbers by points on a plane, through which all doubts were dissipated about the "imaginaries"). For the cubic residues, it was necessary to have recourse to the "integers" $a+bj$, a and b integers, j=the cube root of 1. Gauss recognized as well, and even thought (there is a trace of this in his notes) of studying the domain of the nth roots of unity, at the same time thinking to try to provide a proof of "Fermat's theorem" ($x^n+y^n=z^n$ is impossible), which he suspected would be a simple application (that is what he said) of such a theory. But then he encountered the fact that there was no longer a unique prime decomposition (except for i and j, as 4th and 3rd roots of unity, and I believe also for the 5th roots).

There are many separate threads; it would take 125 years to unravel them and assemble them anew into a new skein. The great names here are Dirichlet (who introduced the zeta functions or L-functions into the theory of quadratic forms, through which he proved among other things that every arithmetic progression contains an infinity of primes; but above all, since that time we have only needed to follow his model in order to apply these functions to the theory of numbers), Kummer (who elucidated the fields generated by roots of unity by inventing "ideal" factors, and went far enough in the theory of these fields in order to obtain some results on Fermat's theorem), Dedekind, Kronecker, Hilbert, Artin. Here is a sketch of the picture that results from their efforts.

I cannot say anything without using the notion of a field, which according to its definition, if one limits oneself to its definition, is simple (it is a set where one has in effect the usual "four elementary {[arithmetic]} operations," these having the usual properties of [247] commutativity, associativity, distributivity); the algebraic extension of a field k (it is a field k', containing k, of which all elements are roots of an algebraic equation $\alpha^n+c_1\alpha^{n-1}+\cdots+c_{n-1}\alpha+c_n=0$ with coefficients $c_1, \ldots, c_n$ in field k); and finally the *abelian* extension of a field k; that means an algebraic extension of k whose *Galois group* is abelian, that is to say commutative. It would be illusory to give a fuller explanation of abelian extensions; it is more useful to say that they are almost the same thing, but not the same thing, as an extension of k obtained by adjoining n-th roots (roots of equations $x^n=a$, a in k); if k contains for whatever integer n, n n-th distinct roots of unity then it is exactly the same thing (but most often one is interested in fields which do not have this property). If k contains n n-th roots of unity (for a given n), then all abelian extensions of degree n (that is to say, having been generated by the adjunction to k of *one* root of an equation of degree n) can be generated by m-th roots (where m is a divisor of n). Abel discovered this idea in his research on equations solvable by radicals (Abel did not know of the notion of the Galois group, which clarifies all these questions). It is impossible to say here how Abel's research was influenced by Gauss's results (see above) on the division of the circle and the n-th roots of unity (which lead to an *abelian* extension of the field of rationals), nor what connections they had with the work of Lagrange, with Abel's own work on elliptic functions (where the division takes place, from Abel's point of view, in the *abelian* equation [the roots generating abelian extensions], results which were already known to Gauss, but not published, at the very least for the particular case of the so-called lemniscate) and abelian functions, just as with Jacobi's work on the same subject (the same Jacobi who invented "abelian functions" in the modern sense and gave them that name, see his memoir "*De transcendentibus quibusdam abelianis*"), nor with Galois's work (which was only understood little by little, and much later; there is *no* trace in Riemann that he had learned from it, *although* (this is most remarkable) Dedekind, Privatdozent in Göttingen and close friend of Riemann, had since 1855 or 6, when Riemann was at the height of his powers, given a course on abstract groups and Galois theory).

To know if a (not a multiple of p) is a residue of p (prime), is to know whether $x^2 - a = py$ has solutions; in passing to the field extension of $\sqrt{a}$, one gets $(x - \sqrt{a})(x + \sqrt{a}) = py$, so in this field p is not prime to $x - \sqrt{a}$, which, nevertheless, it does not divide. In the language of ideals, that is as much to say that in this field p is not prime, but may be decomposed into two prime ideal factors. Thus one is presented with a problem: k being a field (here the field of rationals), k' (here, k' is k adjoined by $\sqrt{a}$) an algebraic extension of k, to know if a prime ideal (here, a number) in k remains prime in k' or if it decomposes into prime ideals, and how: a being given, the law of reciprocity points to those p for which a is the residue, and so resolves the problem for this particular case. Here and in all of what follows, k, k', etc. are fields of algebraic numbers (roots of algebraic equations with rational coefficients).

When it is a question of biquadratic residues, one works with a field generated by $\sqrt[4]{a}$; but such a field is not *in general* an abelian extension of the "base [248] field" k unless the adjunction of a 4th root of a brings along at the same time three others (namely, if α is one of them, the others are $-\alpha$, $i\alpha$, and $-i\alpha$), this requires that k contains $i = \sqrt{-1}$; one would have nothing so simple if one

329

takes as the base field the rationals, but all goes well if one takes (as did Gauss) the field of "complex rationals" $r+si$ (r, s rational). The same is the case for cubic residues. In these cases, one studies the decomposition, in the field k' obtained by the adjunction of a 4th (or, respectively, 3rd) root, starting with a base field k containing i (respectively, j), of an ideal (here, a number) prime in k.

So, this problem of the decomposition in k' of ideals of k is completely resolved when k' is an abelian extension of k, and the solution is very simple and it generalizes the law of reciprocity in a straightforward and direct manner. For the arithmetic progression in which the prime numbers are found, with residue a, one substitutes ideal classes {[*des classes d'idéaux*]}, the definition of which is simple enough. The classes of quadratic forms in two variables, studied by Gauss, correspond to a particular case of these classes of ideals, as was recognized by Dedekind; Dirichlet's analytic methods (using zeta or L-functions) for studying quadratic forms, is translated {[*transportent*]} readily to the more general classes of ideals that had been considered in this theory; for example, for the theorem on arithmetic progressions there corresponds the following result: in each of these ideal classes in k, there is an infinity of prime ideals, therefore an infinity of ideals of k which may be factored in a given fashion in k'. Finally, the decomposition of ideals of k into classes determines k' in a unique way: and, by the theorem called *the law of Artin reciprocity* (because it implicitly contains Gauss's law and all known generalizations), there is a correspondence (an "isomorphism") of the Galois group of k' with respect to k, and the "group" of ideal classes in k. Thus, once one knows what happens in k, one has complete knowledge of *abelian* extensions of k. This does not mean there is nothing more to do about abelian extensions (for example, one can generate these by the numbers $\exp(-2\pi i/n)$ if k is the field of rationals, thus by means of the exponential function; if k is the field generated by $\sqrt{-a}$, a a positive integer, one knows how to generate these extensions by means of elliptic functions or their close relatives; but one knows nothing for all other k). But these questions are well understood and one can say that *everything* that has been done in arithmetic since Gauss up to recent years consists in variations on the law of reciprocity: beginning with Gauss's law; and ending with and crowning the work of Kummer, Dedekind, Hilbert, is Artin's law, *it is all the same law*. This is beautiful, but a bit vexing. We know a little more than Gauss, without doubt; but what we know more (or a bit more) is just that we do not know more.

This explains why, for some time, mathematicians have focused on the problem of the non-abelian decomposition laws (problems concerning k, k', when k' is any nonabelian extension of k; we remain still within the realm of a field of algebraic numbers). What we know amounts to very little; [249] and that little bit was found by Artin. To each field is attached a zeta function, discovered by Dedekind; if k' is an extension of k, the zeta function attached to k' decomposes into factors; Artin discovered this decomposition; when k' is an abelian extension of k, these factors are identical to Dirichlet's L-functions, or rather to their generalization for fields k and classes of ideals in k, and the identity between these factors and these functions *is* (in other words) Artin's reciprocity law; and this is the way Artin first arrived at this law as a bold conjecture (it seems that Landau made fun of him), some time before being able to prove it (a curious fact, his proof is a simple translation {[*transposition*]} of another result by Tchebotareff that had just been published, which he cited; however it is Artin, justly having it bear his name, who had the glory of discovering it). In other words, the law of reciprocity is nothing other than the rule for forming the coefficients of the series that represents the Artin factors (which are called "Artin L-functions"). As the decomposition into factors remains valid if k' is a non-abelian extension, it is these factors, for these "non-abelian L-functions," that it is natural to tackle in order to discover the law of formation of their coefficients. It is worth noting that, in the abelian case, it is known that the Dirichlet L-functions, and consequently the Artin L-functions, which scarcely differ from them, are entire functions. One knows nothing of this sort for the general case: it is there, as already indicated by Artin, that one might find an opening for an attack (please excuse the metaphor): *since* the methods known from arithmetic do not appear to permit us to show that the Artin functions are entire functions, one could hope that in proving it one could open a breach which would permit one to enter this fort (please excuse the straining of the metaphor).

Since the opening is well defended (it had defied Artin), it is necessary to inspect the available artillery and the means of tunneling under the fort (please excuse, etc.). {The reader who has the patience to get to the end will see that as artillery, I make use of a trilingual inscription, dictionaries, adultery, and a bridge which is a turntable, not to speak of God and the devil, who also play a role in this comedy.} And here is where the *analogy* that has been referred to since the beginning finally makes its entrance, like Tartuffe appearing only in the third act.

It is widely believed that there is nothing more to do about algebraic functions of one variable, because Riemann, who had discovered just about all that we know about them (excepting the work on uniformization by Poincaré and Klein, and that of Hurwitz and Severi on correspondences), left us no indication that there might be major problems that concern them. I am surely one of the most knowledgeable persons about this subject; mainly because I had the good fortune (in 1923) to learn it directly from Riemann's memoir, which is one of the greatest pieces of mathematics that has ever been written; there is not a single word in it that is not of consequence. The story is not closed, however; for example, see my memoir in the Liouville Journal (see the introduction to this paper). {["Généralisation des fonctions abéliennes," *Journal de Mathématiques Pures et Appliquées* IX 17 (1938): 47–87, pp. 47–49.]} Of course, I am not foolish enough to compare myself to Riemann; but to add a little bit, whatever it is, to Riemann, that would already be, as they say in [250] Greek, to do something {[*faire quelque chose*]}, even if in order to do it you have the silent help of Galois, Poincaré and Artin.

Be that as it may, in the time (1875 to 1890) when Dedekind created his theory of ideals in the field of algebraic numbers (in his famous "XI Supplements": Dedekind published four editions of Dirichlet's Lectures on the theory of numbers, given at Göttingen during the last years of Dirichlet's life, and admirably edited by Dedekind; among the appendices or "Supplements" of these lectures, which contain nothing indicating they are Dedekind's original work, and which indeed they are only in part, beginning with the 2nd edition there are three entirely different expositions of the theory of ideals, one for each edition), he discovered that an analogous principle permitted one to establish, by purely algebraic means, the principal results, called "elementary," of the theory of algebraic functions of one variable, which were obtained by Riemann by transcendental {[analytic]} means; he published with Weber an account of the consequences of this principle. Until then, when the topic of algebraic functions arose, it concerned a function y of a variable x, defined by an equation $P(x,y)=0$ where P is a polynomial *with complex coefficients*. This latter point was essential in order to apply Riemann's methods; with those of Dedekind, in contrast, those coefficients could come from an arbitrary field (called "the field of constants"), since the arguments were *purely algebraic*. This point will be important shortly.

The analogies that Dedekind demonstrated were easy to understand. For integers one substituted polynomials in x, to the divisibility of integers corresponded the divisibility of polynomials (it is well known, and it is taught even in high schools, that there are other such analogies, such as for the derivation of the greatest common divisor), to the rational numbers correspond the rational fractions, and to algebraic numbers correspond the algebraic functions. At first glance, the analogy seems superficial; to the most profound problems of the theory of numbers (such as the decomposition into prime ideals) there would seem to be nothing corresponding in algebraic functions, and inversely. Hilbert went further in figuring out these matters; he saw that, for example, the Riemann-Roch theorem corresponds to Dedekind's work in arithmetic on the ideal called "the different"; Hilbert's insight was only published by him in an obscure review (Ostrowski pointed me to it), but it was already transmitted orally, much as other of his ideas on this subject. The unwritten laws of modern mathematics forbid writing down such views if they cannot be stated precisely nor, all the more, proven. To tell the truth, if this were not the case, one would be overwhelmed by work that is even more stupid and if not more useless compared to work that is now published in the journals. But one would love it if Hilbert had written down all that he had in mind.

Let us examine this analogy more closely. Once it is possible to translate any particular proof from one theory to another, then the analogy has ceased to be productive for this purpose; it would cease to be at all productive if at one point we had a meaningful and natural way of deriving both theories from a single one. In this sense, around 1820, mathematicians (Gauss, Abel, Galois, Jacobi) permitted themselves, with distress {[*angoisse*]} and delight, to be guided by the analogy between the division of the circle (Gauss's problem) and the division of elliptic functions. Today, we [251] can easily show that both problems have a place in the theory of abelian equations; we have the theory (I am speaking of a purely algebraic theory, so it is not a matter of number theory in this case) of abelian extensions. Gone is the analogy: gone are the two theories, their conflicts and their delicious reciprocal reflections, their furtive caresses, their inexplicable quarrels; alas, all is just one theory, whose majestic beauty can no longer excite us. Nothing is more fecund than these slightly adulterous relationships; nothing gives greater pleasure to the connoisseur, whether he participates in it, or even if he is an historian contemplating it retrospectively, accompanied, nevertheless, by a touch of melancholy. The pleasure comes from the illusion and the far from clear meaning; once the illusion is dissipated, and knowledge obtained, one becomes indifferent at the same time; at least in the Gitâ there is a slew of prayers (slokas) on the subject, each one more defniitive than the previous ones. But let us return to our algebraic functions.

Whether it is due to the Hilbert tradition or to the attraction of this subject, the analogies between algebraic functions and numbers have been on the minds of all the great number theorists of our time; abelian extensions and abelian functions, classes of ideals and classes of divisors, there is material enough for many seductive mind-games, some of which are likely to be deceptive (thus the appearance of theta functions in one or another theory). But to make something of this, two more recent technical contrivances were necessary. On the one hand, the theory of algebraic functions, that of Riemann, depends *essentially* on the idea of birational invariance; for example, if we are concerned with the field of *rational* functions of one variable x, one introduces (initially, I take the field of constants to be the complex numbers) as the *points* corresponding to the various complex values of x, y including the point at infinity, denoted symbolically by $x=\infty$, and defined by $1/x=0$; the fact that this point plays exactly the same role as all the others is essential. Let $R(x)=a(x-\alpha_1) \ldots (x-\alpha_m)/(x-\beta_1) \ldots (x-\beta_n)$ be a rational fraction, with its decomposition into factors as indicated; it will have zeros $\alpha_1, \ldots , \alpha_m$, the poles $\beta_1, \ldots , \beta_n$, and the point at infinity, which is zero if $n > m$, and is infinite if $n < m$. In the domain of rational *numbers*, one always has a decomposition into prime factors, $r = p_1 \ldots p_m/q_1 \ldots q_n$, each prime factor corresponding to a binomial factor $(x-\alpha)$; but nothing apparently corresponds to the point at infinity. If one models the theory of functions on the theory of algebraic numbers, one is forced to give a special role, *in the proofs*, to the point at infinity, sweeping the problem into a corner, if we are to have a definitive statement of the result: this is just what Dedekind-Weber did, this is just what was done by all who have written in algebraic terms about algebraic functions of one variable, until now, I was the first, two years ago, to give (in Crelle's Journal {["Zur algebraischen Theorie der algebraischen Funktionen", 179 (1938), pp. 129–133]}) a purely algebraic proof of the main theorems of this theory, which is as birationally invariant (that is to say, not attributing a special role to any point) as were Riemann's proofs; and that is of more than methodological importance. {Actually, I was not quite the first. The proofs, to be sure very roundabout, of the Italian school (Severi above all) are, in principle, of the same sort, although drafted in classical language.} However fine it is to have these results for the function field, it seems that one has lost sight of the analogy. In order to reestablish [252] the analogy, it is necessary to introduce, into the theory of algebraic *numbers*, something that corresponds to the point at infinity in the theory of functions. That is what one achieves, and in a very satisfactory manner, too, in the theory of "valuations". This theory, which is not difficult but I cannot explain here, depends on Hensel's theory of p-adic fields: to define a prime ideal in a field (a field given *abstractly*) is to represent the field "isomorphically" in a p-adic field: to represent it in the same way in the field of real or complex numbers, *is* (in this theory) to define a "prime ideal at infinity". This latter notion is due to Hasse (who was a student of Hensel), or perhaps Artin, or to both of them. *If one follows it in all of its consequences*, the theory alone permits us to reestablish the analogy at many points where it once seemed defective: it even permits us to discover in the number field simple and elementary facts which however were not yet seen (see my 1939 article in *la Revue Rose* which contains some of the details {["Sur l'analogie entre les corps de nombres algébriques et les corps de fonctions algébriques," *Revue Scientifique* 77 (1939) 104–106, and the comments in the *Oeuvres Scientifiques*, volume 1, pp. 542–543]}). It is not so much this point of view that has been used up to now for giving satisfactory statements of the principal results of the theory of abelian extensions (I forgot to say that this theory is most often called "class field theory"). An important point is that the p-adic field, or respectively the real or complex field, corresponding to a prime ideal, plays exactly the role, in number theory, that the field of power series *in the neighborhood of a point* plays in the theory of functions: that is why one calls it a *local field*.

With all of this, we have made great progress; but it is not enough. The purely algebraic theory of algebraic functions in any *arbitrary* field of constants is not rich enough so that one might draw useful lessons from it. The "classical" theory (that is, Riemannian) of algebraic functions over the field of constants of the complex numbers is infinitely richer; but on the one hand it is too much so, and in the mass of facts some real analogies become lost; and above all, it is too far from the theory of numbers. One would be totally obstructed if there were not a bridge between the two.

And just as God defeats the devil: this bridge exists; it is the theory of the field of algebraic functions over a finite field of constants (that is to say, a finite number of elements: also said to be a Galois field, or earlier "Galois imaginaries" because Galois first defined them and studied them; they are the algebraic extensions of a field with p elements formed by the numbers 0, 1, 2,... , $p-1$ where one calculates with them modulo p, p = prime number). They appear already in Dedekind. A young student in Göttingen, killed in 1914 or 1915, studied, in his dissertation that appeared in 1919 (work done entirely on his own, says his teacher Landau), zeta functions for certain of these fields, and showed that the ordinary methods of the theory of algebraic numbers applied to them. Artin, in 1921 or 1922, took up the question again, again from the point of view of the zeta function; F. K. Schmidt made the bridge between these results and those of Dedekind-Weber, in the process of providing a definition of the zeta function that was birationally invariant. In the last few years, these fields were a favorite subject of Hasse and his school; Hasse made a number of beautiful contributions.

I spoke of a bridge; it would be more correct to speak of a turntable {[*plaque tournante*]}. On one hand the analogy with number fields is so strict and obvious that there is neither [253] an argument nor a result in number theory that cannot be translated almost word for word to the function fields. In particular, it is so for all that concerns zeta functions and Artin functions; and there is more: Artin functions *in the abelian case* are *polynomials*, which one can express by saying that these fields furnish a *simplified* model of what happens in number fields; here, there is thus room to conjecture that the non-abelian Artin functions are still polynomials: *that is just what occupies me at the moment*, all of this permits me to believe that all results for these fields could inversely, if one could formulate them appropriately, be translated to the number fields.

On the other hand, between the function fields and the "Riemannian" fields, the distance is not so large that a patient study would not teach us the art of passing from one to the other, and to profit in the study of the first from knowledge acquired about the second, and of the extremely powerful means offered to us, in the study of the latter, from the integral calculus and the theory of analytic functions. That is not to say that at best all will be easy; but one ends up by learning to see something there, although it is still somewhat confused. Intuition makes much of it; I mean by this the faculty of seeing a connection between things that in appearance are completely different; it does not fail to lead us astray quite often. Be that as it may, my work consists in deciphering a trilingual text {[cf. the Rosetta Stone]}; of each of the three columns I have only disparate fragments; I have some ideas about each of the three languages: but I know as well there are great differences in meaning from one column to another, for which nothing has prepared me in advance. In the several years I have worked at it, I have found little pieces of the dictionary {[*bouts de dictionaire*]}. Sometimes I worked on one column, sometimes under another. My large study that appeared in the Liouville journal made nice advances in the "Riemannian" column; unhappily, a large part of the deciphered text surely does not have a translation in the other two languages: but one part remains that is very useful to me. At this moment, I am working on the middle column. All of this is amusing enough. However, do not imagine that this work on several columns is a frequent occasion in mathematics; in such a pure form, this is almost a unique case. This sort of work suits me particularly; it is unbelievable at this point that distinguished people such as Hasse and his students, who have made this subject the matter of their most serious thoughts over the years, have, not only neglected, but disdained to take the Riemannian point of view: at this point they no longer know how to read work written in Riemannian (one day, Siegel made fun of Hasse, who had declared himself incapable of reading my Liouville paper), and that they have rediscovered sometimes with a great deal of effort, in their dialect, important results that were already known, much as the ideas of Severi on the ring of correspondences were rediscovered by Deuring. But the role of what I call analogies, even if they are not always so clear, is nonetheless important. It would be of great interest to study these things for a period for which we are well provided with texts; the choice would be delicate.

P.S. I send this to you without rereading ... I fear ... having made more of my research than I intended; that is, in order to explain (following your request) how one develops one's research, I have been focusing on the locks I wish to open. In speaking of analogies between numbers and functions, [254] I do not want to give the impression of being the only one who understands them: Artin has thought profoundly about them as well, and that is to say a great deal. It is curious to note that one work (signed by a student of Artin who is not otherwise known, which without proof to the contrary, allows one to presume that Artin is the real source) appeared 2 or 3 years ago which gives perhaps the only example of a result from the classical theory, obtained by a *double* translation, starting with an number theoretic result (on abelian zeta functions), and which is novel and interesting. And Hasse, whose combination of patience and talent make him a kind of genius, has had very interesting ideas on this subject. Moreover (a characteristic trait, and which would be sympathetic to you, of the school of modern algebra) all of this is spread by an oral and epistolary tradition more than by orthodox publications, so it is difficult to make a history of all of it in detail.

You doubt and with good reason that modern axiomatics will work on difficult material. When I invented (I say invented, and not discovered) uniform spaces, I did not have the impression of working with resistant material, but rather the impression that a professional sculptor must have when he plays by making a snowman. It is hard for you to appreciate that modern mathematics has become so extensive and so complex that it is essential, *if* mathematics is to stay as a whole and not become a pile of little bits of research, to provide a unification, which absorbs in some simple and general theories all the common substrata of the diverse branches of the science, suppressing what is not so useful and necessary, and leaving intact what is truly the specific detail of each big problem. This is the good one can achieve with axiomatics (and this is no small achievement). This is what Bourbaki is up to. It will not have escaped you (to take up the military metaphor again) that there is within all of this great problems of strategy. And it is as common to know tactics, as it is rare (and beautiful, as Gandhi would say) to plan strategy. I shall compare (despite the incoherence of the metaphor) the great axiomatic edifices to communication at the rear of the front: there is not much glory in the commissariat and logistics and transport, but what would happen if these good folks did not devote themselves to secondary work (where, moreover, they readily earn their subsistence)? The danger is only too great that various fronts end up, not by starving (the Council for Research is there for that), but by paying insufficient attention to each other and so waste their time, some like the Hebrews in the desert, others like Hannibal at Capua {[where the troops were said to have been entranced by the place]}. The current organization of science does not take into account (unhappily, for the experimental sciences; in mathematics the damage is much less great) the fact that very few persons are capable of grasping the entire forefront of science, of seizing not only the weak points of resistance, but also the part that is most important to take on, the art of massing the troops, of making each sector work toward the success of the others, etc. Of course, when I speak of troops the term (for the mathematician, at least) is essentially metaphoric, each mathematician being himself his own troops. If, under the leadership given by certain teachers, certain "schools" have notable success, the role of the individual in mathematics remains preponderant. Moreover, it is becoming impossible to apply a view of this sort [255] to science as a whole; it is not possible to have someone who can master enough of both mathematics and physics at the same time to control {[*régler*]} their development alternatively or simultaneously; all attempts at "planning" become grotesque, and it is necessary to leave it to chance and to the specialists.

Notes

PREFACE

1. I have not discussed cluster expansions and Padé approximants (Domb et al); discrete complex analysis (Smirnov and collaborators)/conformal field theory/quantum groups, focused on the critical point; rigorous statistical mechanics in terms of states; or Witten and Frenkel on the Geometric Langlands Program. See chapters 7 and 8 of D.E. Evans and Y. Kawahigashi, *Quantum Symmetries on Operator Algebras* (Oxford: Clarendon Press, 1998), for example. B.M. McCoy, *Advanced Statistical Mechanics* (Oxford: Oxford University Press, 2010), chapters 10–14, is authoritative.

2. Quoted in D. Laugwitz, *Bernhard Riemann* (Boston: Birkhäuser, 1999), p. 302.

3. Frank Quinn has written about this in "Contributions to a Science of Contemporary Mathematics," October 2011, 98pp., available at https://www.math.vt.edu/people/quinn/history_nature/nature0.pdf.

4. J. Palmer, *Planar Ising Correlations* (Boston: Birkhäuser, 2007). p. x

5. Palmer, p. 149

6. J.-M. Maillard makes this point in many places. He says, "the theory of the 2-D Ising model is nothing but the theory of elliptic curves and, more generally, theoretical physics is nothing but effective algebraic geometry with a selected role played by birational transformations." He defines the curve in terms of the intersection of two quadrics (each of which define the symmetries of the problem, such as duality), provides a *j*-function, a gauge group, and an account of the genus of the curve (0 or 1). Yang-Baxter is addition on the curve, and the renormalization group is something like a Landen transformation. See, for example, A. Bostan, S. Boukraa, S. Hassani, M. van Hoeij, J.-M. Maillard, J.-A. Weil and N. Zenine, "The Ising model: from elliptic curves to modular forms and Calabi–Yau equations," *J. Phys. A: Math. Theor.* 44 (2011): 045204.

7. See, D. Bressoud, *Proofs and Confirmations* (New York: Cambridge University Press, 1999), pp. xi–xiii, 257–259. Bressoud's title refers to I. Lakatos's work, in which counterexamples play a central role. G.H. Hardy is quoted for the mountain climbing example; Bressoud offers the archaeologist as an alternative account. See also, W.P. Thurston, "On Proof and Progress in Mathematics," *Bulletin (New Series) of the American Mathematical Society* 30 (1994): 161–177.

8. For a discussion of the connection between analysis and arithmetic, between scaling and combinatorics, in a context rather more related to chapters 3 and 4's discussion of the Ising model, see D. Aldous and P. Diaconis, "Longest Increasing Subsequences: From Patience Sorting to the Baik-Deift-Johansson Theorem," *Bulletin (New Series) of the American Mathematical Society* 36 (1999): 413–432. The significant feature here is a recurrently appearing, highly symmetric (Toeplitz) matrix.

9. The probability distributions are known as the Gaudin-Mehta-Dyson (sine kernel, P_V) and the Tracy-Widom distributions (Airy kernel, P_{II}), respectively the "bulk" and "edge" spectrum, referring as well to the kernel of the associated Fredholm determinant, and to the Painlevé transcendent through which the distribution is expressed as an exponential of an integral. In the original Ising case, a Bessel function and P_{III} played corresponding roles.

10. Kac's formulation will come up shortly in the introduction. More generally, these are called Weyl asymptotics. See also, for an account in terms of the Langlands Program, J. Arthur, "Harmonic Analysis and Group Representations," *Notices of the American Mathematical Society* 47 (2000): 26–34. We are here talking about:

the Hilbert-Polya philosophy: the zeros of the Riemann zeta function are "explained" by an infinite dimensional ("Frobenius") Hermitian operator or unitarizable representation (of some "Galois" group).

and the remarkable story:

F. Dyson, upon hearing about Montgomery's work [on the two-point correlation function for the zeros of the Riemann zeta function], immediately wrote down a formula for the two-point function and asked Montgomery, "Did you obtain this?" Montgomery was astounded, and wanted to know how Dyson knew the answer. "Well," said Dyson, "if the zeros of the zeta function behave like the eigenvalues of a random matrix, then the two point function would have to be . . . Suddenly, the scattering of neutrons off nuclei had something to do with zeros of the Riemann zeta function.

D. Joyner, *Distribution Theorems of L-functions* (Essex: Longmans, 1986), p. ix; P. Deift, "Integrable Systems and Combinatorial Theory," *Notices of the American Mathematical Society* 47 (2000): 631–640, at p. 635.

11. The chart below summarizes some of these questions:

Riemann, Reciprocity	Local Counting, packaged into a Partition Function (L-function, zeta function)	PF's exhibit reciprocity. Scaling. [as in $\zeta(s) \approx \zeta(1-s)$, $\theta(t) \approx \theta(1/t)$]
Langlands TaniyamaShirmura, EichlerShiumura	Number of solutions modulo n, of equation or elliptic curve, E, $= b_n \rightarrow$	$= a_n$: Fourier of automorphic form, f: $E = (\sigma(t), \tau(t))$; $1/\sigma \times d\tau/d\sigma = f = \Sigma\, a_n q^n$, $q = \exp 2\pi i x$
Langlands	Permutations (Galois)	Harmonic analysis
Montroll, Onsager,	Trace V, is equal to $\rightarrow$	$\det A = \rightarrow$Laplace Transform of *Density of States*
Weyl Asymptotics	$\Sigma_i\, (Spectrum_i)^n$, of differential operator $\rightarrow$	Equals area, volume, kinetic energy

12. *The Philosophical Writings of Descartes*, tr. J. Cottingham, R. Stoothoff, D. Murdoch (Cambridge: Cambridge University Press, 1985), volume 1, p. 9.

13. For a survey, see, M. Friedman, "On the Sociology of Scientific Knowledge and its Philosophical Agenda," *Studies in the History and Philosophy of Science* 29 (1998): 239–271; Ian Hacking's introduction to J.Z. Buchwald, ed., *Scientific Practice: Theories and Stories of Doing Physics* (Chicago: University of Chicago Press, 1995). Hacking compares Peter Galison's and Andrew Pickering's theories. As for the studies of mathematics, I refer to particular studies as appropriate throughout. I have found the work of Penelope Maddy comfortable, and Frank Quinn on rigor makes good sense. The work in P. Mancosu, ed. *The Philosophy of Mathematical Practice* (New York: Oxford University Press, 2008) and his introduction (pp. 1–21) poses some of the question I offer here. But what is remarkable about all this work is how the mathematics examples are used for philosophical inquiry, and so less attention is paid to the gritty details of the actual work of the mathematician.

14. *The Princeton Companion to Mathematics* edited by T. Gowers (Princeton: Princeton University Press, 2008) may be useful to some readers. E. Frenkel's *Love and Math* (New York: Basic Books, 2013) is a paean to the Langlands Program. The last chapter in George Mackey's *The Scope and History of Commutative and Noncommutative Harmonic Anaysis* (Providence, RI: American Mathematical Society, 1992) is an early inkling of the analogy and syzygy that informs chapter 5. David Corfield's *Toward a Philosophy of Real Mathematics* (Cambridge: Cambridge University Press, 2003), pp. 91–96, discusses Weil's threefold analogy.

PROLOG

1. G.-C. Rota and F. Polombi, *Indiscrete Thoughts* (Boston: Birkhäuser, 1997).

2. A. S. Besicovitch said, "A mathematician's reputation rests on the number of bad proofs he has given." In J. E. Littlewood, *A Mathematician's Miscellany* (Cambridge University Press, 1986), p. 59.

3. Weyl asymptotics, algebraic geometry, Maxwell's equations and the Standard Model, arithmetic algebraic geometry, C* algebras. The Langlands Program that connects properties of equations with properties of automorphic forms through Galois groups and representation theory.

4. In effect, A says that the trace of the group representation is equal to the partition function (or L-function), and C that the partition function is the Laplace transform of something like a density of states (ρ): $PF = \mathrm{Tr}\, V = \mathrm{Laplace}(\rho)$.

5. These themes are discussed or adumbrated in my earlier work, *Marginalism and Discontinuity*, *Doing Physics*, and *Constitutions of Matter*. Chapter 6 of the second edition of *Doing Physics* provides a nontechnical summary of what I have learned about mathematics.

CHAPTER 1

1. Here I follow Eric Livingston: ". . . a proof of a theorem is the pairing of an account (or description) of the proof with the lived-work (the practical actions and reasoning) of proving such that the proof-account is a practically precise description of the organization of lived-work required to prove that theorem." ("Natural Reasoning in Mathematical Theorem Proving," *Communication & Cognition* 38 (2005): 319–344.

My concern here is to show how a proof is organized to do what it claims to be doing, and to explain that doing. A term of art is that a proof is "surveyable."

Gian-Carlo Rota, in his descriptions of mathematical work, tried to characterize beauty in mathematics. Such a beautiful proof shows in the proof all that is needed to see that it is true. (*Indiscrete Thoughts* (Boston: Birkhäuser, 1997)). Frank Quinn has argued that rigor as a mathematical practice, enables all to appreciate a good proof, and it enables one to display errors more exactly. Quinn argues that modern mathematical practice is much less attached to physics, say, than it once was, less attached to concrete pictures or models, from outside the field. Mathematics is autonomous. See http://www.math.vt.edu/people/quinn /history_nature/nature0.pdf.

2. S.S. Abhyankar, "Galois Theory on the Line in Nonzero Characteristic," *Bulletin (New Series) of the American Mathematical Society* 27 (1992): 68–133, at p. 89, n. 38.

3. See, David Reed, *Figures of Thought: Mathematics and Mathematical Texts* (London: Routledge, 1995), for a more textual literary analysis.

4. Husserl did his dissertation under Weierstrass. Here is a contemporary mathematician's paraphrase of Husserl,

Conversely, we only speak of proof in the strict logical sense when an inference is suffused with insight. Much, to be sure, that is proposed as a proof, as a logical inference, is devoid of such insight, and may even be false. But to propose it is at least to make the claim that a logical relation could hold.

And,

The premises prove the conclusion no matter who may affirm the premises and the conclusion. A rule is here revealed which holds irrespective of the statements, bound by acts of "motivation"; it includes all statements. Such a rule becomes clear when we work out a proof, while we reflect on the contents of the statements and their "motivation" in the processes of inference and proof.

(E. Husserl *First and Second Logical Investigation: The Existence of General Objects, presented in contemporary English by Gian-Carlo Rota* Undated notes for MIT lectures (ca. 1990?), mimeo., section #3, "Two senses of "showing"," pp. 3–4.) Rota would surely have revised this "translation" were he to have published it.

The standard English translation is in E. Husserl, *Logical Investigations*, tr. J.N. Findlay from the Second German Edition (New York: Humanities Press, 1970), volume 1,

p. 271. I give the relevant passage below, largely to suggest how the text becomes rather more concrete when there is a particular audience envisioned (for Rota, MIT students, for whom mathematics is a lingua franca):

> ... conversely, we only speak of demonstration in the strict logical sense in the case of an inference which is or could be informed by insight. Much, no doubt, that is propounded as demonstrative or, in the simplest case, as syllogically cogent, is devoid of insight and may even be false. But to propound it is at least to make the claim that a relation of consequence could be seen to hold.

And,

> The premisses prove the conclusion no matter who may affirm the premisses and the conclusion, or the unity that both form. An ideal rule is here revealed which extends its sway beyond the judgements here and now united by 'motivation'; in supra-empirical generality it comprehends as such all judgements having a like content, all judgements, even, having a like form. Such regularity makes itself subjectively known to us when we conduct proofs with insight, . . .

5. See M. Steiner, *The Applicability of Mathematics as a Philosophical Problem* (Cambridge, Massachusetts: Harvard University Press, 1998). Steiner suggests that the interesting phenomenon is the modern use of mathematical analogies for suggesting and formulating physical laws, the formalism and the notation borrowed from mathematics as a ways of providing a structure for the physicist's theories. Jody Azzouni suggests that mathematics applies only approximately to the physical world, that not all mathematics finds application, and that the "implicational opacity" of mathematics itself (that the consequences of our mathematical theory are not at all so apparent ahead of time) is as important as the applicability questions. Moreover, the connections between fields of mathematics also exhibit that implicational opacity. ("Applying Mathematics: An Attempt to Design a Philosophical Problem," *Monist* 83 (2000): 209–227.)

6. H. Weyl, "David Hilbert and His Mathematical Work," *Bulletin of the American Mathematical Society* 50 (1944): 612–654, p 615. In context it reads:

> In his [Hilbert's] papers one encounters not infrequently utterances of pride in a beautiful or unexpected result, and in his legitimate satisfaction he sometimes did not give to his predecessors on whose ideas he built all the credit they deserved. The problems of mathematics [the quoted passage in the main text] . . . any particular science. Hilbert had the power to evoke this life; against it he measured his individual scientific efforts and felt responsible for it in his own sphere.

I have interpreted the phrase in terms of what follows it, rather than what precedes it.

7. See Mark Kac, "Can One Hear the Shape of a Drum?," *American Mathematical Monthly* 73 (1966): 1–23, who refers to Lipman Bers as well as Salomon Bochner. Weyl's asymptotics set the stage, and the problem has a long history. See, M. Gutzwiller, "The Origins of the Trace Formula," in H. Friedrich and B. Eckhardt, eds., *Classical, Semiclassical and Quantum Dynamics in Atoms*, Lecture Notes in Physics 485 (Berlin: Springer, 1997), pp. 8–28.

8. Hermann Weyl showed in about 1911 that you can surely infer the drum's size or area (larger drum surfaces allow for more and lower tones). And more recently Kac and many others have shown that you can also infer its perimeter (if those surfaces are very asymmetrical, they allow for fewer resonant frequencies), and even whether there are holes in the drumskin (which again allow for fewer resonant frequencies). Still, there are significant exceptional cases, with classes of same-sounding or "isospectral" drums have inequivalent geometries. See H. Weyl, "Über die asymptotische Verteilung der Eigenwerte," *Nachrichten der Königlichen Gesellschaft der Wissenschaften zu Göttingen. Mathematisch-Naturwissenschaftliche Klasse* (1911): 110–117; H. Weyl, "Ramifications, Old and New, of the Eigenvalue Problem," *Bulletin of the American Mathematical Society* 56 (1950): 115–139; Mark Kac, "Can One Hear the Shape of a Drum," who as well provides a history; H.P. McKean, Jr., and I.M. Singer, "Curvature and the Eigenvalues of the Laplacian," *Journal of Differential Geometry* 1 (1967): 43–69; J.P. Solovej, "Mathematical Results on the Structure of Large Atoms," *European Congress of Mathematics* (1996) vol II. (Progress in Mathematics 169, Basel: Birkhäuser, 1998), pp. 211–220, esp. pp. 216–220. As for the exceptional classes, see C. Gordon, D.L. Webb, and S. Wolpert, "One cannot hear the shape of a drum," *Bulletin (New Series) of the American Mathematical Society* 27 (1992): 134–138.

Further details:

(1) The total loudness or energy of a drum is proportional to the number of resonant frequencies. Notionally, this is an aspect the central limit theorem of statistics, that the variance (or an energy) scales with N—the main theme of chapter 2. Physicists would call this fact "equipartition of energy."

As we shall see in chapter 4, for actual situations this needs to be modified. The "uncertainty principle"—namely, one resonant mode per unit volume of phase space—becomes in Charles Fefferman's hands a matter of fitting objects into a volume, fitting boxes into/onto balls. A spiky volume in phase space might not accommodate any objects even if it itself is of large volume.

(2) Let us imagine a sibling or "dual" drum, whose resonant tones are slightly shifted from the original drum's tones—much as we might move a Manhattan street grid a half-block north and a half-block east. It can be shown that such a drum shares many features with its sibling. (McKean and Singer) Similarly, in physicists' models of a permanent magnet ("the Ising model of ferromagnetism," for which, see the appendix to chapter 3), there is a corresponding simple relationship between the thermodynamic free energy of that iron bar at high temperature to its free energy at low temperature. There is as well a related "modular" transformation—that there appears a multiplier or modulus as a result of the

transformation—of the elliptic functions, sn(k,u), connecting the high and low modulus (k and $1/k$) elliptic functions: sn(u,k)=($1/k$) sn(ku,$1/k$).

All three of these phenomena—duality in drums, duality on a lattice, and an inversion symmetry for elliptic functions—relating an object and its inversion, so to speak, are consequences of the Poisson sum rule of fourier analysis.

(3) What is perhaps even more remarkable is that the connection between the shape and the sound of the drum turns out to model as well the balance of forces within ordinary matter, so that such matter does not implode. Namely, the shape represents the potential energy due to the forces among electrons and protons, and the sound represents their kinetic energy, just the energy and motion that is needed to keep matter from becoming unstable and collapsing—the story of chapter 4. (What is crucial is that the total of the $N \times N$ electrical interactions of the particles, a potential energy, can be bounded by a term proportional to N alone, their kinetic energy. One also needs the fact that electrons are fermions, obeying the Pauli exclusion principle, their momentum and energy going up as they fill up the energy levels.)

(4) Finally, there would seem to be a remarkable connection between: (i) the shape of a drum, known topologically and by the kind and pattern of vibrations allowed upon it; (ii) the actual sound itself, accounted for in its so-called theta function or Laplace transform or heat expansion,

$$\theta(t) = \Sigma \; Loudness \times \exp\,(- \, t \times Frequency),$$

t being both a parameter in a generating function and the time, where the sum is over all resonant tones, and by *Loudness* I mean the energy in a particular sound (This is, as well, essentially the formula for the temperature of a hot bar of material as it cools down due to heat conduction, with a suitable interpretation of *Loudness* and *Frequency*.); and, (iii) an accounting or "partition" function which encodes information about its resonant frequencies and their loudness, the zeta function or Mellin transform,

$$\zeta(s)=\Sigma \; Loudness/(Frequency)^s,$$

just the analogy that informs chapter 5. (Weyl, "Ramifications, Old and New, of the Eigenvalue Problem")

A propos of the analogies in chapter 5, " . . . in order to be able to cultivate the fields of analytic number theory [and the partition functions that package those combinatorial facts], as a prerequisite, one must acquire considerable skill for counting integral points [say, intersections in the street grid] inside various regular domains and their borders." H. Iwaniec, "Harmonic Analysis in Number Theory," *Prospects in Mathematics*, ed. H. Rossi (Providence: American Mathematical Society, 1999), pp. 51–68, at p. 53. In fact, Fefferman and Seco's proof, discussed in chapter 4, at a crucial point depends on such counting, to check that an approximation is a good one.

9. Recall that means are arithmetic averages ($\langle x \rangle = \Sigma x_i/N$, where N is the number of observations, and $\langle \; \rangle$ means averaging); variances are averages of the squared-deviation

from the mean (namely, the square of the standard deviation (σ), the variance $= \sigma^2 = \langle((\langle x \rangle - x)^2) \rangle$); and the Gaussian or normal distribution is the bell-shaped curve.

As for taking hold of the world, see Martin H. Krieger, *Doing Physics: How Physicists Take Hold of the World* (Bloomington: Indiana University Press, 1992, 2012).

10. For an example in the physical sciences, see the summary of statistics provided by the Particle Data Group, responsible for evaluating fundamental data in the field of particle physics. J. Beringer, et al, (Particle Data Group), "The Review of Particle Physics," *Physical Review* D86, 010001 (2012). Actual experimental work demands somewhat more sophisticated statistics (yet still within the L_2 realm), and arguments about employing Bayesian or frequentist methods are practical ones. See, F. James, L. Lyons, and Y. Perrin, *Workshop on Confidence Intervals*, Report 2000-005 (Geneva: CERN, 2000) and G.J. Feldman and R.D. Cousins, "Unified approach to the classical statistical analysis of small signals," *Physical Review* D57 (1998): 3873–3889.

11. T. Porter, *The Rise of Statistical Thinking* (Princeton: Princeton University Press, 1986); S. Stigler, *The History of Statistics Before 1900* (Cambridge, Massachusetts: Harvard University Press, 1986).

12. P.J. Huber, "Robust Statistics: A Review," *Annals of Mathematical Statistics* 43 (1972): 1041–1067.

13. F. Mosteller and J.W. Tukey, *Data Analysis and Regression* (Reading, Massachusetts: Addison-Wesley, 1977), p. 10. R. Wilcox, *Fundamentals of Modern Statistical Methods: Substantially Improving Power and Accuracy* (New York: Springer, 2010).

14. B. Efron and R. Tibshirani, "Statistical Data Analysis in the Computer Age," *Science* 253 (1991): 390–395.

15. See, for example, David Donoho, "High-Dimensional Data Analysis: The Curses and Blessings of Dimensionality," ms. (Stanford University, August 8, 2000), or obituaries for John Tukey. Donoho suggests that from some points of view the data analysis movement has been quite successful, but now it is time to get back to proving theorems to enable it even further. On broad distributions, see chapter 2, note 13.

16. For a bit more on "why" algebra and topology are useful for each other, see note 1, chapter 5, and note 96(8), chapter 6.

If a field of mathematics is defined by its methods, then we might say that the methods of topology are at first sight incompatible in style and technique. If a field were defined by its object alone, then this issue might not come up. So number theory uses diverse methods to study numbers and their extensions. These points are made by H. Hida, *Modular Forms and Galois Cohomology* (New York: Cambridge University Press, 2000), p. 1.

I have been encouraged to separate the topological method (whether general or combinatorial-algebraic) from the subject of topology. But I think this separation makes it harder to see the connections between the various subfields, even if one might well teach a course on the methods alone, for example.

17. P. Maddy, in *Naturalism in Mathematics* (Oxford: Clarendon, 1997) describes this strategy for set theory.

It is tricky to take quotes from Wittgenstein out of context, but it may not be too farfetched to believe that he is saying much the same here:

In mathematics there can only be mathematical troubles. There can't be philosophical ones. (p. 369)

If you want to know what the expression "continuity of a function" means, look at the proof of continuity. (p. 374)

The philosophy of mathematics consists in an exact scrutiny of mathematical proofs—not in surrounding mathematics with a vapour. (p. 367)

Nothing is more fatal to philosophical understanding than the notion of proof and experience as two different but comparable methods of verification. (p. 361)

L. Wittgenstein, *Philosophical Grammar*, ed. Rush Rhees, trans. A. Kenny, (Berkeley: University of California Press, 1974).

18. J. Munkres, *Topology: A First Course* (Englewood Cliffs: Prentice-Hall, 1975), is the source of this list of theorems.

19. The crucial figures here are Felix Hausdorff (1914), L.E.J. Brouwer (ca. 1912), and Henri Poincaré (1895). While Weierstrass was one of the earliest users of open sets (his were connected), it is only with Lebesgue and ultimately Hausdorff that the modern notion is codified.

20. There surely are other possible accounts of the fundamental notions of topology, much as textbooks in the field have very different starting points. Munkres begins his text on algebraic topology (*Elements of Algebraic Topology*) with simplicial complexes, W. Fulton (*Algebraic Topology: A First Course* (New York: Springer, 1995)) ends with them. Good textbooks are philosophies of mathematics.

I have not focused on homotopy and covering spaces, explicit combinatorial group theory, or matters of homeomorphism, Betti numbers and torsion, and even homology and invariants under continuous mapping. I have not really discussed low-dimensional topology, or differential topology, nor have I been explicitly concerned with purely combinatorial aspects. And I have only hinted at the machinery typically developed in this field.

21. L.E.J. Brouwer, "Zur Analysis Situs," *Mathematische Annalen* 68 (1910): 422–434. The diagram is reprinted on pp. 354–355 of Brouwer, *Collected Works*, volume 2. If the diagram would seem to suggest an abstract painting, that is perhaps not so farfetched. The art historian Meyer Schapiro relates a story about a lecture on a Picasso drawing in which a human figure becomes a tangle of lines (*Painter and His Model Knitting*, 1927, an illustration for Balzac's *The Unknown Masterpiece*). He said,

A mathematician can't arrive at the tangle of Picasso, there's no familiar geometrical theorem that would transform the appearance of a human body at rest into a tangle.

A mathematician (Dresden) in the audience pointed out that:

there is a projective theorem of the great Dutch mathematician Brouwer, through which he was able to obtain a tangle out of any kind of three-dimensional compact, closed field.

And then Schapiro goes on:

I was delighted later to learn that Brouwer was also the author of an important treatise on painting and aesthetics around 1908 or 1909 [actually 1905, "Life, Art and Mysticism"], which was known to Mondrian.

M. Schapiro, "Philosophy and Worldview in Painting [written, 1958–1968]," in *Worldview in Painting, Art and Society: Selected Papers* (New York: Braziller, 1999), pp. 11–71, at p. 30. It is not clear which theorem is being referred to: possibly, the indecomposable plane; or, perhaps the result of Brouwer's on there being no continuous one-to-one mapping of dimension n to dimension m, if m and n differ. As Mark van Atten and Dirk van Dalen point out to me, the latter is not in accord with Schapiro's account of Dresden's statement, but is in accord with Schapiro's first statement. Dresden was well acquainted with Brouwer's work, so perhaps Schapiro's account of what Dresden said is not quite correct.

22. See H. Freudenthal's note in L.E.J. Brouwer, *Collected Works*, volume 2, ed. H. Freudenthal (Amsterdam: North Holland, 1976), pp. 367–368, which gives Yoneyama's 1917 account of Wada's version of the Brouwer construction. (K. Yoneyama, "Theory of continuous sets of points," *Tôhoku Mathematics Journal* 12 (1917): 43–158, especially pp. 60–62—parts of which pages, including the diagram, are reproduced in the Brouwer volume.) It is interesting that land/water analogies come up again in, L.E.J. Brouwer, "On continuous vector distributions on surfaces, III" *KNAW Proceedings* 12 (1910): 171–186, pp. 303–318 in Brouwer, *Collected Works*, volume 2, where he refers to and illustrates two sorts of configurations: an "irrigation territory" (p. 306) and a "circumfluence territory" (p. 312), very roughly, a source that flows outward, and a source whose set of flows have vorticity and return to the source, respectively. (Brouwer is Dutch.)

23. The phrase "Cantor-Schönflies" is taken from Freudenthal's commentary in Brouwer, *Collected Works*, volume 2, p. 206, where he distinguishes this "point-set topology of the plane" style from slightly later work which employs "set theory topology mixed with homological and homotopical topology such as developed by Brouwer himself." The diagram is from p. 794 of "Continuous one-one transformations of surfaces in themselves," *KNAW Proceedings* 11 (1909): 788–798 (p. 201 in Brouwer, *Collected Works*, volume 2).

24. J. Dieudonné, *A History of Algebraic and Differential Topology* (Boston: Birkhäuser, 1989), p. 161.

25. Brouwer begins with the index of a vector field, a measure of vorticity, and then progresses to the analogous degree of a mapping, again, the n in z^n. See H. Freudenthal in Brouwer, *Collected Works*, volume 2, pp. 422–425.

26. H. Freudenthal in Brouwer, *Collected Works*, volume 2, p. 438, quoting a commonplace of around 1930.

27. J. Rotman, "Review of C.A. Weibel, *Introduction to Homological Algebra*," *Bulletin (New Series) of the American Mathematical Society* 33 (1996): 473–476, at p. 474.

28. F. Dyson, p. vii, in E. Lieb, *The Stability of Matter from Atoms to Stars: Selecta of Elliott H. Lieb*, ed. W. Thirring, edition 2 (Berlin: Springer, 1997).

29. The chemist Lars Onsager took the first major step in each case, in 1939 and in 1944. The modern proofs of the stability of matter, that matter composed of nuclei and electrons won't collapse or explode, started out with M. Fisher and D. Ruelle setting the problem, and then F. Dyson and A. Lenard's "hacking," which was reconceptualized and clarified by E. Lieb and W. Thirring's deeply physical proof, and then C. Fefferman and collaborators sought greater precision in order to show that matter can exist as a gas of atoms. In Fefferman and L. Seco's hundreds of pages of refined classical analysis, greater precision is sought for asymptotic estimates for the binding energy of isolated atoms. The studies of the two-dimensional Ising model were marked by L. Onsager's exact solution and C.N. Yang's derivation of the spontaneous magnetization, both of which are usually taken as "too arduous" to follow; then T. Schultz, D. Mattis, and Lieb's very physical and mathematically sweet account, a now-preferred "translation," so to speak; R. Baxter's remarkable exact but admittedly not always rigorous solutions, perhaps initially seen as ingenious and idiosyncratic; and, T.T. Wu, B. McCoy, C. Tracy and their collaborators' work on the large-distance correlation of atoms (the asymptotic limit, large N, scaling limit), forbidding in its monumental equations. Wu, McCoy, C. Tracy, and E. Barouch's 1976 paper is notable for its length and the complexity of the calculation. And, to be sure, I have not mentioned here many other formidable contributions.

I am told it will be decades before the mathematical community absorbs Fefferman and Seco's techniques. On the other hand, there are simplifications and improvements in estimating the asymptotic form, due to Bach, and to Graf and Solovej. There are by now many simplifications for the various Ising-model proofs or derivations. And many of the combinatorial aspects of that Ising model have resonances with other combinatorial problems (as in playing solitaire, or sorting a deck of cards, or scattering of particles on a line). And many of the analytical aspects have resonances with other analysis problems (the scaling behavior of functions of a complex variable, and the corresponding elliptic curves and topological objects).

30. C.N. Yang, "Journey Through Statistical Mechanics," *International Journal of Modern Physics* B3 (1988): 1325–1329.

31. P.D. Lax, in "Jean Leray (1906–1998)," *Notices of the American Mathematical Society* 47 (March 2000): 350–359, at p. 353.

32. This is taken from the *Oxford English Dictionary* (1929), under "Syzygy," definition 5. The OED also quotes: "[U, V, W] are . . . capable of being connected by integral multipliers U', V', W', such that $UU' + VV' + WW' = 0$. Any number of functions U, V, W, so related I call syzygetic functions." From J.J. Sylvester, in the *Cambridge and Dublin Math. Journal* 5 (1850): 262–282, both quotes at p. 276.

See also, *Collected Mathematical Papers of James Joseph Sylvester* (Cambridge: Cambridge University Press, 1904), vol. 1, pp. 131–132.

33. For a survey, see H. Benis-Sinaceur, "The Nature of Progress in Mathematics: The Significance of Analogy," in E. Grosholz and H. Breger, eds., *The Growth of Mathematical Knowledge* (Dordrecht: Kluwer, 2000) pp. 281–293. The crucial figures here are Hilbert, Poincaré, and Polya.

34. M.H. Krieger, *Constitutions of Matter* (Chicago: University of Chicago Press, 1996), pp. 311–312, which was inserted in proof. The classification of solutions in *Constitutions*, driven by the physical models, cuts across the classifications suggested in chapter 5 of this book, driven as it is in chapter 5 by mathematical formalism. I made this diagram at that time, a precursor to chapter 5:

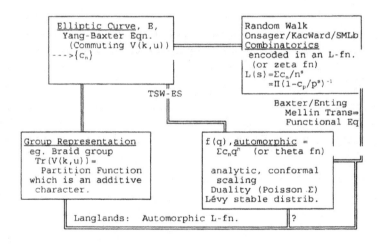

(TSW-ES=Taniyama-Shimura-Weil/Eichler-Shimura)

FIGURE N.1: 12 September 1996 Diagram of Connections.

35. For a history, see, F. Klein, *Vorlesungen Über die Entwicklung der Mathematik im 19. Jahrhundert* (New York: Chelsea, 1956 [1926, 1927, 1950]), pp. 321–335. R. Dedekind and H. Weber, "Theory of Algebraic Functions of One Variable," (1882) wanted to provide

purely algebraic proofs of Riemann's algebraic theorems. (R. Dedekind and H. Weber, "Theorie der algebraische Funktionen einer Veränderlichen," pp. 238–350 in Bd. 1, R. Dedekind, *Gesammelte mathematische Werke*, eds. R. Fricke, E. Noether, and O. Ore (Braunschweig: Vieweg, 1930). Originally, *Journal für reine und angewandte Mathematik ("Crelle")* 92 (1882): 181–290.) They found a suitable algebraic account of the apparently geometrical notion of points on a Riemann surface. And they analogized divisibility and primality in the ring of polynomials over the complex numbers to divisibility and primality in algebraic number fields. J. Dieudonné, *History of Algebraic Geometry* (Monterey: Wadsworth, 1985), pp. 29–30. See *Version H* in chapter 5.

Emmy Noether writes an appreciation of that paper in her commentary in the collection of Dedekind's work. E. Noether, "Erläuterungen zur verstehenden Abhandlung," p. 350, in R. Dedekind, *Gesammelte mathematischen Werke*, vol. 1.

36. This triple was in fact metaphorically called a Trinity by some scientists— a Son of conformal fields, a Holy Spirit of group representations, and a Father of combinatorial exact solutions. E. Date, M. Jimbo, T. Miwa, and M. Okado, "Solvable Lattice Models," in, L. Ehrenpreis and R.C. Gunning, eds., *Theta Functions, Bowdoin 1987*, Proceedings of Symposia in Pure Mathematics 49 (Providence: American Mathematical Society, 1989), pp. 295–331.

37. A. Knapp, *Elliptic Curves* (Princeton: Princeton University Press, 1992), pp. 407–408, recommends the essay by R. Langlands, "Representation Theory: Its Rise and Role in Number Theory," in D.G. Caldi and G.D. Mostow, eds., *Proceedings of The Gibbs Symposium* (Providence: American Mathematical Society, 1990), pp. 181–210.

38. E. Frenkel, *Love and Math* (New York: Basic Books, 2013) indicates the connection of the Langlands Program with quantum field theory and what is called Mirror Symmetry (much as electric and magnetic fields in Maxwell's theory are related). James Arthur, in his talk at the ICM2014, as given in the video, discusses particles and composites and even gauge groups: https://www.youtube.com/watch?v=ljJudppxau4. See also, Langlands, "Representation Theory," where he also talks of particles.

39. Wigner, E. P., "The unreasonable effectiveness of mathematics in the natural sciences. Richard Courant lecture in mathematical sciences delivered at New York University, May 11, 1959," *Communications on Pure and Applied Mathematics* 13 (1960): 1–14. A. De Morgan (in *A Budget of Paradoxes* (Open Court, 1915), vol. 1, p. 285.) earlier told a story similar to Wigner's about the appearance of pi. Most articles discussing Wigner's focus on the title, rather than the text of the article.

40. B.M. McCoy, "Integrable models in statistical mechanics: The hidden field with unsolved problems," *International Journal of Modern Physics A* 14 (1999) 3921-3933.

41. B.M. McCoy, "The Baxter Revolution," *Journal of Statistical Physics* 102 (2001): 375–384.

42. R.P. Langlands, "Representation Theory: Its Rise and Role in Number Theory."

43. Within mathematics, within its abstractions, I shall suggest in chapter 5 that analogy is destiny, some of the time. But mathematics itself, as an activity, cannot be so abstracted.

CHAPTER 2

1. "Tout le monde y croit (la lois des erreurs) parce que les mathématiciens s'imaginent que c'est un fait d'observation, et les observateurs que c'est un théorème de mathématique." Quoted from Poincaré, Preface to *Thermodynamique*, in M. Kac, *Enigmas of Chance* (Berkeley: University of California Press, 1985, 1987), p. 48.

Another version is, ". . . there must be something mysterious about the normal law since mathematicians think it is a law of nature whereas physicists are convinced that it is a mathematical theorem." Quoted in M. Kac, *Statistical Independence in Probability, Analysis and Number Theory* (Providence: American Mathematical Society, 1959), p. 50.

2. Most explicitly, this is seen in E. Durkheim, *Elementary Forms of Religious Life*, tr. K. Fields (New York: Free Press, 1995 [1915]).

3. Imre Lakatos called some of these endeavors "scientific research programs," which have a hard core of assumptions and protective layer or belt of auxiliary assumptions. In Lakatos' terms, in this chapter I am describing the protective belt. For a nice discussion of this in the context of statistics see pp. 457–463 of A. Stuart, J.K. Ord, and S. Arnold, *Kendall's Advanced Theory of Statistics, Vol. 2A, Classical Inference and the Linear Model* (London: Arnold, 1999).

4. See R. Tragesser, *Husserl and Realism in Logic and Mathematics* (Cambridge: Cambridge University Press, 1984) for a nice example from topology. See also the quote from E.H. Moore in chapter 5, at note 36. Moore is especially concerned with when those models are extrapolations, and when they are just summaries of what is known.

5. B. Efron, "Bayesians, Frequentists, and Scientists," *Journal of the American Statistical Association* 100 (2005): 1–5.

6. For a history of the notion of plenitude, see A.O. Lovejoy, *The Great Chain of Being* (Cambridge, Massachusetts: Harvard University Press, 1936). As Kac points out (*Enigmas of Chance*), one also needs an axiom of additivity: if the events are mutually exclusive the probability of at least one of them happening is the sum of their individual probabilities.

7. P. Diaconis, S. Holmes & R. Montgomery, "Dynamical Bias in the Coin Toss," *SIAM Review* 49(2) (2007): 211–235.

8. M.H. Krieger, "Theorems as Meaningful Cultural Artifacts," *Synthese* 88 (1991): 135–154.

9. L. Daston, *Classical Probability in the Enlightenment* (Princeton: Princeton University Press, 1988) ; T. Porter, *The Rise of Statistical Thinking* (Princeton: Princeton University Press, 1986); J. von Plato, *Creating Modern Probability* (Cambridge: Cambridge University Press, 1994); S. Stigler, *The History of*

Statistics Before 1900 (Cambridge, Massachusetts: Harvard University Press, 1986).

10. Stigler, *The History of Statistics Before 1900*.

11. R.A. Fisher, "Theory of Statistical Estimation," *Proceedings of the Cambridge Philosophical Society* 22 (1925): 700–725, at pp. 701–702.

12. Stigler, *The History of Statistics Before 1900*; Porter, *Rise of Statistical Thinking*.

13. For a detailed discussion in the context of magnetization, see T.D. Schultz, D.C. Mattis, and E.H. Lieb, "Two-Dimensional Ising Model as a Soluble Problem of Many Fermions," *Reviews of Modern Physics* 36 (1964): 856–871.

Or, consider a system of atoms in excited states that decay exponentially in time, but their possible decay constants are distributed uniformly: $N_p(t)=N_0\exp{-pt}$, $p \in [0,1]$, say. Then the "typical behavior" for any single atom is an exponential decay. But the average behavior is in fact constant, for small t, while it is $1/t$ for large t (again, where we have now averaged over p). With lots of atoms, we measure the average; with one, we measure the typical behavior. Here, rare anomalous cases can be quite significant; averages do not tell you about individual atoms. Again, peculiar characteristics (what is called "scaling behavior") mark off this sort of situation, when in fact the mean is so very different than typical behavior.

See D.S. Fisher, "Critical behavior of random transverse-field Ising spin chains," *Physical Review* B51 (1995): 6411–6461. Fisher is quite aware he is up to something out of the ordinary, since he is concerned with situations where "the distributions of various physical quantities are extremely broad [say, over many energy scales, as in renormalization group calculations] with rare anomalous values dominating averages of them, but just such averages are what would be measured in macroscopic experiments."(p. 6411) In effect one goes from τ to $\ln \tau$ as one's physical variable (that is, one is in the scaling regime, τ^α). The physics of rare events might well be very different than that prescribed by the statistical ergodic assumptions usual in standard statistical mechanics and by the Central Limit Theorem.

See also, F. Bardou, J.-P. Bouchaud, A. Aspect, and C. Cohen-Tannoudji, *Lévy Statistics and Laser Cooling* (Cambridge: Cambridge University Press, 2002).

14. Stigler, *History of Statistics Before 1900*, pp. 36–91.

15. Stigler, *History of Statistics Before 1900*, chapter 8 on Galton.

16. The frontispiece of volume 5 of *Collected Papers of R.A. Fisher* has a photograph of Fisher at his electromechanical calculator. Recall that these calculators (manufactured by Marchant, Monroe, or Friden) had full keyboards, typically for ten digit numbers, and an accumulator and a multiplier display.

17. B. Efron and R. Tibshirani, "Statistical Data Analysis in the Computer Age," *Science* 253 (1991): 390–395.

18. R.A. Fisher, "Frequency distribution of the values of the correlation coefficient in samples from an indefinitely large population," *Biometrika* 10 (1915): 507–521; R.A. Fisher, "On the Mathematical Foundations of Theoretical Statistics," *Philosophical Transactions A222* (1922): 309–368, at p. 313; R.A. Fisher, "On the interpretation of chi-square from contingency tables, and the calculation of P," *Journal of the Royal Statistical Society* 85 (1922): 87–94; J.F. Box, *R.A. Fisher, the Life of a Scientist* (New York: Wiley, 1978), pp. 122–129; R.J. Wonnacott and T.H. Wonnacott, *Econometrics* (New York: Wiley, 1979), chapter 14.

19. V. Chvatal, *Linear Programming* (New York: Freeman, 1983), chapter 14.

20. R.A. Fisher, "A Mathematical Examination of the Methods of Determining the Accuracy of an Observation by the Mean Error, and by the Mean Square Error," *Monthly Notices of the Royal Astronomical Society* 80 (1920): 758–770; Fisher, "On the Mathematical Foundations of Theoretical Statistics;" Fisher, "On the interpretation of chi-square from contingency tables, and the calculation of P;" R.A. Fisher, "Theory of Statistical Estimation," *Proceedings of the Cambridge Philosophical Society* 22 (1925): 700–725.

21. R.A. Fisher, *Statistical Methods for Research Workers*, p. 15. Reprinted in R.A. Fisher, *Statistical Methods, Experimental Design, and Scientific Inference, A Re-issue of Statistical Methods for Research Workers [1925] (1973), The Design of Experiments [1935] (1971), and Statistical Methods and Scientific Inference [1956] (1973)*. Edited by J.H. Bennett. Oxford: Oxford University Press, 1990.

22. Fisher, "Theory of Statistical Estimation," p. 712.

23. Fisher, "Theory of Statistical Estimation," p. 714.

24. K. Baclawski, G.-C. Rota, and S. Billey, *An Introduction to the Theory of Probability* (Cambridge, Massachusetts: Department of Mathematics, MIT, 1989), pp. 151–154. Rota, in an essay "Twelve Problems in Probability No One Likes to Bring Up" (Fubini Lectures, Torino, June 3–5, 1998, ms.), has suggested the need for such a justification of the normal distribution and the Central Limit Theorem, perhaps as maximum entropy, with an account of speed of convergence; or, an axiomatics of confidence intervals. The essay is reprinted in H. Crapo and D. Senato, *Algebraic Combinatorics and Computer Science* (New York: Springer, 2001) pp. 57 ff.

25. Say we employ the simplex algorithm, and measures of tradeoff prices from optimality in the dual formulation. See, Chvatal, *Linear Programming*, chapter 14.

26. So R. Wilcox, *Fundamentals of Modern Statistical Practice* (New York: Springer, 2010) begins with the conventional story, pointing out its limits, and then goes on to describe such robust and resistant methods. They increase one's capacity not to miss an effect ("power").

27. Here I have followed David Donoho, who writes a confession of his past work in data analysis (ala Tukey) and future work on mathematical statistics. D. Donoho, "High–Dimensional Data Analysis: The Curses and Blessings of Dimensionality" (Stanford University, August 8, 2000, ms.). http://statweb.stanford.edu/~donoho/Lectures/AMS2000/Curses.pdf

28. Stigler, *History of Statistics Before 1900*, chapter 8.

29. W. Feller, *An Introduction to Probability Theory and Its Applications* (New York: Wiley, 1968 (volume I); 1971 (volume II)), volume 1, pp. 261–262. A similar role is played by the Wigner semicircle theorem for random matrices and related phenomena. http://terrytao.wordpress.com/2010/01/05/254a-notes-2-the-central-limit-theorem/ ; http://terrytao.wordpress.com/2010/02/02/254a-notes-4-the-semi-circular-law/.

30. M. Loève, *Probability Theory* (Princeton: Van Nostrand, 1960), p. 335.

31. Lévy's student Mandelbrot tries to invert this orthodoxy. See B. Mandelbrot, *Fractal Geometry of Nature* (San Francisco: Freeman, 1982). And the literature on chaos is an attempt to change the emphasis from variances to how measures of fluctuation scale with size.

32. See M.H. Krieger, "Theorems as Meaningful Cultural Artifacts," *Synthese* 88 (1991): 135–154; also, M.H. Krieger, *Marginalism and Discontinuity: Tools for the Crafts of Knowledge and Decision* (New York: Russell Sage Foundation, 1989).

33. P. Samuelson, "The Fundamental Approximation Theorem of Portfolio Analysis in Terms of Means, Variances and Higher Moments," *Review of Economic Studies* 37 (1970): 537–542; R.C. Merton, *Continuous-Time Finance* (Cambridge, Massachusetts: Blackwells, 1990), pp. 467–491.

34. Feller, *An Introduction to Probability Theory*, volume 2, p. 172.

35. Feller, *An Introduction to Probability Theory*, volume 2, pp. 169–176.

36. Feller, *An Introduction to Probability Theory*, volume 2, section VI.2; Bardou et al, *Lévy Statistics and Laser Cooling*. In the latter case one has distributions with $P(t) \approx 1/t^{1+\alpha}$.

37. Feller, *An Introduction to Probability Theory*, volume 1, section VIII.5.

38. Technically, it would seem that the Gaussian character of the sum distribution is crucial, or at least that its upper tail fall off as does a Gaussian's. For when one examines the proof of the law of the iterated logarithm, what is crucial is the finiteness or not of a sum of probabilities—so being able to draw conclusions

from the Borel-Cantelli lemma. That summability turns out to be equivalent to the finiteness or not of the sum of a harmonic series, $1/n^\beta$. Now, one of the logarithms is immediately needed to reexpress the sum of probabilities as the sum of a harmonic series. The other logarithm is needed in an approximation of the integral of the tail of the Gaussian. Namely, the integral of the tail of the Gaussian goes as $(1/x)\exp{-x^2/2}$ which is less than $\exp{-x^2/2}$ for x large, and if $x=\sqrt{(2 \beta \log \log N)}$ then the tail equals $\exp(-\beta \log \log N)$. If N is set equal to μ^r, r integer, then $(\log N)^{-\beta}$ becomes $\approx 1/r^\beta$, a harmonic series.

39. R.A. Fisher, "On the Mathematical Foundations of Theoretical Statistics," p. 313.

40. F. Mosteller and D.L. Wallace, *Inference and Disputed Authorship: The Federalist*. Reading, Massachusetts: Addison-Wesley, 1964. They estimate the odds of a particular event to be 60,000 to 1, based on the quality of their model and multiple modes of estimation. Still, they might be subject to "outrageous events":

> Clearly, frauds, blunders, poor workmanship, and coincidences snip away at very high odds, and the reader has to supply the adjustment for these. The chance of being brought to what we call "roguish ruin" [namely, large negative expectations] by one or another sort of outrageous event is sufficient large that final odds of millions to one cannot be supported, but these long odds can be understood in their own place within the mathematical model. (p. 91)

They trust themselves, but not millions to one.

41. Or, that we are seeing a blip in a Poisson process, and once we know about the correct average rate or probability, there is nothing special here or in other such extreme-order statistics; or, that there is some clumping in this Poisson process, or a combination of several different Poisson processes; or, given a small sample, that confidence intervals have to be much wider than we would initially expect; or, that the approximation we are using, say the Gaussian, is systematically too low, and were we to conceive of the longshot differently, say as a large deviation measured by an entropy, we might estimate its probability more correctly. These observations are elaborated on in, M.H. Krieger, "Apocalypticism, One-Time Events, and Scientific Rationalty," in S.S. Hecker and G.-C. Rota, eds., *Essays on the Future, In Honor of Nick Metropolis* (Boston: Birkhäuser, 2000), pp. 135–151. As for coincidences, see W. Kruskal, "Miracles and statistics: the casual assumption of independence," *Journal of the American Statistical Association* 83 (1988): 929–940; P. Diaconis and F. Mosteller, "Methods for studying coincidences," *Journal of the American Statistical Association* 84 (1989): 853–861.

42. L. Dubins and L.J. Savage, *Inequalities for Stochastic Processes* (New York: Dover, 1976). Otherwise you will be surely killed by asymptotics.

43. Paul Cohen's notion of a generic real, the foundation of his forcing method in his work on the Continuum Hypothesis, is much like this number.

44. As for what we might mean by "real:" When physicist measure the mass and lifetime (or width in the mass distribution) of a particle, both are actual physical facts, not statistical, although each will have both systematic and statistical errors. Those measurements will determine whether a particle has a nonzero mass (as for the neutrino), or whether two particles have a mass difference (as in the K^0 system).

But for much of social science and economics, whatever is measured is an artifactual property of the combination of observations (the average height of a population), but as far as I can tell has no other claim to real-ness. When the means of two populations are compared using the standard error of the mean, you may well get a significant difference. But if the width of the distributions is much larger than the standard error of the means, we might be much less confident in the claim of a difference. And that width of a distribution may be part of an analysis of variance, but until you have a theoretical structure where the exact value of those measurements matter (and I do not mean a regression equation or an economic model), you are in a very different realm than the physicist's.

45. It is R. A. Fisher (1925) who is crucial here. H. Scheffé, "Alternative Models for the Analysis of Variance," *Annals of Mathematical Statistics* 27 (1956): 251–271, at pp. 255–256; C. Eisenhart, "The Assumptions Underlying the Analysis of Variance," *Biometrics* 3 (1947): 1–21.

46. ". . . the separation of the variance ascribable to one group of causes from the variance ascribable to other groups." Fisher, *Statistical Methods for Research Workers*, p. 213, also p. 120. See also, Fisher, *The Design of Experiments*, pp. 52–58.

47. Fisher, *Statistical Methods for Research Workers*, p. 120.

48. R. Levins and R. Lewontin, *The Dialectical Biologist* (Cambridge, Massachusetts: Harvard University Press, 1985), chapter 4.

49. P. Kennedy, *A Guide to Econometrics* (Cambridge, Massachusetts: MIT Press, 1985), p. 188; Eisenhart, "The Assumptions Underlying the Analysis of Variance."

50. We might think of fractals as being when typical and average differ, although different measures of central tendency can correct this (albeit it is not the average).

51. All are described by a ("parabolic") second order partial differential equation. Newtonian mechanics and Maxwellian electrodynamics are first order systems, ordinary and partial.

52. Feynman R.P. and A.R. Hibbs, *Quantum Mechanics and Path Integrals* (New York: McGraw Hill, 1965); R.P. Feynman, R.B. Leighton, and M. Sands, *The Feynman Lectures on Physics*, volume 1 (Reading, Massachusetts: Addison Wesley, 1963); J.A. Wheeler and R.P. Feynman. "Interaction with the Absorber as the Mechanism of Radiation," *Reviews of Modern Physics* 17 (1945): 157–181, at pp. 170–171; D. Chandler, *Introduction to Modern Statistical Mechanics* (New York: Oxford University Press, 1987), chapter 8.

M.H. Krieger, *Doing Physics: How Physicists Take Hold of the World* (Bloomington: Indiana University Press, 1992, 2012), p. 169 n. 22 in 2012 edn; M.H. Krieger, *Constitutions of Matter: Mathematically Modeling the Most Everyday of Physical Phenomena* (Chicago: University of Chicago Press, 1996), pp. 168–171.

53. L. Onsager, "Reciprocal Relations in Irreversible Processes, I," *Physical Review* 37 (1931): 405-426.

54. F. Black, "Noise," in *Business Cycles and Equilibrium*. New York: Blackwell's, 1987, pp. 152–172.

55. Some of these insights came from an essay on social variation and fit, by Harrison White, of perhaps thirty-plus years ago; A.E. Roth and M. Sotomayor, *Two-Sided Matching* (New York: Cambridge University Press, 1990), on matching medical students with residencies.

56. Of course the distribution of sizes need not be Gaussian, and so a different measure of width would be relevant.

57. I leave out here such emblematic figures as Wiener, Langevin and Doob (stochastic differential equations), Borel and Steinhaus (series of random functions), Bachelier and Kolmogorov (diffusion), and Pearson and Polya (random walk), to use the list in J.-P. Kahane, "Le mouvement brownien," in *Matériaux pour L'Histoire des Mathématiques au XXe Siècle*. Séminaires et Congrès, 3 (Paris: Société Mathématique de France, 1998), pp. 123–155.

58. A. Pais, *"Subtle is the Lord . . .": The Science and Life of Albert Einstein* (New York: Oxford University Press, 1982), p. vii. Einstein applied his analysis of fluctuations to electromagnetic radiation in 1909. In 1917 he analyzed transitions between atomic states (stimulated emission and absorption, as well as spontaneous emission) in terms of thermodynamic equilibrium.

59. Fisher, "On the Mathematical Foundations of Theoretical Statistics," pp. 311–312.

60. R.A. Fisher and W.A. MacKenzie, "Studies in Crop Variation I: The Manurial Response of Different Potato Varieties," *Journal of Agricultural Science* 13 (1923): 311–320; R.A. Fisher, "On a distribution yielding the error functions of several well known statistics," in *Proceedings of the International Mathematical Congress, Toronto, 1924*, ed. J.C. Fields (Toronto: University of Toronto Press, 1928), volume 2, pp. 805–813.

Fisher, in "A Mathematical Examination of the Methods of Determining the Accuracy of an Observation by the Mean Error, and by the Mean Square Error," an article on the various mean errors, works out a lovely example of sufficiency, comparing the mean and the median of the Gaussian (L_2 vs. L_1 measures), the former being sufficient, possessing all the information in a sample, the latter being less than sufficient. What was innovative here was not the calculation but rather the observation about information. See also, "Fisher, R.A.," *Biographical Memoirs of Fellows of the Royal Society of London* 9 (1963): 91–120.

61. Now, if we continuously keep observing or measuring some thing, then fluctuations are precluded and quantum mechanical transitions that depend on fluctuations are prevented. And if we attend to the net payoff of gaining statistical information, as in decision-theoretic formulations, then Fisher's criteria are no longer so absolute. In each case, variation becomes real in a new way. For variation is no longer so untouched by our inquiry, as it once might have seemed. Of course, Einstein and Fisher in their developments of quantum ideas and small sample statistics, respectively, may be said to have laid the foundations for our modern conceptions.

CHAPTER 3

1. H. Weyl, "Topology and Abstract Algebra as Two Roads of Mathematical Comprehension," *American Mathematical Monthly* 102 (1995, originally appearing in 1932), 453–460 and 646–651, at pp. 651 and 454.

2. H. Weyl, "A Half-Century of Mathematics," *American Mathematical Monthly* 58 (1951): 523–553, at p. 548.

3. The seventeenth century mathematical or natural philosophers found the same problems in their descriptions of God's infinity and indivisibility (for which, see chapter 6).

4. A modern version is, N.E. Burkert, *Valentine and Orson* (New York: Farrar, Straus and Giroux, 1989).

5. S. Eilenberg and N. Steenrod, *Foundations of Algebraic Topology* (Princeton: Princeton University Press, 1952), p. xi—quoted in L. Corry, *Modern Algebra and the Rise of Mathematical Structures* (Basel: Birkhäuser, 1996), p. 367.

6. See, for example, R. Palais, "The Visualization of Mathematics," *Notices of the American Mathematical Society* 46 (1999): 647–658.

7. L.A. Steen and J.A. Seebach, Jr., *Counterexamples in Topology* (New York: Dover, [1970, 1978] 1995), p. iv.

8. The origins of rubber sheet topology in complex variable theory, and especially in Riemann's work, is displayed in Hermann Weyl's *The Concept of a Riemann Surface* (Reading, Massachusetts: Addison-Wesley, (English Translation) 1964 [1955, 1913]), and earlier in Felix Klein's *On Riemann's Theory of Algebraic Functions and Their Integrals: A Supplement to the Usual Treatises* (New York: Dover, 1963 [1882, 1893]). This is perhaps most vivid in Weyl's later edition, since the revision incorporates much of the earlier material but with a more modern topological viewpoint guiding the exposition.

9. J. Barrow-Green, *Poincaré and the Three Body Problem* (Providence: American Mathematical Society, 1997), pp. 30–39 and 166–167.

10. Namely, the laws of Kirchoff, in G. Kirchoff (Studiosus), "Ueber den Durchgang eines elektrischen Stromes durch eine Ebene, insbesondere durch eine kreisförmige," *Annalen der Physik und Chemie* 64 (1845): 497–514, especially, 513–514; and, "Ueber der Auflösung der Gleichungen, auf welche man bei der Untersuchung der linearen Vertheilung galvanisher Ströme gefürt wird," *Annalen der Physik und Chemie* 72 (1847): 497–508.

11. R. Engelking, *General Topology* (Berlin: Heldermann, 1989).

12. J. Munkres, *Topology, A First Course* (Englewood-Cliffs: Prentice Hall, 1975), p. xvi.

13. F.W. Lawvere, "Unity and Identity of Opposites in Calculus and Physics," *Applied Categorical Structures* 4 (1996): 167–174; K. Marx, *Mathematical Manuscripts of Karl Marx* (London: New Park, 1983); C. Smith, "Hegel, Marx and the Calculus," in *Mathematical Manuscripts of Karl Marx*.

14. George Mackey has argued this most eloquently in his *The Scope and History of Commutative and Noncommutative Harmonic Analysis* (Providence: American Mathematical Society, 1992).

15. R. Cooke, "Uniqueness of Trigonometric Series and Descriptive Set Theory, 1870–1985," *Archive for the History of the Exact Sciences* 45 (1993): 281–334; A.S. Kechris and A. Louveau, "Descriptive Set Theory and Harmonic Analysis," *Journal of Symbolic Logic* 57 (1992): 413–441.

16. Grothendieck's program of *dessins d'enfants* is called "un jeu de Légo-Teichmüller." See, L. Schneps, ed., *The Grothendieck Theory of Dessins d'Enfants* (Cambridge: Cambridge University Press, 1994).

17. "Topologie ist Stetigkeitsgeometrie." This is the first sentence of the introduction to P. Alexandroff and H. Hopf, *Topologie* Bd. 1 (Berlin: Springer, 1935). They continue:

sie handelt von denjenigen Eigenschaften geometrischer Gebilde, welche bei *topologischen*, d. h. eineindeutigen und in beiden Richtungen stetigen, Abbildungen erhalten bleiben—von Eigenschaften also, welche jedenfalls nichts mit Grössenverhältnissen zu tun haben—, und sie handelt auch von den stetigen Abbildungen selbst.

18. H. Weyl, *Philosophy of Mathematics and Natural Science* (Princeton: Princeton University Press, [1927] 1949), p. 74.

19. La topologie est la branche des mathématiques qui étudie la continuité: elle ne consiste pas seulement en l'étude de celles des propriétés des figures qui sont invariantes par les représentations topologiques [NOTE: Une représentation topologique est une transformation biunivoque qui est continue dans les deux sens.]: les travaux de MM. Brouwer, H. Hopf, Lefschetz lui ont aussi assigné pour but l'étude des réprésentations (c'est à dire des transformations univoques et continues) et des équations. Elle débute par la définition des espaces topologiques; ce sont les espaces abstraits dans lesquels les notions suivantes ont un sens: ensembles de points ouvertes et fermés, représentations. Elle se poursuit par l'introduction de nouveaux êtres algébrico-géometriques: complexes, groupes et anneaux. On nomme topologie ensembliste la partie de la topologie qui n'utilise que les opérations suivantes: réunion, intersection et fermeture d'ensembles de points. . . On nomme topologie algébrique (ou topologie combinatoire) la partie de la topologie qui utilise des notions algébrico-géométriques; . . .

Mon dessein initial fut d'imaginer une théorie des équations et des transformations s'appliquant directement aux espaces topologiques . . . donner une nouvelle définition de l'anneau d'homologie. . . .

J'introduis, à côté de la notion classique de recouvrement, qui appartient à la topologie ensembliste, une notion beaucoup plus maniable, celle de couverture, qui appartient à la topologie algébrique. . . . aucune subdivision de complexes, . . . seulement les propriétés de ses [de l'espace] réprésentations en lui-même.

J. Leray, "Sur la forme des espaces topologiques et sure les points fixes des représentations," *Journal des mathématiques pures et appliquées* 24 (1945): 96–167, at pp. 96–98. In Appendix C, I have translated the introduction to this paper.

... les méthodes par lesquelles nous avons étudié la topologie d'un espace peuvent être adaptées à l'étude de la topologie d'un réprésentation.

J. Leray, "L'anneau d'homologie d'une réprésentation," *Comptes Rendus Academie Sciences de Paris* 222 (1946): 1366–1368, at p. 1366.

20. F. Hausdorff, *Grundzüge der Mengenlehre* (New York:Chelsea, 1949 [1914]), pp. 213, 225. Weierstrass in about 1860 also defined open sets, but they were connected.

21. Continuity as mapping-of-neighborhoods has proven to be a powerful notion, sometimes with surprising consequences. Brouwer eventually made use of this notion of a neighborhood to prove constructively that every "fully-defined" function, H(x) on the unit interval in the real line, say a function defined by rules or "constructively," is uniformly continuous. On Brouwer, see M.J. Beeson, *Foundations of Constructive Metamathematics* (Berlin: Springer, 1985); W.P. van Stigt, *Brouwer's Intuitionism* (Amsterdam: North Holland, 1990), pp. 379–385.

22. "Die Tendenz der starken Algebraisierung der Topologie auf gruppentheoretischer Grundlage," Alexandroff and Hopf, *Topologie*, p. ix.

23. Čech wants to provide a homology theory for any topological space whatsoever. E. Čech, "Théorie générale de l'homologie dans un espace quelconque," *Fundamenta Mathematicae* 19 (1932): 149–183, at p. 150.

24. Alexandroff and Hopf, *Topologie*, p. vii.

25. Der Besprechung der drei Gebiete der Topologie: allgemeine, n-dimensionale [die Topologie (kontinuierlicher) n-dimensionaler Mannigfaltigkeiten] und kombinatorische Topologie, ist im folgenden, um nicht mit dem abstraktesten zu beginnen, zur Einführung ein I. Abschnitt über Punktmengen in n-dimensionalen Zahlenräumen vorausgeschickt. Bei streng systematischer Anordnung müßte ein solcher Abschnitt hinter den über allgemeine Topologie gestellt werden.

H. Tietze and L. Vietoris, "Beziehungen Zwischen den Verschiedenen Zweigen der Topologie," in *Enzyklopädie der Mathematischen Wissenschaften* (Leipzig: Teubner, 1914–1931), Section III AB 13, at p. 146.

26. Tietze and Vietoris, "Beziehungen Zwischen den Verschiedenen Zweigen der Topologie," §§ 5, 39, 40.

27. P.S. Alexandrov, "In Memory of Emmy Noether" (1935), in *Emmy Noether: Gesammelte Abhandlung*, ed. N. Jacobson (Berlin: Springer, 1983), pp. 1–11.

28. Others argue that Poincaré knew about such groups in his earlier work. In any case, see L. Vietoris, "Uber den höheren Zusammenhang kompakter Raüme und eine Klasse von zusammenhangstreusten Abbildungen," *Mathematische Annalen* 97 (1927): 454–472. (An earlier version appeared in *Proc. Amsterdam* 29 (1926): 443–453 and 1009–1013. I was not able to locate this source.); S. MacLane, "Topology Becomes Algebraic With Vietoris and Noether," *Journal of Pure and Applied Algebra* 39 (1986):

305–307; and, J. Dieudonné, *A History of Algebraic and Differential Topology, 1900–1960* (Boston: Birkhäuser, 1989).

29. R. Engelking, *General Topology* (Berlin: Heldermann, 1989), p. 408.

30. J.G. Hocking and G.S. Young, *Topology* (Reading, Massachusetts: Addison-Wesley, 1961), p. 218.

31. So Čech advocated for his own homology theory, its simplices being collections of such sets that have nonzero intersection, the "nerve" of a covering. Another potential unification is algebraic notions of categories and toposes, in which the general structural features of the mappings of objects, which are derived historically from algebraic topology, can be used as well to do point-set topology. For an account, see C. McLarty, *Elementary Categories, Elementary Toposes* (Oxford: Clarendon Press, 1995). The Stone representation of boolean algebras in terms of topology (1936) is perhaps another source of this impulse. M.H. Stone, "Theory of Representations for Boolean Algebras," *Transactions of the American Mathematical Society* 40 (1936): 37–111. McLarty tells me he believes that more important were covering spaces in homotopy, sheaf theory, and derived functor cohomology.

32. Alexandroff and Hopf, *Topologie*, pp. 10–11.

33. He employs Lebesgue's *Pflastersatz*, a paving or pigeonhole principle (in effect that if you have $N+1$ objects and N holes you must have at least two objects in one hole). See also the Epilog.

Freudenthal gives an extensive account of the priority dispute between Lebesgue and Brouwer, Brouwer continuing to doubt the realizability of Lebesgue's scheme as proposed, in, L.E.J. Brouwer, *Collected Works* vol. 2, ed. H. Freudenthal (Amsterdam: North Holland, 1976), pp. 435–445.

34. J. Dieudonné, "Recent Developments in Mathematics," *American Mathematical Monthly* 71 (1964): 239–248, at p. 243–244. He continues a bit further on: "[T]he introduction of sheaves by Leray . . . [gave] a workable mathematical formalism for the intuitive concept of "variation" of structures. . . . doing away with the cumbersome triangulations of former methods . . . " So the algebraic and analytic notions displace the combinatorial ones.

35. The use of "soft" and "hard" is never very precise, but rather reflects what might be called an aesthetic judgment. Gromov says, "Intuitively, 'hard' refers to a strong and rigid structure of a given object, while 'soft' suggests some weak general property of a vast class of objects." (M. Gromov, "Soft and Hard Symplectic Geometry," *Proceedings of the International Congress of Mathematicians* (Berkeley, 1986), pp. 81–97, at p. 96.)

36. "Désignons par δ, ε deux nombres très petits, le premier choisi de telle sorte que, . . . " A. Cauchy, "Résumé des Leçons donné a L'École Royale Polytechnique Sur Le Calcul Infinitésimal," *Oeuvres* series II, vol. 4 (Paris: Gauthier-Villars, 1899[1823]), p. 44.

37. W. Thirring, "Introduction," in E.H. Lieb, *The Stability of Matter: From Atoms to Stars, Selecta of Elliott H. Lieb*, ed. W. Thirring, 2nd edn (Berlin: Springer, 1997), p. 1. See also, W. Thirring, "Preface [to a festshrift issue devoted to Elliott Lieb]," *Reviews*

in Mathematical Physics 6 #5a (1994): v–x; E. Lieb and M. Loss, *Analysis* (Providence: American Mathematical Society, 1997).

38. For a history of early usage, with references, see W. Felscher, "Bolzano, Cauchy, Epsilon, Delta," *American Mathematical Monthly* 107 (2000): 844–862.

39. Dieudonné, *History of Algebraic and Differential Topology*, p. 16.

40. This list is inspired by Lieb and Loss, *Analysis*.

41. C. Fefferman, "The Uncertainty Principle," *Bulletin (New Series) of the American Mathematical Society* 9 (1983): 129–206.

42. Kirchoff's second law says that the currents at a node must balance out, and the first law says that the potential drops in a circuit add up to zero.

43. C.W. Misner, K.S. Thorne, and J.A. Wheeler, *Gravitation* (San Francisco: Freeman, 1973), pp. 367–368.

44. D. Aldous and P. Diaconis, "Hammersley's Interacting Particle Process and Longest Increasing Subsequences," *Probability Theory and Related Fields* 103 (1995): 199–213.

45. D.A. Klain and G.-C. Rota, *Introduction to Geometric Probability* (Cambridge: Cambridge University Press, 1997), p. 1.

46. "Analysis" as Lieb and Loss (*Analysis*) call it.

47. From Laugwitz, *Bernhard Riemann*, pp. 302, 304, 305.

Ich habe versucht, den grossen rechnerischen Apparat von X zu vermeiden, damit auch hier der Grundsatz von Riemann verwirklicht würde, demzufolge man die Beweise nich durch Rechnung, sondern lediglich durch Gedanken zwingen soll.

The original quote refers to Kummer.

48. See B. Simon, *The Statistical Mechanics of Lattice Gases*, volume 1 (Princeton: Princeton University Press, 1993). M.H. Krieger, *Constitutions of Matter* (Chicago: University of Chicago Press, 1996), discusses many of the solutions.

49. B.M. McCoy, "The Connection Between Statistical Mechanics and Quantum Field Theory," in V.V. Bazhanov and C.J. Burden, eds., *Statistical Mechanics and Field Theory* (Singapore: World Scientific, 1995), pp. 26–128.

50. F. Dyson, *Selected Papers of Freeman Dyson* (with Commentary) (Providence: American Mathematical Society, 1996), p. 32. Walter Thirring says,

Everyone knows that two liters of gasoline contain only twice as much energy as one liter but to deduce from the Schrödinger equation that it is a general property of matter that there is a universal bound for the binding energy per particle (=stability) is not so easy.

Thirring points out how important are the details of the Coulomb singularity $(1/r)$ for stability, that this is not a problem about the long-range character of the Coulomb force, and that the finite size of the nucleus is not crucial. W. Thirring, *Selected Papers of Walter E. Thirring with Commentaries* (Providence: American Mathematical Society, 1998), "Commentary," p. 4; W. Thirring, *Quantum Mechanics of Large Systems*, A Course in Mathematical Physics, vol. 4 (New York: Springer, 1983), pp. 17–18, 258.

51. E.H. Lieb and J.L. Lebowitz, "The Constitution of Matter: Existence of Thermodynamics for Systems Composed of Electrons and Nuclei," *Advances in Mathematics* 9 (1972): 316–398.

52. Dyson, *Selected Papers of Freeman Dyson*, p. 32.

53. Fefferman, "The Uncertainty Principle"; and P. Federbush, "A new approach to the stability of matter problem, I and II," *Journal of Mathematical Physics* 16 (1975): 347–351, 706–709.

54. See Lieb and Simon's rigorous account of Thomas-Fermi theory, which also includes careful consideration of breaking up space into cubes, with either Dirichlet or Neumann boundary conditions. (E.H. Lieb and B. Simon, "Thomas-Fermi Theory of Atoms, Molecules and Solids," *Advances in Mathematics* 23 (1977): 22–116, at p. 31). Lieb and Simon point out that Dirichlet-Neumann bracketing of the true value, one being above, the other below, is how R. Courant and D. Hilbert prove the Weyl theorem for the asymptotic distribution of eigenvalues, and allows us to treat the problem as particles in a box (*Methods of Mathematical Physics* (New York: Interscience 1953), volume 1, pp. 429–445).

55. Lieb, *The Stability of Matter* (1991, 1997, 4th edn 2004), surveys the state of the art in Lieb and collaborators' work.

56. C. Fefferman, "The N-Body Problem in Quantum Mechanics," *Communications on Pure and Applied Mathematics* 39 (1986): S67–S109, at p. S88.

57. A sharp enough E_* allows for a one-particle approximation, one atom rather than many within each space-dividing ball. Fefferman's hope was that it is easier to estimate E_* than to prove an inequality that Lieb and Thirring need, the inequality that gives the kinetic energy bound in terms of an integral of the density to the 5/3-power or, more precisely, as an integral of the potential to the 5/2-power.

In the Fefferman-Seco papers, they achieve in effect the Lieb-Thirring "classical" bound through their estimate of the sum of the eigenvalues. This will not be enough, however. (Fefferman-Seco is for an isolated atom, not atoms in bulk.) Giving up rigor for the moment, we *might* employ the Fefferman-Seco asymptotics for the Thomas-Fermi energy and substitute that *number*, for the Thomas-Fermi energy of an atom of a particular atomic number, Z, into the Lieb-Thirring estimate for E_*. (I am not at all sure this is even formally legitimate, given how the Thomas-Fermi energy appears in the derivation. Namely, it would appear that the derivation employs truncated values for the Thomas-Fermi kinetic energy.) Even then, for $Z=1$, the proportionality constant is a bit more than two times too large. Even with the surprisingly good agreement with the actual value of the binding energy (-1.08 Ry ≈ -1 Ry), it is not clear if the asymptotic series is correct or merely fortuitously accurate for atoms in bulk. If the Thomas-Fermi asymptotic energy is correct for $Z \geq 4$ we might have a good enough proportionality constant (that is, $(1+Z^{-2/3})^2$ is less than 2). See also, chapter 4, note 68.

58. At one point, they call what is eventually to be ignored in the correction, since it is suitably bounded, "junk$_s(x,\lambda)$." (C. Fefferman and L.A. Seco, "Eigenvalues and Eigenfunctions of Ordinary Differential Operators," *Advances in Mathematics* 95 (1992): 145–305, at p. 155.) Fefferman and Seco employ a refined optical model—that is, improved WKB or semiclassical asymptotics—for a one-body Schrödinger equation with a (singular) Coulomb potential, $1/r$.

59. For a review of some of Fefferman's earlier work, see P. Martin, "Lecture on Fefferman's Proof of the Atomic and Molecular Nature of Matter," *Acta Physica Polonica* B24 (1993): 751–770.

60. I.M. Sigal, "Review of Fefferman and Seco, 'The density in a one-dimensional potential," *Adv. Math.* 107 (1994) 187–364," *Mathematical Reviews* #96a:81147.

61. Diffusive motion is the archetype of random almost-canceling interactions. The law of the iterated logarithm, which prescribes the fluctuations in that diffusive process, provides a delicate measure of this almost canceling. (See chapter 2, note 38.)

62. Again, I take it that "the subject of topology" is a substantive problem: What can you prove? What do you need to know? What do the mathematical examples, cases, and phenomena demand? How can we systematically organize the facts we discover about spaces understood topologically, and so gain an overview of these diverse phenomena? We might, after these considerations, decide to say that the subject of topology is . . . But new examples, techniques, or problems (and old ones, reconsidered) will force the subject to go in unplanned ways. Dimensionality might seem to be part of general topology, but it needs combinatorial or algebraic means if it is to be understood. Spatial geometrical problems are sometimes best expressed algebraically, or in terms of sets and coverings, and the lattice of intersections and unions of those sets—"best expressed" because then those problems become rather more tractable, and so new notions of geometric intuition become possible and valuable.

63. J. Dieudonné calls them "défaut d'exactitude." See, "Une brève histoire de la topologie," in J.-P. Pier, ed., *Development of Mathematics 1900–1950* (Basel: Birkhäuser, 1994), pp. 35–153, at pp. 68, 71.

64. This sort of theme is not peculiar to topology. Number theorists had to account for the fact that in some number systems, integers did not have a unique prime factorization. Algebraic number theory, with its notion of ideal numbers, explained the obstruction to unique prime factorization and showed how to restore uniqueness by using those ideals.

65. K.G. Wilson, "The Renormalization Group: Critical Phenomena and the Kondo Problem," *Reviews of Modern Physics* 47 (1975): 773–840, pp. 773–775.

66. Engelking, *General Topology*, p. 18.

67. Munkres, *Topology*, pp. 146–147.

68. Lebesgue expliquait la nature de son intégrale par une image plaisante et accessible à tous. "Je dois payer une certaine somme, disait-il; je fouille dans mes poches et j'en sors des pièces et des billets de différentes valeurs. Je les verse à mon créancier dans l'ordre ou elles se présentent jusqu'à atteindre le total de ma dette. C'est l'intégrale de Riemann. Mais je peux opérer autrement. Ayant sorti tout mon argent, je réunis les billets de même valeur, les pièces semblables, et j'effectue le paiement en donnant ensemble les signes monétaires de même valeur. C'est mon intégrale."

A.F. Monna, "The integral from Riemann to Bourbaki," pp. 77–154, at p. 102, in D. van Dalen and A.F. Monna, *Sets and Integration, An outline of the development* (Groningen: Wolters-Noordhoff, 1972). The quote is taken from a biography of Lebesgue by A. Denjoy, L. Félix, and P. Montel.

69. Ampère and Kirchoff played important roles in this story. For a fuller account, see J.D. Jackson and L.B. Okun, "Historical roots of gauge invariance," *Reviews of Modern Physics* 73 (2001): 663–680.

70. I have written more on some of this in M.H. Krieger, *Marginalism and Discontinuity: Tools for the Crafts of Knowledge and Decision* (New York: Russell Sage Foundation, 1989), chapters 1, 2.

71. Very similar issues concern diagrams come up in the use of Feynman diagrams in physics. See D. Kaiser, *Drawing Theories Apart* (Chicago: University of Chicago Press, 2005). In particular, Kaiser discusses how physicists learned to use those Feynman diagrams to calculate, largely at first from personal contact, Dyson being the crucial figure. See also, M. Veltman, *Diagrammatica: The Path to Feynman Diagrams* (Cambridge University Press, 1994).

72. Dieudonné, *History of Algebraic and Differential Topology*, p. 161.

73. T.L. Saaty and P.C. Kainen, *The Four-Color Problem, Assault and Conquest* (New York: Dover, 1986), pp. 60–83. See, H. Heesch, "Chromatic Reduction of the Triangulations T_e, $e=e_5+e_7$," *Journal of Combinatorial Theory* B13 (1972): 46–55. The standard reference, which I have not seen, is H. Heesch, *Untersuchungen zum Vierfarbenproblem* (Mannheim: Bibliographisches Institut, 1969), Hochschulscripten 810/810a/810b.

74. "A picture of A in B," is a phrase due to Mac Lane, I believe.

75. Given a vector field, Stokes', Green's, and Gauss's theorems, with their winding numbers, charges, and indexes, measure vorticity and source strength; and given the sources, then force laws, such as Force = Charge × Field, can be employed to measure field strength.

76. One account of the collaboration is provided by Saunders Mac Lane, "The Work of Samuel Eilenberg in Topology," in A. Heller and M. Tierney, *Algebra, Topology, and Category Theory: A Collection of Papers in Honor of Samuel Eilenberg* (New York: Academic Press, 1976), pp. 133–144. (Technically, they studied the homology of

"$K(\Pi,n)$-complexes" as a function of n, the homotopy or fundamental group, Π, abelian in dimension n.)

77. Feynman then justified it formally, in contrast to its algorithmic power, as would a theoretical physicist in his 1950 paper on operator calculus. Dyson and others proved its correctness, as best can be done.

78. Such inverted redemption sequences are characteristic of much of computer science, with their pushdown stacks. In the Snake Lemma, the prover must first go to the bottom of things and then wend their way back—at each point making an assumption, an *if* that must eventually be redeemed by the next assumption, which must be redeemed by the next.

79. S. Lang, "Review of A. Grothendieck and J. Dieudonné, Éléments de géométrie algébrique," *Bulletin of the American Mathematical Society* 67 (1961): 239–246.

80. And to continue, "Similarly, algebraic K-theory starts from the regrettable fact that not all projective modules are free." (J.F. Adams, "Algebraic Topology in the Last Decade," in, *Algebraic Topology*, Proceedings of Symposia in Pure Mathematics 22 (Providence: American Mathematical Society, 1971), pp. 1–22, at p. 2.) Projectivity is a generalization of freeness (that is, no relations among the objects), a subset of a free module that is a direct sum, so that exact sequences "split."

81. J. Dieudonné calls them "défaut d'exactitude." See note 63 above.

82. This recalls the rhetoric of Ramus (1515–1575), with its binary oppositions and diagrammatic presentation, as an art of memory.

83. Here the usual example quoted is Deligne's (early 1970s) use of Grothendieck's homological/topos-theoretic technology.

84. Let $A(m)$ be the fact that, a union of less than continuum-many measure-zero sets has measure zero; $B(m)$ be, the real line is not the union of less than continuum-many measure-zero sets; $U(m)$ be that, every set of reals of cardinality less than the continuum has measure zero; and $C(m)$ be, there does not exist a family F of measure-zero sets, of cardinality less than the continuum, and such that every measure-zero set is covered by some member of F. $A(c)$, etc., are the same statements with "first Baire category" replacing "measure zero." H. Judah, "Set Theory of Reals: Measure and Category," in H. Judah, W. Just, and H. Woodin, eds., *Set Theory of the Continuum* (New York: Springer, 1992), pp. 75–87, at pp. 75–77.

85. H. Judah, "Set Theory of Reals: Measure and Category." Or, it is argued that one "pattern [of properties in the projective hierarchy of sets] is considered more natural than the one generated by V=L, if only because $\Lambda\Lambda\Lambda\Lambda\Lambda\Lambda\Lambda$ is a more natural continuation of Λ than Λ_____." P. Maddy, *Realism in Mathematics* (New York: Oxford, 1990), p. 139.

86. T. Bartoszynski and H. Judah, *Set Theory: On the Structure of the Real Line* (Wellesley: A.K. Peters, 1995), is devoted to working out the diagram in its details.

87. S. Brush, "History of the Lenz-Ising Model," *Reviews of Modern Physics* 39 (1967): 883–893; C. Domb, *The Critical Point: A Historical Introduction to the Modern Theory of Critical Phenomena* (London: Taylor and Francis, 1996); C. Domb, "The Ising

Model," in P.C. Hemmer, H. Holden, and S. Kjelstrup Ratkje, eds., *The Collected Works of Lars Onsager (with commentary)* (Singapore: World Scientific, 1996), pp. 167–179; C.N. Yang, "Path Crossings with Lars Onsager," *Idem.*, pp. 180–181.

H. V. N. Temperley, "The two-dimensional Ising model," in vol 1 of Domb, C. and M. S. Green, eds., *Phase Transitions and Critical Phenomena*, volume 1 (London, New York: Academic Press, 1972), pp. 227-267, is filled with insights. McCoy's *Advanced Statistical Mechanics* (2010) has more detailed charts for all the various aspects of the Ising model, in chapters 10-14, and provides an historical/chronological account. In the second edition of McCoy and Wu, *The Two-Dimensional Ising Model* (2014) there is a new chapter bringing the situation up to date. See also, "The Romance of the Ising Model," "The Baxter Revolution," and Maillard and McCoy, "The Importance of the Ising Model." As for Toeplitz, see P. Deift, A. Its, and I. Krasnovsky, "Toeplitz Matrices and Toeplitz Determinants Under the Impetus of the Ising Model: Some History and Some Recent Results," *Communications in Pure and Applied Mathematics* 66 (2013): 1360-1438.

88. R.J. Baxter and I. Enting, "399th Solution of the Ising Model," *Journal of Physics A* 11 (1978): 2463–2473.

89. Kramers and Wannier credit Montroll with their general method of solution, derived from Montroll's work on chains of molecules.

90. There is in the literature a number of clarifying expositions, one by Newell and Montroll, and one by Green and Hurst. See, G.F. Newell and E.W. Montroll, "On the Theory of the Ising Model of Ferromagnetism," *Reviews of Modern Physics* 25 (1953): 353–389; H.S. Green and C.A. Hurst, *Order-Disorder Phenomena* (London: Interscience, 1964. A condensed overview is provided in the first paragraph of B. Davies, "Onsager's algebra and superintegrability," *Journal of Physics* A 23 (1990): 2245–2261.

91. H.B. Thacker, "Corner Transfer Matrices and Lorentz Invariance on a Lattice," *Physica* 18D (1986): 348–359. In a Lorentz invariant lattice, the argument u (a measure of lattice anisotropy) plays the role of the rapidity (momentum = mass × sinh u), and Lorentz invariance means that it is the differences in rapidity that matter (what Baxter calls the difference property). The two periods of the elliptic functions correspond to momentum shifts of $2\pi/a$ (the real quarter-period) and a 2π Euclidean rotation (the imaginary quarter-period), a being the lattice spacing. Lorentz invariance means that for a fixed k, transfer matrices commute ($[V(u),V(u')] = 0$).

92. C.A. Tracy, "The Emerging Role of Number Theory in Exactly Solved Models in Lattice Statistical Mechanics," *Physica* 25D (1987): 1–19.

93. Such a matrix is more commonly seen nowadays as an evolution operator, expressed as exp $iH\Delta t/\hbar$, where H is the hamiltonian, and $\hbar$ is Planck's constant over 2π, as in the Heisenberg representation in quantum mechanics. (Baxter's corner transfer matrix is in effect a Lorentz boost operator, shifting the pseudorapidity or spectral parameter u. Baxter's "difference property," that what matters for solvability is that what appears are differences in u, is in effect Lorentz invariance.) But, here we are in the classical and statistical mechanical realm. They compute the partition function by adding

on one more atom's interaction with its neighbors, transferring the interaction down the line, much like computing the quantum mechanical probability amplitude by adding on one more time-unit of interaction.

Smirnov, Chelkak, and Hongler, using discrete complex analysis, show that at criticality, the transfer matrix is in effect analytic continuation, connect it to the SLE process, and rigorously prove conformal invariance. See, D. Chelkak and S. Smirnov, "Universality in the 2D Ising model and conformal invariance of fermionic observables," *Inventiones mathematicae* 189 (2012) 515–580.

94. Onsager, 1944, pp. 121–123, was concerned about the symmetries and group representation of his quaternion algebra. And most of his paper, pp. 124-133, is devoted to finding an appropriate irreducible representation. Similarly, Kaufman, 1949, was concerned with symmetries and group representation of the spinor algebra.

Speculatively: While $\text{Tr}(AB)=\text{Tr}(BA)$, it is not true that $\text{Tr}(AB)=\text{Tr } A \times \text{Tr } B$. To treat the trace of the transfer matrix as a group character, that is, the partition function is a "class function," would seem to be a consequence that there is one eigenvalue that is much larger than the rest. If that is so, then $\text{Tr}(AB)$ is mostly $\Lambda_{A \text{ max}} \times \Lambda_{B \text{ max}}$ and is equal to $\text{Tr } A \times \text{Tr } B$. (On the other hand, if there are two almost equal eigenvalues, $\text{Tr}(AB)$ is about $2 \times \Lambda_{A \text{ max}} \times \Lambda_{B \text{ max}}$, while $\text{Tr } A \times \text{Tr } B$ is about $4 \times \Lambda_{A \text{ max}} \times \Lambda_{B \text{ max}}$. Something does not work here.)

Free energies, F, are additive at the same temperature and pressure, and so we would expect that $\exp -\beta F = PF = \text{Tr } V$, and the partition function (PF) should be multiplicative, as are characters, as we have here. (So thinking physically, since free energies are additive, the PF, however the distribution of the traces, should be multiplicative.)

95. E. W. Montroll, "Statistical Mechanics of Nearest Neighbor Systems," *Journal of Chemical Physics* 9 (1941): 706–721; J. Ashkin and W.E. Lamb, Jr., "The Propagation of Order in Crystal Lattices," *Physical Review* 64 (1943): 169–178.

96. Montroll, "Statistical Mechanics of Nearest Neighbor Systems," pp. 709–710. In Onsager, "Crystal Statistics I," p. 137, this is expressed by the fact that "at" the critical point, one has built up disorder from zero temperature, hence "almost disordered," or built up order from infinite temperature, hence "almost ordered." Formally, this is done by Onsager's Z operators and their multiples, operating on the appropriate of states of order or disorder, respectively. The Z operators introduce spin flips at regular distances.

97. Note that Onsager, following Montroll, uses a row-to-to row transfer matrix, in effect turning a two-dimensional problem into a one-dimensional one. J. Palmer, *Planar Ising Correlations* (Boston: Birkhäuser, 2007), p. 149, says, "...much of what we do in this chapter and the next is to restore a two-dimensional picture of the Ising model (i.e., the Pfaffian formalism) that the transfer matrix formalism has squashed into obscurity."

L. Dolan and M. Grady, "Conserved Charges from Self-Duality," *Physical Review* D25 (1982): 1587–1604; M. Grady, "Infinite Set of Conserved Charges in the Ising Model," *Physical Review* D25 (1982): 1103–1113; M.J. Ablowitz and P.A. Clarkson, *Solitons, Nonlinear Evolution Equations and Inverse Scattering* (London Mathematical Society #149), (Cambridge: Cambridge University Press, 1991).

Those symmetries are of an associated linear partial differential equation, effectively for a Hamiltonian system—much as is the case for the Ising model pseudo-hamiltonian that Onsager discovered, his $B+k^{-1}A$, the symmetries now expressed through an algebra generated by his A and B, kinetic and potential energies, respectively, or disorder and order terms.

98. Notably, the group characters of the spinor representation, the traces of the transfer matrices, $\Sigma\pm\frac{1}{2}\varphi_i$, are formally identical to Onsager's formula for the eigenvalues of his operators. H. Weyl, *The Classical Groups: Their Invariants and Representations* (Princeton: Princeton University Press, 1938, 1946), p. 273; L. Onsager, "Crystal Statistics I," *Physical Review* 65 (1944): 117–149, eqns. 97 and 98. The Onsager paper is reprinted in my *Constitutions of Matter*.

99. T.D. Schultz, D.C. Mattis, and E.H. Lieb, "Two-Dimensional Ising Model as a Soluble Problem of Many Fermions," *Reviews of Modern Physics* 36 (1964): 856–871.

100. Technically, those quasiparticles are not interacting with each other. They are "free" particles or fields. However the interaction of individual spins with each other is mediated through those free quasi-particles, and in effect we add up all the intermediate interactions to know how one spin affects another far away. See chapter 4, note 6.

Alternatively, Nambu pointed out how the Ising lattice might be thought of as being populated by Bloch spin waves (the "kinetic energy" term), which scatter from each other ("potential energy"). This is just how Onsager splits up the transfer matrix into A and B parts. (The spin waves are not the free quasiparticles.) He says, "The [interaction terms] can be neglected when the number of particles are small because the interaction force is of short range, which corresponds to the fact that the notion of a spin wave is a good approximation when the magnetization is nearly complete (low temperature)." Y. Nambu, "A Note on the Eigenvalue Problem in Crystal Statistics," *Progress of Theoretical Physics* 5 (1950): 1–13, at p. 2.

101. McCoy has an illuminating discussion in the section on 'Particles' in "The Connection Between Statistical Mechanics and Quantum Field Theory."

102. I find the following suggestive, although I am sure I do not fully understand it: "These parameters seem to be playing the role of fugacities [reaction constants, as in chemistry] and the [mathematical technology of] Bailey pairs seem to be building up complex systems by gluing these more elementary fermions [for example, Majorana fermions and parafermions] together." A. Berkovich, B.M. McCoy, and A. Schilling, "N=2 Supersymmetry and Bailey Pairs," *Physica* A 228 (1996): 33–62, at p. 61.

103. The Yang-Baxter equation is an "elliptic curve." Namely, it leads to an equation for the Boltzmann weights which can be put in the form:

$$y^2 = 4x^3 - g_2 x - g_3.$$

The integral for the length along an ellipse is expressed as $\int 1/\sqrt{(4x^3 - g_2 x - g_3)}$, and hence the name elliptic curve. The Yang-Baxter relation is also one of the numerous addition formulae for elliptic theta functions, reflecting their home on elliptic curves.

104. On the one- and zero- dimensional analogy, in the Kondo effect, see p. 189.

As for braids, at least for $k=1$ (and then we have a conformal field theory); for $k\neq1$ we have what are called quantum groups, referring to noncommutativities, as in quantum mechanics, or "q-deformations".

Some further speculative considerations:

—The braid group, another form of the Yang-Baxter relation, is much the same as a "pants decomposition" of a hyperbolic space. (R. Penner with J.L. Harer, *Combinatorics of Train Tracks*, Annals of Mathematics Studies 125, (Princeton: Princeton University Press, 1992); E. Martinec, "Quantum Field Theory: A Guide for Mathematicians," in, *Theta Functions, Bowdoin 1987* (Providence: American Mathematical Society, 1989), pp. 406–412.) See also, W. Thurston, *Three-Dimensional Geometry and Topology* (Princeton: Princeton University Press, 1997) for pictures.

And a pair of pants appropriately slit becomes (two of) Onsager's right-angle hyperbolic hexagon(s), and hence the star-triangle relation, equivalent to the Yang-Baxter equation. (M.J. Stephen and L. Mittag, "New Representation of the Solution to the Ising Model," *Journal of Mathematical Physics* 13 (1972): 1944–1951.)

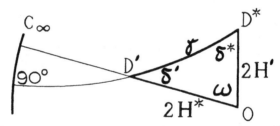

FIG. 4. Hyperbolic triangle. Stereographic projection, conformal. Circles are represented by circles, geodetics by circles invariant towards inversion in the limiting circle $C\infty$ of the projection. See F. Klein, reference 15, pp. 293–299.

FIGURE N.2: Triangle of Ising model horizontal and vertical couplings (H [H^* is the dual of H], H'), with quasiparticle momentum and energy and phases (ω, $\gamma(\omega)$, and δ^* and δ'). From Onsager, "Crystal Statistics. I," (1944).

—Hyperbolic geometry builds in the fact that the Ising model quasiparticle energies are in effect shifted from their one-dimensional values because of lattice interactions, seen in the hyperbolic triangle which defines the eigenvalues of Onsager's problem.

Note that γ is the quasiparticle energy, and ω is its momentum, the formula from hyperbolic geometry relating sides and angles turning out to have direct correspondence with the energy-momentum formula from relativistic mechanics. (Thacker, "Corner Transfer Matrices and Lorentz Invariance on the Lattice")

—Transfer matrices are translation operators in hyperbolic space. (D.B. Abraham and A. Martin-Löf, "The Transfer Matrix for a Pure Phase in the Two-dimensional Ising Model," *Communications in Mathematical Physics* 32 (1973): 245–268.)

105. K.G. Wilson and J. Kogut, "The Renormalization Group and the ε Expansion," *Physics Reports* 12C (1974): 75–200, at pp. 90–94, give as examples of the search for analyticity the mean field model, and Kadanoff's account of the Ising model coupling constant's smooth behavior as one scales up in size.

106. Wilson, "The Renormalization Group: Critical Phenomena and the Kondo Problem," pp. 773–775.

107. R.J. Baxter, *Exactly Solved Models in Statistical Mechanics* (London: Academic Press, 1982).

108. L. Onsager, "The Ising Model in Two Dimensions," in *Critical Phenomena in Alloys, Magnets, and Superconductors*, ed. R.E. Mills, E. Ascher, and R.I. Jaffee (New York: McGraw-Hill, 1971), pp. 3–12; Stephen and Mittag, "New Representation of the Solution to the Ising Model."

109. The symmetries are those of the S-matrix for particle scattering, crossing symmetry corresponding to Onsager's duality. The Ising problem can be expressed in scattering terms, so this is no coincidence.

110. Baxter (*Exactly Solved Models*, chapter 7) derived a functional equation for the Ising model partition function, giving an equation for the eigenvalues, Λ, of the transfer matrix: $\Lambda(u) \times \Lambda(u+I') = (-2/(k \text{ sn } iu))^n + (-2 \text{ sn } iu)^n r$. k and u are as in the text, I' is the imaginary quarter period of the elliptic functions. Baxter and Enting, "399th Solution," derive the following equation for a correlation function, f, where $b(k)$ is given (as "scattering data"), $\{K_i\}$ are coupling constants ($=J/k_B T$, J being the coupling in the hamiltonian): $\Sigma f(K_i,k) = b(k) (1+k^{-1}\Pi \text{ sech } 2K_i)$. See also, J.-M. Maillard, "The inversion relation," *Journal de Physique* 46 (1985): 329–341, and J.-Ch. Anglès d'Auriac, S. Boukraa, and J.-M. Maillard, "Functional Relations in Lattice Statistical Mechanics, Enumerative Combinatorics, and Discrete Dynamical Systems," *Annals of Combinatorics* 3 (1999) 131–158.

There could be an iterative scheme, using functional equations connecting the partition functions for k, $1/k$, and $-q/k$. The solution is in terms of q-series (perhaps not surprisingly, the same sort of series used to define the elliptic theta functions); typically the series look like: $\Pi(1-q^{2n-1}z)$, z being the other independent variable. One makes assumptions about regions of analyticity of the partition function, assumptions that are reminiscent of the Wiener-Hopf factorization that plays a role elsewhere in solutions for the correlation and permanent magnetization. See also chapter 5, note 22.

It is hard to be sure, in this area, when one has exhausted the symmetries: "all these conditions constitute an overlapping set of constraints and it is quite difficult to combine them all together to get some new nontrivial property for the model." (Maillard, "The inversion relation," at p. 340.)

111. Contemporaneous with Wu's first paper, Kadanoff provides an account of the Onsager solution "in the language of thermodynamic Green's functions."(p. 276) What I find most valuable in this paper is his reworking of Yang's 1952 paper on the spontaneous magnetization, making clearer the meaning of some of Yang's mathematical moves—in particular, Yang's in effect employment of Wiener-Hopf methods (pp. 289–293). L.P. Kadanoff, "Spin-Spin Correlations in the Two-Dimensional Ising Model," *Il Nuovo Cimento* B44 (1966): 276–305.

112. Painlevé transcendents so to speak outdo elliptic functions, much as elliptic functions generalize the trigonometric or circular sines and cosines. Asymptotically, at least some of the Painlevé transcendents become elliptic functions. For example, for the first transcendent, defined by the nonlinear equation: $P_I'' = P_I^2 + \lambda \psi$, we find $P_I(v,k;\lambda) = sn(v,k) + \lambda \psi(v,k;\lambda)$, where λ is small. (H.T. Davis, *Introduction to Nonlinear Differential and Integral Equations* (Washington, D.C.: United States Atomic Energy Commission, 1960), chapter 8.)

113. C.A. Tracy and H. Widom, "Fredholm determinants, differential equations and matrix models," *Communications in Mathematical Physics* 163 (1994): 33–72. Also, H. Widom, "On the Solution of a Painlevé III Equation," (solv-int/9808015 24 Aug 1998), which connects Fredholm to Painlevé directly.

Interestingly, Paul Painlevé (1863–1933) was also premier of France (Président du Conseil, 1917, 1925). He was "not an outstanding political leader, [but] he was a brilliant mathematician."(*Encyclopedia Britannica*, 1996) A memoir about him by R. Garnier that prefaces his collected works argues otherwise, including his leadership during the Great War. (P. Painlevé, *Oeuvres*) As for the transcendents, at first Poincaré believed they were an isle isolated from the rest of the mathematical continent. Garnier argues they are a New World linked broadly with the Old World of mathematics, a New World whose exploration is promising.

For a review of Ising solutions, and a list of Painlevé applications subsequent to their use by Wu, McCoy, Tracy, and Barouch, see J. Palmer and C. Tracy, "Two-Dimensional Ising Correlations: Convergence of the Scaling Limit," *Advances in Applied Mathematics* 2 (1981): 329–388, at pp. 329–330; "Two-Dimensional Ising Correlations: The SMJ [Sato, Miwa, Jimbo] Analysis," *Advances in Applied Mathematics* 4 (1983): 46–102, at pp. 46–49.

For more on Toeplitz matrices in this context, see P. Deift, A. Its, and I. Krasovsky, "Toeplitz Matrices and Toeplitz Determinants under the Impetus of the Ising Model: Some History and Some Recent Results."

CHAPTER 4

1. H. Weyl, "Topology and Abstract Algebra as Two Roads of Mathematical Comprehension," *American Mathematical Monthly* 102 (1995, originally appearing in 1932): 453–460, 646–651, at p. 651.

2. Kuhn inveighs against the sleepwalker theory, insisting that a good scientist knows what she is doing, even if in retrospect we might feel that it is partly wrong or confused. The work I am discussing here is in fact correct. The scientist may still feel that the proof is insufficiently satisfying, hoping that it could be rendered transparent. T.S. Kuhn, "Revisiting Planck," *Historical Studies in the Physical Sciences* 14 (1984): 231–252.

3. I.M. Sigal, "Lectures on Large Coulomb Systems," *CRM Proceedings and Lecture Notes* 8 (1995): 73–107, at p. 102 on Fefferman; M. Beals, C. Fefferman, R. Grossman, "Strictly Pseudoconvex Domains in C^n," *Bulletin (New Series) of the American Mathematical Society* 8 (1983): 125–322, at p. 294; B.M. McCoy and T.T. Wu, *The Two-Dimensional Ising Model* (Cambridge, Massachusetts: Harvard University Press, 1973), pp. 97–98 ("the resulting expression for the specific heat contains a forest of hyperbolic functions"). Fefferman and Seco at one point refer to "the hypotheses in the preceding section have become too baroque." (C. Fefferman and L. Seco, "Eigenvalues and Eigenfunctions of Ordinary Differential-Operators," *Advances in Mathematics* 95 (1992): 145–305, at p. 235.) Emmy Noether is supposed to have referred to her dissertation as *Formelnstruppe*, colloquially translated as jungles of formulas.

4. The quote is from T.T. Wu, "Theory of Toeplitz Determinants and the Spin Correlations of the Two-Dimensional Ising Model. I," *Physical Review* 149 (1966): 380–401. R.J. Baxter, *Exactly Solved Models in Statistical Mechanics* (London: Academic Press, 1982); D. Ruelle, *Statistical Mechanics: Rigorous Results* (New York: Benjamin, 1969); B. Simon, *Statistical Mechanics of Lattice Gases*, vol. 1 (Princeton: Princeton University Press, 1993).

5. See B. Davies, "Onsager's algebra and superintegrability," *Journal of Physics* A 23 (1990): 2245–2261; "Onsager's algebra and the Dolan-Grady condition in the non-self-dual case," *Journal of Mathematical Physics* 37 (1991): 2945–2950; B.M. McCoy, "The Connection Between Statistical Mechanics and Quantum Field Theory," in V.V. Bazhanov and C.J. Burden, eds., *Statistical Mechanics and Field Theory* (Singapore: World Scientific, 1995), pp. 26–128. Also, the articles in M. Kashiwara, T. Miwa, eds., *MathPhys Odyssey 2001, Integrable Models and Beyond, In Honor of Barry McCoy* (Boston: Birkhäuser, 2002).

6. (a) R.P. Feynman, in his *QED* (Princeton: Princeton University Press, 1985), presents this sort of argument in pellucid detail, there about the interference of electric fields that define the strength and polarization of light. The Ising model is classical statistical mechanics; the analogy to quantum mechanics is a formal mathematical one.

(b) The *usual* analogy is the computation of the partition function as a path integral. For this generic analogy of statistical mechanics to quantum mechanics, see, for example, R.P. Feynman and A.R. Hibbs, *Quantum Mechanics and Path Integrals* (New York: McGraw-

Hill, 1965). Using field-theoretic techniques and path-integrals (from his diagrammatic method in quantum electrodynamics), Feynman derived the Ising model partition function. (*Statistical Mechanics* (Reading, Massachusetts: Benjamin, 1972), pp. 136–150; also, L. Landau and E. Lifshitz, *Statistical Physics*, Part 1 (Oxford: Pergamon, 1980), pp. 498–506.) The path here can also be related to the series expansion solutions, as in quantum field theory.

In the actual combinatorial (Pfaffian) calculation of the partition function and the correlation function, the lattice is given a supplementary decoration or "bathroom tiling" at the vertices to ensure that the combinatorial polygons are counted correctly. The minimal size of the Pfaffian matrix is related to the shortest path between two spins. See R.E. Shrock and R.K. Ghosh, "Off-axis correlation functions in the isotropic d=2 Ising model," *Physical Review B* 31 (1985): 1486–1489, at p. 1486.

See B. Kaufman and L. Onsager, "Crystal Statistics, III: Short Range Order in a Binary Ising Lattice," *Physical Review* 76 (1949): 1244–1252, at pp. 1246–1250. Also, T.D. Schultz, D.C. Mattis, and E.H. Lieb, "Two-Dimensional Ising Model as a Soluble Problem of Many Fermions," *Reviews of Modern Physics* 36 (1964): 856–871, at p. 865, where they invoke Wick's theorem. For the lattice decoration, see E.W. Montroll, R.B. Potts, and J.C. Ward, "Correlations and Spontaneous Magnetization of the Two-Dimensional Ising Model," *Journal of Mathematical Physics* 4 (1963): 308–322; C.A. Hurst and H.S. Green, "New Solution of the Ising Problem from a Rectangular Lattice," *Journal of Chemical Physics* 33 (1960): 1059–1062; P. Kastelyn, "Dimer Statistics and Phase Transitions," *Journal of Mathematical Physics* 4 (1963): 281–293; McCoy and Wu, *The Two-Dimensional Ising Model*.

(c) As a consequence of these matrix computations of the correlation function, one ends up with the determinant of a Toeplitz matrix. I am suggesting that it is here that we might have something of a formal analogy with quantum mechanics (but not the usual one). In taking the determinant of Toeplitz matrix, we would appear to multiply probability amplitudes along a single "path," and then add up the amplitudes for each of the possible paths. I am unsure if these "paths" can be mapped onto actual paths in the lattice. (Recall that, while for the Pfaffian there is a precise analogy, here we have a very different matrix derived from it.)

Technically, and speculatively, if a_n is an amplitude or probability of correlation for a part of a path, then we are concerned with a total amplitude or probability, that is:

$$\Sigma_{\text{``all'' paths between two points}} \Pi_{\text{subpaths of a ``path,''p}} \, a_{i(p)},$$

which, with attention to minus signs, might be an interpretation of the determinant of a (Toeplitz) matrix $A_{ij}=a_{i-j}$. Notionally, a_n might represent a path distance of n steps (it is the nth fourier coefficient). (Note that the determinant here, derived from a Pfaffian of a matrix, to compute the correlation functions, is not the same matrix and determinant employed by Kac and Ward in their derivation of the Ising model partition function. See (f) below.)

Again, as far as I can tell, in the particular case of the Toeplitz determinant, the analogy does not follow, at least directly, from Feynman and Hibbs's generic account.

(d) The amplitudes that are being added, the products of elements of the Toeplitz matrix, can be identified with products of fourier components of $\exp i\delta'$, where the phase angle, δ', is from Onsager's 1944 paper. δ^* and δ' are two of the angles of Onsager's hyperbolic triangle, the third angle being ω, and $\omega + \delta^* + \delta' < \pi$. See Figure N.2.

(i) Again, as for the δs: $\exp i\delta^*$ is the amplitude in the disorder operator, namely, the disorder operator introduces disorder into the perfectly ordered low temperature ground state. $\exp i\delta'$ is the amplitude of the order operator, δ' being δ^*'s hyperbolic triangle complement, that is, $\delta'(\omega) < \pi-\omega-\delta^*(\omega)$, the order operator introducing order into the perfectly (randomly) disordered high temperature ground state. (Onsager, "Crystal Statistics, I," pp. 136–137.) See notes 152, 153. See our discussion of the meaning of the Wiener-Hopf method, in the main text.

(ii) δ' is "the angle of rotation involved in transforming the transfer matrix into diagonal form."(L. Onsager, "The Ising Model in Two Dimensions," in *Critical Phenomena in Alloys, Magnets, and Superconductors*, ed. R.E. Mills, E. Ascher, and R.I. Jaffee (New York: McGraw-Hill, 1971), pp. 3–12, at p. 10.). See, B. Kaufman, "Crystal Statistics, II: Partition Function Evaluated by Spinor Analysis," *Physical Review* 76 (1949): 1232–1243, at p. 1241.

(iii) δ^* is the phase angle associated with the fermion quasiparticles of momentum ω (the Cooper pairs), energy γ, that may be said to make up the Ising lattice in Schultz, Mattis, and Lieb, "Two-Dimensional Ising Model as a Soluble Problem of Many Fermions," p. 866. Their q corresponds to ω in Onsager, ε_q to γ, and $\varphi(l)$, from the Bogoliubov-Valatin transformation, is simply related to $\delta^*(\omega)$ (= $\varphi(l)+l$, where l refers to a notional ω_l). Note that their Toeplitz matrix's symbol is $\exp i\delta^*$. Presumably δ' could index them just as well.

Kaufman and Onsager formulate their analysis in terms of their angle δ'. Montroll, Potts, and Ward show, in their appendix, that the formulation they find, in terms of δ^*, is equivalent to Kaufman and Onsager's. Kaufman and Onsager actually calculate (p. 1251) the elements, Σ_a, of their Toeplitz matrices Δ_k and Δ_{-k}, and in figure 4, p. 1248 of their paper (our Figure 4.1) they plot the values and so explain the spontaneous magnetization's appearance. At very high or low temperatures, only one matrix element (that is, only one diagonal row) is nonzero. [Σ_a corresponds to a_n, the letter a being different in each case: in the first it is a numerical index, in the second it is the symbol.]

(e) The phase angle, $\delta^*(\omega)$, either rotates 360 degrees, or remains small and returns to zero, as ω goes from $-\pi$ to $+\pi$, depending on whether we are well above or below the critical point. These would be reversed if we worked in terms of the symbol $\exp i\delta'$ and the angle δ'. See also, note 7 below.

(f) Kaufman and Onsager employ Kaufman's spinor representation. Potts and Ward used the determinant techniques of Kac and Ward, which jury-rigs in the needed minus signs; while Montroll, Potts, and Ward use the Pfaffian method of Kastelyn, and of Hurst and Green, which formally builds in the needed minus signs, just what is adopted by Wu and McCoy in their subsequent work. (Kaufman, "Crystal Statistics, II"; M. Kac and J.C.

Ward, "A Combinatorial Solution of the Two-Dimensional Ising Model," *Physical Review* 88 (1952): 1332–1337; R. B Potts and J.C. Ward, "The Combinatorial Method and the Two-Dimensional Ising Model," *Progress of Theoretical Physics* (Kyoto) 13 (1955): 38–46, especially p. 45 on the different angles; Montroll, Potts, and Ward, "Correlations and Spontaneous Magnetization of the Two-Dimensional Ising Model"; Hurst and Green, "New Solution of the Ising Problem from a Rectangular Lattice"; Kastelyn, "Dimer Statistics and Phase Transitions"; McCoy and Wu, *The Two-Dimensional Ising Model*.)

7.

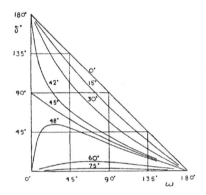

FIGURE N.3: The disorder phase angle, δ^*, which determines the amplitude of the disorder operator for the Ising lattice, Figure 5 from Onsager, "Crystal Statistics. I," (1944). $T = \infty$, T_c, ~0, for curves labeled 0°, 45°, 75° respectively.

For example, since δ' goes from π to 0 for low temperatures, for exp $i\delta'(\omega)$ the zeroth cosine fourier component (Σ_0) will be small or vanishing, and only the first positive component, Σ_1, will be significantly nonzero. For high temperatures, Σ_0 is the only significantly nonzero term. (Kaufman and Onsager, "Crystal Statistics, III," p. 1250 at equation 46, and p. 1251 under the subhead "The Correlation Curves.") In formulations in terms of δ^*, the nonzero matrix elements would be interchanged. My impression from reading some of Onsager's papers (here p. 1251) is that these numerical facts are crucial to his own understanding of his work.

8. For a nice discussion, see G. 't Hooft, "Renormalization of Gauge Theories," in L. Hoddeson, L. Brown, M. Riordan, and M. Dresden, eds., *The Rise of the Standard Model: Particle Physics in the 1960s and the 1970s* (New York: Cambridge University Press, 1997), pp. 178–198.

9. C.N. Yang, "The Spontaneous Magnetization of a Two-Dimensional Ising Model," *Physical Review* 85 (1952): 808–816.

10. C.N. Yang, *Selected Papers 1945–1980, with Commentary* (San Francisco: Freeman, 1983), p. 12; also, C.N. Yang, "Journey Through Statistical Mechanics,"

International Journal of Modern Physics B2 (1988): 1325–1329, p. 1327. Elsewhere, he writes to Mark Kac about "recovering" from this computation.

11. G. Benettin, G. Gallavotti, G. Jona-Lasinio, A.L. Stella, "On the Onsager-Yang-Value of the Spontaneous Magnetization," *Communications in Mathematical Physics* 30 (1973): 45–54.

12. See Yang, "The Spontaneous Magnetization," p. 811, Equation 34. Yamada suggests another perspective on the "artificial limiting process." He introduces another such process which then expresses the correlations as a ratio of determinants, but which determinants are zero in the limit (columns are equal). So he applies L'Hôpital's rule, and takes the derivatives of both numerator and denominator. But such a derivative, $d/d\theta$, is in effect an infinitesimal rotation. See, K. Yamada, "Pair Correlation Function in the Ising Square Lattice, Generalized Wronskian Form," *Progress of Theoretical Physics* 76 (1986): 602–612, at pp. 603, 608.

Baxter claims to be following Yang in his more recent algebraic derivation. (R. J. Baxter, "Algebraic Reduction of the Ising Model, *Journal of Statistical Physics* 132 (2008): 959-982.) The artificial limiting process is given by Baxter's $\exp(-\gamma J)$ $\gamma \rightarrow \infty$, where J is in effect the one dimensional hamiltonian. The transfer matrix is represented by a Hamiltonian, $\exp -\alpha H$, that H much like Onsager's $B + k^{-1}A$. As for Kaufman and Onsager's work, Baxter has unearthed the original manuscripts. He describes the two ways they did the correlation function (Szegő limit theorem, integral equation), and provides a third way of his own invention. (Baxter, R. J., "Onsager and Kaufman's calculation of the spontaneous magnetization of the Ising model," *Journal of Statistical Physics* 145 (2011): 518-548; "Onsager and Kaufman's calculation of the spontaneous magnetization of the Ising model: II," *Journal of Statistical Physics* 149 (2012): 1164-1167; "Some comments on developments in exact solutions in statistical mechanics since 1944," *Journal of Statistical Mechanics: Theory and Experiment* (2010) P11037, 26pp.

13. Kadanoff indicates Yang's in-effect employment of Wiener-Hopf methods:

Yang evaluated the trace of [an expression for the zero-field magnetization, Kadanoff's 3.30, see Figure 4.3 in our text] by finding all the eigenvalues of the operator inside the log [of 3.30] and then performing the indicated summation. The calculation is possible because the difference equation that appears in the eigenvalue problem is soluble through the use of the Wiener-Hopf method.

L.P. Kadanoff, "Spin-Spin Correlations in the Two-Dimensional Ising Model," *Il Nuovo Cimento* B44 (1966): 276–305, especially pp. 289–293, at p. 292. Yang actually solves the equation that gives the eigenvalues "by inspection" in IIIC. He never needs to refer to Wiener-Hopf. (Note that Yang's $\Theta(z)$ at Equation 61 is Onsager's $\exp i\delta'$. Also, the product that appears becomes a sum when one takes the logarithm, as for the Szegő limit theorem.)

14. Onsager got a ratio of theta functions. Onsager, "The Ising Model in Two Dimensions," pp. 11–12.

15. C.N. Yang, "Journey Through Statistical Mechanics," p. 1326. Yang read the paper when he was a graduate student at Chicago (1945–1949). He reports that when Luttinger told him about Kaufman's work (November 1949), "Being familiar with Onsager's difficult maneuvers, I quickly grasped the new perspective." See also, C.N. Yang, "Path Crossing with Lars Onsager," in L. Onsager, *The Collected Works of Lars Onsager*, ed. by P. Hemmer, H. Holden, S. Kjelstrup Ratkje (Singapore: World Scientific, 1996), pp. 180–181.

16. See also, Schultz, Mattis, and Lieb, "Two-Dimensional Ising Model as a Soluble Problem of Many Fermions," pp. 870–871, for an account of Yang's limiting process for defining the spontaneous magnetization. (This is *not* Yang's artificial limiting process that restores rotation symmetry.) The quasiparticles in this system form Cooper pairs of fermions.

Technically, the spontaneous magnetization is an off-diagonal matrix element, between even and odd unperturbed eigenfunctions. (Recall, as a formal analogy, Montroll's resonance analogy between ordered and disordered states to characterize the system at the critical point, in E.W. Montroll, "The Statistical Mechanics of Nearest Neighbor Systems," *Journal of Chemical Physics* 9 (1941): 706–721.)

At least metaphorically, these ideas are generalized by Penrose, Onsager, and Yang into a notion of "off-diagonal long range order." For this generalization, the order is quantum mechanical in origin. Presumably, if the analogy is correct, it works in this statistical mechanical, classical, Ising case because there is a formal analogy between such statistical mechanical systems and a suitably chosen quantum mechanical system. C.N. Yang, "Concept of Off-Diagonal Long-Range Order and the Quantum Phases of Liquid Helium and of Superconductors," *Reviews of Modern Physics* 34 (1962): 694–704.

17. G.F. Newell and E.W. Montroll, "On the Theory of the Ising Model of Ferromagnetism," *Reviews of Modern Physics* 25 (1953): 353–389, at p. 376. In effect, Yang introduces a notional magnetic field, h, and then sends it to zero. Let F be the free energy, N the number of rows, say: Yang is computing $\partial F/\partial h = (F_{h/N} - F_0)/h/N$. See note 21.

18. Yang did not draw from Onsager's unpublished work on the spontaneous magnetization.

19. See Montroll, Potts, and Ward, "Correlations and Spontaneous Magnetization," pp. 308–309.

20. Newell and Montroll, in their 1953 article explaining the Onsager solution and the combinatorial solution of Kac and Ward, "On the Theory of the Ising Model," refer to Yang's solution as, "Although the method is quite obvious, the detailed calculations are both tricky and tedious. . . . The magnetization is thus described by a single matrix element [the off-diagonal matrix element] which formally looks quite simple. The evaluation of it is, however, rather complicated . . . " (pp. 376, 377) Our concern here has been with this tricky, tedious, and complicated derivation.

21. Schultz, Mattis, and Lieb, "Two-Dimensional Ising Model as a Soluble Problem of Many Fermions," pp. 870–871, show how Yang's analysis works to get a term linear in the magnetic field.

22. It seems that Kaufman, at one point, had shown Yang her notes on Onsager's work. (Yang, "The Spontaneous Magnetization of a Two-Dimensional Ising Model," at p. 809 n. 10.) See also, note 145.

Kuhn, "Revisiting Planck," suggests that scientists will often be unreliable in recalling their motivations, retrospectively suggesting that they were "confused" in some paper because it turned out to be right for the wrong reason or it was just wrong, when in fact they actually knew enough of what they were doing and it did made sense then (even if the argument turned out to be wrong). Still, I am convinced that Yang's testimony surely reflects how others, at least, took the paper, and probably how he did too: not sleepwalking, but surely filled with twists and turns in its being worked out.

23. D recalls the Toeplitz matrix of Kaufman and Onsager (1949), their Δs (Equation 45); and, eventually, of Potts and Ward (1955), and Montroll, Potts, and Ward (1963); and Wu and McCoy and collaborators (1966–1981).

24. Here, the sum is expressed as a product since in the Szegő theorem one is dealing with the logarithm of the Toeplitz matrix.

25. Montroll, Potts, and Ward, "Correlations and Spontaneous Magnetization," p. 309; Kaufman and Onsager, "Crystal Statistics III," p. 1252 (Equation 72).

26. A.E. Ferdinand and M.E. Fisher, "Bounded and Inhomogeneous Ising Models: Part I, Specific-Heat Anomaly of a Finite Lattice," *Physical Review* 185 (1969): 832–846.

27. See note 14, above, and Figure 4.3. Given my focus in this chapter, I could not survey many other contributions to understanding the Ising lattice, each within their own context. For a chronology, see pp. 278-280, 328-329, 408-412, 481-484 of B.M. McCoy, *Advanced Statistical Mechanics* (Oxford: Oxford University Press, 1910). Besides McCoy's and collaborators' work, I am thinking of the work of Kadanoff, Fisher, and Abraham. As for Kadanoff, see also, L.P. Kadanoff, "Correlations along a Line in the Two-Dimensional Ising Model," *Physical Review* 188 (1969): 859–863 (on operator product expansions and scaling); L.P. Kadanoff and H. Ceva, "Determination of an Operator Algebra for the Two-Dimensional Ising Model," *Physical Review* B3 (1971): 3918–3938 (order and disorder operators); L.P. Kadanoff and M. Kohmoto, "SMJ's Analysis of Ising Model Correlation Functions," *Annals of Physics* 126 (1980): 371–398. The "SMJ" paper [Sato- Miwa-Jimbo] is again a deliberate attempt to clarify earlier work by others by rederiving it through a method having more familiar origins. I have not discussed the Bethe Ansatz and Quantum Inverse Scattering approaches, nor the work on conformal field theory and perturbations of that theory to encompass magnetic fields and when the temperature is not T_c. On the Bethe Ansatz, see B. Sutherland, *Beautiful Models: Seventy Years of Exactly Solved Models in Statistical Mechanics* (Singapore: World Scientific, 2004). See also, note 167.

28. D.B. Abraham, "On the Correlation Functions of the Ising and X-Y Models," *Studies in Applied Mathematics* 51 (1972): 179–209, quote at p. 195.

29. One-dimensional particle scattering (scattering on a line) provides a host of other associations for understanding the Ising model. See, for example, E.H. Lieb and D.C. Mattis, *Mathematical Physics in One Dimension* (New York: Academic Press, 1966); E.H. Lieb, "Some of the Early History of Exactly Soluble Models," *International Journal of*

Modern Physic B11 (1997): 3–10; C.N. Yang and C.P. Yang, "One-Dimensional Chain of Anisotropic Spin-Spin Interactions," *Physical Review* 150 (1966): 321–327; Sutherland, *Beautiful Models*. Other crucial figures were McGuire and Sutherland. The key words are the Bethe Ansatz, Hopf algebras, and the Yang-Baxter relation, expressing the fact that we may think of this problem in terms of scattering, and just a sequence of two-body scatterings, in fact.

30. For an extension of Yang's method to correlation functions, see K. Yamada's papers: "On the Spin-Spin Correlation Function in the Ising Square Lattice," *Progress of Theoretical Physics* 69 (1983): 1295–1298; "On the Spin-Spin Correlation Function in the Ising Square Lattice and the Zero Field Susceptibility," *Progress of Theoretical Physics* 71 (1984): 1416–1418; "Pair Correlation Function in the Ising Square Lattice, Generalized Wronskian Form."

31. Eric Livingston has shown how this demand shapes actual proofs.

32. "We believe that $(4\pi)^{-2/3}$ (=0.185, 1/5.41) does not belong in [the kinetic energy bound formula] and hope to eliminate it someday." E.H. Lieb, "The stability of matter," *Reviews of Modern Physics* 48 (1976): 553–569, at p. 556 (p. 594 in E.H. Lieb, *The Stability of Matter: From Atoms to Stars, Selecta of Elliott H. Lieb,* ed. W. Thirring, edition 2 (Berlin: Springer, 1997). The good reason for the expectation is that Lieb and Thirring have a "classical" derivation (that is, a semiclassical approximation to quantum mechanics) in which the 4π factor does not appear. One would expect to improve the bound by a factor of 5.4. See note 66 below for more recent attempts to control this factor.

33. V. Bach, "Error bound for the Hartree-Fock energy of atoms and molecules," *Communications in Mathematical Physics* 147 (1992): 527–548; V. Bach, "Accuracy of mean field approximation for atoms and molecules," *Communications in Mathematical Physics* 155 (1993): 295–310; G.M. Graf and J.-P. Solovej, "A correlation estimate with application to quantum systems with Coulomb interactions," *Reviews in Mathematical Physics* 6 #5a (1994): 977–997.

34. The first term in the expansion was rigorously derived by Lieb and Simon, using Thomas-Fermi theory; Scott provided a heuristic or physical derivation of the second term (and Schwinger an improved physical one); Siedentop and Weikard, and Hughes, provided a rigorous derivation; Dirac and Schwinger each provided a physical account of the third term. The physical accounts demanded the imagination of the strongest of physicists. But that is far from a mathematically rigorous derivation of the form or the coefficient. For a survey, see L. Spruch, "Pedagogic notes on Thomas-Fermi theory (and on some improvements): atoms, stars, and the stability of bulk matter," *Reviews of Modern Physics* 63 (1991): 151–209; Lieb, *Stability of Matter*, has the relevant articles by Lieb and collaborators on Thomas-Fermi theory.

35. W. Thirring, *Quantum Mechanics of Large Systems, A Course in Mathematical Physics 4* (New York: Springer, 1983), pp. 256–258. This is not quite a three-page proof, since much needed material is developed earlier in the textbook. The same observation applies to L. Spruch, "Pedagogic notes on Thomas-Fermi theory," pp. 191–194. E. Lieb and R. Seiringer, *The Stability of Matter in Quantum Mechanics* (New York: Cambridge

University Press, 2009), is able eliminate any reference to the Thomas-Fermi model, by developing suitable inequalities. So the three derivations they provide stand on the shoulders of earlier chapters.

36. Spruch points out that Schwinger's conceptual analysis of the Z^2 term (the "Scott correction") is illuminating (although it is deliberately not a rigorous derivation, again that derivation having been later provided by Siedentop and Weikard, and Hughes); this also applies to Schwinger's derivation of the $Z^{5/3}$ term (which is rigorously derived by Fefferman-Seco). Spruch, "Pedagogic notes on Thomas-Fermi theory," pp. 195–202. See also, J. Mehra and K.A. Milton, *Climbing the Mountain, The Scientific Biography of Julian Schwinger* (New York: Oxford University Press, 2000), pp. 538–543; or, J. Schwinger, *A Quantum Legacy, Seminal Papers of Julian Schwinger*, K.A. Milton, ed. (Singapore: World Scientific, 2000), pp. 761–763, which has roughly the same commentary as in the biography.

37. C. Fefferman, "The Uncertainty Principle," *Bulletin (New Series) of the American Mathematical Society* 9 (1983): 129–206, at p. 164.

38. Spruch, "Pedagogic notes on Thomas-Fermi theory," at p. 192.

39. E.H. Lieb and W. Thirring, "Bound for Kinetic Energy Which Proves the Stability of Matter," *Physical Review Letters* 35 (1975): 687–689, at p. 688, equation 13; E.H. Lieb, "The Stability of Matter," *Reviews of Modern Physics* 48 (1976): 553–569, at p. 562, Theorem 10; P. Martin, "Lecture on Fefferman's Proof of the Atomic and Molecular Nature of Matter," *Acta Physica Polonica* B24 (1993): 751–770, at p. 765; J. Schwinger, "Thomas-Fermi model: The leading correction," *Physical Review* A22 (1980): 1827–1832, uses the virial theorem [at p. 1828, at equation 13] to get "the average electrostatic potential, produced by the electrons, at the position of the nucleus."

40. In doing the integral, the integral's limits are determined by the ball (z,R) and that brings down factors of R. The quote is from C. Fefferman, "The N-Body Problem in Quantum Mechanics," *Communications on Pure and Applied Mathematics* 39 (1986): S67–S109, pp. S89-S90.

41. Fefferman, "The N-Body Problem in Quantum Mechanics," p. S106.

42. "Phase space" as it is used here is not to be confused with a phase factor or its phase angle, or with a phase transition and phases of matter, or with "stationary phase." Each is a different use of "phase."

43. L. van Hove, "Quelques Propriétés Générales de L'Intégrale de Configuration d'un Système de Particules Avec Intéraction," *Physica* 15 (1949): 951–961; "Sur L'Intégrale de Configuration Pour les Systèmes de Particules à Une Dimension," *Physica* 16 (1950): 137–143; M.E. Fisher, "The Free Energy of a Macroscopic System," *Archive for Rational Mechanics and Analysis* 17 (1964): 377–410; E.H. Lieb and J.L. Lebowitz, "The Constitution of Matter: Existence of Thermodynamics for Systems Composed of Electrons and Nuclei," *Advances in Mathematics* 9 (1972): 316–398, at p. 351. I survey some of the various divisions of space into cubes in M.H. Krieger, *Constitutions of Matter* (Chicago: University of Chicago Press, 1996), pp. 41–45.

44. Technically, distorted plane waves are the eigenfunctions; they are the simple "plane waves" for this space. Instead of the plane wave approximation for a solution, exp $-ikx$, we have exp $-iS(x)$, and we might expand the eikonal function, $S(x)$, in a Taylor series.

45. Fefferman, "The Uncertainty Principle," pp. 176–179; V. Bach, "Approximative Theories for Large Coulomb Systems," in J. Rauch and B. Simon, eds., *Quasiclassical Methods* (Berlin: Springer, 1997), pp. 89–97. I use the term "reasonably" because for low-Z the asymptotic formula is not yet good enough.

46. J. Moser, "A Broad Attack on Analysis Problems," *Science* 202 (1978): 612–613.

47. L. Carleson, "The work of Charles Fefferman," in *Proceedings of the International Congress of Mathematicians* (Helsinki: Academy Sci. Fennica, 1978), pp. 53–56.

48. S. Mac Lane, *Mathematics, Form and Function* (New York: Springer, 1986), p. 452, on being good at approximations.

49. This is from a posted message on a mathematics website, December 1998. There have been improvements that increased the chance of understanding, but Appel's concern still stands, I believe.

50. L. Onsager, "Electrostatic Interaction of Molecules," *Journal of Physical Chemistry* 43 (1939): 189–196, at p. 189.

51. The paper is reproduced as an appendix in Krieger, *The Constitutions of Matter*.

52. Spruch, "Pedagogic notes on Thomas-Fermi theory," p. 191.

53. F. Dyson and A. Lenard, "Stability of Matter. I," *Journal of Mathematical Physics* 8 (1967): 423–434, at p. 427.

54. In subsequent work by van Hove ("Quelques Propriétés" and "Sur L'Intégrale") the hard sphere assumption is retained.

55. E.H. Lieb and R. Seiringer, *The Stability of Matter in Quantum Mechanics*. New York: Cambridge University Press, 2009. The quote is on pp. 125–126.

56. A. Lenard and F. Dyson, "Stability of Matter. II," *Journal of Mathematical Physics* 9 (1968): 698–711, at p. 698.

57. Lenard and Dyson, "Stability of Matter II," p. 711.

58. See an interview with Dyson at http://www.webofscience/play/freeman.dyson/ 106 and 107.

59. I have adjusted these numbers so that they count the number of fermions or electrons (again, the nuclei are treated as bosons) rather than the number of charged particles, and include the fact of electron spin. Dyson and Lenard, "Stability of Matter I," p. 424.

60. Lenard and Dyson, "Stability of Matter II," p. 699.

61. A. Lenard, "Lectures on the Coulomb Stability Problem," in A. Lenard, ed., *Statistical Mechanics and Mathematical Problems* (Lecture Notes in Physics 20) (Berlin: Springer, 1971), pp. iii–v, 114–135.

62. F. Dyson, "Stability of Matter," in *Statistical Physics, Phase Transitions, and Superfluidity (Brandeis 1966)*, M. Chrétien, E.P. Gross, and S. Deser, eds. (New York: Gordon and Breach, 1968), pp. 177–239, at p. 238.

63. Lenard and Dyson, "Stability of Matter II," p. 698.

64. P. Federbush, "A new approach to the stability of matter problem, I and II," *Journal of Mathematical Physics* 16 (1975): 347–351, 706–709, at p. 347.

65. Federbush, "A new approach to the stability of matter problem," p. 709.

66. Federbush, "A new approach to the stability of matter problem," p. 706.

67. Federbush, "A new approach to the stability of matter problem," p. 706.

68. Lieb and collaborators, and others, have worked at sharpening the proportionality constant from about −33 to −22 to −20 to perhaps −18 Rydbergs/electron by 1990 (the Dyson-Lenard number was about minus-10^{14}). (In natural units, 1 Rydberg is ¼, and Fefferman needs to get his proportionality constant to be −½ if he is to prove the gaseous state of hydrogenic matter, and in general it must be less than two times the binding-energy/electron.) E.H. Lieb, "The Stability of Matter," *Reviews of Modern Physics* 48 (1976): 553–569, p. 563, and E. Lieb, "The Stability of Matter: From Atoms to Stars," *Bulletin (New Series) of the American Mathematical Society* 22 (1990): 1–49, p. 25. See p. 149 of the text for the current best estimates of the proportionality constant.

More recent, better estimates of the Lieb-Thirring constant, L (= $2/(3\sqrt{3})$ = $\pi/\sqrt{3}$ × $2/(3\pi)$, the latter being the semiclassical value), have sharpened the proportionality constant for the stability of matter, although for the best value the Lieb-Oxford inequality is crucial. But L is still about a factor of two "too large" compared to the classical value for the Lieb-Thirring constant. The "classical value" for the proportionality constant is about −4.9 Ry, so for Fefferman we still have a way to go to get to −2 Ry. See, D. Hundertmark, A. Laptev, and T. Weidl, "New Bounds for the Lieb-Thirring Constants," *Inventiones Mathematicae* 140 (2000): 693–704; J. Dolbeault, Laptev, A., and M. Loss, "Lieb-Thirring inequalities with improved constants," *Journal of the European Mathematics Society* 10 (2008) 1121–1126. In all of these estimates I have assumed N is the number of electrons, and electron spin has been folded in already. The Lieb-Thirring constants have to be taken to the 2/3 power to get their effect on the proportionality constant.

See chapter 3, note 57, where I point out that, if we take Fefferman's estimates of the Thomas-Fermi binding energy at face value: for $Z=1$, the proportionality constant is −4.3 Ry, more than two times too large for Fefferman's atomic gas proof, which needs the proportionality constant to be less than two times larger than the ground state energy of an atom; for $Z > 4$ or so, we might have stability with a good enough constant.

69. For part B, I have followed Spruch's account in "Pedagogic notes on Thomas Fermi theory" (pp. 192–193).

70. E. Teller, "On the Stability of Molecules in the Thomas-Fermi Theory," *Reviews of Modern Physics* 34 (1962): 627–631.

71. Lieb, "Stability of Matter," p. 557, paraphrase.

72. Here, recall Onsager's basic insight, an electrostatic inequality: that, for the purposes of a bound, for each electron, the $N-1$ other electrons can be substituted for by one dummy-particle suitably located.

73. For more on the detailed structure of this proof, see Krieger, *Constitutions of Matter*, chapters 2, 3.

74. I review much of this in Krieger, *Constitutions of Matter*, chapters 1–3.

75. Spruch, "Pedagogic notes on Thomas Fermi theory," p. 172.

76. F.J. Dyson, "Ground state energy of a finite system of charged particles," *Journal of Mathematical Physics* 8 (1967): 1538–1545.

77. A. Lenard, in correspondence with the author, 28 August 1998.

78. See also Spruch, "Pedagogic notes on Thomas-Fermi theory" pp. 205–208.

79. Fefferman, "Uncertainty Principle," pp. 142–144. See note 68 above for better estimates of the Lieb-Thirring constant.

80. W. Thirring, "Preface," *Reviews of Mathematical Physics* 6 #5A (1994): v–x.

81. Thirring, "Preface," p. vi.

82. Fefferman, "The *N*-Body Problem," p. S72. Fefferman's 1986 paper in fact does not achieve what is needed for hydrogen, but does give bounds (off by a factor of two, roughly) for the binding energy if Z is between 40 and 86.

83. Of course, in actual nature, the potential is not of this form as r approaches 0, for the nucleus is not a point but is of finite size. But, since the energies here are low there is little sensitivity to nuclear-size effects. Moreover, Thirring points out, the relevant mass in the Rydberg (which is measured in electron-volts) is that of the electron:

> Since the Rydberg, which is measured in electronvolts (eV), is determined by the mass of the electron, it is the kinetic energy of the electrons rather than the size of the nuclei that matters most for stability. The lower bound from the size of the nucleus alone would be ~ N MeV [Million electron volts]. (Thirring, *Quantum Mechanics of Large Systems*, p. 18)

The relevant observation lies elsewhere:

> The important property of the Coulomb potential for stability is that $1/r$ is a function of positive type. . . . [T]he $1/r$ singularity is not the only danger for stability; even regular potentials with energies . . . that take on both signs can lead to instability. This shows the superficiality of the common opinion that stability is not a real physical problem, since actual potentials do not become singular. (p. 258)

And,

> [I]f one uses r exp–μr so that the potential is even zero at the origin and short range the system is not stable neither for bosons nor fermions. (p. 4, *Selected Papers of Walter E. Thirring*)

And Thirring also shows that "the problem of the stability of matter has nothing to do with the long range of the Coulomb potential."(*Quantum Mechanics of Large Systems*, p. 258)

84. "The essence of a proof of the stability of matter should be a demonstration that an aperiodic arrangement of particles cannot give greater binding than a periodic arrangement." (Lenard and Dyson, "Stability of Matter," p. 711.)

85. Fefferman, "Uncertainty Principle," p. 130.

86. Fefferman, "Uncertainty Principle," p. 165.

87. These are his Theorems 9 and 10. Theorem 11 says a bit more about why Lieb-Thirring's use of the Thomas-Fermi model actually works so well: if the Thomas-Fermi and quantum kinetic energies are comparable, then we would expect to find one particle in each of the cubes. (Fefferman, "Uncertainty Principle," p. 176, my paraphrase.)

88. Fefferman, "Uncertainty Principle," p. 165.

89. Fefferman, "Uncertainty Principle," pp. 201–202.

90. Fefferman, "Uncertainty Principle," pp. 202–204.

91. Fefferman, "Uncertainty Principle," pp. 132–139.

92. W. Thirring, "Preface," p. vii. Again, see note 66 for work on the Lieb-Thirring constant.

93. Here I am thinking of Bach's density matrix formalism and the use of second quantization. Sigal, "Lectures on Large Coulomb Systems," pp. 96–99 gives a nice summary of the Bach and Graf-Solovej work. What is notable is that certain steps appear in each derivation (Lieb-Thirring, Fefferman-Seco, Bach, Graf-Solovej) and so they resonate with each other: they have to go from $N \times N$ interactions to N interactions, and they have to have fermions.

Sigal points out that a main contribution to the binding energy comes from the exchange term (that is, identical particles), that the Dirac calculation assuming a Fermi gas on a torus gives the right coefficient, that Fefferman-Seco start out with Dirac's idea, that there is a more general formulation for the exchange term, and that the task that remains is to go from two-particle to one-particle operators. Bach and Graf-Solovej need to decouple the potential into the product of two one-particle factor spaces. To do so they "complete the square," "manage the waste" or the fluctuation term (these two steps resonate with the Fefferman-Seco paper); and then they invoke the fermionic character of the electrons.

94. Krieger, *Constitutions of Matter*, chapter 3, reviews some of this work. The crucial paper is E.H. Lieb and J.L. Lebowitz, "The Constitution of Matter: Existence of Thermodynamics for Systems Composed of Electrons and Nuclei," *Advances in Mathematics* 9 (1972): 316–398. See also, Lieb, *Stability of Matter*, Part VI.

95. C. Fefferman, "The Thermodynamic Limit for a Crystal," *Communications in Mathematical Physics* 98 (1985): 289–311.

96. Fefferman, "The N-Body Problem," p. S88.

97. For a very different kind of account, albeit in a semi-relativistic context, see V. Bach, J. Fröhlich, and I.M. Sigal, "Quantum Electrodynamics of Confined Nonrelativistic Particles," *Advances in Mathematics* 137 (1998): 299–395.

98. Fefferman, "The N-Body Problem," p. S71.

99. C. Fefferman, "The Atomic and Molecular Nature of Matter," *Revista Matematica Iberoamericana* 1 (1985): 1–44, at p. 29.

100. Fefferman, "Atomic and Molecular Nature of Matter," p. 5. For another summary, see Fefferman, "The *N*-Body Problem," pp. S78–S88.

101. C. Fefferman, "Some Mathematical Problems Arising from the Work of Gibbs," *Proceedings of the Gibbs Symposium, Yale University, May 15–17, 1989*, eds. D.G. Caldi, G.D. Mostow (Providence: American Mathematical Society, 1990), p. 159.

102. Fefferman, "Some Mathematical Problems," p. 158.

103. See Martin, "Lecture on Fefferman's Proof of the Atomic and Molecular Nature of Matter," pp. 758–799, for a nice exposition.

104. Fefferman, "The *N*-Body Problem, p. S72.

105. Fefferman, "The *N*-Body Problem," p. S89.

106. Fefferman, "The *N*-Body Problem," p. S89.

107. H.F. Trotter, "Use of symbolic methods in analyzing an integral operator," in E. Kaltofen and S.M. Watt, eds., *Computers and Mathematics* (New York: Springer, 1989), pp. 82–90, at p. 82. Fefferman does say,

> Our attack on many-body Coulomb systems is based on the following philosophy. To each ball we shall associate a potential and a kinetic energy. We shall then write the energy of a Coulomb system as an integral over all possible balls *B* of the energy associated to *B*. The potential energy for a single ball *B* will be very elementary—much easier than the Coulomb potential. Similarly, the kinetic energy for *B* will be much easier to understand than the true kinetic energy for all particles. In particular, we will be very happy whenever *B* contains lots of particles, because then the energy for *B* will be obviously positive. This is in marked contrast with the original Hamiltonian, which looks opaque for large *N*. (Fefferman, "The *N*-Body Problem," p. S89.)

Surely, this is helpful, but I think the Trotter quote is more physical, and gives away more about what is going on.

108. Fefferman, "The *N*-Body Problem," p. S105, Equation 77.

109. For summaries see C. Fefferman, "Atoms and Analytic Number Theory," in F. Browder, ed., *Mathematics into the Twenty-First Century, Proceedings of the AMS Centennial Symposium*, volume 2 (Providence: American Mathematical Society, 1992), pp. 27–36; and, C. Fefferman, V.Ja. Ivrii, L.A. Seco, and I.M. Sigal, "The Energy Asymptotics of Large Coulomb Systems," in E. Balsev, ed., *Schroedinger Operators* (Berlin: Springer, 1992), pp. 79–99.

110. See J. Schwinger's (1980, 1981) work on the strongly bound electrons (the Scott correction to the Thomas-Fermi binding energy, the latter $\approx Z^{7/3}$), $Z^{7/3}/Z^{1/3} = Z^2$; and on the semiclassical asymptotics or quantum corrections and exchange effects, $Z^{7/3}/Z^{2/3} = Z^{5/3}$. Spruch has a nice review of this. Englert and Schwinger then took on shell effects or atomic binding energy oscillations, $Z^{7/3}/Z^{3/3} = Z^{4/3}$, actually $Z^{4/3} \times Z^{1/3}r$, in the end actually $Z^{3/2}$. (B.-G. Englert and J. Schwinger, "Semiclassical atom," *Physical Review* 32A (1985): 26–35, at p. 33.) The semiclassical analysis, a refined WKB method, stood between Thomas-

Fermi averaging and the Hartree-Fock shell model. B.-G. Englert provides a complete survey in *Semiclassical Theory of Atoms* (Berlin: Springer, 1988). Fefferman and Seco's task was to derive as much of this as possible with suitable mathematical rigor. In the case of the binding energy oscillations at the $Z^{4/3}$ level, they get a series in the angular momentum, ℓ, in integrals to the ½ power of the potential including the $\ell(\ell+1)/r^2$ term, rather than a $Z^{4/3}$ term.

Schwinger's Airy integrals appear as well in Fefferman-Seco.

A. Córdoba, C.L. Fefferman, and L.A. Seco, "A Trigonometric Sum Relevant to the Nonrelativistic Theory of Atoms," *Proceedings of the National Academy of Science* 91 (1994): 5776–5778; "Weyl Sums and Atomic Energy Oscillations," *Revista Matematica Iberoamericana* 11 (1995): 165–226; and, "A Number-Theoretic Estimate for the Thomas-Fermi Density," *Communications in Partial Differential Equations* 21 (1996): 1087–1102, show that the oscillating term is of order $Z^{3/2}$, and that corrections are of order Z^γ, with $\gamma < 3/2$.

111. An illuminating survey is provided by V. Bach, "Approximative Theories for Large Coulomb Systems."

112. C. Fefferman and L.A. Seco, "On the Dirac and Schwinger Corrections to the Ground-State Energy of an Atom," *Advances in Mathematics* 107 (1994): 1–185, announced $a=2835$. Bach apparently gets $a=231/2$. None of these represent focused attempts to get a as small as possible.

113. Fefferman and Seco, "On the Dirac and Schwinger Corrections to the Ground-State Energy of an Atom," p. 13.

114. See the Introduction for further discussion of Mark Kac's famous question, a follow-on of Hermann Weyl's earlier work on the asymptotic eigenvalues of the differential equation of a stretched membrane.

115. Fefferman, "Atoms and Analytic Number Theory," pp. 27–28.

116. Spruch, "Pedagogic notes on Thomas-Fermi theory," pp. 203–204.

117. Again, as Thirring, "Preface," p. vii, puts it, "despite all the wisdom on pseudodifferential operators," the suspected bounds have yet to be reached.

118. By the Raylitz-Ritz variational principle, the lowest value, and the true energy, is with the exact wavefunctions; and so you get an upper bound, however poor, with other wavefunctions.

119. Fefferman and Seco, "On the Dirac and Schwinger Corrections to the Ground-State Energy of an Atom," pp. 140–141. They refer to the work on the density and the eigenvalue sum as being "long and complicated."

120. Again, Bach, and Graf and Solovej, subsequently provided an improved derivation. But my purpose here is to attend to the first effort, what might be least transparent.

121. Fefferman and Seco, "On the Dirac and Schwinger Corrections to the Ground-State Energy of an Atom," p. 11.

122. Fefferman and Seco, "On the Dirac and Schwinger Corrections to the Ground-State Energy of an Atom," pp. 7–8. Here I have filled in with ordinary verbiage an interpretation of the symbols, with the attendant risks of imprecision.

123. C. Fefferman, V.Ja. Ivrii, L.A. Seco, and I.M. Sigal, "The Energy Asymptotics of Large Coulomb Systems," at pp. 89–91.

124. Sato, Miwa, and Jimbo's holonomic quantum field methods, various cluster expansions (Nickel), perturbed conformal field theory, and work on Toeplitz matrices (see P. Deift, A. Its, and I. Krasovsky, "Toeplitz Matrices and Toeplitz Determinants under the Impetus of the Ising Model: Some History and Some Recent Results," *Communications on Pure and Applied Mathematics* 116 (2013): 1360-1438, at pp. 1366-1372 and 1386-1390.) are a range of different presentations, and one would hope for "an identity in that manifold presentation of profiles."

125. A. Lenard, "Momentum distribution in the ground state of the one-dimensional system of impenetrable bosons," *Journal of Mathematical Physics* 5 (1964): 930-943. See Deift, Its, Kasovsky for more on Lenard and on Dyson on the boson system and Toeplitz matrices, and random random matrices in the case of Dyson.

126. B.M. McCoy, "Integrable models in statistical mechanics: The hidden field with unsolved problems," *International Journal of Modern Physics* 14 (1999): 3921-3933.

127. One of Palmer and Tracy's purposes in writing their papers were to make rigorous the various improved derivations, which as they say have "holes" of positive measure in their rigor or statements. J. Palmer and C.A. Tracy, "Two-Dimensional Ising Correlations: Convergence of the Scaling Limit," *Advances in Applied Mathematics* 2 (1981): 329–388; J. Palmer and C.A. Tracy, "Two-Dimensional Ising Correlations: The SMJ Analysis," *Advances in Applied Mathematics* 4 (1983): 46–102. The quote is from the 1981 paper, at p. 330. See J. Palmer, *Planar Ising Correlations* (Boston: Birkhäuser, 2007).

128. H. Au-Yang and J.H.H. Perk, "Onsager's Star-Triangle Equation: Master Key to Integrability," in M. Jimbo, T. Miwa, and A. Tsuchiya, eds., *Integrable Systems in Quantum Field Theory and Statistical Mechanics*, Advanced Studies in Pure Mathematics, Tokyo, Japan, 19 (Boston: Academic Press, 1989; Tokyo: Kinokuniya), pp. 57–94, at p. 58.

129. M. Bander and C. Itzykson, "Quantum field theory calculation of the 2d Ising model correlation function," *Physical Review* D15 (1977): 463–469, at p. 463, call it "monumental."

130. H. Cheng and T.T. Wu, "Theory of Toeplitz Determinants and the Spin Correlations of the Two-Dimensional Ising Model. III," *Physical Review* 164 (1967): 719–735, at p. 732. For a history, see Deift, Its, and Krasovsky, "Toeplitz Matrices and Toeplitz Determinants under the Impetus of the Ising Model: Some History and Some Recent Results," at pp. 1366–1372 and 1386–1390.

131. Deift, Its, and Krasovsky, "Toeplitz Matrices and Toeplitz Determinants under the Impetus of the Ising Model: Some History and Some Recent Results;" McCoy, "Integrable models in statistical mechanics: The hidden field with unsolved problems;" B. M. McCoy, "The Baxter Revolution," *Journal of Statistical Physics* 102 (2001): 375-384; B. M. McCoy, "The Romance of the Ising model," *Symmetries, integrable systems*

and representations, 263–295, Springer Proc. Math. Stat., 40, *Springer, Heidelberg,* 2013; B. M. McCoy and J.-M. Maillard, "The Importance of the Ising model," *Progress in Theoretical Physics* 127 (2012): 791–817. Also, B. M. McCoy and T. T. Wu, *The Two-Dimensional Ising Model* has a new final chapter in its second edition summarizing developments since the first edition. And McCoy, *Advanced Statistical Mechanics* (2010), chapters 10-14, surveys those developments as well. As for elliptic curves and the two-dimensional Ising model, on Maillard, see note 6 to the Preface.

132. C.A. Tracy and H. Widom, "Fredholm determinants, differential equations and matrix models," *Communications in Mathematical Physics* 163 (1994): 33–72. Also, H. Widom, "On the Solution of a Painlevé III Equation," (solv-int/9808015 24 Aug 1998), which connects Fredholm to Painlevé directly.

133. Note, however, that there are ways of achieving this infinite volume limit without the Toeplitz matrix, for example Baxter's corner transfer matrices, which in this case emphasize that the solution is a q-series, namely a modular form. Baxter, *Exactly Solved Models,* chapter 13.

134. The quote is from C.A. Tracy and H. Widom, "The Thermodynamic Bethe Ansatz and a Connection with Painlevé Equations," *International Journal of Modern Physics* B11 (1997): 69–74, at p. 71. It reads: "Painlevé functions and generalizations arise in statistical mechanics in the following way: Correlation functions or closely related quantities are shown to be expressible in terms of Fredholm determinants (or minors) and these determinants (or minors) are then explicitly evaluated in terms of solutions to total systems of differential equations. The Painlevé equations arise as the simplest examples of these total systems."

135. Craig Tracy has pointed out to me that if the underlying field theory is not that of free fermions, one won't get the Toeplitz matrix even if the problem is exactly solvable with a Bethe Ansatz. On the other hand, one might still end up with a Painlevé-type solution or at least the Tracy-Widom distributions. Similarly, the $\langle \sigma \sigma_{M,N} \rangle$ formulas do not have Toeplitz form, but we would expect the results to be much the same as for those that do ($\langle \sigma \sigma_{N,N} \rangle$).

136. See also, p. 205 of Krieger, *Constitutions of Matter.* There is always the hope that the zeros of the zeta function or other such would be the eigenvalues of an infinite-dimension hermitian matrix. For the Tracy-Widom distribution, the associated Toeplitz matrix's symbol, also called a generating function, is $\exp \sqrt{t} \times (z + z^{-1})$. For a survey of some of the connections, see, for example, P.M. Bleher and A.R. Its, eds., *Random Matrix Models and Their Applications* (Cambridge: Cambridge University Press, 2001); or, T. Bronson, ed., *Spectral Problems in Geometry and Arithmetic* (Providence: American Mathematical Society, 1999), as well as Deift, Its, and Krasovsky, "Toeplitz Matrices and Toeplitz Determinants under the Impetus of the Ising Model: Some History and Some Recent Results." Pervasive themes are integrable systems and combinatorics. Parts of some of the connections I adumbrate in chapter 5 (zeta functions and statistical mechanical partition functions) are developed in greater detail in the Bronson volume.

137. See pp. 1422–1425 of Deift, Its, and Krasovsky, "Toeplitz Matrices and

Toeplitz Determinants under the Impetus of the Ising Model: Some History and Some Recent Results," for further discussion. The quote comes from p. 1422. In the context of orthogonal polynomials on the unit circle, I began to see a more coherent story, albeit the experts would have to fill it in and correct it. See P. Deift, "Riemann-Hilbert Methods in the Theory of Orthogonal Polynomials," in F. Gesetzsky, et. al., *Spectral Theory and Mathematical Physics*. Providence: American Mathematical Society, 2007, pp. 715–740, and B. Simon, *Orthogonal Polynomials on the Unit Circle*. Providence: American Mathematical Society, 2005.

138. P. Deift, "Integrable Systems and Combinatorial Theory," p. 635, explaining Okounkov's work on random surfaces.

139. See chapters 10 and 17 in McCoy and Wu, *The Two-Dimensional Ising Model*, 2 edition (2014).

140. B. M. McCoy, M. Assis, S. Boukrass, S. Hassani, J.-M. Maillard, W. P. Orrick, and N. Zenine, "The saga of the Ising susceptibility," *New trends in quantum integrable systems* (Hackensack, NJ: World Scientific, 2011): 287–306.

141. D. Wilkinson, "Continuum derivation of the Ising model two-point function," *Physical Review* D17 (1978): 1629–1636, at p. 1630.

142. And there are other ways of going from the determinant to the solution in terms of Painlevé transcendents, through the sinh-Gordon equation, or more directly as do Tracy and Widom.

143. Wilkinson, "Continuum derivation of the Ising model," p. 1634.

144. There is an error in the original paper (in equations 6.109, 6.107, and 6.110) as noted by Tracy (C.A. Tracy, "Asymptotics of a τ-Function Arising in the Two-Dimensional Ising Model," *Communications in Mathematical Physics* 142 (1991): 297–311.), p. 298, note 2), which, however, does not affect the numerical check of equation 6.104, which is correct.

145. The quotes are from Deift, "Riemann-Hilbert Methods," p. 723. As indicated earlier, I am not satisfied with my understanding of the connection, albeit its being central to the mathematics. Put differently, what is it about the physics of Toeplitz and the physics of Wiener-Hopf method and Riemann-Hilbert that makes them connected. That the Wiener-Hopf method was developed to solve an equation that is the continuous analog of the Toeplitz matrix (as its kernel, a convolution operator) does not help me, at least now. See N. Wiener, N. and E. Hopf, "Über eine Klasse Singulärer Integralgleichungen," *Sitzberichte der Preussischen Akademie der Wissenschaften, Phys.-Math. Klasse* (1931): 696–706 (note that they do not use what we would call a Wiener-Hopf sum equation); H. Widom, "Commentary," in C. Morawetz, J. Serrin, and Ya.G. Sinai, eds., *Selected Works of Eberhard Hopf with Commentaries*. Providence: American Mathematical Society, 2002, pp. 47–48; G. Szegő, "Ein Grenzwertsatz über die Toeplitzschen Determinanten einer reellen positive Funktion," Mathematische Annalen 76 (1915): 490–503; B.M. McCoy, "Introductory Remarks on Szegő's Paper, "On Certain Hermitian Forms Associated with the Fourier Series of a Positive Function"," in R. Askey, ed., *Gabor Szegő: Collected Papers, Volume 1*, Boston:

Birkhäuser, 1982, pp. 47–52. For applications, see N.L. Hills, N.L. and S.N. Karp, "Semi-Infinite Diffraction Grating-I," *Communications on Pure and Applied Mathematics* 18 (1965): 203–233; A.N. Ezin, A.N. and A.L. Samgin, "Kramers problem: Numerical Wiener-Hopf-like model characteristics," *Physical Review E* 82 (2010): 056703-1-15; H. Dai, Z. Geary, and L.P. Kadanoff, "Asymptotics of eigenvalues and eigenvectors of Toeplitz matrices," *Journal of Statistical Mechanics: Theory and Experiment* (2009): P05012-1-25, who compute the eigenvalues for a Toeplitz matrix, and so one might compute the determinant by multiplying them.

146. Technically and mathematically, for nice symbols, $a = \exp b$, the Toeplitz matrix, $S(a)$, will prove to be invertible. If a can be expressed as $a_- \times a_+$ (in essence, the Wiener-Hopf factorization), where a_- is expressed in terms of a_n, $n \leq 0$, and similarly for a_+ for $n \geq 0$, then S can be split into two parts, one upper triangular, the other lower triangular, sharing the diagonal. (Also, $S(a)^{-1} = S(1/a_+) \times S(1/a_-)$.) In general $S(a) \times S(b)$ does not equal $S(ab)$, and their difference may be expressed in terms of the product of highly symmetric (Hankel) matrices of the same symbols, but now upper and lower triangular. Now the asymptotic expression for the logarithm of the determinant of the Toeplitz matrix has a term $\Sigma[l \times b(l) \times b(-l)]$, where $b(l)$ is the fourier transform of the logarithm of the generating function, a. This curious term comes from the Hankel matrices. For more on Toeplitz and Ising, see Deift, Its, and Krasovsky, "Toeplitz Matrices and Toeplitz Determinants under the Impetus of the Ising Model: Some History and Some Recent Results,"

In these particular Ising model cases, the logarithm of the determinant turns out to be expressible as *just* $\Sigma \, l \times b(l) \times b(-l)$. In the earlier, non-operator-theoretic derivation, the sum appears in a power series for $\ln (1 - \alpha \exp i\omega)$, say in McCoy and Wu's derivation, in a seemingly arbitrary device in the derivation of the Szegő limit theorem used to calculate the determinant—although a $(1 - k \exp -i\omega)$ term does appear in the actual generating function, $\exp i\delta^*$.

The $\Sigma \, [l \times b(l) \times b(-l)]$ term (divided by n) expresses the difference between the logarithm of the average value of the symbol and the average value of the logarithm of the symbol: here, $\log\langle \exp i\delta^* \rangle - \langle [\log \exp] i\delta^* \rangle$, the difference between the logarithm of average value of the probability amplitude and the average value of its phase angle. In this particular case, the latter average is zero. Formally, and this is surely stretching things, this generic form can be shown to be much like a variance (see below).

See A. Böttcher and B. Silbermann, *Introduction to Large Truncated Toeplitz Matrices* (New York: Springer 1999), pp. 9–18, 121–136; McCoy and Wu, *The Two-Dimensional Ising Model*, pp. 223–224; H. Widom, "Wiener-Hopf Integral Equations," *The Legacy of Norbert Wiener: A Centennial Symposium, Proceedings of the Symposia in Pure Mathematics* 60 (1997): 391–405, p. 402.

By a "generic" Szegő limit theorem: the logarithm of the correlation function is given by:

$$\ln \det S \sim n/2\pi \times \int_{-\pi \text{ to } \pi} [\ln \exp] \, i\delta^*(\omega) \, d\omega \; - \sum q \, \delta^*(q) \, \delta^*(-q) + o(1),$$

where n is the size of the Toeplitz matrix, S, and $\delta^*(q)$ is the fourier transform of $\delta^*(\omega)$. (The minus sign on the sum term is due to the $i\delta^*$s.) By symmetry of $\delta(\omega)$, the first term on the right cancels out. As for the phase angle, see Schultz, Mattis, and Lieb, "Two-Dimensional Ising Model as a Soluble Problem of Many Fermions," pp. 863, 865–866, Equations 3.30, 4.14 and 4.18. My reference to a "variance" comes from an analogy offered by Widom, pp. 402–404 (also Böttcher and Silbermann, pp. 132–136) who also gives an interpretation in terms of spectral asymptotics. If the Toeplitz operator is S, the Σ-term represents:

$$2 \text{ trace } [\log S(g) - S(\log g)],$$

and the analogy is to replace the "log" by the "square" (2 trace $[(S(g))^2 - S(g^2)]$).

In general, a trace of a matrix (divided by its dimension, n) "is" an arithmetic average; the nth root of a determinant is a geometric average. The normed trace is also the average value of the "Rayleigh quotient," $\langle Ax \ x\rangle/\langle x \ x\rangle$.

The Toeplitz determinant can also be expressed in terms that recall distributions of eigenvalues of random matrices, with the distributions' Jacobians expressed in terms of a discriminant, Π |exp $i\theta_l$ – exp $i\theta_m$| $d\theta$. Johansson points out that the contribution to the integral is maximal when the angles are uniformly spaced around the circle (as in the Coulomb gas analog), and with a reasonable symbol the maximal contribution comes from an almost uniform spacing. K. Johansson, "On Szegő's asymptotic formula for Toeplitz determinants and generalizations," *Bullétin des Sciences Mathematiques* 112 (1988): 257–304; "On random matrices from the compact classical groups," *Annals of Mathematics* 145 (1997): 519–545.

147. I believe that these techniques were employed by Onsager for his first announcement of the result, in about 1948. We have a letter from Onsager to Kaufman dated April 12, 1950, where he refers to "a general formula for the value of an infinite recurrent determinant" (reproduced in *Journal of Statistical Physics* 78 (1995): 585–588), presumably what becomes the relevant Szegő limit theorem. See also, Baxter, "Some comments on developments in exact solutions in statistical mechanics since 1944."

148. Onsager, "The Ising Model in Two Dimensions," p. 11. See Schultz, Mattis, and Lieb, "Two-Dimensional Ising Model as a Soluble Problem of Many Fermions," section V, for a careful discussion of the definition of the spontaneous magnetization.

149. Onsager, "The Ising Model in Two Dimensions," pp. 10–12. He concludes with, "before I knew what sort of conditions to impose on the generating function, we talked to Kakutani and Kakutani talked to Szegő and the mathematicians got there first."

150. W. Feller, *An Introduction to Probability Theory and Its Applications*, volume II, edn. 2 (New York: Wiley, 1971), pp. 403–404. Feller first sets up separate solvable equations for each of the two problems, and then shows how a Wiener-Hopf form can be developed by combining the equations. As for the Kramers Problem, see Ezin and Samgin, "Kramers problem: Numerical Wiener-Hopf-like model characteristics," p. 0567036. They refer to the work on the spontaneous magnetization in the 2-d Ising model (Wu, McCoy and Wu), the Wiener-Hopf sum equation that appears in both their

context and in Wu, and then draw the following analogy: "Since the function S_N [the correlation in Ising] is given in the form of a Toeplitz determinant, we ascribe to $P(\Delta)$, thus defined, a similar sense of a correlation function which is capable of reducing the total ionic conductivity a real super ionic conductor."

151. T.T. Wu, "Theory of Toeplitz Determinants and the Spin Correlations of the Two-Dimensional Ising Model. I," *Physical Review* 149 (1966): 380–401, quotation at pp. 380–381. As far as I can tell, such a device might well have been part of conventional practice. As for its mathematical foundation, see Deift, "Riemann-Hilbert Methods," pp. 715–716. But see Szegő, "Ein Grenzwertsatz über die Toeplitzschen Determinanten einer reellen positive Funktion," of 1915, p. 494.

152. Accounts employ either the disorder phase angle δ^* or the order angle δ'. (See note 6, above.) People show that the two accounts give the same results, so that, for example, Potts and Ward, "The Combinatorial Method and the Two-Dimensional Ising Model, p. 45, discuss this. But note that the angles are notionally complementary in that they are the angles of a hyperbolic triangle (see Figure N.2), and $\delta^*+\delta'+\omega < \pi$. For a lattice in which the horizontal and vertical couplings are equal, they represent high and low temperatures respectively (or vice versa).

153. Onsager, "Crystal Statistics. I," p. 137. The operator that introduces disorder is of the form: $\exp -i[Z(\omega)\times\delta^*(\omega)]/2$, operating on the totally ordered state (Equation 96b). The quote in the text begins, "A detailed inspection of the operators $Z(\omega)$ can be made to show that those for which ω is small . . . " $Z(\omega)$ is defined on p. 128, Eq. 62, of the article. In effect, it is a raising or lowering operator.

Correspondingly, the operator that introduces order into the totally random state is of the form: $\exp -i[Z(\omega)\times(\pi -\omega -\delta^*(\omega))/2]$. Either the totally ordered state $(|\uparrow\rangle \pm |\downarrow\rangle)/\sqrt{2})$ or the totally random state (all configurations equally likely) can be used as a basis to build up the lattice at a particular temperature.

154. Wu, "Theory of Toeplitz Determinants and the Spin Correlations of the Two-Dimensional Ising Model. I," p. 381, Equations 1.10–1.14.

155. Wiener-Hopf techniques are needed as well to solve the Kondo problem, another such phase transition. See also, A.C. Hewson, *The Kondo Problem to Heavy Fermions* (Cambridge: Cambridge University Press, 1997).

156. For (5) in the text, see note 6 above on "paths." Also, Böttcher and Silbermann, *Introduction to Large Truncated Toeplitz Matrices*, pp. 127–131.

Say we took the perspective of the Schultz, Mattis, and Lieb ("Two-Dimensional Ising Model as a Soluble Problem of Many Fermions") account of the lattice as being populated by quasiparticles. Then we might say that the specific heat of the lattice is due to the excitation of the quasiparticles (Krieger, *Constitutions of Matter*, pp. 146–149). The correlations are also due to contributions of the quasiparticles. (Schultz, Mattis, and Lieb, p. 865.) This should lead to an interpretation of the a_n and the Toeplitz determinant.

For (6) in the text, see note 6[d(i)] above for more on building up the lattice from above and below.

157. See H. Dai, Z. Geary, and L.P. Kadanoff, "Asymptotics of eigenvalues and

eigenvectors of Toeplitz matrices," for further development of the particle interpretation, and the eigenvectors and eigenvalues derived from the translational symmetry of the Toeplitz matrix as plane waves.

158. Deift, Its, and Krasovsky, "Toeplitz Matrices and Toeplitz Determinants under the Impetus of the Ising Model: Some History and Some Recent Results," begins to do exactly this.

159. In the analogy between statistical mechanics and quantum field theory, the connections are rather more formally made. McCoy, "The Connection Between Statistical Mechanics," indicates these in some detail.

As I have indicated earlier, in the realm of analogy there may be some distance between suggestive evidence and proven theorems.

160. Here I paraphrase and translate R. Garnier, pp. 14 and 16, in his preface to *Oeuvres de Paul Painlevé*, volume I (Paris: Éditions du Centre National de la Récherches Scientifique, 1972).

161. Quoted on p. 301 of Simon, *Orthogonal Polynomials*.

162. But "no method is algebraically trivial," at least so far. Nickel, "Addendum to 'On the singularity structure of the 2D Ising model susceptibility'," *Journal of Physics A, Mathematical and General* 33 (2000): 1693–1711, at p. 1696. But see McCoy, "Introductory Remarks on Szegő's Paper, "On Certain Hermitian Forms Associated with the Fourier Series of a Positive Function"," for a quick survey going from Toeplitz, to nonlinear partial difference equations relating the various correlations, and then going to the scaling limit, and then one has the Painlevé III differential equation.

163. Palmer and Tracy, "Two-Dimensional Ising Correlations: The SMJ Analysis," at p. 46.

164. V.E. Korepin, N.M. Bogoliubov, A.G. Izergin, *Quantum Inverse Scattering Method and Correlation Functions* (New York: Cambridge University Press, 1993); D.B. Abraham, "Some Recent Results for the Planar Ising Model," in D.E. Evans and M. Takesaki, *Operator Algebras and Applications, volume II: Mathematical Physics and Subfactors*, London Mathematical Society Lecture Note Series, 136 (Cambridge: Cambridge University Press, 1988), pp. 1–22.

165. For a survey, see the Introduction in U. Grimm and B. Nienhuis, "Scaling limits of the Ising model in a field," *Physical Review* E55 (1997): 5011–5025, at pp. 5011–5012. For the work on conformal field theory and perturbations to include nonzero magnetic field, see P. Fonseca and A. Zamolodchikov, "Ising Field Theory in a Magnetic Field: Analytic Properties of the Free Energy, *Journal of Statistical Physics* 110 (2003): 527-590, and references therein for Zamolodchickov's pioneering work. According to McCoy, there are three approaches to nonzero magnetic field: perturbation theory, scaling, and conformal field theory.

166. The Schultz, Mattis, and Lieb derivation of the Ising model partition function is one of the cruxal points in such a history. I have devoted much of *Constitutions of Matter* to showing how this naturalness and physical-ness has emerged.

167. A nice summary of the argument of WMTB is provided on p. 1060 of B.M. McCoy, C.A. Tracy, and T.T. Wu, "Painlevé functions of the third kind," *Journal of Mathematical Physics* 18 (1977): 1058–1092.

168. Cheng and Wu, "Theory of Toeplitz Determinants," Appendix A, addresses these convergence issues.

169. Wilkinson, "Continuum derivation of the Ising model," p. 1633.

170. E.L. Ince, *Ordinary Differential Equations* (New York: Dover, 1956 [1926]), p. 351.

171. B.M. McCoy, C.A. Tracy, and T.T. Wu, "Painlevé functions of the third kind," at p. 1060. See also, B. Nickel, "On the singularity structure of the 2D Ising model susceptibility," *Journal of Physics A, Mathematical and General* 32 (1999): 3889–3906. Nickel begins with WMTB's equation 7.43 (at the end of their paper), and shows that it is in the form of a dispersion series, as well as provides evidence of the fermionic nature of the problem (p. 3890). Nickel also provides a dictionary, connecting Onsager-Yang elliptic function parametrizations with trigonometric function parametrizations (Appendix C, p. 1710). Yamada, "On the Spin-Spin Correlation Function in the Ising Square Lattice and the Zero Field Susceptibility," derives the same series, it would seem. Yamada, "Pair Correlation Function in the Ising Square Lattice," refers to Baxter's extension of the Montroll, Potts, and Ward formula, and its connection with his work (R.J. Baxter, "Solvable Eight-Vertex Model on an Arbitrary Planar Lattice," *Philosophical Transactions of the Royal Society* 289A (1978): 315–344, at pp. 342–343.).

172. In this case, it would seem that the functional equation embodies the basic symmetries most directly.

173. Newell and Montroll, "On the Theory of the Ising Model of Ferromagnetism," provide a motivated account of Onsager's proof, in 1953, writing down what was then part of the oral tradition among the cognoscenti.

CHAPTER 5

1. A nice presentation is in D. Cox, J. Little, and D. O'Shea, *Ideals, Varieties, and Algorithms* (New York: Springer, 1997). Earlier in the twentieth century, Emmy Noether developed the abstract or formal algebra needed to encompass what might be called these ideals of geometry, in effect how they may be decomposed into a finite number of smaller or simpler parts or a basis (as in Hilbert's Syzygy or Basis Theorem), much as ordinary integers have unique prime factors. That same commutative algebra is the device that is needed to make topology algebraic. Again, the problem is one of decomposing a (topological) structure into a sequence of "smaller" or simpler parts, much as in algebra, we may decompose an object into an (exact) sequence of parts.

This remarkable utility of the same body of algebra, from the same source, for two rather different fields, algebraic geometry and algebraic topology, is striking. As Weil puts it:

Whether it be class fields, multiple integrals on an algebraic variety, or homology properties of a space, one finds everywhere commutative groups . . . it is not a matter of chance that abelian functions have the same family name as abelian groups. [Qu'il s'agisse du corps de classes, d'intégrales multiples sur les variétés algébriques, ou des propriétés d'homologie d'un espace, ce sont partout des groups commutatifs qu'on fait intervenir, . . . ce n'est pas un effet du hasard que les fonctions abélienne portent le même nom que les groupes abéliens, . . .]

A. Weil, "Généralisation des fonctions abéliennes," *Journal des mathematiques pures et appliquées* 9 (1938): 47–87, at p. 47; also, *Oeuvres Scientifiques, Collected Papers*, volume I (New York: Springer, 1979), pp. 185–225.

The recurrent problem is to account for why algebra is so effective in topology, or why topological models are so effective for algebraic computations. It haunts chapter 3, although I have not made it so thematic there. As Mac Lane puts it:

Algebra (the manipulation of formulas according to rules) and topology (the classification of spaces under continuous deformations) seem intrinsically different, but they interact repeatedly.

Mac Lane argues that perhaps the connection here between the discrete and the continuous is part of a "network of formal ideas." See S. Mac Lane, "Group Extensions for 45 Years, *The Mathematical Intelligencer* 10 (1988): 29–35, at pp. 29, 35. Also, his "Origins of the Cohomology of Groups," *L'Enseignement Mathématique* 24 (1978): 1–29; and, *Mathematics, Form and Function*. New York: Springer, 1986. I earlier quoted from H. Weyl on the same problem, although he characterizes it as a tension. A historical answer would show how the machinery and ideas in one area influenced what people thought they might do in another (with that machinery and those ideas) and vice versa, to some extent what I have tried to do in chapter 3. Presumably topology or algebra might make new alliances, and new questions of this sort will arise. And of course algebra was remade for its topological work.

D. Corfield, *Toward a Philosophy of Real Mathematics* (Cambridge: Cambridge University Press, 2003) pp. 88–90, reviews the literature on analogy, and discusses Stone duality and the connection between Boolean algebra and topology.

See also, note 96(8), chapter 6.

2. For a review that encompasses Polya, Hilbert, and Poincaré see H. Benis-Sinaceur, "The Nature of Progress in Mathematics: The Significance of Analogy," in E. Grosholz and H. Breger, *The Growth of Mathematical Knowledge* (Dordrecht: Kluwer, 2000), pp. 281–293. Benis-Sinaceur is concerned with model theory, which describes formal analogies mathematically. More generally, she discusses the role of analogy for

invention and prediction in mathematics. Again, on analogy, D. Corfield, *Toward a Philosophy of Real Mathematics*, pp. 91–95.

3. The conception Robert Langlands proposed in about 1967, connecting automorphic forms to Galois groups, let us say, came to be known by others as the "Langlands Program," surely by 1984 (Gelbart, A. Borel). In time the notion had a life of its own, out of Langlands' control. In his programmatic statements over the last 25+ years he has tried to sketch fruitful directions for further research. His sentences are wonderful to read, complex and filled with qualifiers. Langlands described some work that he thought well of as, "unlikely to be emphemeral." His basic themes are *functoriality*, a form of universality, and *reciprocity* (echoing quadratic reciprocity and Artin reciprocity).

4. For a nice list of geometric objects connected to spectral objects, see J. Arthur, "Harmonic Analysis and Group Representations," *Notices of the American Mathematical Society* 47 (2000): 26–34, at p. 27, including: sum of diagonal entries of a square matrix vs. sum of eigenvalues of the matrix; conjugacy classes of finite groups vs. their irreducible characters; lengths of closed geodesics vs. eigenvalues of the laplacian; logarithms of powers of prime numbers vs. zeros of the zeta function; rational conjugacy classes vs. automorphic forms. The Langlands Program connects two spectral objects: "motives" and automorphic representations. "[T]he arithmetic information wrapped up in motives comes from solutions of polynomial equations with rational coefficients." (p. 26). As for a dynamical system, see A. Knauf, "Number Theory, Dynamical Systems, and Statistical Mechanics," *Reviews in Mathematical Physics* 11 (1999): 1027–1060. Also, J.-Ch. Anglès d'Auriac, S. Boukraa , and J.-M. Maillard, "Functional Relations in Lattice Statistical Mechanics, Enumerative Combinatorics, and Discrete Dynamical Systems," *Annals of Combinatorics* 3 (1999) 131–158. B.M. McCoy, *Advanced Statistical Mechanics* (Oxford: Oxford University Press, 2010), pp. 281–283 reviews the literature on zeros and symmetries for the Ising model.

Note that since the addends are all positive real numbers, the zeroes are either at complex values of s or they pinch the real line.

5. "The Trace Formula and its Applications, An Introduction to the Work of James Arthur," *Canadian Mathematics Bulletin* 20 (2001): 160–209. Arthur describes the Langlands Program as an "organizing scheme for fundamental arithmetic data in terms of highly structured analytic data." See Arthur, "Harmonic Analysis and Group Representations," p. 26. The structuring of the analytic data is due to symmetries it possesses (automorphy and "the rigid structure of Lie theory"), dramatically cutting down the obvious degrees of freedom. And one gets at the representation of the Lie group through the action of the group on a manifold mediated by the automorphic form, and then decomposing it into irreducible parts. See note 30 below and the main text it is attached to.

6. "yoga" in *Webster's Ninth New Collegiate Dictionary* (Springfield, Massachusetts: Merriam-Webster, 1986).

7. The preface to the paper is clear about their programmatic intent. It is pp. 181–184 in the original paper, pp. 238–241 in Dedekind's collected works, and there is a translation into English in Laugwitz, pp. 154–156. (R. Dedekind and H. Weber, "Theorie der algebraischen Funktionen einer Veränderlichen," *Journal für reine und angewandte Mathematik ("Crelle")* 92 (1882): 181–290; R. Dedekind, *Gesammelte mathematische Werke*, ed. R. Fricke, E. Noether, and O. Ore (Braunschweig: Vieweg, 1930); D. Laugwitz, *Bernhard Riemann 1826–1866, Turning Points in the Conception of Mathematics*, tr. A. Shenitzer (Boston: Birkhäuser, 1999).)

8. D. Hilbert, *The Theory of Algebraic Number Fields*, tr. I.T. Adamson (Berlin: Springer, 1998), pp. viii–ix, "Hilbert's Preface." Originally, D. Hilbert, "Die Theorie der algebraischen Zahlkörper," *Jahresbericht der Deutschen Mathematiker-Vereinigung* 4 (1897): 175–546.

9. A. Weil, "Review of 'The Collected Papers of Emil Artin,'" *Scripta Mathematica* 28 (1967): 237–238; in Weil, *Oeuvres Scientifiques, Collected Papers, Volume III (1964–1978)* (New York: Springer, 1979), pp. 173–174.

10. R.J. Baxter and I. Enting, "399th Solution of the Ising Model," *Journal of Physics A* 11 (1978): 2463–2473. As for a possible "400th" solution, see R.J. Baxter, "On solving the Ising model functional relation without elliptic functions," *Physica* A177 (1991): 101–108.

11. More generally, Hopf algebras are useful, with their natural representation of putting-together and taking-apart.

12. Automorphic *functions* are invariant to birational transformations, while automorphic or modular *forms* obtain a factor or modulus under such transformations. Modular functions also obtain a factor. The partition function in the Ising model is a modular form reflecting some of its symmetries; it has a functional equation (which Baxter uses to great effect). The partition function is also the trace of a matrix, the transfer matrix, devised to do the counting work. That matrix may be expressed as a product of a transfer matrix for each row of the lattice. The eigenvalues of the transfer matrix also have a functional form.

These are just the properties one will want for the number theorists' partition function, the L-function (thereby the modularity of an associated function gives one analytic continuation, for example) and a group representation. But unlike for the Ising model, the modular form related to L-functions, that associated function, f, is *not* the partition or L-function. f itself may well be a partition function (here I am thinking of the theta function).

L-functions are used to package combinatorial information, much as the partition function does for the physicist. If you have sufficiently many functional equations for your L-function, then there exists an automorphic form, f, whose fourier coefficients are the relevant combinatorial numbers. In fact, the connection is formal, so that $L(s)=\Sigma\ c_n B_n$ and the automorphic form $f=\Sigma\ c_n q^n$ (where $B_n=\exp -s \ln n$ and $q=\exp 2\pi i z$), n moving from subscript to superscript, from index to power.

13. C.W. Curtis, *Pioneers of Representation Theory: Frobenius, Burnside, Schur, and Brauer* (Providence: American Mathematical Society, 1999). Whether you can find such a matrix representation is open; it is not available in all cases, I believe. See notes 24–26 about the peculiar features of the transfer matrix eigenvalues and traces.

14. In the case of the physicists' problems, one can also define a counting matrix, whose Pfaffian leads to the combinatorial information. As mentioned in the text, I remain unclear if this has anything *directly* to do with the group representations and traces (characters) and symmetries I refer to here, other than its being another way of obtaining that same partition function. (Notionally, the counting matrix and the transfer matrix might be seen to be related, much as is the lagrangian in the path integral formulation (the Pfaffian of A_{Pf} in effect is the sum of the paths' amplitudes, each path amplitude defined by the product of the amplitudes of the steps in the path) related to the exponentiated hamiltonian as an evolution operator (that is, $V_\tau = \exp H\tau$, where H is the pseudo-hamiltonian).)

In any case, we end up with a volume (or determinant derived from the Pfaffian) related to the sum-of-the-eigenvalues (a trace), the two quite different matrices being about the same system. Some combinatorial problems, such as graph coloring, can be put in the transfer matrix form, and in that case that form does reflect symmetries or constraints built into the problem. See, R.J. Baxter, *Exactly Solved Models in Statistical Mechanics* (New York: Academic Press, 1982), p. 165–168 for an account of three colorings of a square lattice due to Lenard and Lieb. See note 25 below.

15. H.P. McKean and V. Moll, *Elliptic Curves: Function Theory, Geometry, and Arithmetic* (Cambridge: Cambridge University Press, 1997), as indicated by its title, is particularly clear about all these connections. See also, A.W. Knapp, *Elliptic Curves* (Princeton: Princeton University Press, 1992).

Technically, consider the cubic equation, $y^2 = 4x^3 - ax - b$, where a and b are rational numbers. Combinatorially, one can form a partition function or L-function that packages into one function the number of rational solution modulo a prime number, p, for each p.

Moreover, the rational solutions form a mathematical group. Namely, one can define addition, "+", for points on the curve (its solutions), so that z_1 "+" $z_2 = z_3$, where if z_1 and z_2 are rational solutions, so is z_3. There is a function, labeled by Weierstrass by a small capital script pee, $\wp$, so that for a solution, $z_i = (x_i, y_i)$, $(x_i, y_i) = (\wp(t_i), \wp'(t_i))$, for some complex t_i; namely, if we set $y = \wp'(t)$ and $x = \wp(t)$, then $\wp(t)$'s differential equation is just the cubic equation above. t and $\wp(t)$ are said to parametrize the cubic equation, and the equation is called an elliptic curve. (The distance along an ellipse is measured by an integral of the square root of a cubic or quartic, and hence $y^2 = 4x^3 - ax - b$ came to be called an elliptic curve.) And, reflecting the group property, $t_1 + t_2 = t_3$ (within what is called the period parallelogram in the complex plane). Also, $\wp(t_3)$ may be expressed rationally in terms of $\wp(t_1)$, $\wp(t_2)$, and their derivatives.

$\wp(t)$ is a periodic function, with two periods. So to each cubic equation we might associate a torus, with its two radii, on which the $\wp(t)$ associated to that equation has a natural home—go around one of the circumferences and you get back to the same value of the function. If the torus is cut and flattened, it becomes the period parallelogram.

16. The Rosetta Stone's three languages were Egyptian hieroglyphics, Greek, and a demotic (cursive, and simplified, less pictorial than hieroglyphics). I first read of Weil's analogy in D. Reed, *Figures of Thought: Mathematics and Mathematical Texts* (London: Routledge, 1995).

In 1944, Lars Onsager spoke of a "conspiracy," albeit in a different context: "The three integrals are infinite at the critical point, otherwise finite. The singularity results from a conspiracy." (p. 250, in L. Onsager, "Crystal Statistics I," *Physical Review* 65 (1944): 117–149). But I think he captures the sense that more is going on than can be accounted for by the calculation itself. In any case, Onsager's paper touches on all three of Weil's columns or languages.

17. R.P. Langlands, "Automorphic Representations, Shimura Varieties and Motives, Ein Märchen" in *Automorphic Forms, Representations, and L-Functions*, Proceedings of Symposia in Pure Mathematics, volume 33, part 2 (Providence: American Mathematical Society, 1979), pp. 205–246, at p. 206.

18. Frenkel, in *Love and Math*, p. 222, is explicit in invoking the analogy:

Number Theory and Curves/finite fields	Riemann surfaces X	Quantum Physics
Langlands relation	geometric L. relation	electromagnetic duality mirror symmetry
Galois group	fundamental group of X	fundamental group of X
representation of the Galois group in ^LG	representation of the fundamental group in ^LG	zero-brane on $M(^LG,X)$
automorphic function	automorphic sheaf	A-brane on $M(G,X)$

His first column corresponds to Weil's Columns Three and Two, and his second column corresponds to Weil's Column One.

19. I.G. Enting, "Series Expansions from the Finite Lattice Method," *Nuclear Physics* B (Proceedings Supplement) 47 (1996): 180–187; A.J. Guttmann and I. Enting, "Solvability of Some Statistical Mechanical Systems," *Physical Review Letters* 76 (1996): 344–347. For further support for their conclusions, see B. Nickel, "On the singularity structure of the 2D Ising model susceptibility," *Journal of Physics A, Mathematical and General* 32 (1999): 3889–3906.

20. In another context, these features have been dubbed a Trinity: the exact combinatorial solution is the Father, the conformal invariance (in effect, automorphy) of the field is the Son, and the group representation is the Holy Spirit. E. Date, M. Jimbo, T.

Miwa, and M. Okado, "Solvable Lattice Models," in, L. Ehrenpreis and R.C. Gunning, eds., *Theta Functions, Bowdoin 1987*, Proceedings of Symposia in Pure Mathematics 49 (Providence: American Mathematical Society, 1989), pp. 295–331.

21. M.H. Krieger, *Constitutions of Matter: Mathematically Modeling the Most Everyday of Physical Phenomena* (Chicago: University of Chicago Press, 1996) surveys some of the import of the infinite volume limit. There is a substantial literature on the loci of the zeros of the partition function, beginning with Lee and Yang's proof that the zeros lie on a line or a unit circle, depending on the variable employed. I review some of this in *Constitutions of Matter*, pp. 112–116. More generally, the zeros can be distributed over the complex plane, if the lattice is of finite size. See also, D.A. Lavis and G.M. Bell. *Statistical Mechanics of Lattice Systems* (Berlin: Springer, 1999), volume 2, pp. 148–163 for a review. For zeroes, see B. Nickel, "On the singularity structure of the 2D Ising model susceptibility," *Journal of Physics* A 32 (1999): 3889–3906. See also, McCoy's survey articles.

22. J.M. Maillard, "The inversion relation," *Journal de Physique* 46 (1985): 329–341, provides an extensive survey. See chapter 3, note 111. See also, J.-Ch. Anglès d'Auriac, S. Boukraa , and J.-M. Maillard, "Functional Relations in Lattice Statistical Mechanics, Enumerative Combinatorics, and Discrete Dynamical Systems," *Annals of Combinatorics* 3 (1999) 131–158, especially pp. 136–137.

One can solve for Q by iterating a pair of functional equations, something like $Q(k,u) \times Q(k,-u) = -i \sinh (I' + u)$, and $Q(k,u) = Q(k, I' + u)$. These are analogous to S-matrix unitarity and crossing relationships. Namely, $S(\theta) \times S(-\theta) = F(\theta)$, F being known; and $S(\theta) = S(\lambda-\theta)$ (Note that the Ising duality relationship is $Q(k,u) = k^N Q(1/k,ku)$.)

23. More precisely, there is an equation for the curve in an ambient space (in projective coordinates), for which see C.A. Tracy, "Embedded Elliptic Curves and the Yang-Baxter Equation," *Physica 16D* (1985): 202–220. In fact, as we shall see, for more complicated models, that curve need not be an elliptic or a cubic curve, and may be of higher degree and genus (or complicatedness). See H. Au-Yang, B.M. McCoy, J.H.H. Perk, S. Tang, and M.-L. Yan, "Commuting Transfer Matrices in the Chiral Potts Models: Solution of Star-Triangle Equations with Genus > 1," *Physics Letters* A123 (1987): 219–223.

24. For an illustration of these patterns and their connection with the partition function, see p. 153 of J.M. Dixon, J.A. Tuszynski, and M.L.A. Nip, "Exact eigenvalues of the Ising Hamiltonian in one-, two- and three-dimensions in the absence of a magnetic field," *Physica A* 289 (2001): 137–156. A similar numbering scheme is in the original papers of Kramers and Wannier (that is, a row pattern is known by a binary number encoding the ups and downs of the spins, say up is 1 and down is 0). Duality is a matter of recoding the pattern by asking if each adjacent pair of spins is like (coded by 1) or unlike (coded by 0). Then invert that coding (interchanging the 1s and 0s), and then decoding that in terms of up and down spins. Thereby, an arrangement with no spin flips becomes one with the maximal number of spin flips.

I should note that in the cases of interest, it is the largest eigenvalue of the transfer matrix that effectively determines the trace and the partition function, unless there are two roughly equal largest values—as below the critical point. The others are exponentially smaller; they are approximately exp–N smaller when we multiply the row-to-row transfer matrices to get the transfer matrix of a lattice of N rows.) It is for this formal reason that in the case of building up a lattice row by row (rectangular or diagonal), not only is the product of the transfer matrices of the rows the transfer matrix of the larger system (which is true by definition): $V=(V_r)^N$, where there are N rows; but the trace of the product is the product of the traces (which is not the case in general). The partition function of the system of subsystems (the rows) is the product of the partition functions of the subsystems, the free energies then adding. The maximum eigenvalues of the matrices are a good approximation to the trace, they have the appropriate multiplication for a group character, *and* they have the correct physical interpretation. One still needs to establish the irreducible representations or the good particles of the system.

In the actual solution of the Ising model, one solves the problem using spinor rotations, diagonalizing just one row matrix, as does Kaufman. (See note 99 of chapter 3.) Then the group characters are the phases, but this group representation is not my main concern here.

25. Besides showing how the combinatorial method and the transfer matrix method may be derived from the same expression for the partition function, one might try to see how the algebra did the combinatorial work (as did Kac and Ward), or to go from the trace of the transfer matrix to the Pfaffian of the combinatorial matrix. The latter can be done, starting out with the transfer matrix expressed in terms of fermion operators and reexpressing it in terms of Grassmann variables (ones that totally anticommute), using "coherent states." See, for example, K. Nojima, "The Grassmann Representation of the Ising Model," *International Journal of Modern Physics B* 12 (1998): 1995–2003.

Again, as far as I can tell, matrices though they be, the As are not a group representation. They are not multiplied (they seem not to have an algebra), and their Pfaffians rather than their traces are what is of interest. Moreover, the Pfaffian form of the partition function would appear to be occasional, not working for many models (as pointed out by partisans of Grassmann integration, Samuel in particular). The transfer matrix is generic, even if the model is not exactly solvable. The properties of the Vs are in fact like those of a scattering or S-matrix in quantum mechanics, albeit here we are in the classical realm. (I suspect that the Bethe Ansatz method connects with the As, the rules for the Bethe Ansatz reminding one of the rules for taking the Pfaffian of a matrix.)

26. Those products were similar to the product formulas for the theta functions.

If we think of the products of V_d's as akin to adding up random variables, the commutation of the V_d's represents the statistical independence of the variables or particles. This relationship of commutation to independence is suggested in D.S. Fisher, "Critical behavior of random transverse-field Ising spin chains," *Physical Review* B51 (1995): 6411–6461.

27. See note 32 on theta functions.

28. Essentially you know all about a rectangle, its sides representing the two periods of the theta functions, whether you put its long side horizontally or vertically. The "imaginary" part comes in because one side of the rectangle is said to have an imaginary length, the other a real length (the rectangle is in the complex plane), and the transformation interconverts them, namely it inverts the period ratio, τ: $\tau \rightarrow -1/\tau$. For more on thetas, see note 32.

29. See note 42.

30. I am sure I do not fully appreciate the content of the Langlands Program, and especially the notion of an automorphic group representation. My source here is S. Gelbart, "An elementary introduction to the Langlands Program," *Bulletin (New Series) of the American Mathematical Society* 10 (1984): 177–219. As for irreducible representations, see Lavis and Bell, *Statistical Mechanics of Lattice Systems*, vol. 2, pp. 137–138, 358–360. Onsager, in his original paper, is also deriving an irreducible representation, in terms of his A_i and G_k matrices.

31. Again, what I call the "Onsager Program" has never been announced as a program (except perhaps for Baxter's and Lieb's work). It is my impromptu rubric for what I take to be a fairly coherent strand of work in mathematical physics.

32.

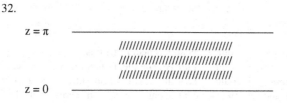

$z = \pi$

$z = 0$

FIGURE N.4: Slab of material, "infinite" in the x and y directions, with $z=0$ at the bottom of the slab. At the initial time the temperatures at the surfaces of the slab and in its interior are given.

When Joseph Fourier (1807, 1822) solved the equation for the propagation of heat in a solid sphere his solution was expressed in terms of a series for the temperature, T, times the radius, z:

$$T(z,t) \times z = \Sigma \, a_i \exp(-4\kappa n_i^2 t) \sin n_i z,$$

where t is the time, and the n_i are determined from solutions of a transcendental equation. (J. Fourier, *The Analytical Theory of Heat* , tr. A. Freeman (New York: Dover Publications, 1955 [1878, 1822]), p. 275.) Those functions are now called theta (θ) functions (where in this case, $T(z,t) \times z = \theta(z,t)$), and were originally discovered by Jakob Bernoulli and Euler. For the infinite slab, the temperature fixed at its surfaces (a heat sink is external), with a given initial temperature or heat source inside (say at the center of the slab), and where z is now the distance from the bottom surface: then, the temperature inside the slab is expressed simply in terms of a theta function. For other boundary conditions, one of the other three theta functions provides a solution. (D.F. Lawden, *Elliptic functions and applications* (New York: Springer, 1989), pp. 1–3.)

The time dependence is expressed by the "nome" (as in binomial, but here it is a single variable), q, of the theta function, that is, the $e^{-4\kappa t}$ factors, to the n^2 power, where κ, the diffusivity, equals the thermal conductivity divided by the specific heat times the density. So we have $\theta(z,q)$. (If we define the variable τ by $\ln q = i\pi\tau$, then τ turns out to be the ratio of the two periods of the elliptic functions. By convention the imaginary part of τ is positive, and for the heat equation $\tau=it$, t being the time.) The elliptic modulus, k, the nome, q, and the period ratio, τ, are directly related to each other.)

Now, the threefold analogy of analyticity, automorphy, and associativity, has a natural meaning for theta functions. We now know that heat flow is the consequence of diffusion processes—namely, of many and multiple molecular collisions, in effect a random walk. Perhaps we should not be too surprised that if τ is fixed the value of a theta function at a point is an average of its value in the neighborhood (the temperature is a harmonic function, whose value is an average of the temperatures in the neighborhood, as we might expect for the heat equation), that theta functions exhibit the scaling symmetry we might recall from the Central Limit Theorem, and that their fourier coefficients are of combinatorial significance (products of thetas are a way of adding up the squared-displacements in a random walk).

It may be useful to group together some facts about theta functions:

Automorphy and Periodicity and Modularity: The theta function, $\theta(z,\tau)$, is said to be automorphic in τ and doubly-periodic in z. Using the Poisson sum formula of fourier analysis—Σ(series-coefficients)=Σ(fourier-coefficients)—one can show that the theta functions, $\theta(z,\tau)$, have that lovely property of automorphy—for $z=0$, $\theta(0,-1/\tau) = (-i\tau)^{1/2} \times \theta(0,\tau)$. So we have a functional equation for the thetas.

Elliptic Functions: Ratios of those theta functions act much as do trigonometric or circular functions, and in honor of this fact those ratios are called, for example, *sn* and *cn*: the Jacobian elliptic functions. Corresponding to the automorphy of theta functions, $\mathrm{sn}(u,k) = 1/k\,\mathrm{sn}(ku,1/k)$, for example, where u is an angle or argument, and k is a function of the ratio, τ, of the two periods of the elliptic function (that is, $k=k(\tau)$) and is called the elliptic modulus. For $k=0$, when there is just one nontrivial period, $\mathrm{sn}\,u = \sin u$.

These Jacobian elliptic functions (the theta functions are also elliptic functions) now parametrize an ellipse much as the trigonometric functions parametrize a circle, giving $x=\mathrm{sn}(u,k)$, $y=\mathrm{cn}(u,k)$ for an ellipse of eccentricity k, for example, rather than $x=\sin u$ for a circle, where u is the radial angle for a circle and is the argument or mean anomaly for an ellipse. (See Krieger, *Constitutions of Matter*, pp. 187–188.)

Note that elliptic functions, and sines too are elliptic functions, have analytic, automorphic, and associative properties. They are known by their poles and zeros; they possess modularity; and, there is the "chord and tangent construction" for adding elliptic functions. This turns out to define an additive group on an elliptic curve.

Moreover, elliptic substitutions are employed to produce single-valued expressions for square roots. If $x=\mathrm{sn}(u)$, then $\sqrt{(1+k^2x^2)}$ is well defined as $\mathrm{cn}(u)$, as in Onsager's, Yang's, and Baxter's work.

33. For a popular version, see E. Frenkel, *Love and Math* (New York: Basic Books, 2013).

34. Much of this is suggested by the 1941 work of Kramers and Wannier and of Montroll. H.A. Kramers and G. Wannier, "Statistics of the Two-Dimensional Ferromagnet," *Physical Review* 60 (1941): 252–276; E.W. Montroll, "The Statistical Mechanics of Nearest Neighbor Systems," *Journal of Chemical Physics* 9 (1941): 706–721.

35. I am not at all sure of this correspondence. For me, it is hinted at in R. Langlands, "Representation Theory: Its Rise and Role in Number Theory," in *Proceedings of The Gibbs Symposium* (Providence: American Mathematical Society, 1990), pp. 181–210 and in, R. Langlands, "Where Stands Functoriality Today?," *Proceedings of Symposia in Pure Mathematics* 61 (1997): 457–471.

Langlands provides a particle scattering interpretation for his automorphic representations—much as group representations in physics are seen as elementary particles of given spin, parity, etc., to be combined with or scattered off other particles. In the particle metaphor, an automorphic representation is a collection of particles with structure, moving at various speeds, whose scattering amplitudes are quotients of L-functions. (Langlands, "Where Stands Functoriality Today?," p. 461.)

James Arthur, in his ICM 2014 address on the Langlands program, suggests the analogies: Laplacian (or Schroedinger) and Hecke operators to energy and momentum, the families of simultaneous eigenvalues to families of elementary particles, and the Langlands dual to the gauge group. And automorphy connects families at prime p to those at other primes. https://www.youtube.com/watch?v=ljJudppxau4.

In the Ising model, in the diagonal-to-diagonal matrix representation of the transfer matrix, $V_d(k,u)$, the rapidity or momentum-like variable in scattering is the argument u of the elliptic functions, the "energy" being the modulus k, and the Yang-Baxter relations are a way of phrasing conservation of momentum or rapidity. The Bethe Ansatz says that in the particle interpretation scattering, or interaction is a cumulation of nondiffrative two-body processes, their order not mattering: namely, that Yang-Baxter relation. (Note that these "particles" associated with the group representation, $V_d(k,u)$, are not the quasiparticles in the Schultz, Mattis, and Lieb fermion field theory account—the latter quasiparticles known by their energy and momentum, γ and ω to use Onsager's variables, for a fixed k and u.)

In suggesting the analogy of primes and elementary particles, all I am confident of here is the idea of their being "elementary."

36. E.H. Moore, "On a Form of General Analysis with Applications to Linear Differential and Integral Equations," in *Atti del IV Congresso Internazionale dei*

Matematici (Roma, 6–11 Aprile 1908) (Roma: Tipografia della R. Accademia dei Lincei, 1909), volume 2.

37. R.P. Langlands, M.-A. Lewis, and Y. Saint-Aubin, "Universality and Conformal Invariance for the Ising Model in Domains with Boundary," *Journal of Statistical Physics* 98 (2000): 131–244, at p. 132. Smirnov points out how their numerical studies are a prelude to his work. See Chelkak and Smirnov, "Universality in the 2D Ising…"

38. R. Langlands, P. Pouliot, and Y. Saint-Aubin, "Conformal Invariance in Two-Dimensional Percolation," *Bulletin (New Series) of the American Mathematical Society* 30 (1994): 1–61, at p. 29.

39. M.E. Fisher, "Renormalization group theory: Its basis and formulation in statistical physics," *Reviews of Modern Physics* 70 (1998): 653–681, at p. 655.

40. Some of the relevant names here are Hellegouarch, Serre, Frey, Ribet, Taniyama, Shimura, Weil, Wiles, and Taylor. A technical survey is provided in G. Cornell, J.H. Silverman, and G. Stevens, eds., *Modular Forms and Fermat's Last Theorem* (New York: Springer, 1997). I learned what I understand of this from the expository literature, including: F.Q. Gouvêa, "'A Marvelous Proof'," *American Mathematical Monthly* 101 (1994): 203–222; K. Ribet, "Galois Representations and Modular Forms," *Bulletin (New Series) of the American Mathematical Society* 32 (1995): 375–402. I found Knapp, *Elliptic Curves*, to be especially helpful. (And it got me to read Langlands.)

41. See note 32 for a survey.

42. Technically, the appropriate parallel for connecting combinatorial numbers is a product of two zeta functions, and it is associated with an Eisenstein series (which is a modular form). So the curiousness of the connection in the text is not fully explained by the conventional strategy. See, for example, H. Iwaniec, *Topics in Classical Automorphic Forms* (Providence: American Mathematical Society, 1997), pp. 203–204. He points out that the zeta function itself has "the wrong gamma factor in its functional equation" and instead of having local factors of degree 2 in its Euler product, they are of degree 1 (for example, $(1-p^{-s})^{-1}$). Hence the need for a product of zeta functions.

What is crucial is that the dimension of the vector space of modular forms is finite and often small. So there are many connections among the combinatorial numbers that make up the L-function and which form the fourier coefficients of the modular forms.

43. Klein, *Entwicklung der Mathematik*, p. 324. Translation from p. 305 of F. Klein, *Development of Mathematics in the 19th Century*, tr. M. Ackerman (Brookline: Math Sci Press, 1979). See p. 409, top, for the English translation.

44. Klein, *Entwicklung der Mathematik*, p. 324.

45. Dedekind and Weber employ Kummer's notion of an ideal and prime divisors, and more generally the notion of algebraic extensions by an irrational number $(a+b\sqrt{2})$; the geometric notion of a place on a curve, which then takes on a certain value; and the notion of a divisor. Klein, *Entwicklung der Mathematik*, pp. 321–335.

46. "Es geht dann wie bei dem Turnbau von Babel, dass sich die verschiedenen Sprachen bald nicht mehr verstehen." Klein, *Entwicklung der Mathematik*, p. 327.

Starting point: an irreducible polynomial equation $f(x) = 0$	Starting Point: an irreducible equation $f(w,z) = 0$ that contains z rationally (so that its coefficients, after being multiplied by a common denominator, are polynomials in z with arbitrary coefficients that are of no interest)
Field of all $R(x)$	Field of all $R(w,z)$, i.e. of all algebraic functions that are single valued on the Riemann surface.
Selection of the integral algebraic numbers of the field.	Selection of the integral algebraic functions of the field, i.e., the functions that become infinite only for $z = \infty$.
Decomposition into real and ideal prime factors and units.	Ideal decomposition into functions $G(w,z)$ that vanish at one or at none of the points of the Riemann surface.

47. "... le mémoire [de Riemann, 1851] est certes l'une des plus grandes choses que mathématicien ait jamais écrites; il n'y en a pas un seul mot qui ne soit considérable." André Weil, "Une lettre et un extrait de lettre à Simone Weil," (1940), in his *Oeuvres Scientifiques, Collected Papers*, volume I, pp. 244–255, p. 249.

48. "Mais les meilleurs spécialistes des théories arithmétique et 'galoisienne' ne savaient plus lire le riemannien, ni a plus forte raison l'italien; et il fallut vingt ans de recherches avant que le traduction fût mise au point et que la démonstration de l'hypothèse de Riemann dans la seconde colonne fût complètement déchiffrée [by Weil]." A. Weil, "De la métaphysique aux mathématiques," pp. 408–412 in A. Weil, *Oeuvres Scientifiques, Collected Papers* II (New York: Springer, 1979), at p. 412; originally in *Science* (Paris) (1960): 52–56, at p. 56.

49. Weil, "Une lettre et un extrait de lettre à Simone Weil." See also, A. Weil, "De la métaphysique aux mathématiques," whose text is substantially taken from the 1940 letter. I have included some of the additional 1960 material in my discussion of the 1940 letter. For "un texte trilingue," see "Une lettre," p. 253. For "une inscription trilingue," and "la pierre de Rosette," see "De la métaphysique aux mathématiques" and Weil's commentary on a 1939 paper, in *Oeuvres Scientifiques* I, p. 542. He provides more commentary in volume I, pp. 540–544, and volume III, pp. 454–458.

50. " ... j'ai quelques notions sur chacune des trois langues: mais je sais aussi qu'il y a de grandes différences de sens d'une colonne à l'autre, et dont rien ne m'avertit à l'avance. Depuis quelques années que j'y travaille, j'ai des bouts de dictionnaire. Quelquefois c'est sur une colonne que je fais porter mes efforts, quelquefois sur l'autre. Mon grand travail du Journal de Liouville a fait avancer beaucoup la colonne en "Riemannien"; par maleur, une grande partie du texte ainsi déchiffré n'a sûrement pas de traduction dans les deux autres langues;" Weil, "Une lettre et un extrait de lettre à Simone Weil," (1940), p. 253.

51. Weil, "De la métaphysique aux mathémathiques," at p. 408.

52. Conformal transformations connect similar configurations. Eventually, all of this becomes enshrined in algebraic topology, where cohomology is a general form of Gauss's law connecting fields to charges.

53. As in the other columns, in the middle column we have zeta or L-functions and their functional equation. A propos of Weil's notion of learning to read Riemannian, in this realm Weil (1948) resolves the appropriate version of the Riemann hypothesis concerning the location of the zeros of the zeta function.

54. Weil, "Une lettre et un extrait de lettre à Simone Weil," pp. 249–253. Corfield, *Toward a Philosophy of Real Mathematics*, pp. 91–96, provides a more technical account of the relationship of number and function as Weil describes it.

55. Hilbert's Problem XII, from his 1900 address, presents the analogy in this form. (D. Hilbert, "Mathematical Problems," *Bulletin of the American Mathematical Society* 8 (1902): 437–479, reprinted in *Mathematical Developments Arising From Hilbert Problems*, Proceedings of Symposia in Pure Mathematics 28. (Providence: American Mathematical Society, 1976), part 1, pp. 1–34, with Problem XII on pp. 18–20.) See, also, R. Langlands, "Some Contemporary Problems with Origins in the Jugendtraum," in part 2, pp. 401–418, in which the version of the analogy that concerns me earlier in this chapter is emphasized (p. 408). Namely, partition functions defined from "diophantine data" (the number of rational solutions to an equation in y and x) are the same as the partition function defined from "analytic data" (an automorphic form), and this is a natural extension of the notion of "reciprocity" (from Gauss, from Artin).

Dedekind and Weber ("Theorie der algebraischen Funktionen einer Veränderlichen, p. 29):

Versteht man, analog der Zahlentheorie, unter einem Körper algebraische Funktionen . . . der vier Spezies [+−×÷] . . . so deckt sich dieser Begriff vollständig mit dem der Riemannschen Klasse algebraischer Funktionen.

56. Weil, "Une lettre et un extrait de lettre à Simone Weil," p. 252.

57. "Finie l'analogie [of straightedge-and-compass division of a circle, to division of elliptic functions]: finies les deux théories, finis ces troubles et délicieux reflets de l'une à l'autre, ces caresses furtives, ces brouilleries inexplicables; nous n'avons plus, hélas, qu'une seul théorie, dont la beauté majestueuse ne saurait nous émouvoir. Rien n'est plus fécond que ces attouchements quelque peu adultères; rien ne donne plus de plaisir au connaisseur, soit qu'il y participe, soit même qu'en historien il les contemple rétrospectivement, ce qui ne va pas néamoins sans un peu de mélancolie. . . . Main revenons à nos fonctions algébriques." Weil, "Une lettre et un extrait de lettre à Simone Weil," p. 251.

58. A functor is a transformation between categories, transforming both mappings and objects, so that the diagram "commutes:" you can map and then functor, or functor and then map and you end up in the same place. If we have category *One(f,O)* and

category $Two(g,T)$, where f, g are mappings, and O and T are sets, then a functor $F(One,Two)$ acts as follows: $F(O)$ is a member of Two, $F(f)$: $F(O_1) \rightarrow F(O_2)$ is a member of Two, and F preserves the identity and $F(f_1 \circ f_2) = F(f_1) \circ F(f_2)$.

59. Without claiming to understand many of these connections myself, it may still be useful to mention the following: G. Shimura writes that Eichler's mid-1950s diagram was a hexagon, its vertices labeled by: automorphic forms, modular forms, quadratic forms, quaternion algebras, Riemann surfaces, and algebraic functions. Physicists and mathematicians may try to find a quantum field theory explanation for the appearance of modular forms in the study of the L-functions of elliptic curves. Witten (pp. 430, 585, 588) suggests a potential connection of conformal field theory to Langlands' realm of automorphic representations of adèle groups, that scaling and congruences matter.

Ruelle points out with respect to the Lee-Yang circle theorem (a statement about the location of the zeros of the partition function), "One might think of a connection with zeta functions (and the Weil conjectures); the idea of such a connection is not absurd, . . . But the miracle has not happened; one still does not know what to do with the circle theorem."(p. 263)

Mackey provides a cardinal set, rather than a triple or a sextet: (1) Kac-Moody Lie algebras; (2) analyticity of the S-matrix embodied in the Veneziano model, strings as harmonic oscillators, and Virasoro algebras; (3) the Kortweg-deVries equation, solitons, integrable models, and inverse scattering methods; and, (4) the Lieb and Sutherland Bethe-Ansatz (plane wave plus phase shift) exact solution to the ice model. E. Frenkel (1995) connects affine algebras, D-modules on curves, and integrable systems. M. Kapranov (1995) looks into the Langlands duality or correspondence between automorphic representations and the combinatorics represented by Galois groups: ". . . the formal structure of the Langlands correspondence is not unlike that of topological quantum field theory" (p. 120). Drinfeld's quantum groups are everpresent.

Katz and Sarnak, in their work on zeros of the zeta function, connect the distribution of the eigenvalues of random matrices, zeros of L-functions, Fredholm determinants, monodromy groups, and scaling—reminiscent of the themes in chapter 4 concerning the correlations in the Ising model.

See, G. Shimura, in "Steele Prizes," *Notices of the American Mathematical Society* 43 (November 1996): 1340–1347); E. Witten, "Quantum Field Theory, Grassmannians, and Algebraic Curves," *Communications in Mathematical Physics* 113 (1988): 529–600; D. Ruelle, "Is Our Mathematics Natural?, The Case of Equilibrium Statistical Mechanics," *Bulletin (New Series) of the American Mathematical Society* 19 (1988): 259–268; G. Mackey, "Final Remarks," in *The Scope and History of Commutative and Noncommutative Harmonic Analysis* (Providence: American Mathematical Society, 1992), pp. 325–369; E. Frenkel, "Affine algebras, Langlands duality and Bethe Ansatz," in, *Proceedings of the XIth International Congress of Mathematical Physics, Paris, 1994*, ed. D. Iagolnitzer (Cambridge, Massachusetts: International Press, 1995), pp. 606–642; M.M. Kapranov, "Analogies between the Langlands Correspondence and Topological Quantum Field Theory," in *Functional Analysis on the Eve of the 21st*

Century, Gelfand Festschrift, eds. R. Wilson et al., volume 1, Progress in Mathematics 131 (Boston: Birkhäuser, 1995), pp. 119–151; N.M. Katz and P. Sarnak, "Zeros of zeta functions and symmetry," *Bulletin (New Series) of the American Mathematical Society* 36 (1999): 1–26.

M.H. Krieger, *Constitutions of Matter* (Chicago: University of Chicago Press, 1996), chapter 8, includes a number of quotes from other attempts to indicate the range of topics unified under these columns. This closing chapter in fact forced me to look further, and hence in part the current book.

60. "Dès qu'elle s'est traduite par la possibilité de transporter une démonstration telle quelle d'une théorie à l'autre, elle a déjà cessé sur ce point d'être féconde; elle l'aura cessé tout à fait si un jour on arrive d'une manière sensée et non artificielle, à fondre les deux théories en une seule." Weil, "Une lettre et un extrait de lettre à Simone Weil," p. 250.

61. D. Bump, *Automorphic Forms and Representations* (New York: Cambridge University Press, 1997), p. 94: "Langlands functoriality says that operations on Artin L-functions correspond to operations on automorphic forms." Another way functoriality is described is as the relationship of automorphic representation of different groups, G and its Langlands dual, ^{L}G.

Metaphorically, as in our shape of the drum problem, the Poisson sum formula of fourier analysis (or, I believe, the Selberg trace formula) connects sums of objects with sums of those objects' properties when expressed as fourier coefficients (connecting geometric to spectral data).

62. C. N. Yang, "Some Exact Results for the Many-Body Problem in One Dimension with Repulsive Delta-Function Interaction, *Physical Review Letters* 19 (1967): 1312–1315, at p. 1312, Equation 8. B. Sutherland, "Exact Solution of a Two-Dimensional Model for Hydrogen-Bonded Crystals," *Physical Review Letters* 19 (1967): 103, and *Beautiful Models: 70 Years of Exactly Solved Quantum Many-Body Problems* (Singapore: World Scientific, 2004). J. B. McGuire, "Study of Exactly Soluble One-Dimensional N-Body Problem," *Journal of Mathematical Physics* 5 (1964): 622–636, expresses this graphically. And the Bethe Ansatz expresses two-bodyness as well.

63. Baxter, *Exactly Solved Models in Statistical Mechanics*, p. 85, Eq. 6.4.27.

64. Tracy, "Embedded Elliptic Curves," p. 203.

65. So Emmy Noether says in commenting on the Dedekind-Weber 1882 paper:

Für den allgemeinen Fall, welcher sich zu dem eben genannten ähnlich verhält, wie der Fall der allgemeinsten algebraischen Zahlen zu demjenigen der rationalen Zahlen, wiesen die mit bestem Erfolg in der Zahlentheorie angewandten Methoden, die sich an Kummers Schöpfung der idealen Zahlen anschließen, und der Übertragung auf die Theorie der Funktionen fähig sind, auf den richtigen Web.

E. Noether, "Erläuterungen zur vorstehenden Abhandlung," in R. Dedekind, *Gesammelte mathematische Werke*, p. 350, referring to R. Dedekind and H. Weber, "Theorie der algebraischen Funktionen einer Veränderlichen."

So also does Gerd Faltings describe his proof of the Mordell-Shafarevich-Tate conjecture: ". . . our proof is an adaptation of Zarhin's method to the number field case. The dictionary necessary for the translation is supplied by Arekelov, and the author built his proof on Arekelov's method." [. . . unser Beweis ist eine Übertragung seiner [Zahrins] Methoden auf den Zahlkörper-Fall. Das zur Übersetzung notwendige Wörterbuch hat Arekelov geliefert, und seine Methoden sind vom Verfasser ausgebaut worden. Kurz gesagt handelt es sich darum, "alles" mit hermiteschen Metriken zu versehen.] "Endlichkeitssätze für abelsche Varietäten über Zahlkörpern," *Inventiones mathematicae* 73 (1983): 349–366, at p. 349; also, in English translation, "Finiteness Theorems for Abelian Varieties over Number Fields," in G. Cornell and J.H. Silverman, eds., *Arithmetic Geometry* (New York: Springer, 1986), p. 9.

He goes on, "The basic idea for this was communicated to me by a referee . . . and I then had only to translate it from Hodge theory into étale cohomology." (p. 10)

66. "De son côté, Langlands s'était tracé un vaste programme qui tendait avant tout à établir le lien entre produits eulériens et réprésentations de groupes. . . . le lien entre ses recherches et les miennes ne m'apparut que plus tard. . . . il y a une sorte de réciprocité, généralisant . . ." Weil, *Oeuvres Scientifiques*, volume III, pp. 455, 458.

67. S. Lang. *Undergraduate Algebra* (New York: Springer, 1987), pp. 196–197. The section (VIII § 6) is entitled, "Where Does It All Go? Or Rather, Where Does Some of It Go?," and tries to indicate questions that are unresolved about Galois groups.

68. R. Langlands, "Representation Theory: Its Rise and Role in Number Theory," at p. 209.

69. Langlands, Lewis, and Saint-Aubin, "Universality and Conformal Invariance for the Ising Model in Domains with Boundary." (Langlands and collaborators started out with a "noninteracting" Ising model, percolation.) Langlands, Pouliot, Saint-Aubin, "Conformal Invariance in Two-Dimensional Percolation." Langlands has expressed some discomfort with the "Russians'" use of sheaves and automorphic sheaves, in their work on the geometric Langlands program; he is concerned that reciprocity is not present, while the main theme there is functoriality, his other main theme. (https://publications.ias.edu/ sites/default/files/message-to-Peter-Sarnak_0.pdf) Smirnov has made major mathematical advances in studying these models rigorously, as Langlands might want.

It is interesting that in Baxter's studies of more complex lattices, and in McCoy's work, they end up using crucial number-theoretic facts: the Rogers-Ramanujan identities and, more generally, properties of a natural extension of binomial expansions, q-series.

70. See D.S. Fisher, "Critical behavior of random transverse-field Ising spin chains," for a situation of randomness and scaling when the order of the transfer matrices matters (and what is typical is not in fact the average).

71. "Ces textes sont l'unique source de nos connaissances sur les langues dans lequel ils sont écrits;" Weil, "De la métaphysique aux mathématiques," p. 411.

72. Somewhat modified from the T.M. Knox translation. The reference is to Goethe's *Faust*. Mephistopheles says:

> My worthy friend, gray are all theories
> And green alone life's golden tree.

CHAPTER 6

1. J.C. Maxwell, *Treatise on Electricity and Magnetism* edn. 3 (Oxford: Clarendon Press, 1891, Dover reprint, 1954), vol. 1, pp. 16–20, and section 421 in vol. 2. (If I understand it correctly, the technical way of putting all of this nowadays is that "the cokernels in Sheaves(X) are not equal to the cokernels in Presheaves(X).") There are sources and holes that make for inexactness.

2. The "discharging rules" of Heesch (1969) that are crucial to the Appel-Haken solution (1976) of the Four Color Problem, replace Kirchoff's laws and Coulomb's with another prescription for the flow of "charges," in order to show the topological relationship of the local to the global. Namely, in a candidate non-fourcolorable configuration, the local rules for the flow of charge prove inconsistent with global facts (Euler's formula) about edges, vertices, and faces of a network or map. See T. L. Saaty and P.C. Kainen, *The Four-Color Problem, Assault and Conquest* (New York: Dover, 1986).

3. For mathematicians, this is the deRham or sheaf cohomology; for physicists, it is Faraday-Maxwell electromagnetism. Technically, I have been describing what is called the Riemann-Roch theorem, and modern versions of it are called index theorems—recalling the Poincaré-Brouwer index of a vector field, in effect a multipole moment, the n of z^n.

Technically, cohomology is a mapping or "functor" from topology to algebra, from charges and shapes to a vector space of solutions to a differential equation (the fields): If we continuously transform the topological space, the vector space comes along nicely. What we want is information about these vector spaces (as groups) directly, especially their finiteness.

Put differently, if A "divided" by B equals C, and given B and C, what could be A? Or, given various integrals of the field, what can we know about the field itself.

4. Quoted in M. Monastyrsky, *Riemann, Topology and Physics* (Boston: Birkhäuser, 1987), p. 42, from which I have taken the translation. Originally, A. Sommerfeld, "Klein, Riemann und die mathematische Physik," *Die Naturwissenschaften* 17 (1919): 300–303, at p. 301. Sommerfeld quotes Klein, "Riemann im Gebiete der Mathematik und Faraday

im Gebiete der Physik stehen parallel." See also, F. Klein, *Vorlesungen Über die Entwicklung der Mathematik*, vol. 1 (New York: Chelsea, 1956), pp. 72, 264–265.

5. "Was Maxwell hingegen in hohem Masse auszeichnet, das ist eine starke Intuition, die sich bis zur Divination steigert und mit einer phantasievollen Anschauungskraft Hand in Hand geht. . . . seine Vorliebe für Diagramme, die Verwendung von Rollkurven, stereoskopischen Figuren, reziproken Kräfteplänen." (F. Klein, *Vorlesungen Über die Entwicklung der Mathematik*, p. 245.)

6. N. Swerdlow, *The Babylonian Theory of the Planets* (Princeton: Princeton University Press, 1998).

7. Quinn argues for the truth and objectivity of mathematics, much more universally than I allow here. See F. Quinn, "Contributions to a Science of Contemporary Mathematics," October 2011, 98pp, available at https://www. math.vt.edu/people/quinn/history_ nature/nature0.pdf.

8. H. Weyl, "David Hilbert and His Mathematical Work," *Bulletin of the American Mathematical Society* 50 (1944): 612–654, at p. 615.

9. J.-L. Lagrange, "Réflections sur la résolution algébrique des équations," (1771), in *Oeuvres de Lagrange*, ed. J.-A. Serret, G. Darboux (Paris: Gauthier-Villars, 1869). He says: ". . . je donnerai à cette occasion les vrais principes, pour ainsi dire, la métaphysique de la résolution des équations du troisième et du quatrième degré [p. 357] . . . tout se réduit, comme on voit, à une espèce de calcul des combinaisons. [p. 403]"

10. When Paul Gordan said something to the effect, "It is not mathematics, it is theology," he was referring to an abstract proof by Hilbert of the kind of result (Gordan's) that until then was proved by exhaustive calculation, the kind of calculation that Gordan excelled at. And it was after further study, that he added, "I have convinced myself that even theology has its merits." He then, in 1899, simplified Hilbert's proof. See pp. 330–331 in F. Klein, *Vorlesungen Über die Entwicklung der Mathematik im 19. Jahrhundert* (New York: Chelsea, 1956 [1926, 1927, 1950]).

Gordan felt that Hilbert's proofs often did not explicitly deal with all the technical niceties, although he was quite willing to believe that Hilbert himself had satisfied himself about them. He says,

Sie verlangen meine Ansicht über die Hilbert'sche Arbeit. Leider muss ich Ihnen sagen, dass ich sehr unzufrieden mit derselben bin. Die Dinge sind ja sehr wichtig und auch richtig, also darauf geht mein Tadel nicht. Derselbe bezieht sich vielmehr auf den Beweis seines Fudamentalsatzes, welcher den bescheidensten Anforderungen, welche man an einen mathematischen Beweis macht, nicht entspricht. Es genügt keineswegs, dass sich der Verfasser selbst die Sache klar macht, sondern man verlangt, dass er den Beweis nach festen Regeln aufbaut.

Gordan to Klein, Sign.: Klein 9,461, 24 February, 1890, in G. Frei, ed., *Der Briefwechsel David Hilbert-Felix Klein (1866–1918)* (Göttingen: Vandenhoeck & Ruprecht, 1985), p. 65.

11. I have surveyed the introductory passages of many of the papers of Gian-Carlo Rota (1932-1999). A fine book could be created out of those prefatory historical and contextual remarks, stopping just when each paper says something like, "Let X be a Hopf algebra . . ."

12. H. Weyl, *The Philosophy of Mathematics and Natural Science* (Princeton: Princeton University Press, 1949 [1927]), p. vi.

13. On Weyl and Husserl, see D. van Dalen, "Hermann Weyl's Intuitionistic Mathematics," *Bulletin of Symbolic Logic* 1 (1995): 145–169.

14. Responding to criticism by André Weil, Weyl the historically acute thinker says: "Who says to us that we have already reached the end of the methodical development of topology?" (P. ix in the 1964 translation of the 1955 edition of *The Concept of a Riemann Surface* (Reading, Massachusetts: Addison-Wesley, 1964 [1955, 1913])).

15. Lagrange, "Réflections sur la résolution algébrique des équations."

16. In this vein: Combinatorial (simplicial) topology is the piecewise linear image of topology into algebra. Finding exact sequences is algebraic elimination theory (as in solving simultaneous equations), projecting out what will then be annihilated.

17. M. Atiyah, "Mathematics in the 20th Century," *American Mathematical Monthly* 108 (2001): 654–666, at pp. 659–660. Atiyah says that K-theory finds commutative representations of non-commutative algebras by providing abelian or linear invariants (trace, dimension, determinant).

18. Still, Weyl, by his own admission, is an impossible act to follow. "[H]ow could I hope to recapture the faith and spirit of that epoch of my life when I first composed it [*Philosophy of Mathematics*]—after due literary preparations dashing off the manuscript in a few weeks?" Weyl, *Philosophy of Mathematics and Natural Science*, pp. v–vi.

19. I do not mean by this that we need to account for mathematical practice in terms of a formal logic, as has been the goal of some philosophizing.

An interpretive account of the specific tasks of some of mathematics is not what is conventionally called philosophy of mathematics, a field that would appear to be about traditional philosophical problems as they may be represented in the mathematical realm. Mathematics has been used to consider traditional philosophic questions about objects and knowledge: What are mathematical objects?, How do we acquire mathematical knowledge?, and, How do we distinguish correct from incorrect knowledge? See, D.M. Jessep, *Berkeley's Philosophy of Mathematics* (Chicago: University of Chicago Press, 1993), p. 4.

20. So, Brouwer's and Weyl's philosophical analyses of the continuum in the first part of this century lead to some interesting mathematics, as such. But if they did not lead to interesting mathematics, their philosophical analyses would have much less force, I believe. See P. Mancosu, ed., *From Brouwer to Hilbert: The Debate on the Foundations of Mathematics in the 1920s* (New York: Oxford University Press, 1998).

21. I have also found some support for what I am trying to do in P. Maddy's call for a naturalistic philosophy of mathematics, where one pays attention to what mathematicians actually do. P. Maddy, "Set Theoretic Naturalism," *Journal of Symbolic Logic* 61 (1996): 490–514.

22. I am thinking of the essay by George Mackey, "Harmonic Analysis as the Exploitation of Symmetry," *Rice University Studies* 64 (1978): 73–228 (also in *Bulletin (New Series) of the American Mathematical Society* 3 (1980): 543–699; reprinted in *The Scope and History of Commutative and Noncommutative Harmonic Analysis* (Providence: American Mathematical Society, 1992)). Mackey deliberately covers a very wide swath of mathematics, displaying the meaning of symmetry, and of decomposition under symmetry's varied regimes. These connections are to be demonstrated mathematically. See also, Jean Dieudonné's *A History of Algebraic and Differential Topology, 1900–1960* (Boston: Birkhäuser, 1989).

23. Here, I hear the voice of Gian-Carlo Rota, always looking for models of special functions in terms of combinatorial distribution-and-occupancy, balls into boxes.

24. C. Fefferman, "The Uncertainty Principle," *Bulletin (New Series) of the American Mathematical Society* 9 (1983): 129–206.

25. F. Klein, *Development of Mathematics in the 19th Century*, tr. M. Ackerman (Brookline, Massachusetts: Math Sci Press, 1979), p. 3. William Thurston has emphasized the social and community aspects of mathematics, asking "How do mathematicians advance the human understanding of mathematics?" And the answer is not just by proving theorems. An advance in understanding is a collective endeavor. W.P. Thurston, "On Proof and Progress in Mathematics," *Bulletin (New Series) of the American Mathematical Society* 30 (1994): 161–177.

26. It has been argued that standards of proof and rigor do change, even after we have thought we have got it down right. J. Dieudonné insists otherwise, "what constitutes a correct proof [in each sector of mathematics] has remained unchanged ever since [they got it right, and did not depend on intuition]." (*History of Algebraic and Differential Topology*, pp. 15–16.) See also Quinn, "Contributions to a Science of Contemporary Mathematics."

27. Much of this is a commonplace of Marxist analysis, most famously in Boris Hessen's discussion (1931) of the social and economic origins of the calculus. Marx himself studied the calculus and he saw the process of taking a derivative as dialectical, in the algebra going from a finite to a zero-limit differential, and that "the abstract forms express the relationships of material processes." (C. Smith, "Hegel, Marx and the Calculus," in Karl Marx, *Mathematical Manuscripts of Karl Marx* (London: New Park, 1983), p. 269.) This is also a commonplace of Darwinian thought about natural selection, the fit of mathematics and the world being an historical artifact—none of which insists that mathematics is either arbitrary in its truths or "relative."

28. D. A. Martin, "Mathematical Evidence," in H. G. Dales and G. Oliveri, eds., *Truth in Mathematics* (New York: Oxford University Press, 1998), pp. 215–231, at pp. 215–216.

29. See, for example, P. Mancosu, ed., *Philosophy of Mathematical Practice* (Oxford: Oxford University Press, 2008), and Mancosu's Introduction, pp. 1–21. He discusses the work of Batterson, Butterfield, Corfield, Kitcher, Maddy, and others who contributed to his volume.

30. Quinn, "Contributions to a Science of Contemporary Mathematics." Kitcher points to the 19th century, in P. Kitcher, *The Nature of Mathematical Knowledge* (Oxford: Oxford University Press, 1985).

31. "Does Mathematics Need New Axioms?," Solomon Feferman, Harvey M. Friedman, Penelope Maddy and John R. Steel, *The Bulletin of Symbolic Logic* 6 (2000): 401–446.

32. Mit jedem einfachen Denkakt tritt etwas Bleibendes, Substantielles in unsere Seele ein. Dieses Substantielle erscheint uns zwar als ein Einheit, scheint aber . . . ein Innere Mannigfaltigkeit zu enthalten; ich nenne es daher "Geistesmasse" . . .

Die sich bildenden Geistesmassen verschmelzen, verbinden oder compliciren sich in bestimmtem Grade, theils unter einander, theils mit älteren Geistesmassen.

"Zur Psychologie und Metaphysik," pp. 509–520 in B. Riemann, *Gesammelte Mathematische Werke (Collected Works)*, edition 2, ed. H. Weber (New York: Dover, 1953 [1892, 1902]), p. 509.

33. H. Weyl, *The Concept of a Riemann Surface*, p. vii. Originally, in 1913, pp. iv–v, "Die Riemannsche Flaeche . . . muss durchaus als das prius betrachtet werden, als der Mutterboden, auf dem die Funktionen allererst wachsen und gedeihen koennen." ["The Riemann surface must be considered as the prius, as the virgin soil, whereupon the functions above all can grow and prosper."] See, on Riemann, D. Laugwitz, *Bernhard Riemann 1826–1866, Turning Points in the Conception of Mathematics*, tr. A. Shenitzer (Boston: Birkhäuser, 1999).

34. The tropes of mixture and layering have been adopted even in cases where they are less pervasive, as in medieval cities. So Michael Camille, in *Image on the Edge: Margins of Medieval Art* (Cambridge, Massachusetts: Harvard University Press, 1992), says: "Like the famous cries of the street-vendors, the discourse of the city [Paris, 1325] was the vernacular babble after Babel. Polysemous and multicoded, the city was the site of exchange, of money, goods and people, creating a shifting nexus rather than a stable hierarchy."(p. 130) And, " . . . a world that contrasts as much with the ecclesiastical as with the courtly model of space—one of horizontal multiplicity rather than vertical hierarchy, a mingling of rich, poor, free, unfree, peasants, knights and clergy as well as the rising bourgeoisies, all rubbing shoulders within the same walls. The distinction between high and low was, however, still retained in *representations* of urban spaces." (p. 129) Perhaps the Industrial Revolution's effect is a secondary effect.

35. See our Figure P.1. C. Neumann, *Vorlesungen über Riemann's Theorie der Abel'schen Integrale* (Leipzig: Teubner, edn. 2, 1884 [1865]). It portrays a surface associated with the function "the square root of z." Since there are two values, $\pm\sqrt{z}$, there are two layers on which the function lies (except at $z=0$). At any point on the surface, only one of the values is being referred to. (This is not a graph of the function, $f(z)=\pm\sqrt{z}$;

rather, it is the underlying space in which $f(z)$ lies.) In general, over any point, z, there may be many values associated with an $f(z)$: the function is "ramified" at z, and there is a z-layer for each of the possible values of $f(z)$. And even more generally, as Hermann Weyl pointed out, the surface may be freed from the complex plane, the freed surface now to be known by its topology.

36. An interview with Pierre Cartier that appeared in *The Mathematical Intelligencer* 1 (1998): 22–28.

37. I have left out the improvements needed to assure unique prime factorization.

38. M.H. Stone, "Theory of Representations for Boolean Algebras," *Bulletin of the American Mathematical Society* 44 (1938): 807–816.

39. This became embodied in homology theories, and so we might speak of the sheaf cohomology of a city.

40. H. Adams, "The Rule of Phase Applied to History," in E. Stevenson, ed., *A Henry Adams Reader* (Garden City: Anchor, 1958).

41. W. Benjamin, *Gesammelte Schriften*, volume 5, part 1 (Frankfurt am Main: Suhrkamp, 1982), p. 137.

> Paris steht über einem Höhlensystem, . . . durchkreuzt sich mit den altertümlichen Gewölben, den Kalksteinbrüchen, Grotten, Katakomben, die seit dem frühen Mittelalter Jahrhunderte hindurch gewachsen sind. Noch heute kann man gegen zwei Franken Entgelt sich seine Eintrittskarte zum Besuche dieses nächtlichsten Paris lösen, das so viel billiger und ungefährlicher als das der Oberwelt ist. Das Mittelalter hat es anders gesehen.

In 21st century Beijing, there is a large population living underground, serving vital needs but residing illegally in the city.

42. Eventually, as railroads and transit develops, the city differentiates horizontally, workers sometimes relegated to the outskirts, the middle classes to the suburbs. Racial and ethnic segregations become entrenched, in part because ready transportation brought diverse groups from far away into a city, and made possible more distant trips to work. The vertical differentiation perhaps becomes less significant, although I can think of obvious counterexamples.

43. Derivatives of potentials expressed the actual flows and fields we experience. And, the second derivatives expressed the conservation laws. Eventually, these become the chain complexes of algebraic topology, the sequences of derivatives or derived entities, and the exact sequences of conserved flows—formally, where the second derivatives, suitably understood, are zero—so employed by the mathematicians to decompose things into their presumably natural parts.

44. Für die folgenden Betrachtungen beschränken wir die Veränderlichkeit der Grössen x, y auf ein endlichen Gebiet, indem wir als Ord des Punktes O nicht mehr die Ebene A selbst, sondern eine über dieselbe angebreitete Fläche T

betrachten. Wir wählen diese Einkleidung, bei der es anstössig sein wird, von auf einander liegenden Flächen zu reden, . . . dass der Ort des Punktes O über denselben Theil der Ebene sich mehrfach erstreke . . . 1851, § V) [In the following considerations we shall restrict the variables x and y to a finite region, with points to be thought of not located on the complex plane itself but rather on a surface covering of it. . . .]

B. Riemann, *Collected Works*, p. 5—from Riemann's Inaugural Dissertation of 1851 ("Grundlagen für eine allgemeine Theorie der Funktionen einer veränderlichen complexen Grösse"), section 5.

45. If $f(z)=u(z)+iv(z)$ is analytic, that is, possessing well defined derivatives, u and v are to be harmonic. For harmonic functions, their value at a point is an average of their values in the neighborhood of that point (as long as there are no infinities of the function in that neighborhood).

46. Buchwald argues that Maxwell had a fluidic (rather than particulate) notion of charge. J.Z. Buchwald, *From Maxwell to Microphysics* (Chicago: University of Chicago Press, 1985).

47. Maxwell Equation's relationships are eventually expressed through Gibbs' vector calculus, as electromagnetic theory's gradients, divergences, and curls (∇, $\nabla\cdot$, $\nabla\times$), the flows and circulations in a place—which in fact for Maxwell were fluidic and material concepts.

The Yang-Mills theories that are the foundation of the Standard Model in modern physics and geometry, are founded in a tradition that has explicitly mixed Maxwellian electricity and mathematics, tensors and topology, charges and fields, poles and holes, functions and spaces: Riemann (1851, 1857) and Gustav Roch (1865), Maxwell (1873), Felix Klein (1890), Hermann Weyl (1913), and W.V.D. Hodge (1940). The textbook, *Gravitation* (1973), by Charles Misner, Kip Thorne, and John Archibald Wheeler is perhaps the most beautiful product of this tradition. See also, J.D. Jackson and L.B. Okun, "Historical Roots of Gauge Invariance," *Reviews of Modern Physics* 73 (2001): 663–680.

48. T.P. Hughes, *Networks of Power* (Baltimore: Johns Hopkins University Press, 1983).

49. . . . il semble bien que certains architectes encore oublient parfois l'électricité. . . . il s'en trouve ou va s'en trouver (de plus en plus nombreux) dans chaque bâtiment, dans chaque demeure, et qu'il faut donc prévoir, au moment que l'on conçoit ce bâtiment, l'arrivée, la "circulation" et le débouché au maximum d'endroits possible, du ou des courants mis à la disposition de chacun.

. . . pour demander à mon architecte de prévoir, dans la maison qu'il me bâtit, des interrupteurs près des fenêtres (et non seulement près des portes ou des lits), pour me permettre de mieux goûter la nuit.

. . . Il paraît que le fameux Tabernacle des Juifs, l'Arche Sante construite par Moïse, étant donné sa description comme elle figure au chapitre xxv du livre de l'Exode, pourrait être considérée comme un très savant condensateur. Faite, selon les ordres du Seigneur, en bois de sétim (isolant) recouvert sur ses deux faces, intérieure et extérieure, de feuille d'or (conductrices), surmontée encore d'une couronne d'or destinée peut-être, grâce au classique "pouvoir des pointes," à provoquer la charge spontanée de l'appareil dans le champ atmosphérique, lequel, dans ces régions sèches, peut atteindre paraît-il jusqu'à des centaines de volts à un ou deux mètres du sol—il n'est étonnant que cette Arche Sainte, toute prête à foudroyer les impies, ait pu être approchée sans danger seulement par les grands prêtres, tels Moïse et Aaron, dont l'Écriture nous apprend par ailleurs qu'ils portaient des vêtements "entièrement tissés de fils d'or et orné de chaînes d'or traînant jusqu'aux talons." Comme l'ajoute le patient commentateur auquel nous empruntons cette hypothèse, cette ingénieuse "mise à la terre" leur permettant de décharger le condensateur sans dommage pour leur personne.

F. Ponge, *Lyres* (Paris: Gallimard, 1967, 1980), pp. 75, 79, 83–84. Also, the text is available with an English translation in F. Ponge, *The Power of Language*, tr. S. Gavronsky (Berkeley: University of California Press, 1979), pp. 160, 166, 172, 174.

50. "Zur Psychologie und Metaphysik," pp. 509–520 in Riemann, *Collected Works*. See also, U. Bottazzini and R. Tazzioli, "Naturphilosophie and its Role in Riemann's Mathematics," *Revue d'histoire des mathématiques* 1 (1995): 3–38; K.L. Caneva, "Physics and Naturphilosophie," *History of Science* 35 (1997): 35–106.

51. Continuity, for Herbart, is "union in separation and separation in union." The experienced world is a sequence of representations (*Reiheformen*).

52. As in C. Smith and M.N. Wise, *Energy and Empire: A Biographical Study of Lord Kelvin* (Cambridge: Cambridge University Press, 1989).

53. L. Kronecker, *Leopold Kronecker's Werke*, ed. K. Hensel (New York: Chelsea, 1968), volume 5, p. 457:

Es handelt sich um meinen liebsten Jugendtraum, nämlich um den Nachweis, dass die Abel'schen Gleichungen . . . grade so erschöpft werden, wie die ganzzahligen Abel'schen Gleichungen durch die Kreistheilungsgleichungen. Dieser Nachweis ist mir, wie ich glaube, nun vollständig gelungen, und ich hoffe, dass sich bei der Ausarbeitung, auf die ich nun allen Fleiss verwenden will, kein neuen Schwierigkeiten zeigen werden. Aber nicht bloss das—wie mich dünkt—werthvolle Resultat, auch die Einsicht die mir auf dem Wege geworden ist, hat mir mannigfache Befriedigung meiner mathematischen Neugierde gewährt, und ich habe auch die Freude gehabt, mit meinen bezüglichen Mittheilungen das mathematische Herz meines Freundes Kummer vielfach zu erfreuen, da auch Aussichten für Erledigung seiner Lieblingsfragen sich zeigen.

54. This is transcribed from a NOVA television program on Wiles' proof.
55. A. Grothendieck, "Cohomology Theory of Abstract Algebraic Varieties," in J.A. Todd, ed., *Proceedings of the International Congress of Mathematicians 1958* (Cambridge: Cambridge University Press, 1960), pp. 103–118, at pp. 105–106.
56. R. Williams, *The Country and the City* (New York: Oxford University Press, 1973); W. Cronon, *Nature's Metropolis* (New York: Norton, 1991).
57. Riemann articulates various notions of holes in a plane on which a function resides and the connectedness of that plane. He defines cross-cuts (*Querschnitt*) and circular paths (*Rückkehrschnitt*). We might recall branch cuts (the *lignes d'arrêt*). (Some of these distinctions are due to V. Puiseux. See also, A. Weil, "Riemann, Betti and the Birth of Topology," *Archive for the History of the Exact Sciences* 20 (1979): 91–96, p. 95, where Betti says that Riemann got his ideas for cuts from Gauss. The letters are also, in French translation, in J.C. Pont, *La topologie algébrique dès origine à Poincaré* (Paris: Presses Universitaires de France, 1974).)
58. It is a *uni*formization, since there is now one variable, t, a parameter. D. Hilbert, "Mathematical Problems," *Bulletin of the American Mathematical Society* 8 (1902): 437–479. Reprinted in, *Mathematical Developments Arising from Hilbert Problems*, Proceedings of Symposia in Pure Mathematics 28, part 1 (Providence: American Mathematical Society, 1976), pp. 1–34.
59. A. Grothendieck and J.A. Dieudonné, "Préface," *Eléments de Géométrie Algébrique I* (Berlin: Springer, 1971).
60. S. Buck-Morss, *The Dialectics of Seeing, Walter Benjamin and the Arcades Project* (Cambridge, Massachusetts: MIT Press, 1989), p. 102. Some other passages from Benjamin (*Gesammelte Schriften*, volume 5, part 1):

Die ersten Eisenbauten dienten transitorischen Zwecken: Markthallen, Bahnhöfe, Ausstellungen. Das Eisen verbindet sich also sofort mit funktionalen Momentum im Wirtschaftsleben. Aber was damals funktional und transitorisch war, beginnt heute in verändertem Tempo formal un stabil zu wirken. (p. 216)

Jedes der 12 000 Metallstücke [of the Eiffel Tower] ist auf Millimeter genau bestimmt, jeder der 2½ Millionen Niete . . Auf diesem Werkplatz ertönte kein Meißelschlag, der dem Stein die Form entringt; selbst dort herrschte der Gedanke über die Muskelkraft, die er auf sichere Gerüste und Krane übertrug. (A.G. Meyer, quoted by Benjamin, p. 223)

61. M. Rudwick, *The Great Devonian Controversy* (Chicago: University of Chicago Press, 1985).
62. J.C. Maxwell, *Scientific Letters and Papers of James Clerk Maxwell*, vol. 1, 1846–1862, ed. P.M. Harman (Cambridge: Cambridge University Press, 1990), articles and manuscripts beginning on pages 267, 284, 633, 654, 675. Notes 6 and 7 on pp. 269–270 provide a brief history of color vision. Harman suggests that the three-color theory

comes out of seventeenth century painting. Young is crucial for the three-color theory of vision.

63. A. Cornwell-Clyne, *Colour Cinematography* (London: Chapman and Hall, 1951).

64. Colin McLarty has cautioned me that Noether was not abstracting or reducing what she saw on the blackboard, so much as seeing or comprehending it in her terms.

65. We might think of the precision of craftsmanship yet the roomy-enough tolerance of fit among nested Russian *matryoshka* dolls, one inside the other. By analogy, the archetypal problem of algebraic topology is that an ant might march around the surface of one such doll, noticing there is enough room to move in-between two differently sized dolls. But it would not know whether the dolls can really be separated with any particular slice—for a limb might be wedged within a larger limb or intertwined with it.

In effect, local information (that there is enough room in-between) does not tell us global facts (whether you can separate the dolls by using a particular cut). What mathematicians call "exactness" is a claim that local information is enough to tell you global facts, and that your division of an object is into nice independent parts.

Obstructions to exactness turn out to be just those charges in electricity or those holes in a doughnut, which then mean that there is path dependence in integrals of functions, that local information is not enough to tell you global facts. For example, an integral from a to b of a function, $\int_a^b f(x)dx=I(a,b)$, can be expressed as $J(b)-J(a)$, where the derivative of J is the function f itself. It is the indefinite integral or antiderivative. But if the function has jumps or discontinuities, the particular path from a to b will in part determine the integral, and $J(b)-J(a)$ needs to be supplemented by some measure of the obstructions or discontinuities crossed by that path.

The Snake Lemma shows one way that exactness propagates between two parallel sequences of objects, each sequence decomposing related objects.

Imagine multiplying nonzero diagonal matrices so that the product of two of them is zero. They might look like the following,

$$\begin{bmatrix} a & 0 & 0 \\ 0 & b & 0 \\ 0 & 0 & 0 \end{bmatrix} \text{ and } \begin{bmatrix} 0 & 0 & 0 \\ 0 & 0 & 0 \\ 0 & 0 & c \end{bmatrix},$$

each matrix in effect taking charge of a part of the larger vector space. But if we are in a modular number system, as is the minute hand of a clock, in which $n \times p$ gets you back to zero (just as 60×1 minute gets you back to 0), then in fact that division of the vector space may not be nicely separated. For both matrices may have only nonzero diagonal elements yet the product could be zero.

Technically, for example, we might have two matrices, g and f, such that $gf=0$, where $U \!-\! f \!\rightarrow\! V \!-\! g \!\rightarrow\! W$. Say we have a τ, a member or set of V, such that $g\tau=0$. We might ask if there is a u in U such that $fu=\tau$: namely, that f delivers to g what g will then annihilate. In so far as that is not the case there is a defect. Now, dim U is the dimension of U, and the rank of a matrix is the number of independent rows or columns, in effect its dimension. Then the defect is: dim U–(rank f)–(rank g). The defect is the dimension of the homology module. For a nice statement, see C.A. Weibel, *An Introduction to Homological Algebra* (New York: Cambridge University Press, 1994), p. 1.

66. Yu. Manin, *A Course in Mathematical Logic* (Berlin: Springer, 1977), p. 19.

67. J. Valley, "It's My Turn," *Rolling Stone* #331 (27 November 1980): 54–57, at p. 57.

68. In effect, we have a naturalistic philosophy of science exemplified, where ". . . the methodologist's job is to account for [mathematics] as it is practiced." P. Maddy, "Set Theoretic Naturalism," at p. 490. Maddy refers to set theory rather than mathematics more generally.

69. W. Thurston, *Three-Dimensional Geometry and Topology*, vol. 1 (Princeton: Princeton University Press, 1997), p. 259.

70. Thurston, *Three Dimensional Geometry and Topology*, p. 272.

71. W.P. Thurston, "On Proof and Progress in Mathematics," p. 174.

72. R. Penner with J.L. Harer, *Combinatorics of Train Tracks*, Annals of Mathematics Studies 125 (Princeton: Princeton University Press, 1992), pp. 3–6, 13, 204.

If all of this might appear to be something of a fantasy, the phallic diagram of a "CW complex having one open cell in each dimension 0, 1, 2," from Munkres' textbook (Figure 38.3, p. 217) does not appear to deny that. J. Munkres, *Elements of Algebraic Topology* (Reading, Massachusetts: Addison-Wesley, 1984).

73. W. Thurston, "On the Construction and Classification of Foliations," *Proceedings of the International Congress of Mathematicians* (Vancouver, 1974), p. 547.

74. My discussion of the body is quite specific and concrete. Brian Rotman has written about the body and mathematics in a very different light, in *Ad infinitum . . . Taking God out of mathematics and putting the body back in: an essay in corporeal semiotics* (Stanford: Stanford University Press, 1993).

It is perhaps irrelevant, but still interesting to note that it was Walt Disney's fascination with model trains, one of his hobbies, that led him to create monorails and the like at his theme parks.

75. http://mathnexus.wwu.edu/Archive/mathematician/detail.asp?ID=30. See Jon Rogawski, *Calculus* (New York: Freeman, 2008), p. 107. Also, http://www-groups.dcs.st-and.ac.uk/history/Printonly/Fefferman.html. I do not have the original source.

76. Mathematics provides as well a model of apparently eternal truths and method, one that may be viewed as echoing theological goals. And mathematics' capacity to prove the uniqueness (or categoricity) of objects should be of interest to theological concerns about monotheism; its capacity to show irreducible variety, to polytheism.

77. Descartes, "Second Set of Replies to the *Meditations*," in R. Descartes, *Philosophical Writings of Descartes*, vol. 2, tr. J. Cottingham, R. Stoothoff, and D. Murdoch (Cambridge: Cambridge University Press, 1984), p. 100.

78. Descartes, "Meditations, Third Meditation," *Philosophical Writings*, p. 31 (original text page 45).
79. Descartes, "Meditations, Fifth Meditation," *Philosophical Writings*, p. 46 (original text page 66).
80. *Ethics*, Book I, II/57 in B. Spinoza, *A Spinoza Reader*, tr. and ed. by E. Curley (Princeton: Princeton University Press, 1994), p. 95.
81. Spinoza, *Ethics*, Book I, D2, D6, P13 (pp. 85, 86, 93).
82. "Discourse on Metaphysics" (1786), section 6, in G.W. Leibniz, *Philosophical Essays*, ed. and tr. R. Ariew and D. Garber (Indianapolis: Hackett, 1989), p. 39.
83. And then there is Leopold Kronecker's (1880) remark (reported by Heinrich Weber, but we do not have any published articles by Kronecker saying this) which is actually a claim about how one should do mathematics and what is fundamental.

Die [ganzen, natürlichen] Zahlen hat der liebe Gott gemacht, alles andere ist Menschenwerk. [The integers were made by our dear God, the rest is man's work.]

Of course, we might argue that the integers and the polynomials with integer coefficients, which Kronecker also included, are the great construction developed out of our experience of a continuum, God having actually given us ongoing experience or a continuum already.

84. Funkenstein, so writing about the 17th century, calls this connection "metabasis," transferring a model from its original realm to another. A. Funkenstein, *Theology and the Scientific Imagination* (Princeton: Princeton University Press, 1986), p. 346.
85. Weyl, *Philosophy of Mathematics and Natural Science*, p. 9: "mathematics always accomplishes the extension of a given domain of operation through the introduction of ideal elements. Such an extension is made in order to force the validity of simple laws." See the quote from Hilbert a bit further on in the main text.
86. For the physicists, there is a length scale when in fact that self-similarity fails (in effect, one then encounters a fundamental particle or a new strength of interaction). P. Dehornoy ("Another Use of Set Theory," *Bulletin of Symbolic Logic* 2 (1996): 379–391.) shows how self-similarity as an idea leads to important ideas about finite objects (algorithms for the word problem for braids). Albeit, after much further invention, one can now do without any reference to set theory and the infinite in proofs of these algorithms.
87. M. Hallett, *Cantorian Set Theory and Limitation of Size* (New York: Oxford, 1984). Dauben discusses Cantor's interest in the theological consistency of his mathematical ideas, and his relationship to Catholic priests in this regard. J.W. Dauben, *Georg Cantor* (Cambridge, Massachusetts: Harvard University Press, 1979), pp. 140–148. It should be noted that the mathematics came first, and then the theological inquiries.

88. Quoted in translation in H. Wang, *From Mathematics to Philosophy* (New York: Humanities Press, 1974), p. 210. It is from Cantor's collected works, p. 400 (1886), which I have not seen. Wang is skeptical of Cantor's theological speculations.

89. David Hilbert, "On the Infinite" (1925), in J. van Heijenoort, ed., *From Frege to Gödel* (Cambridge, Massachusetts: Harvard University Press, 1967), pp. 376, 379, 392.

90. Maddy, *Realism in Mathematics*, p. 141.

91. There is also an orderly articulation of the smaller measure-zero large cardinals, by Shelah.

92. Complementation corresponds to a universal quantifier, while union corresponds to an existential quantifier. And so sets in the hierarchies will correspond to particular sequences of quantifiers in formulae.

93. J.R. Steel, "Mathematics Needs New Axioms," *Bulletin of Symbolic Logic* 6 (2000): 422–433. See note 96(5).

94. P.J. Cohen, "Comments on the Foundations of Set Theory," in, *Axiomatic Set Theory*, Proceedings of Symposia in Pure Mathematics 13, part 1 (Providence: American Mathematical Society, 1967), pp. 9–15, at p. 12.

95. W.H. Woodin, "The Continuum Hypothesis, Part I and Part II," *Notices of the American Mathematical Society* 48 (2001): 657–676, 681–690.

96. L'Analyse eût porté en elle-même un principe de limitation. . . . La démonstration [by Lebesgue] était simple, courte, mais fausse. . . . Ainsi, l'Analyze ne porte pas en elle-même un principe de limitation. L'étendue de la famille des fonctions de Baire était vaste à donner le vertige, le champ de l'Analyse est plus vaste encore. Et combien plus vaste! . . . Le champ de l'Analyse est peut-être limité, il embrasse en tout cas cette infinité de types de fonctions!

H. Lebesgue, "Préface," in N. Lusin, *Leçons sur les Ensembles Analytiques et leurs Applications* (New York: Chelsea, 1972 [1930]), pp. vii–viii.

97. Moschavakis' chapter headings in Y. Moschovakis, *Descriptive Set Theory* (Amsterdam: North Holland, 1980) include "The Playful Universe," when he is talking about determinacy.

Some possible technical models of God's properties, some of which overlap:

(1) God as Unspecifiable. Our axiomatic conceptions of mathematical objects, such as the real numbers or infinite sets, often prove to have a variety of very different inequivalent models or realizations (they are not "categorical" conceptions, to use a term from John Dewey).

(2) God as Incommensurable and Inaccessible. Our mode of specification of an object can be unending, an infinitude of specifications needed to distinguish it from another. Set theoretic technology, in model theory, demonstrates the everpresence of problematic plenitude and nonuniqueness. We have an infinity of specifications, as in an unending binary or decimal expansion (0.011001010 . . . "tail"). These sorts of sets have very peculiar properties: Tail sets exhibit the Kolmogorov zero-one law; free-choice sequences

lead to functions that are necessarily uniformly continuous (Brouwer). (Lawvere describes general properties of these "variable sets," connecting them to Marxian dialectics. F.W. Lawvere, "Unity and Identity of Opposites in Calculus and Physics," *Applied Categorical Structures* 4 (1996): 167–174.)

The tail of that infinity is of course unknowable as of yet, since it is not yet given, never fully given in any finite time. But say that we set things up so that "so far" in the list of specifications that number (or world or set) is by its explicit specification undeniably not at all like any one we know. Hence, it is other and different, in a specified way, so far—"forced" to be so different, to use Paul Cohen's (1963) term of art. So Cohen constructed a set whose size or cardinality is strictly larger than the integers and strictly smaller than the reals. Within this model world, we do not have the resources to count up these intermediate-sized sets. Were we outsiders, we might be able to count them up. (Technically, one says that these "nonstandard" sets are not countable "internally," but they are countable or denumerable "externally.") So we have created a model of the infinite world, a world that is in one sense inaccessible, but transcendently it is accessible.

(3) God as External. Following Hilbert's suggestion, say we append something not at all everyday or standard such as the Absolute or God onto our everyday conceptions. The "transfer principle" of nonstandard analysis and model theory says that what is true in our language in the nonstandard universe, the one in which God is now a good notion, is as well true in the standard one, in which God is incommensurable, and vice versa. (This is in first order language, no quantification over functions.) So such a God is legitimately present in ordinary life. God is "external," not expressible in standard language, much as i or dx are present but not expressible "standardly" in the real number system or in ordinary analysis, respectively.

(4) God as Ineffable: We might model God in terms of mathematical measure or weight. We stipulate that two sets are equivalent as long as their difference were of zero measure (as in "Boolean valued models"). For ineffability, God might be taken to be of measure zero in our world, hence present but unavailable. Or, perhaps it is the other way around, and we are of measure zero. Or, we might conceive of a world made up of worlds, a Heaven that incorporates all of space and time, as did Augustine in *The City of God* (420 CE), again its total specification beyond us, but "up to the current point" different than ours. (This is in effect an "ultrapower" or reduced powers construction, where sets disagree in a comparatively small number of places.)

(5) God as a Game. Consider a "determined" set, one defined by its containing infinite sequences of natural numbers, such that every sequence represents a game in which a particular one of the two players has a winning strategy. (For a nice review, see D.A. Martin and J.R. Steele, "Projective Determinacy," *Proceedings of the National Academy of Science* 85 (1988): 6582–6586.) While Einstein may have said that God does not play dice, and that God's providence is difficult to decode even if God is not wicked, we might imagine God playing a game of chance in figuring out or playing out this world in order to construct such an interesting set. What is again remarkable is the existence of a

linear hierarchy (by set inclusion) of these worlds or sets or cardinal numbers, each new mode of specification finding a home within the linear hierarchy.

(6) God as Indefinitely Large. Technically, we might have a concept of "zillions," finite but very very large or indefinitely large sets, as one such model of the availability of the transfinite. An "Omnipotent Mathematician" is actually combinatorially counting out a set or iteratively building-up a set as a collection through repeated applications of basic rules, rather than specifying it by some logical or formal property. (See, for example, S. Lavine, *Understanding the Infinite* (Cambridge, Massachusetts: Harvard University Press, 1994). It is, in part, derived from J. Mycielski's work.)

(7) God as Compact. God may have an accessible model. Gödel demonstrated that, so to speak, if every finite aspect of an object or world has an actual model, or in this case, every aspect of God has a model in our world (namely, we can imagine it concretely, let us say), then the object or world itself has a model (technically, God's putative "compactness"). And, "having a model" is to give us access to the world. Löwenheim and Skolem showed that for any higher transfinite picture or model of the world, there is as well a lower transfinite picture or model, albeit still countably infinite, which fulfills all our (first order) axiomatized expectations. (This latter is the basis for Cohen's constructions.) Hence, no matter how distant is the Absolute or God's realm, we have some access to it; albeit, our model is not unique or categorical; and our transfinite countable model is perhaps not quite what we want to mean by the Absolute or God.

(8) God as a Sequence of Decisions. There are models or interpretations of the infinite as it might be present in everyday life, models that provide us with an intuition for what is at stake in the Absolute. Potentially infinite sequences of decisions as in steering an automobile or in cybernetics may be modeled by a real or rational number, each digit representing a choice, let us say independent of the previous choices. The probability that such a number is not a rational fraction is zero or one, and in fact it is one. (M.H. Krieger, "Could the Probability of Doom Be Zero-or-One?," *Journal of Philosophy* 92 (1995): 382–387.)

Kevin Kelly has constructed a theory of inquiry and decision using topological notions, where various sets in the Borel/Kleene hierarchy represent different sorts of reliability claims. He begins with (a Baire space of) sequences of natural numbers,

> . . . a standard topological space in which points correspond to infinite data streams, sets of points become empirical hypotheses, elements of a countable basis correspond to finite data sequences, and continuity reflects the bounded perspective of the scientist.

What is remarkable is how deeply applicable is his epistemological model in instantiating all the basic topological notions and quite sophisticated theorems from topology and set theory. One might construct a philosophy of topology based on this analogy between epistemology and topology, between decision analysis and his account of descriptive set theory. (K. Kelly, *The Logic of Reliable Inquiry* (New York: Oxford, 1996), p. 78.) Kelly's work may well be a significant philosophical advance, since it encourages the philosopher to borrow notions from descriptive set theory and analysis. See also, S. Vickers, *Topology via Logic* (Cambridge: Cambridge University Press, 1989), where in the preface he says *"open sets are semidecidable properties"* to begin to account for the power of topological notions in computer science.

98. J. Freccero, "Foreword," to Dante, *The Inferno of Dante*, tr. R. Pinsky (New York: Noonday, 1996), pp. ix–xvii, at p. ix. Freccero argues that much of Dante's thought is founded in Augustine.

99. E. Durkheim, *Elementary Forms of the Religious Life*, tr. K. Fields (New York: Free Press, 1995).

100. D. Fowler, *The Mathematics of Plato's Academy* (New York: Oxford University Press, 1999).

101. As for Newton, Chandrasekhar's recent reconstruction was criticized for its deliberate focus on Chandrasekhar's understanding of what Newton was doing rather than Newton's and his contemporaries' understanding. (S. Chandrasekhar, *Newton's Principia for the Common Reader* (Oxford: Clarendon Press, 1995).) As for the Italian algebraic geometers, they are often described as not being up to modern standards of rigor, Zariski's redoing of their insights being the crucial move historically. But to appreciate how they were thinking might well lead one to new insights. C. Parikh, *The Unreal Life of Oscar Zariski* (Boston: Academic Press, 1991).

Or, as André Weil puts it, referring in part to the Italian school, "the so-called 'intuition' of earlier mathematicians, reckless as their use of it may sometimes appear to us, often rested on a most painstaking study of numerous special examples, from which they gained an insight not always found among modern exponents of the axiomatic creed." But, he then says, "it is the duty, as it is the business, of the mathematician to prove theorems, and that this duty can never be disregarded for long without fatal effects." A. Weil, *Foundations of Algebraic Geometry* (New York: American Mathematical Society, 1946), p. vii.

102. K. Kuratowski, *A Half Century of Polish Mathematics* (Oxford: Pergamon, 1980).

EPILOG

1. From A. Kanamori, "Review of S. Shelah's *Proper and Improper Forcing*," *Mathematical Reviews* #98m:03002.

2. M. Beals, C. Fefferman, R. Grossman, "Strictly Pseudoconvex Domains in C^n," *Bulletin (New Series) of the American Mathematical Society* 8 (1983): 125–322, at p. 294. Or,

Indeed, the spherically symmetric case is the only case in which Classical Mechanics yields reasonable information about periodic orbits. In fact, it is only the case when it yields any information. Furthermore, the quasiclassical, or WKB, techniques in one dimension [Fefferman-Seco] use extensively explicit computations.

I.M. Sigal, et al, "Lectures on Large Coulomb Systems," *CRM Proceedings and Lecture Notes* 8 (1995): 73–107, at p. 102.

3. G. Mackey, *The Scope and History of Commutative and Noncommutative Harmonic Analysis* (Providence: American Mathematical Society, 1992); C. Fefferman, "The Uncertainty Principle," *Bulletin (New Series) of the American Mathematical Society* 9 (1983): 129–206.

4. See note 33 of chapter 3.

5. Technically, phase space is the realm of Hamilton's equations, which preserve phase space volume (Liouville's theorem). And harmonic analysis is those nice action-angle variables, J (angular momentum) and ω (frequency), the energy or hamiltonian being $J\omega$. Those volume-preserving canonical transformations, such as Hamilton's equations, are employed to express the hamiltonian in this simple fashion.

Speculatively, I believe this is also the realm of inverse scattering transforms, the Kortweg-deVries equation, and the Ising model. In effect, it is the mathematics of spinning tops. So those equations are effectively diagonalized into amplitudes and frequencies—perhaps for each box of phase space separately, "microlocally." And one obtains, some of the time, an infinity of conserved quantities and associated symmetries.

APPENDIX D

1. Lang has indicated that Weil's potted history is tendentious, in part because Weil would seem to be less that respectful of Artin's contribution. S. Lang, "On the AMS *Notices* Publication of Krieger's Translation of Weil's 1940 Letter," *Notices of the American Mathematical Society* 52 (2005): 612–614.

Note added in proof: In the quantum mechanical realm and in statistical mechanics, the largest eigenvalue is the ground state or equilibrium. Speculatively, we might think of the ordered state of the Ising lattice as such a ground state or equilibrium; perhaps it is not so surprising that the scaled correlation is associated with largest eigenvalues. So, for example, the Tracy-Widom distribution describes the extremes (largest) of a set of N *dependent* random variables, and it is centered at $[\sqrt{(2N)} - \sim 2]$ with a width that goes as $N^{1/6}$. If the dependent random variables fall off as x^{-2}, then $N \times$ {falloff evaluated at the center-of-extreme-value-distribution} is on the order of 1, a substantial probability.

Bibliography

Abbott, B., *Berenice Abbott, Photographs*. New York: Horizon Press, 1970.

Abelson, H., and G. Sussman, *Structure and Interpretation of Computer Programs*. Cambridge, Massachusetts: MIT Press, 1996.

Abhyankar, S.S., "Galois Theory on the Line in Nonzero Characteristic," *Bulletin (New Series) of the American Mathematical Society* 27 (1992): 68–133.

Ablowitz, M.J., and P.A. Clarkson, *Solitons, Nonlinear Evolution Equations and Inverse Scattering* (London Mathematical Society #149). Cambridge: Cambridge University Press, 1991.

Abraham, D.B., "On Correlation Functions of the Ising and X-Y Models," *Studies in Applied Mathematics* 51 (1972): 179–209.

———, "*n*-Point Functions for the Rectangular Ising Model," *Physics Letters* 61A (1977): 271–274.

———, "Susceptibility of the Rectangular Ising Ferromagnet," *Journal of Statistical Physics* 19 (1978): 349–358.

———, "Odd Operators and Spinor Algebras in Lattice Statistics: *n*-Point Functions for the Rectangular Ising Model," *Communications in Mathematical Physics* 59 (1978): 17–34.

———, "Some Recent Results for the Planar Ising Model," in D.E. Evans and M. Takesaki, eds., *Operator Algebras and Applications* (volume II: Mathematical Physics and Subfactors), London Mathematical Society Lecture Note Series, 136. Cambridge: Cambridge University Press, 1988, pp. 1–22.

Abraham, D.B., and A. Martin-Löf, "The Transfer Matrix for a Pure Phase in the Two-dimensional Ising Model," *Communications in Mathematical Physics* 32 (1973): 245–268.

Adams, H., "The Rule of Phase Applied to History," in E. Stevenson, ed., *A Henry Adams Reader*. Garden City: Anchor, 1958.

Adams, J.F., "Algebraic Topology in the Last Decade," in, *Algebraic Topology* (Proceedings of Symposia in Pure Mathematics 22). Providence: American Mathematical Society, 1971, pp. 1–22.

Aharony, A. and M. E. Fisher, "Universality in Analytic Corrections to Scaling for Planar Ising Models," *Physical Review Letters* 45 (1980): 679-682.

Aldous, D., and P. Diaconis, "Hammersley's Interacting Particle Process and Longest Increasing Subsequences," *Probability Theory and Related Fields* 103 (1995): 199–213.

———, "Longest Increasing Subsequences: From Patience Sorting to the Beik-Deift-Johansson Theorem," *Bulletin (New Series) of the American Physical Society* 36 (1999): 413–432.

Alexandroff, P., and H. Hopf, *Topologie* Bd. 1. Berlin: Springer, 1935.

Alexandrov, P.S., "In Memory of Emmy Noether" (1935), in N. Jacobson, ed., *Emmy Noether: Gesammelte Abhandlung*. Berlin: Springer, 1983, pp. 1–11.

Andrews, G., *q-series*. Providence: American Mathematical Society, 1986.

Anglès d'Auriac, J.-Ch., S. Boukraa , and J.-M. Maillard, "Functional Relations in Lattice Statistical Mechanics, Enumerative Combinatorics, and Discrete Dynamical Systems," *Annals of Combinatorics* 3 (1999) 131–158.

Appel, K.I., W. Haken, and J. Koch, "Every Planar Map is Four Colorable, Part I: Discharging, Part II: Reducibility," *Illinois Journal of Mathematics* 21 (1977): 421–490, 491–567.

Arnett, R., *Supernovae and Nucleosynthesis*. Princeton: Princeton University Press, 1996.

Arratia, R., and L. Gordon, "Tutorial on Large Deviations for the Binomial Distribution," *Bulletin of Mathematical Biology* 51 (1989): 125–131.

Arthur, J., "Harmonic Analysis and Group Representations," *Notices of the American Mathematical Society* 47 (2000): 26–34.

_____, "L-functions and automorphic representations," address to the ICM 2014, https://www.youtube.com/watch?v=ljJudppxau4

Ashkin, J., and W.E. Lamb, Jr., "The Propagation of Order in Crystal Lattices," *Physical Review* 64 (1943): 169–178.

Assis, M., S. Boukraa, S. Hassani, M. van Hoeij, J.-M. Maillard, and B. M. McCoy, "Diagonal Ising susceptibility: elliptic integrals, modular forms and Calabi-Yau equations," *J. Phys. A* 45 (2012), no. 7, 075205, 32 pp.

Assis, M., J.-M. Maillard, and B. M. McCoy, Factorization of the Ising model form factors. *J. Phys. A* 44(2011), no. 30, 305004, 35 pp.

Atiyah, M., "Mathematics in the 20th Century," *American Mathematical Monthly* 108 (2001): 654–666.

Au-Yang, H., B.M. McCoy, J.H.H. Perk, S. Tang, and M.-L. Yan, "Commuting Transfer Matrices in the Chiral Potts Models: Solution of Star-Triangle Equations with Genus > 1," *Physics Letters* A123 (1987): 219–223.

Au-Yang, H., and J.H.H. Perk, "Onsager's Star-Triangle Equation: Master Key to Integrability," in M. Jimbo, T. Miwa, and A. Tsuchiya, eds., *Integrable Systems in Quantum Field Theory and Statistical Mechanics* (Advanced Studies in Pure Mathematics, Tokyo, Japan, 19). Boston: Academic Press, 1989; Tokyo: Kinokuniya, pp. 57–94.

Azzouni, J., *Metaphysical Myths, Mathematical Practice: The Ontology and Epistemology of the Exact Sciences*. New York: Cambridge University Press, 1994.

_____, Applying Mathematics: An Attempt to Design a Philosophic Problem," *The Monist* 83 (2000): 209–227.

Bach, V., "Error bound for the Hartree-Fock energy of atoms and molecules," *Communications in Mathematical Physics* 147 (1992): 527–548.

_____, "Accuracy of mean field approximation for atoms and molecules," *Communications in Mathematical Physics* 155 (1993): 295–310.

_____, "Approximative Theories for Large Coulomb Systems," in J. Rauch and B. Simon, eds., *Quasiclassical Methods*. Berlin: Springer, 1997, pp. 89–97.

Bach, V., J. Fröhlich, and I.M. Sigal, "Quantum Electrodynamics of Confined Nonrelativistic Particles," *Advances in Mathematics* 137 (1998): 299–395.

Baclawski, K., G.-C. Rota, and S. Billey, *An Introduction to the Theory of Probability*. Cambridge, Massachusetts: Department of Mathematics, MIT, 1989.

Baik, J., Borodin, A., Deift, P., and T Suidan, "A model for the bus system in Cuernavaca (Mexico)," *Journal of Physics A* 39 (2006): 8965–8975.

Bander, M., and C. Itzykson, "Quantum field theory calculation of the 2d Ising model correlation function," *Physical Review* D15: 463–469.

Bardou, F., J.-P. Bouchaud, A. Aspect, and C. Cohen-Tannoudji, *Lévy Statistics and Laser Cooling*. Cambridge: Cambridge University Press, 2002.

Barrow-Green, J., *Poincaré and the Three Body Problem*. Providence: American Mathematical Society, 1997.

Bartoszynski, T. and H. Judah, *Set Theory: On the Structure of the Real Line*. Wellesley: A. K. Peters, 1995.

Batchelor, M. T., and K.A. Seaton, "Magnetic correlation length and universal amplitude of the lattice E_8 Ising model," *Journal of Physics A* 30 (1997): L479–L484.

Batterman, R., *The devil in the details: asymptotic reasoning in explanation, reduction, and emergence*. New York: Oxford University Press, 2002.

Baxter, R. J., "Corner Transfer Matrices of the Eight-Vertex Model. I. Low-Temperature Expansions and Conjectured Properties," *Journal of Statistical Physics* 15 (1976): 485–503.

_____, "Corner Transfer Matrices of the Eight-Vertex Model. II. The Ising Model Case," *Journal of Statistical Physics* 17 (1977): 1–14.

_____, "Solvable Eight-Vertex Model on an Arbitrary Planar Lattice," *Philosophic Transactions of the Royal Society A* 289 (1978): 315–346.

_____, "Variational Approximation for Square Lattice Models in Statistical Mechanics," *Journal of Statistical Physics* 19 (1978): 461–478.

_____, "Solvable Eight-Vertex Model on an Arbitrary Planar Lattice," *Philosophical Transactions of the Royal Society* 289A (1978): 315–344.

_____, "Corner Transfer Matrices," *Physica A* 106 (1981): 18–27.

_____, *Exactly Solved Models in Statistical Mechanics*. London: Academic Press, 1982.

_____, "Chiral Potts Model: Eigenvalues of the Transfer Matrix," *Physics Letters A* 146 (1990): 110–114.

_____, "On solving the Ising model functional relation without elliptic functions," *Physica* A177 (1991): 101–108.

_____, "Corner Transfer Matrices of the Chiral Potts Model. II. The Triangular Lattice," *Journal of Statistical Physics* 70 (1993): 535–582.

_____, "Solvable Models in Statistical Mechanics, from Onsager Onward," *Journal of Statistical Physics* 78 (1995): 7–16.

_____, "Functional Relations for the Order Parameter in the Chiral Potts Model," *Journal of Statistical Physics* 91 (1998): 499–524.

_____, "Functional relations for the order parameters of the chiral Potts model: low-temperature expansions," *Physica A* 260 (1998): 117–130.

_____, "Some hyperelliptic function identities that occur in the Chiral Potts model," *Journal of Physics A* 31 (1998): 6807–6818.

_____, "Corner transfer matrices in statistical mechanics," *Journal of Physics A* 40 (2007): 12577–12588.

_____, "A Conjecture for the Superintegrable Chiral Potts Model," *Journal of Statistical Physics* 132 (2008): 983–1000.

_____, "Algebraic Reduction of the Ising Model, *Journal of Statistical Physics* 132 (2008): 959–982.

_____, "Some comments on developments in exact solutions in statistical mechanics since 1944," *Journal of Statistical Mechanics: Theory and Experiment* (2010) P11037, 26pp.

_____, "Onsager and Kaufman's calculation of the spontaneous magnetization of the Ising model," *Journal of Statistical Physics* 145 (2011): 518–548,

_____, "Onsager and Kaufman's calculation of the spontaneous magnetization of the Ising model: II," *Journal of Statistical Physics* 149 (2012): 1164–1167.

Baxter, R.J., and I. Enting, "399th Solution of the Ising Model," *Journal of Physics A* 11 (1978): 2463–2473.

_____, "Series Expansions from Corner Transfer Matrices: The Square Grid Ising Model," *Journal of Statistical Physics* 21 (1979): 103–123.

Bazhanov, V.V., B. Nienhuis and S.O. Warnaar, "Lattice Ising model in a field: E_8 scattering theory," *Physics Letters* B322 (1994): 198–206.

Beals, M., C. Fefferman, and R. Grossman, "Strictly Pseudoconvex Domains in C^n," *Bulletin (New Series) of the American Mathematical Society* 8 (1983): 125–322.

Beeson, M.J., *Foundations of Constructive Metamathematics*. Berlin: Springer, 1985.

Bell, J.L., and M. Machover, *A Course in Mathematical Logic*. Amsterdam: North Holland, 1977.

Benettin, G., G. Gallavotti, G. Jona-Lasinio, and A.L. Stella, "On the Onsager-Yang-Value of the Spontaneous Magnetization," *Communications in Mathematical Physics* 30 (1973): 45–54.

Benis-Sinaceur, H., "The Nature of Progress in Mathematics: The Significance of Analogy," in E. Grosholz, and H. Breger, eds., *The Growth of Mathematical Knowledge*. Dordrecht: Kluwer, 2000, pp. 281–293.

Benjamin, W., *Gesammelte Schriften*, vol. 5, part 1. Frankfurt am Main: Suhrkamp, 1982.

Berkovich, A., B.M. McCoy, and A. Schilling, "N=2 Supersymmetry and Bailey Pairs," *Physica A* 228 (1996): 33–62.

Beringer, J., et al., (Particle Data Group), "The Review of Particle Physics," *Physical Review D* 86 (2012): 010001.

Black, F., "Noise," in *Business Cycles and Equilibrium*. New York: Blackwells, 1987, pp. 152–172.

Blackwell, D., "Infinite games and analytic sets," *Proceedings of the National Academy of Science* 58 (1967): 1836–1837.

Bleher, P.M., and A.R. Its, eds., *Random Matrix Models and Their Applications*. Cambridge: Cambridge University Press, 2001.

Boi, L., "Le concept de Variété et la Nouvelle Géométrie de l'Espace dans le Pensée de Bernhard Riemann," *Archives internationales d'histoire des sciences* 45 (1995): 82–128.

Borel, A., "Jean Leray and Algebraic Topology," in J. Leray, *Selected Papers, Oeuvres Scientifiques* volume 1, Topology and Fixed Point Theorems. Berlin and Paris: Springer, 1998, pp. 1–21.

Bostan, A., S. Boukraa, S. Hassani, M. van Hoeij, J.-M. Maillard, J.-A. Weil and N. Zenine, "The Ising model: from elliptic curves to modular forms and Calabi–Yau equations," *J. Phys. A: Math. Theor.* 44 (2011): 045204.

Bott, R., "On induced representations," in R.O. Wells, ed., *The Mathematical Heritage of Hermann Weyl* (Proceedings of Symposia in Pure Mathematics 48). Providence: American Mathematical Society, 1988, pp. 1–13.

———, "The Topological Constraints on Analysis," in *A Century of Mathematics in America II*. Providence: American Mathematical Society, 1989, pp. 527–542.

Bottazzini, U., and R. Tazzioli, "*Naturphilosophie* and its Role in Riemann's Mathematics," *Revue d'histoire des mathématiques* 1 (1995): 3–38.

Böttcher, A., and B. Silbermann, *Introduction to Large Truncated Toeplitz Matrices*. New York: Springer, 1999.

Boukraa, S., Hassani, S., Maillard, J.-M., McCoy, B. M., Orrick, W. P., and N. Zenine, Holonomy of the Ising model form factors," *J. Phys. A* 40 (2007): 75–111.

Boukraa, S., Hassani, S., Maillard, J.-M., McCoy, B. M., Weil, J.-A., and N. Zenine, "Fuchs versus Painlevé," *J. Phys. A* 40 (2007): 12589–12605.

Boukraa, S., and J.-M. Maillard, "Let's Baxterise," *Journal of Statistical Physics* 102 (2001): 641–700.

Box, J.F., *R. A. Fisher, the Life of a Scientist*. New York: Wiley, 1978.

Breger, H., "Tacit Knowledge in Mathematical Theory," in J. Echeverria, A. Ibarra, and T. Mormann, eds., *The Space of Mathematics*. Berlin: de Gruyter: 1992, pp. 79–90.

Bressoud, D.M., *Proofs and Confirmations, The Story of the Alternating Sign Matrix Conjecture*. Cambridge: Cambridge University Press, 1999.

Bronson, T., ed., *Spectral Problems in Geometry and Arithmetic*. Providence: American Mathematical Society, 1999

———, "Continuous one-one transformations of surfaces into themselves," *KNAW Proceedings* 11 (1909): 788–798.

———, "Continuous vector distribution of surfaces," *KNAW Proceedings* 12 (1910): 171–186.

Brouwer, L.E.J., *Collected Works* vol. 2 (ed. H. Freudenthal). Amsterdam: North Holland, 1976. Included are all the following, which appeared in *Mathematische Annalen*: "Zur Analysis Situs," 68 (1910): 422–434; "Beweis des Jordanschen Kurvensatzes," 69 (1910): 169–175; "Beweis der Invarianz der Dimensionenzahl," 70 (1911): 161–165; "Über Abbildung von Mannigfaltigkeiten," 71 (1911): 97–115). Also, "On continuous vector distributions on surfaces, III" *KNAW Proceedings* 12 (1910): 171–186, pp. 303–318; "Continuous one-one transformations of surfaces in themselves," *KNAW Proceedings* 11 (1909): 788–798.

Brush, S., "History of the Lenz-Ising Model," *Reviews of Modern Physics* 39 (1967): 883–893.

Buchwald, J.Z., *From Maxwell to Microphysics*. Chicago: University of Chicago Press, 1985.

_____, ed., *Scientific Practice: Theories and Stories of Doing Physics*. Chicago: University of Chicago Press, 1995.

Buck-Morss, S., *The Dialectics of Seeing, Walter Benjamin and the Arcades Project*. Cambridge, Massachusetts: MIT Press, 1989.

Bullough, R.K., and J.T. Timonen, "Quantum and Classical Integrable Models and Statistical Mechanics," in, V.V. Bazhanov and C.J. Burden, eds., *Statistical Mechanics and Field Theory*. Singapore: World Scientific, 1995.

Bump, D., *Automorphic Forms and Representations*. New York: Cambridge University Press, 1997.

Burkert, N.E., *Valentine and Orson*. New York: Farrar, Straus and Giroux, 1989.

Butterfield, J., "Reduction, Emergence, and Renormalization," *Journal of Philosophy* 111 (2009): 5-49.

Camille, M., *Image on the Edge: The Margins of Medieval Art*. Cambridge, Massachusetts: Harvard University Press, 1992.

Caneva, K.L., "Physics and *Naturphilosophie*," *History of Science* 35 (1997): 35–106.

Cardy, J.L., "Operator content of two-dimensional conformally invariant theories," *Nuclear Physics B* 270 (1986): 186–204.

Carleson, L., "The work of Charles Fefferman," in *Proceedings of the International Congress of Mathematicians*. Helsinki: Academy Sci. Fennica, 1978, pp. 53–56.

Cartier, P., "A Mad Day's Work: From Grothendieck to Connes and Kontsevich, The Evolutions of Concepts of Space and Symmetry," *Bulletin (New Series) of the American Mathematical Society* 38 (2001): 389–408.

Cauchy, A., "Résumé des Leçons donné a L'École Royale Polytechnique Sur Le Calcul Infinitésimal," *Oeuvres,* series II, vol. 4. Paris: Gauthier-Villars, 1899, p. 44 [1823].

Čech, E., "Théorie générale de l'homologie dans un espace quelconque," *Fundamenta Mathematicae* 19 (1932): 149–183.

Chandler, D., *Introduction to Modern Statistical Mechanics*. New York: Oxford University Press, 1987.

Chandrasekhar, S., *Newton's Principia for the Common Reader*. Oxford: Clarendon Press, 1995.

Chang, C.C., and H.J. Keisler, *Model Theory*. Amsterdam: North Holland, 1990.

Chelkak, D. and S. Smirnov, "Universality in the 2D Ising model and conformal invariance of fermionic observables," *Inventiones mathematicae* 189 (2012) 515–580.

Cheng, H., and T.T. Wu, "Theory of Toeplitz Determinants and the Spin Correlations of the Two-Dimensional Ising Model. III," *Physical Review* 164 (1967): 719–735.

Chvatal, V., *Linear Programming*. New York: Freeman, 1983.

Cohen, P.J., "The Independence of the Continuum Hypothesis," *Proceedings of the National Academy of Science* 50 (1963): 1143–1148.

_____, "Comments on the Foundations of Set Theory," in, *Axiomatic Set Theory* (Proceedings of Symposia in Pure Mathematics 13, part 1). Providence: American Mathematical Society, 1967, pp. 9–15.

Cooke, R., "Uniqueness of Trigonometric Series and Descriptive Set Theory, 1870–1985," *Archive for the History of the Exact Sciences* 45 (1993): 281–334.

Córdoba, A., C.L. Fefferman, and L.A. Seco, "A Trigonometric Sum Relevant to the Nonrelativistic Theory of Atoms," *Proceedings of the National Academy of Science* 91 (1994): 5776–5778.

_____, "Weyl Sums and Atomic Energy Oscillations," *Revista Matematica Iberoamericana* 11 (1995): 165–226.

_____, "A Number-Theoretic Estimate for the Thomas-Fermi Density," *Communications in Partial Differential Equations* 21 (1996): 1087–1102.

Corfield, D., "Assaying Lakatos's Philosophy of Mathematics," *Studies in the History and Philosophy of Science* 28 (1997): 99–121.

_____, *Toward a Philosophy of Real Mathematics*. Cambridge: Cambridge University Press, 2003.

Cornell, G., J.H. Silverman, and G. Stevens, eds., *Modular Forms and Fermat's Last Theorem*. New York: Springer, 1997.

Cornwell-Clyne, A., *Colour Cinematography*. London: Chapman and Hall, 1951.

Corry, L., *Modern Algebra and the Rise of Mathematical Structures*. Basel: Birkhäuser, 1996.

Courant, R., and D. Hilbert, *Methods of Mathematical Physics*. New York: Interscience, 1953.

Cox, D., J. Little, and D. O'Shea, *Ideals, Varieties, and Algorithms*. New York: Springer, 1997.

Cronon, W., *Nature's Metropolis*. New York: Norton, 1991.

Curtis, C.W., *Pioneers of Representation Theory: Frobenius, Burnside, Schur, and Brauer*. Providence: American Mathematical Society, 1999.

da Silva, J.J., "Husserl's Phenomenology and Weyl's Predicativism," *Synthese* 110 (1997): 277–296.

Dai, H., Z. Geary, and L.P. Kadanoff, "Asymptotics of eigenvalues and eigenvectors of Toeplitz matrices," *Journal of Statistical Mechanics: Theory and Experiment* (2009): P05012-1-25.

Dante, *The Inferno of Dante*, tr. R. Pinsky. New York: Noonday, 1996.

Daston, L., *Classical Probability in the Enlightenment*. Princeton: Princeton University Press, 1988.

Date, E., M. Jimbo, T. Miwa, and M. Okado, "Solvable Lattice Models," in, L. Ehrenpreis and R.C. Gunning, eds., *Theta Functions, Bowdoin 1987* (Proceedings of Symposia in Pure Mathematics 49). Providence: American Mathematical Society, 1989, pp. 295–331.

Dauben, J.W., *Georg Cantor*. Cambridge, Massachusetts: Harvard University Press, 1979.

Davies, B., "Onsager's algebra and superintegrability," *Journal of Physics* A 23 (1990): 2245–2261.

_____, "Onsager's algebra and the Dolan-Grady condition in the non-self-dual case," *Journal of Mathematical Physics* 37 (1991): 2945–2950.

Davis, H.T., *Introduction to Nonlinear Differential and Integral Equations*. Washington, D.C.: United States Atomic Energy Commission, 1960.

Dedekind, R., and H. Weber, "Theorie der algebraische Funktionen einer Veränderlichen," pp. 238–350 (with a comment by E. Noether, "Erläuterungen zur verstehenden Abhandlung," p. 350, on the threefold analogy) in Bd. 1, R. Dedekind, *Gesammelte mathematische Werke* (eds. R. Fricke, E. Noether, and O. Ore) Braunschweig: Vieweg, 1930. Originally, *Journal für reine und angewandte Mathematik ("Crelle")* 92 (1882): 181–290.

Dehornoy, P., "Another Use of Set Theory," *Bulletin of Symbolic Logic* 2 (1996): 379–391.

Deift, P., "Integrable Systems and Combinatorial Theory," *Notices of the American Mathematical Society* 47 (2000): 631–640.

_____, "Riemann-Hilbert Methods in the Theory of Orthogonal Polynomials," in F. Gesetzsky, et al., *Spectral Theory and Mathematical Physics*. Providence: American Mathematical Society, 2007, pp. 715–740.

Deift, P., Its, A., and I. Krasovsky, "Toeplitz Matrices and Toeplitz Determinants under the Impetus of the Ising Model: Some History and Some Recent Results," *Communications on Pure and Applied Mathematics* 116 (2013): 1360–1438.

Delfino, G., "Integrable field theory and critical phenomena: the Ising model in a magnetic field," *Journal of Physics A* 37 (2004): R45-R78.

Descartes, R., *Philosophical Writings of Descartes*, vol. 2, tr. J. Cottingham, R. Stoothoff, and D. Murdoch. Cambridge: Cambridge University Press, 1984.

Diaconis, P., and F. Mosteller, "Methods for studying coincidences," *Journal of the American Statistical Association* 84 (1989): 853–861.

Dieudonné, J., *Foundations of Modern Analysis*. New York: Academic Press, 1960.

_____, "Recent Developments in Mathematics," *American Mathematical Monthly* 71 (March 1964): 239–248.

_____, *History of Algebraic Geometry*. Monterey: Wadsworth, 1985.

_____, J., *A History of Algebraic and Differential Topology, 1900–1960*. Boston: Birkhäuser, 1989.

_____, "Une brève histoire de la topologie," in J.-P. Pier, ed., *Development of Mathematics 1900–1950*. Basel: Birkhäuser, 1994, pp. 35–153.

Dirac, P.A.M., *The Principles of Quantum Mechanics*. New York: Oxford University Press, 1982 (1930).

Dirichlet, P.G.L., *Vorlesungen über Zahlentheorie* (ed. R. Dedekind). New York: Chelsea, 1968 [1863, 1893].

Dolan, L., and M. Grady, "Conserved Charges from Self-Duality," *Physical Review* D25 (1982): 1587–1604.

Dolbeault, J., Laptev, A., and M. Loss, "Lieb-Thirring inequalities with improved constants," *Journal of the European Mathematics Society* 10 (2008) 1121–1126.

Domb, C., "The Ising Model," in P.C. Hemmer, H. Holden, and S. Kjelstrup Ratkje, eds., *The Collected Works of Lars Onsager (with commentary)*. Singapore: World Scientific, 1996, pp. 167–179.

_____, *The Critical Point: A Historical Introduction to the Modern Theory of Critical Phenomena*. London: Taylor and Francis, 1996.

Donoho, D., "High-Dimensional Data Analysis: The Curses and Blessings of Dimensionality," ms (August 8, 2000). Stanford: Stanford University.

Dubins, L., and L.J. Savage, *Inequalities for Stochastic Processes*. New York: Dover, 1976.

Durkheim, E., *Elementary Forms of Religious Life*, tr. K. Fields. New York: Free Press, 1995.

Dyson, F.J., "Stability of Matter," in *Eastern Theoretical Physics Conference*. Providence: Brown University, 1966, pp. 227–239.

_____, "Ground state energy of a finite system of charged particles," *Journal of Mathematical Physics* 8 (1967): 1538–1545.

_____, "Stability of Matter," in M. Chrétien, E.P. Gross, and S. Deser, eds., *Statistical Physics, Phase Transitions, and Superfluidity (Brandeis 1966)*. New York: Gordon and Breach, 1968, pp. 177–239.

_____, *Selected Papers of Freeman Dyson* (with Commentary). Providence: American Mathematical Society, 1996.

Dyson, F., and A. Lenard, "Stability of Matter. I," *Journal of Mathematical Physics* 8 (1967): 423–434.

Efron, B., and R. Tibshirani, "Statistical Data Analysis in the Computer Age," *Science* 253 (1991): 390–395.

Eilenberg, S., and S. MacLane, "General Theory of Natural Equivalences," *Transactions of the American Mathematical Society* 58 (1945): 231–294.

Eilenberg, S., and N. Steenrod, *Foundations of Algebraic Topology*. Princeton: Princeton University Press, 1952.

Eisenhart, C., "The Assumptions Underlying the Analysis of Variance," *Biometrics* 3 (1947): 1–21.

Engelking, R., *General Topology*. Berlin: Heldermann, 1989.

Englert, B.-G., *Semiclassical Theory of Atoms*. Berlin: Springer, 1988.

Englert, B.-G., and J. Schwinger, "Semiclassical Atom," *Physical Review* A32 (1985): 26–35.

Enting, I.G., "Series Expansions from the Finite Lattice Method," *Nuclear Physics B (Proceedings Supplement)* 47 (1996): 180–187.

Evans, D.E. and Y. Kawahigashi, *Quantum Symmetries on Operator Alge*bras. Oxford: Clarendon Press, 1998.

Ezin, A.N. and A.L. Samgin, "Kramers problem: Numerical Wiener-Hopf-like model characteristics," *Physical Review E* 82 (2010): 056703-1-15.

Faltings, G., "Endlichkeitssätze für abelsche Varietäten über Zahlkörpern," *Inventiones mathematicae* 73 (1983): 349–366; "Finiteness Theorems for Abelian Varieties over Number Fields," (English translation), in G. Cornell and J.H. Silverman, eds., *Arithmetic Geometry*. New York: Springer, 1986, pp. 9ff.

Federbush, P., "A new approach to the stability of matter problem, I and II," *Journal of Mathematical Physics* 16 (1975): 347–351, 706–709.

Feferman, S., H.M. Friedman, P. Maddy and J.R. Steel, "Does Mathematics Need New Axioms?," *The Bulletin of Symbolic Logic* 6 (2000): 401–446.

Fefferman, C., "The Uncertainty Principle," *Bulletin (New Series) of the American Mathematical Society* 9 (1983): 129–206.

———, "The Thermodynamic Limit for a Crystal," *Communications in Mathematical Physics* 98 (1985): 289–311.

———, "The Atomic and Molecular Nature of Matter," *Revista Matematica Iberoamericana* 1 (1985): 1–44.

———, "The N-Body Problem in Quantum Mechanics," *Communications on Pure and Applied Mathematics* 39 (1986): S67–S109.

———, "Some Mathematical Problems Arising from the Work of Gibbs," in D.G. Caldi and G.D. Mostow, eds., *Proceedings of the Gibbs Symposium, Yale University, May 15–17, 1989*. Providence: American Mathematical Society, 1990, pp. 155–162.

———, "Atoms and Analytic Number Theory," in F. Browder, ed., *Mathematics into the Twenty-First Century, Proceedings of the American Mathematical Society Centennial Symposium*, volume 2. Providence: American Mathematical Society, 1992, pp. 27–36.

Fefferman, C. and V.Ja. Ivrii, L.A. Seco, and I.M. Sigal, "The Energy Asymptotic of Large Coulomb Systems," in E. Balsev, ed., *Schroedinger Operators*. Berlin: Springer, 1992, pp. 79–99.

Fefferman, C., and L.A. Seco, "On the Energy of a Large Atom," *Bulletin (New Series) of the American Mathematical Society* 23 (1990): 525–530.

———, "Eigenvalues and Eigenfunctions of Ordinary Differential Operators," *Advances in Mathematics* 95 (1992): 145–305.

———, "Aperiodicity of the Hamiltonian Flow in the Thomas-Fermi Potential," *Revista Matematica Iberoamericana* 9 (1993): 409–551.

_____, "On the Dirac and Schwinger Corrections to the Ground-State Energy of an Atom," *Advances in Mathematics* 107 (1994): 1–185.

_____, "The Eigenvalue Sum for a One-Dimensional Potential," *Advances in Mathematics* 108 (1994): 263–335.

_____, "The Density in a One-Dimensional Potential," *Advances in Mathematics* 107 (1994): 187–364.

_____, "The Density in a Three-Dimensional Potential," *Advances in Mathematics* 111 (1995): 88–161.

_____, "The Eigenvalue Sum for a Three-Dimensional Potential," *Advances in Mathematics* 119 (1996): 26–116.

Feldman, G.J., and R.D. Cousins, "Unified approach to the classical statistical analysis of small signals," *Physical Review* D57 (1998): 3873–3889.

Feller, W., *An Introduction to Probability Theory and Its Applications.* New York: Wiley, 1968 (vol. I); 1971 (vol. II).

Felscher, W., "Bolzano, Cauchy, Epsilon, Delta," *American Mathematical Monthly* 107 (2000): 844–862.

Ferdinand, A.E., and M.E. Fisher, "Bounded and Inhomogeneous Ising Models: Part I, Specific-Heat Anomaly of a Finite Lattice," *Physical Review* 185 (1969): 832–846.

Feuer, L.S., "America's First Jewish Professor: James Joseph Sylvester at the University of Virginia," *American Jewish Archives* 36 (1964): 151–201.

Feynman, R.P., *Statistical Mechanics.* Reading, Massachusetts: Benjamin, 1972.

_____, *QED.* Princeton: Princeton University Press, 1985.

Feynman R.P., and A.R. Hibbs, *Quantum Mechanics and Path Integrals.* New York: McGraw-Hill, 1965.

Feynman, R.P., R.B. Leighton, and M. Sands, *The Feynman Lectures on Physics*, vol. 1. Reading, Massachusetts: Addison Wesley, 1963.

Fischer, G. *Mathematical Models* (2 vols.). Braunschweig: Vieweg, 1986.

Fisher, D.S., "Critical behavior of random transverse-field Ising spin chains," *Physical Review* B51 (1995): 6411–6461.

Fisher, M.E., "The Free Energy of a Macroscopic System," *Archive for Rational Mechanics and Analysis* 17 (1964): 377–410.

_____, "Renormalization group theory, its basis and formulation in statistical mechanics," *Reviews of Modern Physics* 70 (1998): 653–681.

Fisher, M.E., and D. Ruelle, "The Stability of Many-Particle Systems," *Journal of Mathematical Physics* 7 (1966): 260–270.

Fisher, R.A., "Frequency distribution of the values of the correlation coefficient in samples from an indefinitely large population," *Biometrika* 10 (1915): 507–521.

_____, "A Mathematical Examination of the Methods of Determining the Accuracy of an Observation by the Mean Error, and by the Mean Square Error," *Monthly Notices of the Royal Astronomical Society* 80 (1920): 758–770.

_____, "On the interpretation of chi-square from contingency tables, and the calculation of P," *Journal of the Royal Statistical Society* 85 (1922): 87–94.

_____, "On the Mathematical Foundations of Theoretical Statistics," *Philosophical Transactions* A222 (1922): 309–368.

_____, On a distribution yielding the error functions of several well known statistics," in *Proceedings of the International Mathematical Congress, Toronto, 1924*, ed. J. C. Fields. Toronto: University of Toronto Press, 1928, volume 2, pp. 805–813.

_____, "Theory of Statistical Estimation," *Proceedings of the Cambridge Philosophical Society* 22 (1925): 700–725.

_____, *Statistical Methods for Research Workers*. Reprinted in R.A. Fisher, *Statistical Methods, Experimental Design, and Scientific Inference, A Re-issue of Statistical Methods for Research Workers [1925] (1973), The Design of Experiments [1935] (1971), and Statistical Methods and Scientific Inference [1956] (1973)* (ed. J.H. Bennett). Oxford: Oxford University Press, 1990.

"Fisher, R.A." (Biographical Memoir on Fisher), *Biographical Memoirs of Fellows of the Royal Society of London* 9 (1963): 91–120.

Fisher, R.A., *Collected Papers of R. A. Fisher*, 5 vols., ed. J.H. Bennett. Adelaide: University of Adelaide, 1971–1974.

Fisher, R.A., and W.A. MacKenzie, "Studies in Crop Variation I: The Manurial Response of Different Potato Varieties," *Journal of Agricultural Science* 13 (1923): 311–320.

Flaschka, H., and A.C. Newell, "Monodromy- and Spectrum-Preserving Deformations I," *Communications in Mathematical Physics* 76 (1980): 65–116.

Fonseca, P., and A. Zamolodchikov, "Ising Field Theory in a Magnetic Field: Analytic Properties of the Free Energy, *Journal of Statistical Physics* 110 (2003): 527–590.

Forrester, P.J., "Random walks and random permutations," *Journal of Physics* A 34 (2001): L417–L423.

Fourier, J., *The Analytical Theory of Heat*, tr. A. Freeman. New York: Dover Publications, 1955 [1878, 1822]).

Fowler, D., *The Mathematics of Plato's Academy*. New York: Oxford University Press, 1999.

Freccero, J., "Foreword," in R. Pinsky, trans., *The Inferno of Dante*. New York: Noonday, 1995, pp. ix–xvii.

Frei, G., ed., *Der Briefwechsel David Hilbert-Felix Klein (1866–1918)*. Göttingen: Vandenhoeck & Ruprecht, 1985.

Frenkel, E., "Affine algebras, Langlands duality and Bethe Ansatz," in, *Proceedings of the XIth International Congress of Mathematical Physics, Paris, 1994*, ed. D. Iagolnitzer. Cambridge, Massachusetts: International Press, 1995, pp. 606–642.

_____, *Love and Math*. New York: Basic Books, 2013.

Friedan, D., Z. Qiu, and S. Shenker, "Superconformal invariance in two dimensions and the tricritical Ising model," *Physics Letters B* 51 (1985): 37–43.

Friedman, M., "On the Sociology of Scientific Knowledge and its Philosophical Agenda," *Studies in the History and Philosophy of Science* 29 (1998): 239–271.

Fröhlich, J., "The Electron is Inexhaustible," in H. Rossi, ed., *Prospects in Mathematics*. Providence: American Mathematical Society, 1999.

Fulton, W., *Algebraic Topology, A First Course*. New York: Springer, 1995.

Funkenstein, A., *Theology and the Scientific Imagination*. Princeton: Princeton University Press, 1986.

Garnier, R., "Préface," *Oeuvres de Paul Painlevé*, R. Gérard, G. Reeb, and A. Sec, eds., Tome I. Paris: Éditions du Centre National de la Récherche Scientifique, 1972, pp. 11–17.

Gelbart, S., "An Elementary Introduction to the Langlands Program," *Bulletin (New Series) of the American Mathematical Society* 10 (1984): 177–219.

Geroch, R., *Mathematical Physics*. Chicago: University of Chicago Press, 1985.

Gibbs, J.W., "A Method of Geometrical Representation of the Thermodynamic Properties of Substances by Means of Surfaces," *Transactions of the Connecticut Academy* 2, Article 14 (1873): 382–404.

Gillies, D., ed., *Revolutions in Mathematics*. Oxford: Clarendon Press, 1992.

Giraud, O., and K. Thas, "Hearing shapes of drums: Mathematical and physical aspects of isospectrality," *Reviews of Modern Physics* 82 (2010): 2213–2255.

Gordon, C., D.L. Webb, and S. Wolpert, "One Cannot Hear the Shape of a Drum," *Bulletin (New Series) of the American Mathematical Society* 27 (1992): 134–138.

Gouvêa, F.Q., "'A Marvelous Proof'," *American Mathematical Monthly* 101 (1994): 203–222.

Grabiner, J.V., *The Origins of Cauchy's Rigorous Calculus*. Cambridge, Massachusetts: MIT Press, 1981.

Grady, M., "Infinite Set of Conserved Charges in the Ising Model," *Physical Review* D25 (1982): 1103–1113.

Graf, G.M., and J.-P. Solovej, "A correlation estimate with application to quantum systems with Coulomb interactions," *Reviews in Mathematical Physics* 6 (1994): 977–997.

Green, H.S., and C.A. Hurst, *Order-Disorder Phenomena*. London: Interscience, 1964.

Griffiths, R. B., "Peirls Proof of Spontaneous Magnetization in a Two-Dimensional Ising Ferromagnet," *Physical Review* 136 (1964): A437–A439.

Grimm, U., and B. Nienhuis, "Scaling limits of the Ising model in a field," *Physical Review* E55 (1997): 5011–5025.

Gromov, M., "Soft and Hard Symplectic Geometry," *Proceedings of the International Congress of Mathematicians*. Berkeley, 1986, pp. 81–97.

Grosholz, E., and H. Breger, *The Growth of Mathematical Knowledge*. Dordrecht: Kluwer, 2000.

Grothendieck, A., "Cohomology Theory of Abstract Algebraic Varieties," in J.A. Todd, ed., *Proceedings of the International Congress of Mathematicians 1958*. Cambridge: Cambridge University Press, 1960, pp. 103–118.

Grothendieck, A., and J.A. Dieudonné, *Eléments de Géométrie Algébrique I*. Berlin: Springer, 1971.

Gurarie, C., "The Inverse Spectral Problem," in S.A. Fulling and F.J. Narcowich, eds., *Forty More Years of Ramifications: Spectral Asymptotics and Its Applications,*

Discourses in Mathematics and Its Applications, Number 1, College Station: Department of Mathematics, Texas A and M University, 1991, pp. 77–99.

Guttmann, A.J., and I. Enting, "Solvability of Some Statistical Mechanical Systems," *Physical Review Letters* 76 (1996): 344–347.

Gutzwiller, M., "The Origins of the Trace Formula," in H. Friedrich and B. Eckhardt, eds., *Classical, Semiclassical and Quantum Dynamics in Atoms,* Lecture Notes in Physics 485. Berlin: Springer, 1997, pp. 8–28.

Hacking, I., *The Taming of Chance.* Cambridge: Cambridge University Press, 1990.

Hallett, M., *Cantorian Set Theory and Limitation of Size.* New York: Oxford, 1984.

Hausdorff, F., *Grundzüge der Mengenlehre.* New York: Chelsea, 1949 [1914].

Heesch, H., *Untersuchungen zum Vierfarbenproblem.* Mannheim: Bibliographisches Institut, 1969, Hochshulscripten 810/810a/810b.

_____, "Chromatic Reduction of the Triangulations T_e, $e=e_5+e_7$," *Journal of Combinatorial Theory* B13 (1972): 46–55.

Hersh, R., "Some Proposals for Reviving the Philosophy of Mathematics," *Advances in Mathematics* 31 (1979): 31–50.

Hessen, B., "The Social and Economic Roots of Newton's 'Principia'," in *Science at the Cross Roads*, ed. N.I. Bukharin. London: Cass, 1971 [1931], pp. 149–212.

Hewson, A.C., *The Kondo Problem to Heavy Fermions.* Cambridge: Cambridge University Press, 1997.

Hida, H., *Modular Forms and Galois Cohomology.* New York: Cambridge University Press, 2000.

Hilbert, D., *The Theory of Algebraic Number Fields*, tr. I.T. Adamson. Berlin: Springer, 1998 [1897].

_____, "Mathematical Problems," *Bulletin of the American Mathematical Society* 8 (1902): 437–479. Reprinted in, *Mathematical Developments Arising from Hilbert Problems* (Proceedings of Symposia in Pure Mathematics 28, part 1). Providence: American Mathematical Society, 1976, pp. 1–34.

_____, "On the Infinite," (1925), in J. van Heijenoort, ed., *From Frege to Gödel.* Cambridge, Massachusetts: Harvard University Press, 1967.

Hills, N.L., and S.N. Karp, "Semi-Infinite Diffraction Grating-I," *Communications on Pure and Applied Mathematics* 18 (1965): 203–233.

Hocking, J.G., and G.S. Young, *Topology.* Reading, Massachusetts: Addison-Wesley, 1961.

Hoddeson, L., E. Braun, J. Teichmann, and S. Weart, *Out of the Crystal Maze, Chapters from the History of Solid State Physics.* New York: Oxford University Press, 1992.

Holley, R., and D.W. Stroock, "Central limit phenomena of various interacting systems," *Annals of Mathematics* 110 (1979): 333–393.

Hongler, C. and S. Smirnov, "The energy density in the planar Ising Model," *Acta Mathematica* 211 (2013): 191–225.

't Hooft, G., "Renormalization of Gauge Theories," in L. Hoddeson, L. Brown, M. Riordan, and M. Dresden, eds., *The Rise of the Standard Model: Particle*

444

Physics in the 1960s and 1970s. New York: Cambridge University Press, 1997, pp. 179–198.

Hoyningen-Huene, P., "Two Letters of Paul Feyerabend to T.S. Kuhn on a Draft of The Structure of Scientific Revolutions," *Studies in the History and Philosophy of Science* 26 (1995): 353–387.

Huber, P.J., "Robust Statistics: A Review," *Annals of Mathematical Statistics* 43 (1972): 1041–1067.

Hughes, T.P., *Networks of Power.* Baltimore: Johns Hopkins University Press, 1983.

Hundertmark, D., A. Laptev, and T. Weidl, "New Bounds on the Lieb-Thirring Constants," *Inventiones Mathematicae* 140 (2000): 693–704.

Hundertmark, D., "Some bound state problems in quantum mechanics," in F. Gesztesy, P. Deift, and C. Galvez, eds, *Spectral Theory and Mathematical Physics.* Providence: American Mathematical Society, 2007, volume I, pp. 463–496.

Hurst, C.A., "Applicability of the Pfaffian Methods to Combinatorial Problems on a Lattice," *Journal of Mathematical Physics* 5 (1964): 90–100.

Hurst, C.A., and H.S. Green, "New Solution of the Ising Problem from a Rectangular Lattice," *Journal of Chemical Physics* 33 (1960): 1059–1062.

Husserl, E., *Logical Investigations*, tr. J.N. Findlay from the second German edition. New York: Humanities Press, 1970.

Ince, E.L., *Ordinary Differential Equations.* New York: Dover, 1956 [1926].

Iranpour, R., and P. Chacon, *Basic Stochastic Processes: The Mark Kac Lectures.* New York: Macmillan, 1988.

Iwaniec, H., *Topics in Classical Automorphic Forms.* Providence: American Mathematical Society, 1997.

———, "Harmonic Analysis in Number Theory," in H. Rossi, ed., *Prospects in Mathematics.* Providence: American Mathematical Society, 1999, pp. 51–68.

Jackson, J.D., and L.B. Okun, "Historical Roots of Gauge Invariance," *Reviews of Modern Physics* 73 (2001): 663–680.

James, F., L. Lyons, and Y. Perrin, *Workshop on Confidence Intervals*, Report 2000-005. Geneva: CERN, 2000.

James, I.M., ed., *History of Topology.* Amsterdam: Elsevier, 1999.

Jessep, D.M., *Berkeley's Philosophy of Mathematics.* Chicago: University of Chicago Press, 1993.

Jimbo, M., "Introduction to Holonomic Quantum Fields for Mathematicians," in, L. Ehrenpreis and R.C. Gunning, eds., *Theta Functions, Bowdoin 1987.* Providence: American Mathematical Society, 1989, pp. 379–390.

Jimbo, M. and T. Miwa, "Studies on Holonomic Quantum Fields. XVII," *Proceedings of the Japanese Academy* 56 (1980): 405–410.

———, "Monodromy Preserving Deformation of Linear Ordinary Differential Equations with Rational Coefficients, II" *Physica* D 2 (1981): 407–448.

Jimbo, M., T. Miwa, and M. Sato, "Holonomic Quantum Fields," in K. Osterwalder, ed., *Mathematical Problems in Theoretical Physics* (Lecture Notes in Physics 116). Berlin: Springer, 1980, pp. 119–142.

Jimbo, M., T. Miwa, and K. Ueno, "Monodromy Preserving Deformation of Linear Ordinary Differential Equations with Rational Coefficients, I," *Physica* D 2 (1981): 306–352.

Johansson, K., "On Szegö's asymptotic formula for Toeplitz determinants and generalizations," *Bullétin des Sciences Mathematiques* 112 (1988): 257–304.

_____, "On random matrices from the compact classical groups," *Annals of Mathematics* 145 (1997): 519–545.

Joyner, D., *Distribution Theorems of L-functions*. Essex: Longmans, 1986.

Judah, H., "Set Theory of Reals: Measure and Category," in H. Judah, W. Just, and H. Woodin, eds., *Set Theory of the Continuum*. New York: Springer, 1992, pp. 75–87.

Kac, M., *Statistical Independence in Probability, Analysis and Number Theory*. Providence: Mathematical Association of America, 1959.

_____, "Can one hear the shape of a drum?," *American Mathematical Monthly* 73 (April 1966): 1–23.

_____, *Enigmas of Chance*. Berkeley: University of California Press, 1985, 1987.

Kac, M., and J.C. Ward, "A Combinatorial Solution of the Two-Dimensional Ising Model," *Physical Review* 88 (1952): 1332–1337.

Kadanoff, L.P., "Spin-Spin Correlations in the Two-Dimensional Ising Model," *Il Nuovo Cimento* 44B (1966): 276–305.

_____, "Correlations Along a Line in the Two-Dimensional Ising Model," *Physical Review* 188 (1969): 859–863.

Kadanoff, L.P., and H. Ceva, "Determination of an Operator Algebra for the Two-Dimensional Ising Model," *Physical Review* B3 (1971): 3918–3938.

Kadanoff, L.P., and M. Kohmoto, "SMJ's Analysis of Ising Model Correlation Functions," *Annals of Physics* 126 (1980): 371–398.

Kahane, J.-P., "Le mouvement brownien," in *Matériaux pour L'Histoire des Mathématiques au XXe Siècle* (Séminaires et Congrès, 3). Paris: Société Mathématique de France, 1998, pp. 123–155.

Kaku, M., *Introduction to Superstrings*. New York: Springer, 1988.

Kanamori, A., "Review of S. Shelah's *Proper and Improper Forcing*," *Mathematical Reviews* #98m:03002.

Kanamori, A., *The Higher Infinite*. Berlin: Springer, 1994, 2003, 2009.

Kapranov, M.M., "Analogies Between the Langlands Correspondence and Topological Quantum Field Theory," in R. Wilson et al., eds., *Functional analysis on the eve of the 21st century* (Gelfand Festschrift), volume 1, Progress in Mathematics 131. Boston: Birkhäuser, 1995, pp. 119–151.

Kashiwara, M., and T. Miwa, eds., *MathPhys Odyssey 2001, Integrable Models and Beyond, In Honor of Barry McCoy*. Boston: Birkhäuser, 2002.

Kastelyn, P., "Dimer Statistics and Phase Transitions," *Journal of Mathematical Physics* 4 (1963): 281–293.

Katz, N.M., and P. Sarnak, "Zeros of Zeta Functions and Symmetry," *Bulletin (New Series) of the American Mathematical Society* 36 (1999): 1–26.

Kaufman, B., "Crystal Statistics, II: Partition Function Evaluated by Spinor Analysis," *Physical Review* 76 (1949): 1232–1243.

Kaufman, B., and L. Onsager, "Crystal Statistics, III: Short Range Order in a Binary Ising Lattice," *Physical Review* 76 (1949): 1244–1252.

Kechris, A., *Classical Descriptive Set Theory*. New York: Springer, 1995.

_____, "New Directions in Descriptive Set Theory," *Bulletin of Symbolic Logic* 5 (1999): 161–174.

Kechris, A.S., and A. Louveau, "Descriptive Set Theory and Harmonic Analysis," *Journal of Symbolic Logic* 57 (1992): 413–441.

Kelly, K., *The Logic of Reliable Inquiry*. New York: Oxford, 1996.

Kennedy, P., *A Guide to Econometrics*. Cambridge, Massachusetts: MIT Press, 1985.

Kirchoff (Studiosus), "Ueber den Durchgang eines elektrischen Stromes durch eine Ebene, insbesondere durch eine kreisförmige," *Annalen der Physik und Chemie* 64 (1845): 497–514.

Kirchoff, G., "Ueber der Auflösung der Gleichungen, auf welche man bei der Untersuchung der linearen Vertheilung galvanisher Ströme gefürt wird," *Annalen der Physik und Chemie* 72 (1847): 497–508.

Kitcher, P., *The Nature of Mathematical Knowledge*. New York: Oxford University Press, 1983.

Klain, D.A., and G.-C. Rota, *Introduction to Geometric Probability*. Cambridge: Cambridge University Press, 1997.

Klein, F., *On Riemann's Theory of Algebraic Functions and Their Integrals: A Supplement to the Usual Treatises*. New York: Dover, 1963 [1882, 1893].

_____, *Vorlesungen Über die Entwicklung der Mathematik im 19. Jarhhundert*. New York: Chelsea, 1956 [1926, 1927, 1950]. Also, *Development of Mathematics in the 19th Century*, tr. M. Ackerman. Brookline: Math Sci Press, 1979 [1928].

Knapp, A.W., *Elliptic Curves*. Princeton: Princeton University Press, 1992.

Knauf, A., "Number Theory, Dynamical Systems, and Statistical Mechanics," *Reviews in Mathematical Physics* 11 (1999): 1027–1060.

Kodaira, K., "Harmonic Fields in Riemannian Manifolds," *Annals of Mathematics* 50 (1949): 587–665.

Koetsier, T., *Lakatos' Philosophy of Mathematics: A Historical Approach*. Amsterdam: North-Holland, 1991.

Kol'man, E., "Karl Marx and Mathematics," in K. Marx, *Mathematical Manuscripts of Karl Marx*. London: New Park, 1983.

Korepin, V.E., N.M. Bogoliubov, and A.G. Izergin, *Quantum Inverse Scattering Mathod and Correlation Functions*. New York: Cambridge University Press, 1993.

Kramers, H.A., and G. Wannier, "Statistics of the Two-Dimensional Ferromagnet," *Physical Review* 60 (1941): 252–276.

Krbalek, M. and P. Seba, "The statistical properties of the city transport in Cuernavaca (Mexico) and random matric ensembles," *Journal of Physics A* 33 (2000): L229–L234.

447

_____, "Headway statistics of public transport in Mexican Cities," *Journal of Physics A* 36 (2003): L7–L11.

Krieger, M.H., *Marginalism and Discontinuity: Tools for the Crafts of Knowledge and Decision.* New York: Russell Sage Foundation, 1989.

_____, Segmentation and Filtering into Neighborhoods as Processes of Percolation and Diffusion: Stochastic Processes (Randomness) as the Null Hypothesis," *Environment and Planning* A 23 (1991): 1609–1626.

_____, "Contingency in Planning: Statistics, Fortune, and History," *Journal of Planning Education and Research 10* (1991): 157–161.

_____, "Theorems as Meaningful Cultural Artifacts: Making the World Additive," *Synthese* 88 (1991): 135–154.

_____, *Doing Physics: How Physicists Take Hold of the World.* Bloomington: Indiana University Press, 1992.

_____, "Could the Probability of Doom be Zero-or-One?," *Journal of Philosophy* 92 (1995): 382–387.

_____, *Constitutions of Matter: Mathematically Modeling the Most Everyday of Physical Phenomena.* Chicago: University of Chicago Press, 1996.

_____, "Apocalypticism, One–Time Events, and Scientific Rationalty," in S.S. Hecker and G.-C. Rota, eds., *Essays on the Future, In Honor of Nick Metropolis* (Boston: Birkhäuser, 2000), pp. 135–151.

Kronecker, L., *Leopold Kronecker's Werke*, ed. K. Hensel, Bd. 5. New York: Chelsea, 1968 [1930].

Kruskal, W., "Miracles and statistics: the casual assumption of independence," *Journal of the American Statistical Association* 83 (1988): 929–940.

Kuhn, T.S., *The Structure of Scientific Revolutions.* Chicago: University of Chicago Press, 1970.

_____, "Revisiting Planck." *Historical Studies in the Physical Sciences* 14 (1984): 231–252.

Kuratowski, K., *Topology*, vol. 1. New York: Academic Press, 1966.

_____, *A Half Century of Polish Mathematics.* Oxford: Pergamon, 1980.

Lagrange, J.-L., "Réflections sur la résolution algébrique des équations," (1771), in *Oeuvres de Lagrange*, eds. J.-A. Serret, G. Darboux. Paris: Gauthier-Villars, 1869.

Landau, L., and E. Lifshitz, *Statistical Physics*, Part 1. Oxford: Pergamon, 1980.

Lang, S., "Review of A. Grothendieck and J. Dieudonné, *Éléments de géométrie algébrique*," *Bulletin of the American Mathematical Society* 67 (1961): 239–246.

_____, *Algebra*, edn. 2. Redwood City: Addison-Wesley, 1984.

_____, *Undergraduate Algebra.* New York: Springer, 1987.

Langlands, R.P "Some Contemporary Problems with Origins in the Jugendtraum," in *Mathematical Developments Arising from Hilbert Problems*, part 2. Providence: American Mathematical Society, 1976, pp. 401–418.

_____, "Automorphic Representations, Shimura Varieties and Motives, Ein Märchen," in *Automorphic Forms, Representations, and L-Functions* (Proceedings of Symposia in Pure Mathematics 33, part 2). Providence: American Mathematical Society, 1979, pp. 205–246.

———, "Representation Theory: Its Rise and Role in Number Theory," in D.G. Caldi and G.D. Mostow, eds., *Proceedings of The Gibbs Symposium.* Providence: American Mathematical Society, 1990, pp. 181–210.

———, "Where Stands Functoriality Today?," *Proceedings of the Symposia in Pure Mathematics* 61 (1997): 457–471.

Langlands, R.P., M.-A. Lewis, and Y. Saint-Aubin, "Universality and Conformal Invariance for the Ising Model in Domains with Boundary," *Journal of Statistical Physics* 98 (2000): 131–244.

Langlands, R., P. Pouliot, and Y. Saint-Aubin, "Conformal Invariance in Two-Dimensional Percolation," *Bulletin (New Series) of the American Mathematical Society* 30 (1994): 1–61.

Langlands, R.P., and Y. Saint-Aubin, "Algebro-Geometric Aspects of the Bethe Equation," in *Strings and Symmetries* (Lecture Notes in Physics 447). New York: Springer, 1995, pp. 40–53.

Laugwitz, D., *Bernhard Riemann 1826–1866, Turning Points in the Conception of Mathematics*, tr. A. Shenitzer. Boston: Birkhäuser, 1999.

Lavine, S., *Understanding the Infinite.* Cambridge, Massachusetts: Harvard University Press, 1994.

Lavis, D.A., and G.M. Bell, *Statistical Mechanics of Lattice Systems* (2 vols.). Berlin: Springer, 1999.

Lawden, D.F., *Elliptic Functions and Applications.* New York: Springer, 1989.

Lawvere, F.W., "Unity and Identity of Opposites in Calculus and Physics," *Applied Categorical Structures* 4 (1996): 167–174.

Lax, P.D., in "Jean Leray (1906–1998)," *Notices of the American Mathematical Society* 47 (March 2000): 350–359. See also, "Jean Leray and Partial Differential Equations," in J. Leray, *Selected Papers, Oeuvres Scientifiques*, volume 2. Berlin: Springer, 1998, pp. 1–9.

Lebesgue, H., "Préface," in N. Lusin, *Leçons sur les Ensembles Analytiques et leurs Applications.* New York: Chelsea, 1972 [1930].

Lee, T.D., and C.N. Yang, "Statistical Theory of Equations of State and Phase Transitions: Part 2, Lattice Gas and Ising Model," *Physical Review* 87 (1952): 410–419.

Leibniz, G., *Philosophical Essays*, ed. and tr. R. Ariew and D. Garber. Indianapolis: Hackett, 1989.

Lenard, A., "Momentum distribution in the ground state of the one-dimensional system of impenetrable bosons," *Journal of Mathematical Physics* 5 (1964): 930–943.

———, "Lectures on the Coulomb Stability Problem," in A. Lenard, ed., *Statistical Mechanics and Mathematical Problems* (Lecture Notes in Physics 20). Berlin: Springer, 1971, pp. iii–v, 114–135.

Lenard, A., and F. Dyson, "Stability of Matter. II," *Journal of Mathematical Physics* 9 (1968): 698–711.

Leray, J., "Sur la forme des espaces topologiques et sur les points fixes des représentations," *Journal des mathématiques pures et appliquées* 24 (1945): 96–167; "Sur la position d'un ensemble fermé de points d'un espace topologique," *Ibid.*, pp. 169–199; "Sur les équations et les transformations," *Ibid.*, pp. 201–248.

_____, "L'anneau d'homologie d'une représentation," *Comptes Rendus, Académie des Sciences de Paris* 222 (1946): 1366–1368.

Levins, R., and R. Lewontin, *The Dialectical Biologist*. Cambridge, Massachusetts: Harvard University Press, 1985.

Levitt, A., "Best Constants in Lieb-Thirring Inequalities: A Numerical Investigation," *Journal of Spectral Theory* 4 (2014): 153–175.

Lévy, P., *Processus Stochastiques et Mouvement Brownien*. Paris: Gauthier-Villars, 1948.

Lieb, E., "Exact Solution of the Problem of the Entropy of Two-Dimensional Ice," *Physical Review Letters* 18 (1967): 692–694; "Exact Solution of the F Model of an Antiferroelectric," *Physical Review Letters* 18 (1967): 1046–1048; "Exact Solution of the Two-Dimensional Slater KDP Model of a Ferroelectric," *Physical Review Letters* 19 (1967): 108–110; "The Residual Entropy of Square Ice, *Physical Review* 162 (1967): 162–172.

_____, "The Stability of Matter," *Reviews of Modern Physics* 48 (1976): 553–569.

_____, "The Stability of Matter: From Atoms to Stars," *Bulletin (New Series) of the American Mathematical Society* 22 (1990): 1–49.

_____, "Some of the Early History of Exactly Soluble Models," *International Journal of Modern Physics* B11 (1997): 3–10.

_____, *The Stability of Matter: From Atoms to Stars, Selecta of Elliott H. Lieb*, ed. W. Thirring, 2nd edn. Berlin: Springer, 1997.

Lieb, E.H., and J.L. Lebowitz, "The Constitution of Matter: Existence of Thermodynamics for Systems Composed of Electrons and Nuclei," *Advances in Mathematics* 9 (1972): 316–398.

Lieb, E.H., and Loss, M., *Analysis*. Providence: American Mathematical Society, 1997.

Lieb, E.H., and D.C. Mattis, *Mathematical Physics in One Dimension*. New York: Academic Press, 1966.

Lieb, E.H., T.D. Schultz, and D.C. Mattis, "Two Soluble Models of an Antiferromagnetic Chain," *Annals of Physics* 16 (1961): 407–466.

Lieb, E.H. and R. Seiringer, *The Stability of Matter in Quantum Mechanics*. New York: Cambridge University Press, 2009.

Lieb, E. H, and B. Simon, "Thomas-Fermi Theory Revisited," *Physical Review Letters* 31 (1973): 681–683.

_____, "Thomas-Fermi Theory of Atoms, Molecules and Solids," *Advances in Mathematics* 23 (1977): 22–116.

Lieb, E.H., and W. Thirring, "Bound for Kinetic Energy Which Proves the Stability of Matter," *Physical Review Letters* 35 (1975): 687–689.

Littlewood, J. E., *A Mathematician's Miscellany*. Cambridge: Cambridge University Press, 1986.

Livingston, E., *Ethnomethodological Foundations of Mathematics*. London: Routledge, 1986.

_____, "Cultures of Proving," *Social Studies of Science* 29 (December 1999): 867–888.

_____, "Natural Reasoning in Mathematical Theorem Proving," *Communication & Cognition* 38 (2005): 319–344.

Loève, M., *Probability Theory*. Princeton: Van Nostrand, 1960.

Loss, M., "Stability of Matter," version 1.0, 168pp. ms., August 4, 2005.

_____, "Stability of Matter," 28pp ms., November 14, 2005.

Lovejoy, A.O., *The Great Chain of Being*. Cambridge, Massachusetts: Harvard University Press, 1936.

Mac Lane, S., "Duality for Groups," *Bulletin of the American Mathematical Society* 56 (1950): 485–516.

_____, "The Work of Samuel Eilenberg in Topology," in A. Heller, M. Tierney, eds., *Algebra, Topology, and Category Theory: A Collection of Papers in Honor of Samuel Eilenberg*. New York: Academic Press, 1976, pp. 133–144.

_____, "Origin of the Cohomology of Groups," *L'Enseignement Mathématique* 24 (1978): 1–29.

_____, "Topology Becomes Algebraic With Vietoris and Noether," *Journal of Pure and Applied Algebra* 39 (1986): 305–307.

_____, *Mathematics, Form and Function*. New York: Springer, 1986.

_____, "Group Extensions for 45 Years, *The Mathematical Intelligencer* 10 (1988).

Mackey, G., *The Scope and History of Commutative and Noncommutative Harmonic Analysis*. Providence: American Mathematical Society, 1992. Includes a reprint of "Harmonic Analysis as the Exploitation of Symmetry," *Rice University Studies* 64 (1978): 73–228 (also in *Bulletin (New Series) of the American Mathematical Society* 3 (1980): 543–699).

Maddy, P., *Realism in Mathematics*. New York: Oxford University Press, 1990.

_____, "Set Theoretic Naturalism," *Journal of Symbolic Logic* 61 (1996): 490–514.

_____, *Naturalism in Mathematics*. New York: Oxford University Press, 1997.

_____, "How applied mathematics became pure," *Review of Symbolic Logic* 1 (2008): 16–41.

Maillard, J.M., "The inversion relation," *Journal de Physique* 46 (1985): 329–341.

Mancosu, P., ed., *From Brouwer to Hilbert: The Debate on the Foundations of Mathematics in the 1920s*. New York: Oxford University Press, 1998.

_____, *Philosophy of Mathematical Practice*. Oxford: Oxford University Press, 2008.

Mandelbrot, B., *Fractal Geometry*. San Francisco: Freeman, 1982.

Mangazeev, V., M.Yu. Dudalev, V. Bazhanov, and M. T. Batchelor, "Scaling and universality in the two-dimensional Ising model with magnetic field," *Physical Review B* 81 (2010): 060103-1 – 060103-4.

Manin, Yu., *A Course in Mathematical Logic*. Berlin: Springer, 1977.

Martin, D.A., "Mathematical Evidence," in H.G. Dales and G. Oliveri, eds., *Truth in Mathematics*. Oxford: Clarendon Press, 1998, pp. 215–231.

Martin, D.A., and J.R. Steele, "Projective Determinacy," *Proceedings of the National Academy of Science* 85 (1988): 6582–6586.

Martin, Ph., "Lecture on Fefferman's Proof of the Atomic and Molecular Nature of Matter," *Acta Physica Polonica* B24 (1993): 751–770.

Martinec, E., "Quantum Field Theory: A Guide for Mathematicians," in *Theta Functions, Bowdoin 1987*. Providence: American Mathematical Society, 1989, pp. 406–412.

Marx, K., *Mathematical Manuscripts of Karl Marx*, ed. S.A. Yanovskaya. London: New Park, 1983.

Massey, W.S., *Algebraic Topology: An Introduction*. New York: Springer, 1967.

Maxwell, J.C., "Experiments on Colour as Perceived by the Eye, with Remarks on Colour-Blindness," *Proceedings of the Royal Society of Edinburgh* 3 (1855): 299–301.

_____, *Treatise on Electricity and Magnetism*, 3rd edn. New York: Dover, 1954 [1891].

_____, *Scientific Letters and Papers of James Clerk Maxwell*, ed. P.M. Harman, vol. 1, 1846–1862. Cambridge: Cambridge University Press, 1990.

McCoy, B.M., "Introductory Remarks on Szegő's Paper, "On Certain Hermitian Forms Associated with the Fourier Series of a Positive Function," in R. Askey, ed., *Gabor Szegő: Collected Papers, Volume 1*, Boston: Birkhäuser, 1982, pp. 47–52.

_____, "The Connection Between Statistical Mechanics and Quantum Field Theory," in V.V. Bazhanov and C.J. Burden, eds., *Statistical Mechanics and Field Theory*. Singapore: World Scientific, 1995, pp. 26–128.

_____, "Integrable models in statistical mechanics: The hidden field with unsolved problems," *International Journal of Modern Physics A* 14 (1999) 3921–3933.

_____, "The Baxter Revolution," *Journal of Statistical Physics* 102 (2001): 375–384.

_____, *Advanced Statistical Mechanics*. New York: Oxford University Press, 2010.

_____, "The Romance of the Ising model," in *Symmetries, integrable systems and representations*, Springer Proceedings in Mathematics and Statistics, 40. Heidelberg: Springer, 2013, pp. 263–295.

McCoy, B. M., Assis, M, Boukrass, S., Hassani, S., Maillard, J.-M., Orrick, W.P. and N. Zenine, "The saga of the Ising susceptibility," in *New trends in quantum integrable systems*. Hackensack NJ: World Scientific, 2011, pp. 287–306.

McCoy, B. M., and J.-M. Maillard, The Importance of the Ising model," *Progress in Theoretical Physics* 127 (2012): 791–817.

McCoy, B., C.A. Tracy, and T.T. Wu, "Painlevé functions of the third kind," *Journal of Mathematical Physics* 18 (1977): 1058–1092.

McCoy, B. M., and T. T. Wu, "Two-dimensional Ising field theory in a magnetic field: Breakup of the cut in the two-point function," *Physical Review D* 18 (1978): 1259–1267.

_____, *The Two-Dimensional Ising Model*. Cambridge, Massachusetts: Harvard University Press, 1973; second edition, New York: Dover, 2014.

McGuire, J.B., "Study of Exactly Soluble One-Dimensional N-Body Problem," *Journal of Mathematical Physics* 5 (1964): 622–636.

McKean, H.P., "Kramers-Wannier Duality for the 2-Dimensional Ising Model as an Instance of Poisson's Summation Formula," *Journal of Mathematical Physics* 5 (1964): 775–776.

_____, *Stochastic Integrals*. New York: Academic Press, 1969.

McKean, H., and V. Moll *Elliptic Curves: Function Theory, Geometry, Arithmetic*. New York: Cambridge University Press, 1997.

McKean, H., Jr., and I.M. Singer, "Curvature and the Eigenvalues of the Laplacian," *Journal of Differential Geometry* 1 (1967): 43–69.

McLarty, C., *Elementary Categories, Elementary Toposes*. Oxford: Clarendon Press, 1995.

Mehra, J., and K.A. Milton, *Climbing the Mountain, The Scientific Biography of Julian Schwinger*. New York: Oxford University Press, 2000.

Mehta, M.L., *Random Matrices*. San Diego: Academic Press, 1991.

Merton, R.C., *Continuous-Time Finance*. Cambridge, Massachusetts: Blackwells, 1990.

Misner, C., K. Thorne, and J.A. Wheeler, *Gravitation*. San Francisco: Freeman, 1973.

Monastyrsky, M., *Riemann, Topology and Physics*. Boston: Birkhäuser, 1987.

Montroll, E.W., "The Statistical Mechanics of Nearest Neighbor Systems," *Journal of Chemical Physics* 9 (1941): 706–721.

Montroll, E.W., R.B. Potts, and J.C. Ward, "Correlations and Spontaneous Magnetization of the Two-Dimensional Ising Model," *Journal of Mathematical Physics* 4 (1963): 308–322.

Moore, E.H., "On a Form of General Analysis with Applications to Linear Differential and Integral Equations," in *Atti del IV Congresso Internazionale dei Matematici (Roma, 6–11 Aprile 1908)*. Roma: Tipografia della R. Academia dei Lincei, 1909, volume 2.

Moore, R.L., *Foundations of Point Set Theory* (Colloquium Publications, XIII). Providence: American Mathematical Society, 1962.

Moschovakis, Y., *Descriptive Set Theory*. Amsterdam: North Holland, 1980.

Moser, J., "A Broad Attack on Analysis Problems," *Science* 202 (1978): 612–613.

Mosteller, F., and J.W. Tukey, *Data Analysis and Regression: A Second Course in Statistics*. Reading, Massachusetts: Addison Wesley, 1977.

Mosteller, F., and D.L. Wallace, *Inference and Disputed Authorship: The Federalist*. Reading, Massachusetts: Addison-Wesley, 1964.

Munkres, J., *Topology, A First Course*. Englewood-Cliffs: Prentice Hall, 1975.

_____, *Elements of Algebraic Topology*. Reading, Massachusetts: Addison-Wesley, 1984.

Nambu, Y., "A Note on the Eigenvalue Problem in Crystal Statistics," *Progress of Theoretical Physics* 5 (1950): 1–13.

Nettleton, R.E., and M.S. Green, "Expression in Terms of Molecular Distribution Functions for the Entropy Density in an Infinite System," *Journal of Chemical Physics* 29 (1958): 1365–1370.

_____, "Moebius Function on the Lattice of Dense Subgraphs," *Journal of Research of the National Bureau of Standards* 64B (1960): 41–47.

Neumann, C., *Vorlesungen über Riemann's Theorie der Abel'schen Integrale*, 2nd edn. Leipzig: Teubner, 1884 [1865].

Newell, G. F., and E. W. Montroll, "On the Theory of the Ising Model of Ferromagnetism," *Reviews of Modern Physics* 25 (1953): 353–389.

Nickel, B., "On the singularity structure of the 2D Ising model susceptibility," *Journal of Physics* A 32 (1999): 3889–3906.

_____, "Addendum to 'On the Singularity Structure of the 2D Ising model susceptibility,'" *Journal of Physics* A 33 (2000): 1693–1711.

Nojima, K., "The Grassmann Representation of the Ising Model," *International Journal of Modern Physics B* 12 (1998): 1995–2003.

Onsager, L., "Electrostatic Interaction of Molecules," *Journal of Physical Chemistry* 43 (1939): 189–196.

_____, "Crystal Statistics I," *Physical Review* 65 (1944): 117–149.

_____, "The Ising Model in Two Dimensions," in R. E. Mills, E. Ascher, and R.I. Jaffee, eds., *Critical Phenomena in Alloys, Magnets, and Superconductors*. New York: McGraw-Hill, 1971, pp. 3–12.

_____, "Letter to B. Kaufman, April 12, 1950," *Journal of Statistical Physics* 78 (1995): 585–588.

_____, *The Collected Works of Lars Onsager*, ed. P. Hemmer, H. Holden, S. Kjelstrup Ratkje. Singapore: World Scientific, 1996.

Orrick, W.P., Nickel, B., Guttmann, A.J., and J.H.H. Perk, "The Susceptibility of the Square Lattice Ising Model: New Developments," *Journal of Statistical Physics* 102 (2001) 795–841.

Pais, A., *"Subtle is the Lord . . .": The Science and Life of Albert Einstein*. New York: Oxford, 1982.

Palais, R., "The Visualization of Mathematics," *Notices of the American Mathematical Society* 46 (1999): 647–658.

Palmer, J., *Planar Ising Correlations*. Boston: Birkhäuser, 2007.

Palmer, J., and C.A. Tracy, "Two-Dimensional Ising Correlations: Convergence of the Scaling Limit," *Advances in Applied Mathematics* 2 (1981): 329–388.

_____, "Two-Dimensional Ising Correlations: The SMJ Analysis," *Advances in Applied Mathematics* 4 (1983): 46–102.

Parikh, C., *The Unreal Life of Oscar Zariski*. Boston: Academic Press, 1991.

Particle Data Group, "Review of Particle Properties." *Physical Review* D 54, part 1 (1996): 1–720.

Penner, R., and J.L. Harer, *Combinatorics of Train Tracks*, Annals of Mathematics Studies 125. Princeton: Princeton University Press, 1992.

Pincock, C., *Mathematics and Scientific Representation*. New York: Oxford University Press, 2011.

Polya, G., "Über den zentralen Grenzwertsatz de Wahrscheinlichkeitsrechnung und das Momentenproblem," *Mathematische Zeitschrift* 8 (1920): 171–181.

Ponge, F., *Lyres*. Paris: Gallimard, 1967, 1980.

————, *The Power of Language*, tr. S. Gavronsky. Berkeley: University of California Press, 1979.

Pont, J.C., *La topologie algébrique dès origine à Poincaré*. Paris: Presses Universitaires de France, 1974.

Porter, T., *The Rise of Statistical Thinking*. Princeton: Princeton University Press, 1986.

Potts, R.B., and J.C. Ward, "The Combinatorial Method and the Two-Dimensional Ising Model," *Progress of Theoretical Physics* 13 (1955): 38–46.

Quinn, F., "Contributions to a Science of Contemporary Mathematics," October 2011, 98pp, available at https://www.math.vt.edu/people/quinn/history_nature /nature0 .pdf.

Reed, D., *Figures of Thought: Mathematics and Mathematical Texts*. London: Routledge, 1995.

Ribet, K., "Galois Representations and Modular Forms," *Bulletin (New Series) of the American Mathematical Society* 32 (1995): 375–402.

Riemann, G.F.B., *Gesammelte Mathematische Werke*. New York: Dover, 1953 [1892, 1902].

Robinson, A., *Non-standard Analysis*. Princeton: Princeton University Press, 1996 [1974].

Rota, G.-C., "E. Husserl, *First and Second Logical Investigation: The Existence of General Objects, presented in contemporary English by Gian-Carlo Rota.*" Cambridge, Massachusetts: Department of Mathematics, mimeo. nd (ca. 1990?), pp. 3–4.

————, *Gian-Carlo Rota on Combinatorics—Introductory Papers and Commentary*, ed. J.P.S. Kung. Boston: Birkhäuser, 1995.

Rota, G.-C. and F. Palombi, *Indiscrete Thoughts*. Boston: Birkhäuser, 1997.

Roth, A.E., and M. Sotomayor, *Two-Sided Matching*. New York: Cambridge University Press, 1990.

Rotman, B., *Ad infinitum . . . Taking God out of mathematics and putting the body back in: an essay in corporeal semiotics*. Stanford: Stanford University Press, 1993.

Rotman, J., "Review of C.A. Weibel, *Introduction to Homological Algebra*," *Bulletin (New Series) of the American Mathematical Society* 33 (1996): 473–476.

Rowe, D., ed., *History of Modern Mathematics* (2 vols.). San Diego: Academic Press, 1989.

Rudwick, M., *The Great Devonian Controversy*. Chicago: University of Chicago Press, 1985.

Ruelle, D., *Statistical Mechanics: Rigorous Results*. Singapore: World Scientific, 1999 [1969, 1974, 1989].

————, "Is Our Mathematics Natural?, The Case of Equilibrium Statistical Mechanics," *Bulletin (New Series) of the American Mathematical Society* 19 (1988): 259–268.

Saaty, T.L., and P.C. Kainen, *The Four-Color Problem, Assault and Conquest*. New York: Dover, 1986.

Samuel, S., "The use of anticommuting variable integrals in statistical mechanics. I. The computation of partition functions, II. The computation of correlation functions, III. Unsolved models," *Journal of Mathematical Physics* 21 (1980): 2806–2833.

Samuelson, P., "The Fundamental Approximation Theorem of Portfolio Analysis in Terms of Means, Variances and Higher Moments," *Review of Economic Studies* 37 (1970): 537–542.

Sato, M., T. Miwa, and M. Jimbo, "Studies on Holonomic Quantum Fields. I," *Proceedings of the Japanese Academy* 53A (1977): 6–10.

Schapiro, M., "Philosophy and Worldview in Painting [written, 1958–1968]," in *Worldview in Painting—Art and Society: Selected Papers*. New York: Braziller, 1999, pp. 11–71.

Scheffé, H., "Alternative Models for the Analysis of Variance," *Annals of Mathematical Statistics* 27 (1956): 251–271.

Schirn, M., ed., *The Philosophy of Mathematics Today*. Oxford: Clarendon Press, 1998.

Schneps, L., ed., *The Grothendieck Theory of Dessins d'Enfants*. Cambridge: Cambridge University Press, 1994.

Schultz, T.D, D.C. Mattis, and E.H. Lieb, "Two-Dimensional Ising Model as a Soluble Problem of Many Fermions," *Reviews of Modern Physics* 36 (1964): 856–871.

Schwarz, A., *Topology for Physicists*. New York: Springer, 1994.

Schwinger, J., "Thomas Fermi model: The leading correction," *Physical Review* A22 (1980): 1827–1832.

_____, "Thomas-Fermi model: The second correction," *Physical Review* A24 (1981): 2353.

_____,, *A Quantum Legacy, Seminal Papers of Julian Schwinger*, ed. K.A. Milton. Singapore: World Scientific, 2000.

Seifert, H., "Topology of 3-Dimensional Fibered Spaces," in H. Seifert and W. Threlfall, *Textbook of Topology*. New York: Academic Press, 1980 [1934].

Shafarevich, I.R., *Basic Algebraic Geometry* (2 vols.). Berlin: Springer, 1994.

_____, *Basic Notions of Algebra*. Berlin: Springer, 1997.

Shimura, G., in "Steele Prizes," *Notices of the American Mathematical Society* 43 (November 1996): 1340–1347.

Shin, Sun-Joo, *The Logical Structure of Diagrams*. New York: Cambridge University Press, 1994.

Shrock, R.E., and R.K. Ghosh, "Off-axis correlation functions in the isotropic d=2 Ising Model," *Physical Review B* 31 (1985): 1486–1489.

Sierpinski, W., *Hypothèse du Continu*. New York: Chelsea, 1956 [1934].

Sigal, I.M., "Lectures on Large Coulomb Systems," *CRM Proceedings and Lecture Notes* 8 (1995): 73–107.

_____, "Review of Fefferman and Seco, 'The density in a one-dimensional potential,'" *Advances in Mathematics* 107 (1994) 187–364," *Mathematical Reviews* #96a:81147.

Simon, B., *Statistical Mechanics of Lattice Gasses*, vol. 1. Princeton: Princeton University Press, 1993.

_____, *Orthogonal Polynomials on the Unit Circle*. Providence: American Mathematical Society, 2005.

Smith, A., *An Inquiry into the Nature and Causes of the Wealth of Nations*. Oxford: Clarendon, 1976 [1776].

Smith, C., "Hegel, Marx and the Calculus," in Karl Marx, *Mathematical Manuscripts of Karl Marx Marx*, ed. S.A. Yanovskaya. London: New Park, 1983.

Smith, C., and M.N. Wise, *Energy and Empire: A Biographical Study of Lord Kelvin*. Cambridge: Cambridge University Press, 1989.

Solovej, J.P., "Mathematical Results on the Structure of Large Atoms," *European Congress of Mathematics*, 1996, vol. II (Progress in Mathematics 169). Basel: Birkhäuser, 1998, pp. 211–220.

Sommerfeld, A., "Klein, Riemann und die mathematische Physik," *Die Naturwissenschaften* 17 (1919): 300–303.

Spencer, T., "Scaling, the Free Field and Statistical Mechanics," *Proceedings of Symposia in Pure Mathematics* 60 (1997): 373–389.

Spinoza, B., *A Spinoza Reader*, tr. and ed. E. Curley. Princeton: Princeton University Press, 1994.

Spruch, L., "Pedagogic notes on Thomas-Fermi theory (and on some improvements): atoms, stars, and the stability of bulk matter," *Reviews of Modern Physics* 63 (1991): 151–209.

Steel, J.R., "Mathematics Needs New Axioms," *Bulletin of Symbolic Logic* 6 (2000): 422–433.

Steen, L.A., and J.A. Seebach, Jr. *Counterexamples in Topology*. New York: Dover, [1970, 1978] 1995.

Steiner, M., *The Applicability of Mathematics as a Philosophical Problem*. Cambridge, Massachusetts: Harvard University Press, 1998.

Stephen, M.J., and L. Mittag, "A New Representation of the Solution of the Ising Model," *Journal of Mathematical Physics* 13 (1972): 1944–1951.

Stigler, S., *The History of Statistics Before 1900*. Cambridge, Massachusetts: Harvard University Press, 1986.

Stone, M.H. "Theory of Representations for Boolean Algebras," *Transactions of the American Mathematical Society* 40 (1936): 37–111.

_____, "Theory of Representations for Boolean Algebras," *Bulletin of the American Mathematical Society* 44 (1938): 807–816.

Stuart, A., J.K. Ord, and S. Arnold, *Kendall's Advanced Theory of Statistics, Vol. 2A, Classical Inference and the Linear Model*. London: Arnold, 1999.

Sutherland, B., "Exact Solution of a Two-Dimensional Model for Hydrogen-Bonded Crystals," *Physical Review Letters* 19 (1967): 103-104.

_____, *Beautiful Models: 70 Years of Exactly Solved Quantum Many-Body Problems*. Singapore: World Scientific, 2004.

Swerdlow, N., *The Babylonian Theory of the Planets*. Princeton: Princeton University Press, 1998.

Sylvester, J.J., *Collected Mathematical Papers of James Joseph Sylvester*. Cambridge: Cambridge University Press, 1904.

Szegő, G., "Ein Grenzwertsatz über die Toeplitzschen Determinanten einer reellen positive Funktion," *Mathematische Annalen* 76 (1915): 490–503.

Teller, E., "On the Stability of Molecules in the Thomas-Fermi Theory," *Reviews of Modern Physics* 34 (1962): 627–631.

Temperley, H. V. N., "The two-dimensional Ising model," in C. Domb and M. S. Green, eds., *Phase Transitions and Critical Phenomena*, volume 1. London, New York: Academic Press, 1972, pp. 227–267.

Thacker, H. B., "Corner Transfer Matrices and Lorentz Invariance on a Lattice," *Physica D* 18 (1986): 348–359.

Thacker, H. B. and H. Bergknoff, "Structure and solution of the massive Thirring model," *Physical Review D* 19 (1979): 3666–3681.

Thirring, W., *Quantum Mechanics of Large Systems*, A Course in Mathematical Physics 4. New York: Springer, 1983.

_____, "Preface [to a festshrift issue devoted to Elliott Lieb]," *Reviews in Mathematical Physics* 6 #5a (1994): v–x.

_____, *Selected Papers of Walter E. Thirring with Commentaries*. Providence: American Mathematical Society, 1998.

Thurston, W., "On the Construction and Classification of Foliations," *Proceedings of the International Congress of Mathematicians*. Vancouver, 1974.

_____, "On Proof and Progress in Mathematics," *Bulletin (New Series) of the American Mathematical Society* 30 (1994): 161–177.

_____, *Three-Dimensional Geometry and Topology*, vol. 1. Princeton: Princeton University Press, 1997.

Tietze, H., and L. Vietoris, "Beziehungen Zwischen den Verschiedenen Zweigen der Topologie," III AB 13 in *Encyklopädie der Mathematischen Wissenschaften*. Leipzig: Teubner, 1914–1931, pp. 141ff.

Tracy, C.A., "Embedded Elliptic Curves and the Yang-Baxter Equation," *Physica* 16D (1985): 202–220.

_____, "The Emerging Role of Number Theory in Exactly Solved Models in Lattice Statistical Mechanics," *Physica* 25D (1987): 1–19.

_____, "Asymptotics of a τ-Function Arising in the Two-Dimensional Ising Model," *Communications in Mathematical Physics* 142 (1991): 297–311.

Tracy, C.A., L. Grove, and M.F. Newman, "Modular Properties of the Hard Hexagon Model," *Journal of Statistical Physics* 48 (1987): 477–502.

Tracy, C.A., and H. Widom, "Fredholm determinants, differential equations and matrix models," *Communications in Mathematical Physics* 163 (1994): 33–72.

_____, "The Thermodynamic Bethe Ansatz and a Connection with Painlevé Equations," *International Journal of Modern Physics* B11 (1997): 69–74.

_____, "Correlation Functions, Cluster Functions and Spacing Distributions for Random Matrices," *Journal of Statistical Physics* 92 (1998): 809–835.

———, "On the distributions of the lengths of the longest monotone subsequences in random words," *Probability Theory and Related Fields* 119 (2001): 350–380.

Tragesser, R., *Phenomenology and Logic*. Ithaca: Cornell University Press, 1977.

———, *Husserl and Realism in Logic and Mathematics*. Cambridge: Cambridge University Press, 1984.

Troelstra, A., *Choice Sequences*. Oxford: Clarendon Press, 1977.

Trotter, H.F., "Use of symbolic methods in analyzing an integral operator," in E. Kaltofen and S.M. Watt, eds., *Computers and Mathematics*. New York: Springer, 1989, pp. 82–90.

Uhlenbeck, K., "Instantons and Their Relatives," in F. Browder, ed., *Mathematics into the Twenty-First Century, Proceedings of the American Mathematical Society Centennial Symposium*. Providence: American Mathematical Society, 1992, pp. 467–477.

Umemura, H., "Painlevé Equations in the Past 100 Years," *American Mathematical Society Translations (2)* 204 (2001): 81–110.

Valley, J., "It's My Turn," *Rolling Stone* #331 (27 November 1980): 54–57.

van Dalen, D., "Hermann Weyl's intuitionistic mathematics," *Bulletin of Symbolic Logic* 1 (1995): 145–169.

van Dalen, D., and A.F. Monna, *Sets and Integration, An outline of the development*. Groningen: Wolters-Noordhoff, 1972.

van der Waerden, B. "Die lange Reichweite der regelmassigen Atomordnung," *Zeitschrift fur Physik* 118 (1941/42): 473–488.

van Hove, L., "Quelques Propriétés Générales de L'Intégrale de Configuration d'un Système de Particules Avec Intéraction," *Physica* 15 (1949): 951–961.

———, "Sur L'Intégrale de Configuration Pour les Systèmes de Particules à Une Dimension," *Physica* 16 (1950): 137–143.

van Stigt, W.P., *Brouwer's Intuitionism*. Amsterdam: North Holland, 1990.

Veltman, M., *Diagrammatica: The Path to Feynman Diagrams*. Cambridge: Cambridge University Press, 1994.

Vickers, S., *Topology via Logic*. Cambridge: Cambridge University Press, 1989.

Vietoris, L., "Über den höheren Zusammenhang kompakter Räume und eine Klasse von zusammenhangstreusten Abbildungen," *Mathematische Annalen* 97 (1927): 454–472. (An earlier version appeared in *Proc. Amsterdam* 29 (1926): 443–453 and 1009–1013. I was not able to locate this source.)

von Plato, J., *Creating Modern Probability*. Cambridge: Cambridge University Press, 1994.

Wang, H., *From Mathematics to Philosophy*. New York: Humanities Press, 1974.

Warnaar, S.O., Nienhuis, and K. A. Seaton, "New Construction of Solvable Lattice Models Including the Ising Model in a Field," *Physical Review Letters* 69 (1992): 710–712.

Wegner, F.J., "Duality in Generalized Ising Models and Phase Transitions without Local Order Parameter," *Journal of Mathematical Physics* 12 (1971): 2259–2272.

Weibel, C.A. *An Introduction to Homological Algebra*. New York: Cambridge University Press, 1994.

Weil, A., "Généralisation des fonctions abéliennes," *Journal de Mathematiques pures et appliquées* 9 (1938): 47–87.

———, *Foundations of Algebraic Geometry*. New York: American Mathematical Society, 1946.

———, "Review: 'The Collected Papers of Emil Artin,'" *Scripta Mathematica* 28 (1967): 237–238 (in Weil, *Collected Papers*, vol. 3, pp. 173–174).

———, Über die Bestimmung Dirichletscher Reihen durch Functionalgleichungen," *Mathematische Annalen* 168 (1967): 149–156; *Dirichlet Series and Automorphic Forms*. New York: Springer, 1971.

———, "Two Lectures on Number Theory, Past and Present," *Enseignement Mathématique* 20 (1974): 87–110 (in Weil, *Collected Papers*, vol. 3, pp. 279–310).

———, "Riemann, Betti, and the Birth of Topology," *Archive for the History of the Exact Sciences*, 20 (1979): 91–96.

———, "Une lettre et un extrait de lettre à Simone Weil," (1940), vol. 1, pp. 244–255; "De la métaphysique aux mathématiques," [*Science* (Paris): 1960, pp. 52–56] vol. 2, pp. 408–412. *Oeuvres Scientifiques, Collected Papers*. New York: Springer, 1979.

———, *Oeuvres Scientifiques, Collected Papers* (3 vols.). New York: Springer, 1979.

———, *The Apprenticeship of a Mathematician*. Basel: Birkhäuser, 1992.

Weinberg, S., *The Quantum Theory of Fields II*. New York: Cambridge University Press, 1996.

Weyl, H., "Über die asymptotische Verteilung der Eigenwerte," *Nachrichten der Königlichen Gesellschaft der Wissenschaften zu Göttingen. Mathematisch-Naturwissenschaftliche Klasse* (1911): 110–117.

———, *The Concept of a Riemann Surface*. Reading, Massachusetts: Addison-Wesley, 1964 [1955, 1913].

———, *Philosophy of Mathematics and Natural Science*. Princeton: Princeton University Press, 1949 [1927].

———, "Topology and Abstract Algebra as Two Roads of Mathematical Comprehension," *American Mathematical Monthly* 102 (1995, originally appearing in 1932): 453–460, 646–651.

———, "David Hilbert and His Mathematical Work," *Bulletin of the American Mathematical Society* 50 (1944): 612–654.

———, *The Classical Groups: Their Invariants and Representations*. Princeton: Princeton University Press, 1946.

———, "Das asymptotische Verteilungsgesetz der Eigenschwingungen eines beliebig gestalteten elastischen Koerpers," *Rend. Cir. Mat. Palermo* 39 (1950): 1–50.

———, "Ramifications, old and new, of the eigenvalue problem," *Bulletin of the American Mathematical Society* 56 (1950): 115–139.

_____, "A Half-Century of Mathematics," *American Mathematical Monthly* 58 (1951): 523–553.

Wheeler, J.A., and R.P. Feynman, "Interaction with the Absorber as the Mechanism of Radiation," *Reviews of Modern Physics* 17 (1945): 157–181.

Whittaker, E.T., and Watson, G.N., *A Course of Modern Analysis*. Cambridge University Press, 1990 [edn. 4, 1927; 1902].

Widom, H., "Wiener Hopf Integral Equations," in *The Legacy of Norbert Wiener* (Proceedings of Symposia in Pure Mathematics 60). Providence: American Mathematical Society, 1997, pp. 391–405.

_____, "On the Solution of a Painlevé III Equation," (solv-int/9808015 24 Aug 1998).

_____, "Commentary," in C. Morawetz, J. Serrin, and Ya.G. Sinai, eds., *Selected Works of Eberhard Hopf with Commentaries*. Providence: American Mathematical Society, 2002, pp. 47–48.

Wiener, N. and E. Hopf, "Über eine Klasse Singulärer Integralgleichungen," *Sitzberichte der Preussischen Akademie der Wissenschaften, Phys.-Math. Klasse* (1931): 696–706.

Wightman, A., "Introduction," in W. Israel, *Convexity and the Theory of Lattice Gases*. Princeton: Princeton University Press, 1979.

Wigner, E.P., "The Unreasonable Effectiveness of Mathematics in the Natural Sciences," in, *Symmetries and Reflections*. Cambridge, Massachusetts: MIT Press, 1970.

Wilcox, R., *Fundamentals of Modern Statistical Methods: Substantially Improving Power and Accuracy*. New York: Springer, 2010.

Wiles, A., "Modular Elliptic Curves and Fermat's Last Theorem," *Annals of Mathematics* 141 (1995): 443–551.

Wilkinson, D., "Continuum derivation of the Ising model two-point function," *Physical Review* D17 (1978): 1629–1636.

Williams, R., *The Country and the City*. New York: Oxford University Press, 1973.

Wilson, K.G., "The renormalization group: Critical phenomena and the Kondo problem," *Reviews of Modern Physics* 47 (1975): 773–840.

Wilson, K.G., and J. Kogut, "The Renormalization Group and the ε Expansion," *Physics Reports* 12C (1974): 75–200.

Wise, M.N., "Work and Waste, Political Economy and Natural Philosophy in Nineteenth-Century Britain," *History of Science* (1989) 27: 263–301, 391–449; (1990) 28: 221–261.

Witten, E., "Quantum Field Theory, Grassmannians, and Algebraic Curves," *Communications in Mathematical Physics* 113 (1988): 529–600.

_____, "Geometry and Quantum Field Theory," in F. Browder, ed., *Mathematics into the Twenty-First Century, Proceedings of the American Mathematical Society Centennial Symposium*. Providence: American Mathematical Society, 1992, pp. 479–491.

_____, "Duality, Spacetime and Quantum Mechanics," *Physics Today* 50 (May 1997): 28–33.

Wittgenstein, L., *Philosophical Grammar*, ed. R. Rhees, tr. A. Kenny. Berkeley: University of California Press, 1974.

Wonnacott, R.J., and T.H. Wonnacott, *Econometrics*. New York: Wiley, 1979.

Woodin, W.H., "The Continuum Hypothesis, Part I and Part II," *Notices of the American Mathematical Society* 48 (2001): 657–676, 681–690.

Wu, T.T., "Theory of Toeplitz Determinants and the Spin Correlations of the Two-Dimensional Ising Model. I," *Physical Review* 149 (1966): 380–401.

Wu, T.T., B.M. McCoy, C.A. Tracy, and E. Barouch, "Spin-spin correlation functions for the two-dimensional Ising model: exact theory in the scaling region," *Physical Review* B13 (1976): 315–374.

Yamada, K., "On the Spin-Spin Correlation Function in the Ising Square Lattice," *Progress of Theoretical Physics* 69 (1983): 1295–1298.

_____, "On the Spin-Spin Correlation Function in the Ising Square Lattice and Zero Field Susceptibility," *Progress of Theoretical Physics* 71 (1984): 1416–1418.

_____, "Pair Correlation Function in the Ising Square Lattice," *Progress of Theoretical Physics* 76 (1986): 602–612.

Yang, C.N., "The Spontaneous Magnetization of a Two-Dimensional Ising Model," *Physical Review* 85 (1952): 808–816. Commented on in C.N. Yang, *Selected Papers 1945–1980, with Commentary*. San Francisco: Freeman, 1983.

_____, "Concept of Off-Diagonal Long-Range Order and the Quantum Phases of Liquid Helium and of Superconductors," *Reviews of Modern Physics* 34 (1962): 694–704.

_____, "S-Matrix for the One-Dimensional N-Body Problem With Repulsive or Delta-Function Interaction," *Physical Review* 168 (1968): 1920–1923.

_____, "Journey Through Statistical Mechanics," *International Journal of Modern Physics* B2 (1988): 1325–1329.

_____, "Path Crossing with Lars Onsager," in *The Collected Works of Lars Onsager*. Singapore: World Scientific, 1996, pp. 180–181.

Yang, C.N., and C.P. Yang, "One Dimensional Chain of Anisotropic Spin-Spin Interaction," *Physical Review* 150 (1966): 321–327.

Yoneyama, K., "Theory of continuous sets of points," *Tôhoku Mathematics Journal* 12 (1917): 43–158.

Thematic Index

See also the list of sections that head each chapter. Names are indexed when they refer to papers (Dyson-Lenard), or styles of work (Baxter), or they are iconic (Onsager). Mentions are not indexed.

Printed in the United States
By Bookmasters